VOLUME ONE HUNDRED AND FOUR

ADVANCES IN BOTANICAL RESEARCH

Lignin and Hydroxycinnamic Acids: Biosynthesis and the Buildup of the Cell Wall

ADVANCES IN BOTANICAL RESEARCH

Series Editor

Jean-Pierre Jacquot
Membre de l'Institut Universitaire de France, Unité Mixte de Recherche INRA, UHP 1136 "Interaction Arbres Microorganismes", Université de Lorraine, Faculté des Sciences, Vandoeuvre, France

VOLUME ONE HUNDRED AND FOUR

ADVANCES IN BOTANICAL RESEARCH

Lignin and Hydroxycinnamic Acids: Biosynthesis and the Buildup of the Cell Wall

Volume Editor

RICHARD SIBOUT

UR1268 BIA (Biopolymères Interactions Assemblages), INRAE, Nantes, France

ELSEVIER

ACADEMIC PRESS

An imprint of Elsevier

Academic Press is an imprint of Elsevier
125 London Wall, London, EC2Y 5AS, United Kingdom
The Boulevard, Langford Lane, Kidlington, Oxford OX5 1GB, United Kingdom
50 Hampshire Street, 5th Floor, Cambridge, MA 02139, United States
525 B Street, Suite 1650, San Diego, CA 92101, United States

First edition 2022

ISBN: 978-0-323-91220-4
ISSN: 0065-2296

For information on all Academic Press publications
visit our website at https://www.elsevier.com/books-and-journals

Publisher: Zoe Kruze
Acquisitions Editor: Sam Mahfoudh
Developmental Editor: Jhon Michael Peñano
Production Project Manager: James Selvam
Cover Designer: Vicky Pearson

Typeset by STRAIVE, India

Contents

Contributors

Marie Baucher
Laboratory of Plant Biotechnology, Department of Biology of the Organisms, Université libre de Bruxelles, Gosselies, Belgium

Marc Behr
Laboratory of Plant Biotechnology, Department of Biology of the Organisms, Université libre de Bruxelles, Gosselies, Belgium

Hua Cassan-Wang
Laboratoire de Recherche en Sciences Végétales, Université de Toulouse, CNRS, UPS, Auzeville-Tolosane, France

Maxime Chantreau
Umeå Plant Science Centre (UPSC), Department of Forest Genetics and Plant Physiology, Swedish University of Agricultural Sciences, Umeå, Sweden

Anne-Laure Chateigner-Boutin
INRAE, UR1268 BIA, F-44316, Nantes, France

José C. del Río
Instituto de Recursos Naturales y Agrobiología de Sevilla (IRNAS), CSIC, Seville, Spain

Marie-Françoise Devaux
INRAE, UR BIA, Nantes, France

Mondher El Jaziri
Laboratory of Plant Biotechnology, Department of Biology of the Organisms, Université libre de Bruxelles, Gosselies, Belgium

Aymerick Eudes
Joint BioEnergy Institute, Emeryville; Lawrence Berkeley National Laboratory, Biosciences Area, Environmental Genomics and Systems Biology, Berkeley, CA, United States

Notburga Gierlinger
Institute of Biophysics, Department of Nanobiotechnology, University of Natural Resources and Life Sciences Vienna (BOKU), Vienna, Austria

Eliana Gonzales-Vigil
Department of Cell & Systems Biology, University of Toronto; Department of Biological Sciences, University of Toronto Scarborough, Toronto, ON, Canada

András Gorzsás
Department of Chemistry, Umeå University, Umea, Sweden

Jacqueline Grima-Pettenati
Laboratoire de Recherche en Sciences Végétales, Université de Toulouse, CNRS, UPS, Auzeville-Tolosane, France

Fabienne Guillon
INRAE, UR BIA, Nantes, France

Ana Gutiérrez
Instituto de Recursos Naturales y Agrobiología de Sevilla (IRNAS), CSIC, Seville, Spain

Ines Hadj Bachir
Laboratoire de Recherche en Sciences Végétales, Université de Toulouse, CNRS, UPS, Auzeville-Tolosane, France

Natalie Hoffmann
Department of Cell & Systems Biology, University of Toronto, Toronto, ON, Canada

Steven G. Hussey
Department of Biochemistry, Genetics and Microbiology, Forestry and Agricultural Biotechnology Institute (FABI), University of Pretoria, Pretoria, South Africa

Hoon Kim
Department of Energy Great Lakes Bioenergy Research Center, the Wisconsin Energy Institute, University of Wisconsin-Madison, Madison, WI, United States

Chang-Jun Liu
Joint BioEnergy Institute, Emeryville, CA; Brookhaven National Laboratory, Biology Department, Upton, NY, United States

Shawn D. Mansfield
Department of Wood Science, University of British Columbia, Vancouver, BC, Canada

Fabien Mounet
Laboratoire de Recherche en Sciences Végétales, Université de Toulouse, CNRS, UPS, Auzeville-Tolosane, France

Raphael Ployet
Department of Biochemistry, Genetics and Microbiology, Forestry and Agricultural Biotechnology Institute (FABI), University of Pretoria, Pretoria, South Africa

John Ralph
Department of Energy Great Lakes Bioenergy Research Center, the Wisconsin Energy Institute; Department of Biochemistry, University of Wisconsin-Madison, Madison, WI, United States

Jorge Rencoret
Instituto de Recursos Naturales y Agrobiología de Sevilla (IRNAS), CSIC, Seville, Spain

A. Lacey Samuels
Department of Botany, University of British Columbia, Vancouver, BC, Canada

Luc Saulnier
INRAE, UR1268 BIA, F-44316, Nantes, France

Chantal Teulières
Laboratoire de Recherche en Sciences Végétales, Université de Toulouse, CNRS, UPS, Auzeville-Tolosane, France

Hannele Tuominen
Umeå Plant Science Centre (UPSC), Department of Forest Genetics and Plant Physiology, Swedish University of Agricultural Sciences, Umeå, Sweden

Preface

Lignin polymers are made up of phenolics that impregnate plant-specific cell walls. Being the second most abundant plant-derived polymer, lignin represents a huge reservoir of carbon in terrestrial ecosystems because, apart from rare exceptions, lignin is exclusively detected in land plants. Interestingly, its composition changes according to plant lineages.

The definition of lignin is often a matter of discussion. A few years ago, a journal editor for whom I have a lot of esteem asked me to change the plural form of lignins into the singular form in my submitted manuscript. I argued that hemicelluloses are often written in plural. The editor was thinking of lignin as a general concept, i.e., "that brown messy polyphenolic glop in the cell wall" as he noted it. Beyond the semantic differences in the use of plurals in French and English languages, he was correct because, once incrusted into the cell wall, lignin polymers do not have a regular repeating sequence or structure and can appear disordered compared with highly structured cell wall polysaccharides. This is due to the polymerization mechanism that randomly governs linkages between monomers following oxidase activities. However, lignin is composed of incredibly diversified structures at the nanometer scale. Composition and structure vary according to position around the cell or according to the cell wall type (primary or secondary cell wall), tissue, and organ.

In this volume, you will discover how much the lignin paradigm needs to be reassessed and that lignins are plural with newly identified compounds found incorporated into the polymers (sometimes in their glycosylated form) as the "conventional" monomers, hydroxyphenyl (H), guaiacyl (G), and syringyl (S) lignin units. In Chapter 1, J. Del Río and colleagues review the recent discovery of flavonoids, hydroxystilbenes, and ferulic acid amides in lignins and describe how they are mechanistically included into the polymers. All of these discoveries indicate that the lignin structure is far more complex than previously thought, and that the definition of "lignin" should not be restricted to a polymer derived solely from the three monolignols: *p*-coumaryl, coniferyl, and sinapyl alcohols.

The ability of lignins to incorporate various molecules in their polymers opens new avenues for cell wall engineering. In the 1990s, the availability of *Arabidopsis thaliana* mutants allowed us to reach the crucial first steps in deciphering the monolignol pathway; but now, the emergence of efficient gene editing methods allows a myriad of prospects for building lignin polymers for dedicated applications and bioengineering of cell wall materials. In Chapter 2, C.-J. Liu and A. Eudes provide a dense and comprehensive review of gene editing approaches in the lignin field and reveal some promising avenues through posttranslational and transcriptional modifications of biosynthetic enzymes, their cofactors, cosubstrates, and metabolic precursors.

Other organisms such as bacteria have the ability to develop a polysaccharide matrix surrounding their cells. However, the formation of lignin polymers is specific to plants. Because lignification occurs in a polysaccharide environment, it is important to understand how glycosylation of the cell wall occurs and how it interacts with lignification. Chapter 3 illustrates the roles of glucan polymerization and of the de/glycosylation of carbohydrates, proteins, and monolignols in the building of the secondary cell wall matrix as well as in the regulation of monolignol homeostasis.

Ironically, lignin, which is one of the most recalcitrant plant polymers to degradation, is known to be constructed through one of the most simple mechanisms. It only requires oxidation of monolignols or related phenolics. It is even surprising at first glance that lignin polymerization *in planta* and the depolymerization of lignin by microorganisms are predominantly under the control of the same type of enzymes, namely peroxidases and/or laccases. Thus, fungal high redox potential laccases are able to break down the complex polymer of lignin into smaller oligomers or related phenolics, probably with the help of natural mediators (small molecules derived from laccase-catalyzed lignin oxidation). In contrast to many fungal laccases, substrate specificity of plant laccase was unknown until very recently. Going beyond this research question, N. Hoffman and colleagues discuss in Chapter 4 how the location and composition of lignin in the secondary cell wall could be fine-tuned by localization and substrate specificity of plant-specific oxidative enzymes.

As aromatic compounds, hydroxycinnamic acids are not only intermediary compounds in the monolignol pathway but also molecules that play multiple important roles *in planta*. In Chapter 5, A.L. Chateigner-Boutin and L. Saulnier outline how ferulic acid and to a lesser extent coumaric acid decorate the cell wall of cereal grains. Hydroxycinnamic acids play a key role in the interactions between the different polymers of the cell wall including lignin polymers. In addition to tackling their biosynthesis *in planta*, the authors also explain how ferulic acid has major effects on the properties of the cell walls and, consequently, on the processing of cereal grain into food products.

Understanding cell wall biosynthesis requires the appropriate wet chemistry methods, and considerable progress was made in the study of lignin polymers that led to identification of new structures. However, vibrational and fluorescence microscopy have the immense advantage to provide details about microstructures and chemical composition *in situ*. Here, F. Guillon and colleagues provide a comprehensive review in Chapter 6 and guide the researcher from sample preparation to image analysis through data acquisition with Raman, FTIR, and fluorescence microscopy in the context of lignin studies. Data acquired from a broad range of plant tissues are also discussed.

Often seen as one the most important components in wood, lignin is also finely and specifically deposited in various tissues conferring hydrophobicity or mechanical resistance. In Chapter 7, M. Chantreau and H. Tuominen review the localization, types, and functions of lignin found in various parts of land plants, from the amazing shaped sclereids to the endodermis. They decipher the coordinated control and fine-tuning of lignification and address the impact of the environmental regulation of lignin deposition.

The coordinated synthesis of the secondary cell wall biopolymers is controlled at a transcriptional level by an interconnected network of transcription factors. This transcriptional level of control can, however, gain complexity in perennial plants. With clear and exciting schematic representations, S. Hussey summarizes in Chapter 8 the current hierarchical transcriptional regulatory network model for secondary cell walls. He elaborates on its expanded structure, discussing key transcription factors underlying lignin formation in *Arabidopsis* and poplar.

Finally, lignin deposition has long been known to be part of the plant response to biotic and abiotic stresses. The identification of transcriptional players involved in the regulation of lignification during stress was poorly investigated until the last decade, and I. Hadj Bachir and colleagues focus their review on the most recent findings on the cross-talk between stress-signaling pathways and lignin biosynthesis (Chapter 9). They show the dual role of secondary cell wall-related transcription factors, which are able to integrate both developmental and environmental signals to reach an increased tolerance to abiotic and biotic stresses.

There are still many open questions regarding lignin biosynthesis and deposition that need to be addressed including its role and versatility across the plant kingdom. Despite the huge progress made in deciphering its composition, the impact of different linkages/units/acylations within the polymers on plant development and on interactions with environmental factors remains elusive. Lignin also represents a huge reserve of carbon, which may lead to production of high-value molecules. In the face of climate change and increasing CO_2 emissions, understanding how lignin is synthesized and regulated may also significantly contribute to reaching net-zero carbon and decarbonization goals with the possibility to produce biofuels, chemicals, or innovative material.

Richard Sibout

Dedication

The volume editor would like to dedicate this book to **Dr. Lise Jouanin** and **Prof. Catherine Lapierre**.

CHAPTER ONE

Unconventional lignin monomers—Extension of the lignin paradigm

José C. del Río[a,]*, Jorge Rencoret[a], Ana Gutiérrez[a], Hoon Kim[b], and John Ralph[b,c]

[a]Instituto de Recursos Naturales y Agrobiología de Sevilla (IRNAS), CSIC, Seville, Spain
[b]Department of Energy Great Lakes Bioenergy Research Center, the Wisconsin Energy Institute, University of Wisconsin-Madison, Madison, WI, United States
[c]Department of Biochemistry, University of Wisconsin-Madison, Madison, WI, United States
*Corresponding author: e-mail address: delrio@irnase.csic.es

Contents

Abstract

Lignin is largely produced by the oxidative radical coupling of the three monolignols, *p*-coumaryl, coniferyl, and sinapyl alcohols, which form the *p*-hydroxyphenyl, guaiacyl, and syringyl lignin units in the lignin polymer. However, it is increasingly appreciated that other "unconventional" phenolic compounds can also behave as lignin monomers participating in oxidative radical reactions during the lignification process and becoming integrated into the lignin polymer in some plants. Various phenolic compounds have been found incorporated into the lignin, including phenolic compounds derived from the monolignol biosynthetic pathway (e.g., monolignol ester conjugates, caffeyl and 5-hydroxyconiferyl alcohols, hydroxycinnamaldehydes, among others). More fascinating still was the discovery in some plants of a series of phenolic compounds arising from other polyphenolic biosynthetic pathways, including flavonoids (tricin, dihydrotricin,

Advances in Botanical Research, Volume 104
ISSN 0065-2296
https://doi.org/10.1016/bs.abr.2022.02.001

naringenin, naringenin chalcone), hydroxystilbenes (resveratrol, piceatannol, and isorhapontigenin, and their *O*-glucosylated counterparts piceid, astringin, and isorhapontin), and ferulic acid amides (feruloyltyramine, and diferuloylputrescine), that are integrated into the lignin polymer. All of these discoveries indicate that the lignin structure is far more complex than previously thought, that the definition of lignin should not be restricted to a polymer derived solely from the three traditional monolignols, and that other unconventional lignin monomers must be considered, thus expanding the established definition of lignin. The overriding factor remains, however, that lignification, the process of polymerization to the lignin polymer, allows this compositional flexibility because it is a simple chemical process.

1. Introduction

Lignin is an aromatic polymer characteristic of vascular plants that provides mechanical rigidity, allows the transport of water and solutes through the vascular system, and plays a role in plant protection against pathogens. Lignin derives primarily from the oxidative radical coupling of three *p*-hydroxycinnamyl alcohols (called "monolignols") that differ in their degree of methoxylation, *p*-coumaryl, coniferyl, and sinapyl alcohols (Boerjan, Ralph, & Baucher, 2003; Freudenberg & Neish, 1968; Ralph et al., 2004; Ralph, Lapierre, & Boerjan, 2019). Unlike other plant biopolymers (e.g., cellulose, hemicelluloses, or proteins), lignin does not have a well-defined structure with a specific sequence of monomeric units and linkages. The inherent complexity, heterogeneity, and wide variability of lignin among plants make the study of the lignin structure particularly challenging. In fact, although the structure of the lignin has been studied for many decades, various features in many plants remain to be elucidated, and new phenolic compounds incorporated into the lignins continue to be discovered. Over the past years, much research on the structural characteristics of the plant lignins, including in transgenics, has indicated the occurrence of unconventional phenolic compounds diverging from the canonical monolignols that are incorporated into the lignin polymer where they behave as true lignin monomers participating in radical coupling reactions during lignification. Most of these unconventional lignin monomers also arise from the monolignol biosynthetic pathway, as is the case of monolignol ester conjugates or phenolic compounds arising from the incomplete biosynthesis of monolignols, such as caffeyl and 5-hydroxyconiferyl alcohols, or hydroxycinnamaldehydes, among others. But more fascinatingly, various investigations have shown that phenolic compounds arising from beyond the canonical monolignol biosynthetic pathway, such as flavonoids, hydroxystilbenes, and ferulic acid amides, also behave as lignin monomers and are integrated into the lignin polymer in many plants (del Río et al., 2020, 2021; Ralph et al., 2019).

In this chapter, we will review the occurrence of unconventional lignin precursors that are derived from both the monolignol biosynthetic pathway, and from other polyphenolic biosynthetic pathways, in the lignin polymer. We will focus exclusively on those lignin precursors that have been described as occurring naturally in wild-type plants, although some of them were previously discovered and described in transgenic and mutant plants.

2. Canonical monolignols and unconventional lignin monomers derived from the monolignol biosynthetic pathway

2.1 Canonical monolignols

Lignin is produced primarily from the oxidative radical coupling of three monolignols, namely the *p*-coumaryl, coniferyl, and sinapyl alcohols (Fig. 1A), that produce the respective *p*-hydroxyphenyl (H), guaiacyl (G), and syringyl (S) lignin units when incorporated into the lignin polymer. Over the last decades, extensive research has been dedicated to clarifying the different steps of the monolignol biosynthetic pathway as well as the mechanisms of gene regulation, which are nowadays largely unraveled (Boerjan et al., 2003; Dixon & Reddy, 2003; Vanholme, De Meester, Ralph, & Boerjan, 2019; Vanholme, Demedts, Morreel, Ralph, & Boerjan, 2010; Vanholme, Morreel, Ralph, & Boerjan, 2008). The biosynthesis of a monolignol starts with the deamination of phenylalanine (or tyrosine), which is derived from the shikimate biosynthetic pathway, and involves sequential hydroxylation reactions of the aromatic ring followed by *O*-methylation; on the sidechain, the carboxylic acid group is ultimately reduced to an alcohol over several steps (Barros et al., 2016; Boerjan et al., 2003; Dixon, Chen, Guo, & Parvathi, 2001; Liu, 2012; Vanholme et al., 2008; Vanholme, De Meester, et al., 2019; Vanholme, Demedts, et al., 2010). The enzymes involved in the pathway are well known, and include phenylalanine ammonia lyase (PAL), tyrosine ammonia lyase (TAL), cinnamate 4-hydroxylase (C4H), 4-coumarate: coenzyme A ligase (4CL), ferulate 5-hydroxylase (F5H), *p*-coumarate 3-hydroxylase (C3H), *p*-hydroxycinnamoyl-CoA: quinate/shikimate hydroxycinnamoyltransferase (HCT), caffeoyl shikimate esterase

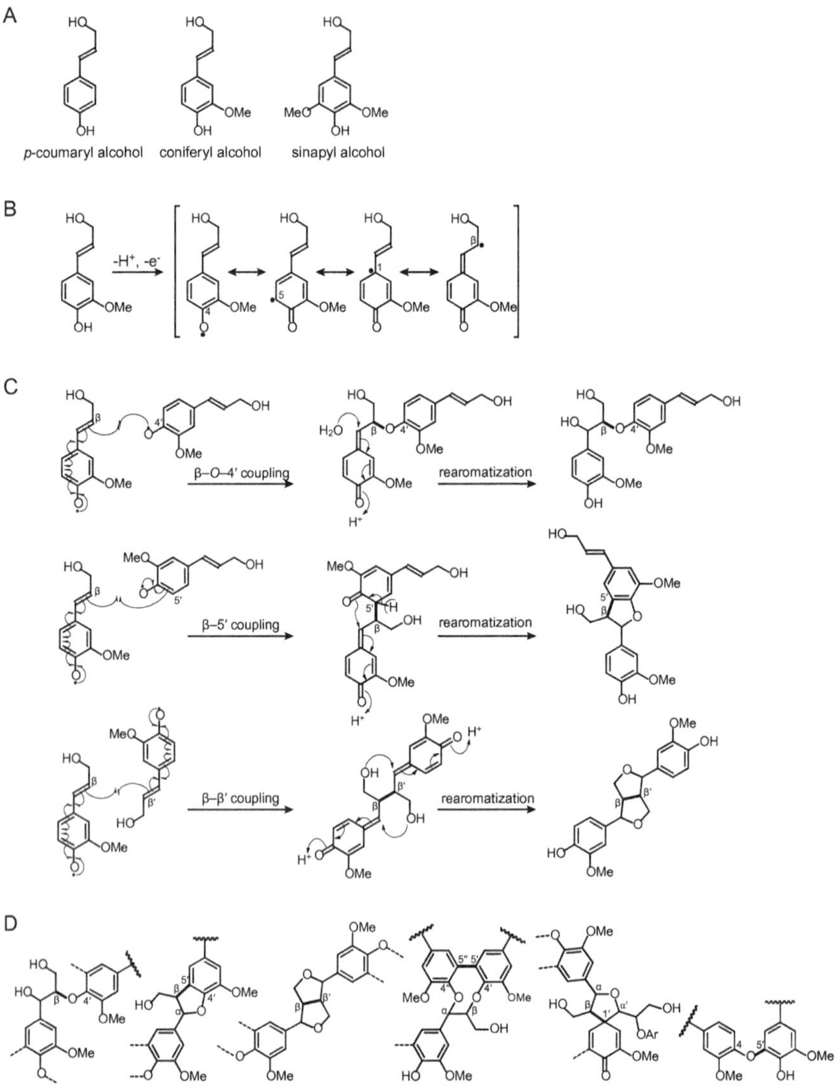

Fig. 1 (A) Structures of the three canonical monolignols, *p*-coumaryl, coniferyl, and sinapyl alcohols. (B) Radical formation from monolignols (shown for coniferyl alcohol here); upon oxidative dehydrogenation, the phenolic radical formed is stabilized via delocalization, as shown by the different resonance forms that allow us to understand the possible coupling modes. (C) Examples of the radical coupling reactions of two coniferyl alcohol radicals through β–*O*–4′, β–5′ and β–β′ coupling. (D) Examples of the major units, with their characteristic inter-unit linkages, in the lignin polymer.

(CSE), caffeoyl-CoA *O*-methyltransferase (CCoAOMT), cinnamoyl-CoA reductase (CCR), caffeic acid *O*-methyltransferase (COMT), and cinnamyl alcohol dehydrogenase (CAD).

After their synthesis, the monolignols are translocated to the cell-wall where they are oxidized by peroxidases and/or laccases producing radicals that are stabilized by resonance (Fig. 1B) and then polymerized in a combinatorial manner through free-radical coupling mechanisms (Fig. 1C). As the radicals are delocalized such that single-electron density is present at various sites, this polymerization mechanism (lignification) generates a wide variety of linkages and structures within the resulting lignin polymer, including β−*O*−4′ alkyl-aryl ethers, β−5′ phenylcoumarans, β−β' resinols, 5–5′ dibenzodioxocins, and β−1′ spirodienones, among others (Fig. 1D). This mechanism of lignin biosynthesis produces a racemic polymer that lacks any regular and ordered repeating units (Boerjan et al., 2003; Ralph et al., 2004, 2008, 2019; Tobimatsu & Schuetz, 2019). The lignin composition and the relative abundance of the different inter-unit linkages are highly dependent on the supply of the individual monolignols to the lignifying zone, which greatly varies among taxa, cell type, environmental conditions or growth stage (Campbell & Sederoff, 1996; Donaldson, 2001; Lourenço et al., 2016; Rencoret et al., 2011, 2018; Terashima, Fukushima, He, & Takabe, 1993; Vermerris & Boon, 2001). In general terms, lignins in gymnosperms are formed primarily from coniferyl alcohol, along with minor amounts of *p*-coumaryl alcohol, lignins in angiosperms are formed from coniferyl and sinapyl alcohols with small amounts of *p*-coumaryl alcohol, whereas there may be slightly higher levels of *p*-coumaryl alcohol incorporation into the lignins from monocotyledonous plants, including grasses (Boerjan et al., 2003; Ralph et al., 2004).

2.2 Unconventional lignin monomers derived from the monolignol biosynthetic pathway

During recent years, extensive research has shown that several phenolic compounds, also deriving from the canonical monolignol biosynthetic pathway (Fig. 2A), behave as lignin monomers in many plants, and are integrally incorporated into the lignin structure via radical coupling reactions during lignification. This is the case of the monolignol ester conjugates with various esters (such acetate, benzoate, *p*-hydroxybenzoate, vanillate, *p*-coumarate, and ferulate), phenolic compounds derived from the incomplete biosynthesis of monolignols such as caffeyl and 5-hydroxyconiferyl alcohols, and the hydroxycinnamaldehydes (coniferaldehyde, and sinapaldehyde), and

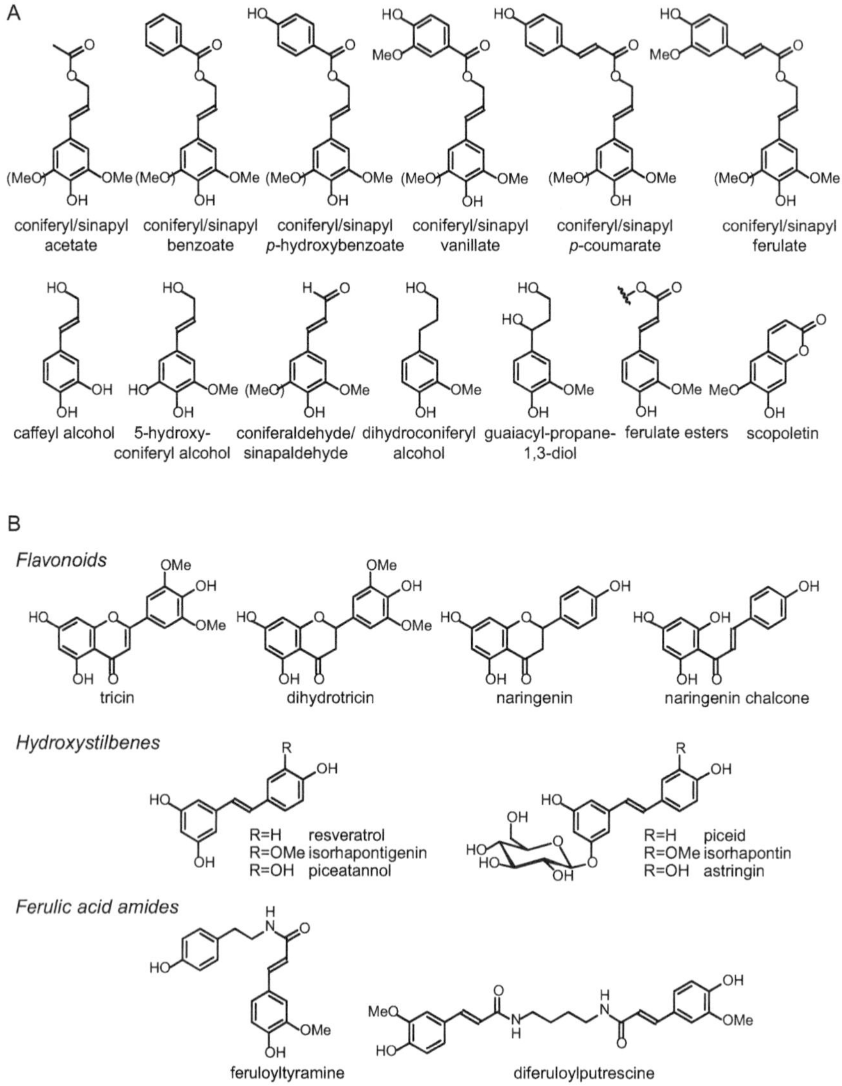

Fig. 2 (A) Unconventional lignin monomers derived from the monolignol biosynthetic pathway. Monolignol ester conjugates (with acetate, benzoate, *p*-hydroxybenzoate, vanillate, *p*-coumarate, and ferulate), the catechols caffeyl alcohol and 5-hydroxyconiferyl alcohol, the hydroxycinnamaldehydes (coniferaldehyde, and sinapaldehyde), along with dihydroconiferyl alcohol, guaiacylpropane-1,3-diol, ferulate esters, and scopoletin. (B) Unconventional lignin monomers derived from beyond the canonical monolignol biosynthetic pathway. Flavonoids (tricin, dihydrotricin, naringenin, and naringenin chalcone). Hydroxystilbenes (resveratrol, isorhapontigenin, and piceatannol), and their *O*-glucosides (piceid, isorhapontin, and astringin). Ferulic acid amides (feruloyltyramine, and diferuloylputrescine).

phenolics from side-reactions of pathway intermediates or products such as dihydroconiferyl alcohol and its related guaiacylpropane-1,3-diol, the hydroxycinnamate esters (particularly ferulate esters on polysaccharides), and the recently discovered coumarin scopoletin.

2.2.1 Monolignol ester conjugates

Monolignol ester conjugates are the most widespread of the unconventional monomers derived from the monolignol biosynthetic pathway found incorporated into the lignin in many plants. They are in fact so prevalent in various classes of plants that they deserve to be regarded as standard monomers, particularly since it was discovered that lignin acylation was a result of lignification with monolignol ester conjugates (Hatfield et al., 2009; Lu & Ralph, 2002, 2008; Lu, Ralph, Morreel, Messens, & Boerjan, 2004; Morreel et al., 2004; Petrik et al., 2014; Wilkerson et al., 2014; Withers et al., 2012). Several groups of esters are found acylating monolignols, including acetates, *p*-hydroxybenzoates, *p*-coumarates, ferulates, and, to a lesser extent, with benzoates and vanillates (Fig. 2A), and have been recognized to act as lignin monomers in a wide variety of natural plants (del Río et al., 2008, 2015; del Río, Marques, Rencoret, Martínez, & Gutiérrez, 2007; Karlen et al., 2016, 2017, 2018; Lu et al., 2015; Martínez et al., 2008; Ralph, 2010).

Monolignol acetates seem to be ubiquitous among angiosperms, usually at low levels in various hardwoods, but in substantial levels (up to 80% acetylation degree in some plants) in the lignins of kenaf (*Hibiscus cannabinus*), sisal (*Agave sisalana*), and abaca (*Musa textilis*), as well as in the lignins of some hardwoods, such as hornbeam (*Carpinus betulus*), in which acetylation reached up to 45% of all S-lignin units (del Río et al., 2008; del Río, Gutiérrez, & Martínez, 2004; del Río, Marques, et al., 2007; Lu & Ralph, 2002; Ralph, 1996). In most plants, acetylation occurred preferentially in S-lignin units, indicating that the monolignol acetate transferase involved presented higher affinities to sinapyl alcohol over coniferyl alcohol. However, in some grasses, such as bamboo, wheat straw, sugarcane, or the pith of elephant grass, acetylation occurred preferentially in G-lignin units (del Río et al., 2012a, 2012b, 2015; del Río, Marques, et al., 2007).

Monolignol *p*-hydroxybenzoates are commonly found in palms, poplars/aspen, and willows (Kuroda, Ozawa, & Ueno, 2001; Landucci, Deka, & Roy, 1992; Lu et al., 2004, 2015; Morreel et al., 2004; Rencoret et al., 2013, 2018; Sun, Fang, & Tomkinson, 1999) and have also been identified at substantial levels in the lignin of the seagrass *Posidonia oceanica*, an endemic species of the Mediterranean Sea (Kaal, Serrano, del Río, & Rencoret, 2018; Rencoret et al., 2020). In most plants, *p*-hydroxybenzoylation occurred overwhelmingly on S-lignin units, with minor levels on G-units (Lu et al., 2015). The acyltransferase responsible for monolignol *p*-hydroxybenzoylation in poplar has been recently identified and characterized, and displays a kinetic preference for sinapyl alcohol *in vitro* (Zhao et al., 2021), which is consistent with the preferential *p*-hydroxybenzoylation of S-lignin units observed in poplar. However, the seagrass *P. oceanica* displays a unique lignin acylation pattern, as it is extensively *p*-hydroxybenzoylated over both G- and S-units (Rencoret et al., 2020). The highest degree of *p*-hydroxybenzoylation was observed in the lignin from *P. oceanica* roots, attaining up to 58% of all G-units and 65% of all S-units (Rencoret et al., 2020).

Monolignol *p*-coumarates are typical of grasses and other commelinid monocots (del Río et al., 2008; Karlen et al., 2018; Lu & Ralph, 1999; Martínez et al., 2008; Ralph, 2010; Ralph et al., 1994; Rosado et al., 2021; Rosado, Rencoret, Marques, Gutiérrez, & del Río, 2021). The extent of lignin acylation with *p*-coumarates is particularly high in some grasses, as is the case of elephant grass (*Pennisetum purpureum*) or papyrus (*Cyperus papyrus*) sedge in which *p*-coumaroylation extent reaches above 50% (del Río, Prinsen, et al., 2012a; Rosado, Bausch, et al., 2021), although in other grasses, as in wheat straw, acylation with *p*-coumarates is rather scarce (del Río, Rencoret, et al., 2012b). In all cases, *p*-coumarates were found acylating the γ-OH of the lignin side-chains overwhelmingly in S-lignin units. The *p*-coumaroyl-CoA: monolignol transferases responsible for lignin *p*-coumaroylation in grasses, and which displayed higher affinities towards sinapyl alcohol than towards coniferyl alcohol, have already been identified and characterized (Marita, Hatfield, Rancour, & Frost, 2014; Petrik et al., 2014; Withers et al., 2012).

Monolignol ferulates were successfully introduced into the lignin of poplar through the introduction of a gene expressing feruloyl-CoA: monolignol transferase (FMT) (Wilkerson et al., 2014), but further detailed analytical studies indicated that small amounts also occurred in the wild-type plant. This discovery prompted the screening of a large number of plant species that concluded that monolignol ferulates were also widely present in the lignins of many plants, albeit in low amounts (Karlen et al., 2016, 2017).

Monolignol benzoates have been described in the lignins of some plants, such as the leaf tissues of the Canary Island date palm (*Phoenix canariensis*) (Karlen et al., 2017), the macaúba (*Acrocomia aculeata*) palm fruit endocarp

(Rencoret et al., 2018), and at trace levels in the wild-type lignin of *Populus trichocarpa*, where it was massively increased in PtrC4H1/ PtrC4H2/ PtrC3H3 transgenic lines (Kim et al., 2020). The data indicated a preferential acylation of S-units in the lignins of both the date palm and *P. trichocarpa*, and a benzoyl-CoA: monolignol transferase candidate was identified in *P. trichocarpa* (Chedgy, Köllner, & Constabel, 2015). However, in macaúba, the monolignol benzoates do not show any particular preference for G- or S-units but were present in a similar proportion as the lignin S/G ratio. The simultaneous occurrence of monolignol benzoates and monolignol *p*-hydroxybenzoates in the same lignins suggests a connection in the biosynthesis of both types of monolignol ester conjugates. Indications are that benzoates and *p*-hydroxybenzoates are derived from the monolignol intermediates, cinnamic and *p*-coumaric acids (Kim et al., 2020).

Finally, monolignol vanillates have also been described, although at trace levels, in the lignin of Canary Island date palm (Karlen et al., 2017). As occurs with other monolignol ester conjugates, vanillates preferentially acylate the S-lignin units.

These monolignol ester conjugates, as the monolignols themselves, can participate in oxidative radical coupling reactions with other monolignol ester conjugates, or with the canonical monolignols, through the same types of linkages that incorporate them into the lignin polymer. Among the different coupling structures, the formation of the β–β' coupled structures is the most influenced by the acylation. In β–β' coupling of conventional monolignols, resinol structures are produced as the free γ-OH groups act as internal nucleophiles to attack and rearomatize the intermediate quinone methides. However, if the γ-OH of the monolignol is acylated, the resinol structure cannot be formed and instead tetrahydrofuran structures are produced following β–β' coupling of two γ-acylated monolignols or from β–β' cross-coupling of a normal (γ-OH) monolignol with a γ-acylated monolignol, as shown in Fig. 3. These characteristic and diagnostic tetrahydrofuran structures were unambiguously identified in the lignins of different plants (del Río et al., 2008, 2015; del Río, Marques, et al., 2007; Lu et al., 2004, 2015; Lu & Ralph, 2002, 2005, 2008; Ralph, 2010), and unequivocally demonstrate that acylated monolignols are true lignin monomers participating in radical coupling reactions during lignification.

2.2.2 Caffeyl alcohol and 5-hydroxyconiferyl alcohol

Caffeyl and 5-hydroxyconiferyl alcohols, which are derived from incomplete methylation in the biosynthetic pathway towards the monolignols,

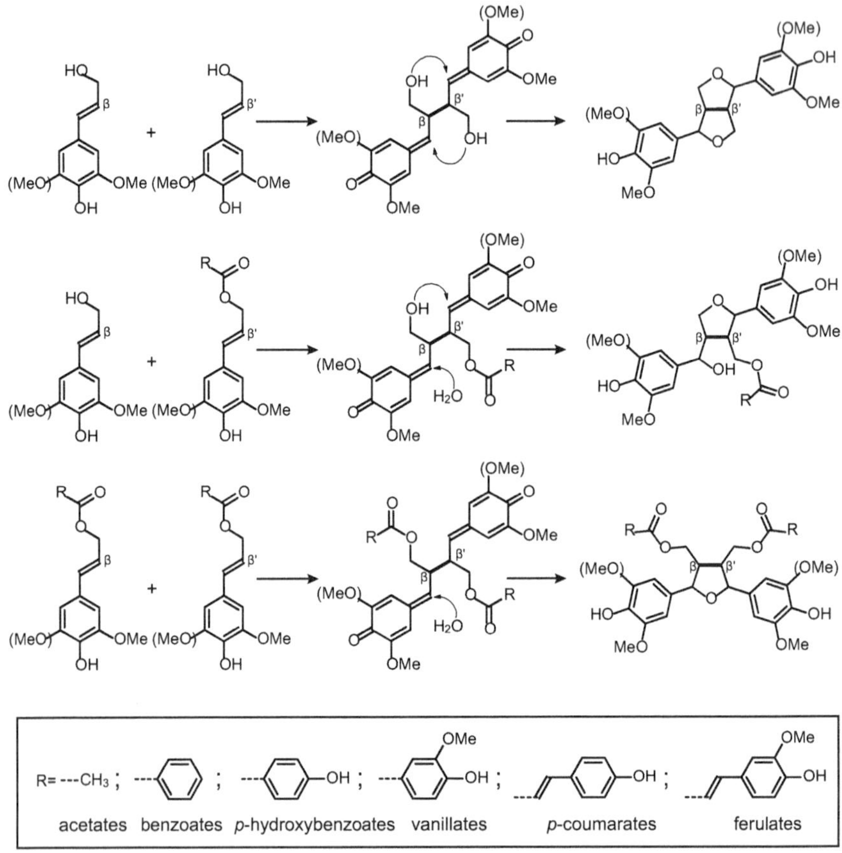

Fig. 3 Structures of the products arising from the β–β′ coupling of normal (γ-OH) and γ-acylated monolignols. β–β′ Coupling of two normal (γ-OH) monolignols produces the typical resinol structure. Coupling of two γ-acylated monolignols, or cross-coupling of a normal (γ-OH) and a γ-acylated monolignol, produces the characteristic tetrahydrofuran structures that are diagnostic for coupling involving acylated monolignols.

have been reported in the lignins of several normal and transgenic plants. These phenolic compounds were first described in the lignins of *OMT*-downregulated transgenic and mutant plants (Lapierre, Tollier, & Monties, 1988; Li, Popko, Umezawa, & Chiang, 2000; Marita et al., 2003; Parvathi, Chen, Guo, Blount, & Dixon, 2001; Ralph et al., 2001, 2001; Vanholme et al., 2010; Wagner et al., 2011; Weng, Mo, & Chapple, 2010). However, later on, lignins produced solely from the polymerization of caffeyl alcohol (called C-lignin) were discovered in the seedcoats of vanilla orchid (*Vanilla planifolia*) and in some members of the Cactaceae, Euphorbiaceae, and Cleomaceae families (Chen et al., 2013; Chen, Tobimatsu,

Havkin-Frenkel, Dixon, & Ralph, 2012; Tobimatsu et al., 2013). Likewise, lignins containing 5-hydroxyguaiacyl units (called 5H-lignin) derived from the incorporation of 5-hydroxyconiferyl alcohol were discovered in the seedcoats of some species of Escobaria (Chen et al., 2013), and in lower amounts in the lignins of other normal plants (del Río, Martínez, & Gutiérrez, 2007).

The incorporation of caffeyl and 5-hydroxyconiferyl alcohols, having a catechol group, into the lignin polymer produces characteristic benzodioxane structures, which are formed by β–*O*–4′ radical coupling of the monomer (at its β-position) with the catechyl end-unit of the polymer (at its 4′–*O*-position) followed by internal trapping of the quinone methide intermediates by the 3-hydroxyl of the catechyl unit to form the benzodioxane ring (Fig. 4). The formation of the characteristic benzodioxane units, therefore, provided evidence for the participation of these phenolic compounds in radical coupling reactions during lignification, and indicates that they behave as true lignin monomers. A recent study examined whether quinones produced by oxidation of these catechyl monomers could allow an additional mechanism to operate in lignification, the Diels-Alder cycloaddition reaction (Ando et al., 2021). Such a reaction even has the potential to allow lignification from unsaturated starting-ends of the polymer (reacting as dienophiles) rather than solely from the phenolic end via radical coupling. Monomeric models produced beautiful Diels-Alder products in which either the diene itself acted as the diene to produce tricyclic products from a monolignol or a 4–*O*-methylated (i.e., etherified) monolignol, or the quinone acted as the diene in a hetero-Diels-Alder reaction to produce benzodioxanes but with the opposite regiochemistry to those from radical coupling. No evidence

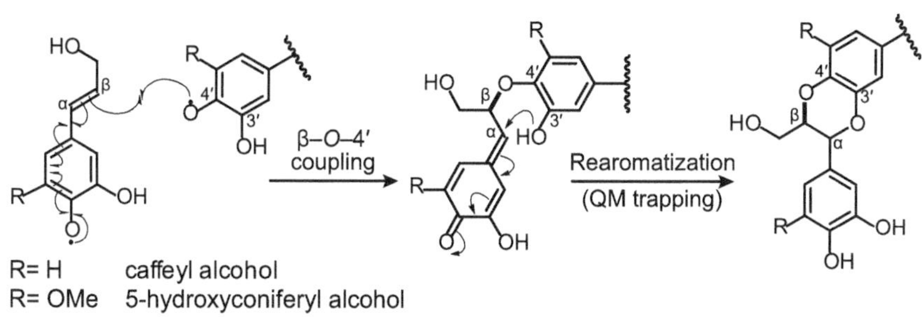

Fig. 4 Scheme of the mechanism of formation of benzodioxane units via β–*O*–4′ radical coupling reactions between caffeyl alcohol (R = H) or 5-hydroxyconiferyl alcohol (R = OMe) and the catechyl polymer followed by internal trapping of the quinone methide (QM) intermediates by the 3′-OH of the catechyl unit.

for such products can be found even in the lignins derived from high or 100% levels of such caffeyl alcohol or 5-hydroxyconiferyl alcohol, however, suggesting that, even for these special monomers, lignification remains firmly in the realm of radical coupling chemistry.

2.2.3 Hydroxycinnamaldehydes

Hydroxycinnamaldehydes (coniferaldehyde and sinapaldehyde) have been found incorporated in considerable amounts into the lignin in various mutant and transgenic plants in which CAD (cinnamyl alcohol dehydrogenase), the enzyme that reduces the hydroxycinnamaldehydes to the monolignols, is deficient (Kim et al., 2002, 2003; Lapierre et al., 2004; Liu et al., 2020; Martin et al., 2019; Ralph et al., 1997, 1998; Ralph, Lapierre, Marita, et al., 2001; Van Acker et al., 2017). Thus, in a CAD-deficient mutant pine, coniferaldehyde units become significant in the polymer (Ralph et al., 1997). Likewise, in a CAD-deficient tobacco transgenic, sinapaldehyde becomes a major component of the lignin, and participates in radical coupling reactions with other aldehyde units as well as with the canonical monolignols to be incorporated into the lignin polymer (Ralph et al., 1998). However, hydroxycinnamaldehydes have also been observed, although at low levels, in the lignins of most normal plants (Chen, 1992; Lüdemann & Nimz, 1974; Van Acker et al., 2017). In fact, the Wiesner test (phloroglucinol/HCl) used to confirm the presence of lignin in cell-walls is based on its specificity for coniferaldehyde residues incorporated into the lignin polymer (Blaschek et al., 2020).

Most structural studies regarding the modes of incorporation of hydroxycinnamaldehydes into the lignin polymer have been carried out on CAD-deficient plants, in which hydroxycinnamaldehydes predominate. As for the monolignols, the hydroxycinnamaldehydes can be incorporated into the lignin polymer via oxidative radical coupling reactions with other hydroxycinnamaldehydes (or with canonical monolignols) through different linkage types, predominantly 8–O–4′ linkages, although other types of linkages, such as 8–5′, 5–5′, 8–8′ or 8–β' linkages also occur (Fig. 5A). In normal plants, with a low content of hydroxycinnamaldehydes, it is likely that they will be incorporated into their lignins in a similar way. In fact, although the units most easily characterized in NMR spectra, for example, are end-unit cinnamaldehydes (and derived benzaldehydes) that arise from the cross-coupling with monolignols early in the process, small peaks associated unambiguously with the cross-coupling of hydroxycinnamaldehyde monomers into the growing polymer, such as the coniferaldehyde- and

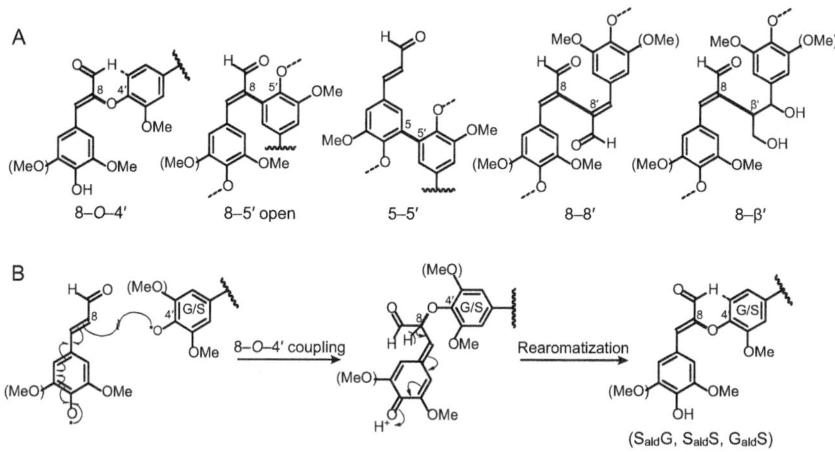

Fig. 5 (A) Major structures produced from the incorporation of hydroxycinnamaldehydes into the lignin polymer through different linkage types, including 8–*O*–4′, 8–5′, 5–5′, 8–8′, and 8–β′- coupling products. (B) Mechanism of 8–*O*–4′ cross-coupling reaction of hydroxycinnamaldehydes with the growing lignin polymer; note the unsaturation that results from rearomatization via loss of the acidic 8-proton. Note that sinapaldehyde (S_{ald}) can cross-couple with both G- and S-lignin units, whereas coniferaldehyde (G_{ald}) can only cross-couple with S-lignin units.

sinapaldehyde-(8–*O*–4′)-guaiacyl/syringyl cross-coupled units, can also be observed. These components (and others) are substantially increased in the lignins in CAD-deficient plants, as nicely illustrated in the recent discovery of the first CAD-deficient hardwood mutant, a plant long known for its interesting red wood (and the superiority of its leaves as a silkworm feed) (Yamamoto et al., 2020). The occurrence of all these linkages illustrates that hydroxycinnamaldehydes incorporate into the lignin polymer by radical coupling reactions in much the same way as normal monolignols do, and indicate that they should also be considered to be authentic lignin precursors (Liu et al., 2020; Martin et al., 2019; Van Acker et al., 2017). The major mechanistic and consequent structural difference arises from the rearomatization step. Following β–*O*–4′ coupling, normal monolignols produce a quinone methide intermediate that is rearomatized by water addition to the benzylic position, as shown in Fig. 1C. The quinone methide resulting from the analogous 8–*O*–4′ coupling of a hydroxycinnamaldehyde, however, has a particularly acidic 8-proton because of the aldehyde at the 9-position of the sidechain; rearomatization is therefore readily achieved via loss of this 8-proton, producing a new unsaturated sidechain that is obviously chemically and spectroscopically different than the conventional β-ethers, as shown in Fig. 5B (Kim et al., 2003; Ralph, Lapierre, Marita, et al., 2001). On the other hand, it was experimentally observed, both *in planta* and *in vitro*, that sinapaldehyde cross-coupled 8–O–4′ with both guaiacyl and syringyl units in the growing polymer, whereas coniferaldehyde cross-coupled 8–O–4′ only with syringyl units but not with guaiacyl units (Kim, Ralph, Yahiaoui, Pean, & Boudet, 2000; Ralph, Lapierre, Marita, et al., 2001). As a result, incorporation of hydroxycinnamaldehydes into the lignin of angiosperms is more substantial than in conifers.

2.2.4 Dihydroconiferyl alcohol and guaiacylpropane-1,3-diol

Dihydroconiferyl alcohol and its derived guaiacylpropane-1,3-diol have also been found in the lignins of wild-type and mutant softwoods (Ralph et al., 1999; Ralph, Kim, Peng, & Lu, 1999; Ralph, Lapierre, Marita, et al., 2001). Substantial amounts of these phenolic compounds have been found incorporated into the lignin in a CAD-deficient pine mutant (Lapierre, Pollet, MacKay, & Sederoff, 2000; Ralph et al., 1997; Ralph, Kim, et al., 1999) but have also been found, although at low levels, into most normal softwood lignins (Lundquist & Stern, 1989; Ralph, Kim, et al., 1999; Sakakibara, 1980), and also in hardwoods (Fukagawa, Meshitsuka, & Ishizu, 1991). It was suggested that dihydroconiferyl alcohol might derive from coniferaldehyde, rather than from coniferyl alcohol, due to the large amounts of coniferaldehyde present in the CAD-deficient pine mutant (Ralph et al., 1997); the 1,4-reduction of coniferaldehyde via a phenylpropenal double-bond reductase (PPDBR) has been demonstrated (Kasahara et al., 2006). On the other hand, it was demonstrated that guaiacylpropane-1,3-diol derives directly from dihydroconiferyl alcohol by peroxidase-mediated reactions (Ralph, Kim, et al., 1999) now recognized to be via disproportionation (Ralph et al., 2009).

As for the monolignols, dihydroconiferyl alcohol and guaiacylpropane-1,3-diol can be incorporated into the lignin via the typical oxidative radical coupling reactions. These phenolic compounds can undergo homo-coupling but also cross-coupling reactions with the normal monolignol, coniferyl alcohol, to incorporate them into the lignin polymer. However, it is important to note that dihydroconiferyl alcohol and guaiacylpropane-1,3-diol do not possess unsaturated sidechains, and therefore coupling is restricted to 4–O-, and 5-positions of the aromatic ring (Ralph et al., 2004, 2008; Ralph, Lapierre, Marita, et al., 2001). Hence, these compounds are mostly involved in 5–5′ biphenyl units forming dibenzodioxocins, as well as in biphenyl 4−O−5′ ether linkages, or are found at the starting end of the polymer that arise from initial lignification reactions in which they cross-couple with coniferyl alcohol, the monolignol invariably coupling at its β-position, as usual.

2.2.5 Polysaccharide hydroxycinnamate esters

Polysaccharide hydroxycinnamate esters (principally ferulates, but also *p*-coumarates to a lower extent) are also incorporated into the lignin in some plants, particularly in grasses and other commelinid monocots (Eugene, Lapierre, & Ralph, 2020; Hatfield, Ralph, & Grabber, 1999; Ralph, 2010; Ralph, Grabber, & Hatfield, 1995). In grasses, ferulates predominantly acylate polysaccharides, generally at the C-5 hydroxyl of arabinosyl residues in arabinoxylans (Fig. 6A). Ferulate esters can be oxidized by peroxidases and/or laccases forming radicals that are stabilized by resonance and can participate in radical coupling reactions that may occur at the 4−O-, 5- or 8-positions forming ferulate dehydrodimers and higher dehydro-oligomers through different linkages (such as 8−O−4′-, 4−O−5′-, 8–8′-, 5–5′- and 8–5′- linkages, as depicted in Fig. 6B) and creating cross-links between the arabinoxylan chains (Bunzel, Ralph, Funk, & Steinhart, 2005; Bunzel, Ralph, Lu, Hatfield, & Steinhart, 2004; Grabber, Ralph, & Hatfield, 2000, 2002; Hatfield et al., 1999; Ralph et al., 1995; Ralph, Quideau, Grabber, & Hatfield, 1994). In addition, ferulates and dehydro-diferulates can also cross-couple with monolignols (or oligolignols) through the same types of linkages (such as β−O−4′-, β−5′-, 8–5′-, or 8−β’- linkages, as depicted in Fig. 6C), thus cross-linking the lignin to the polysaccharide matrix (Bunzel et al., 2004; Eugene et al., 2020; Grabber et al., 2000, 2002; Ralph et al., 1995; Ralph, Quideau, et al., 1994; Schatz, Ralph, Lu, Guzei, & Bunzel, 2006; Yang et al., 2021; Zhang, Lu, Sun, & Ralph, 2009).

2.2.6 Scopoletin

Scopoletin is a coumarin that is naturally present in some plants. It is produced after 6-hydroxylation of ferulate by the feruloyl-CoA 6-hydroxylase (F6H1), and subsequent ring-closure after *cis-trans* isomerization by newly discovered coumarin synthase (COSY) (Vanholme et al., 2019). Scopoletin was successfully introduced into the lignin in Arabidopsis plants in which the F6H1 was expressed (Sakamoto et al., 2020). More recently, additional research indicated that scopoletin may also be incorporated into the lignin

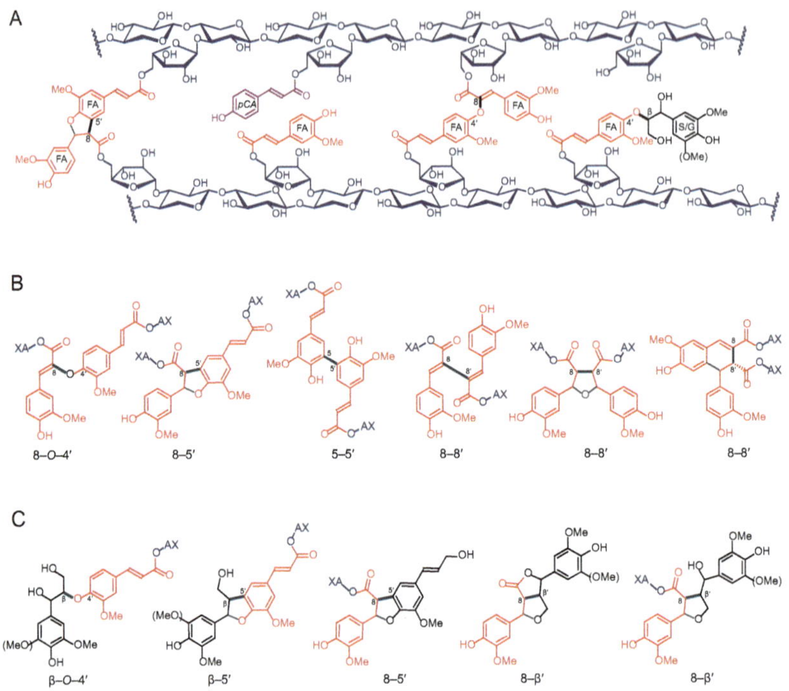

Fig. 6 (A) Scheme of the arabinoxylan chains showing the ferulates, and *p*-coumarates acylating the C-5 hydroxyl of arabinosyl residues. Two representative diferulates (8–*O*–4′ and 8–5′) and a monolignol ferulate conjugate formed via 4′–*O*–β coupling with a monolignol, are also shown. (B) Ferulate dehydrodimerization products produced from 8–*O*–4′, 8–5′, 5–5′, and 8–8′ homo-coupling; note that different structures can be formed following 8–8′ cross-coupling. (C) Different modes of incorporation of ferulates into the lignin polymer through cross-coupling reactions with monolignols, the β–*O*–4′, β–5′, 8–5′, and 8–β′ cross-coupled products are shown; note that 8–β′ cross-coupling produces more than one structure. AX: arabinoxylan chains. *Panel A: Adapted from Eugene, A., Lapierre, C., & Ralph, J. (2020). Improved analysis of arabinoxylan-bound hydroxycinnamate conjugates in grass cell walls.* Biotechnology for Biofuels, 13, *2020.*

in the wild-type plants, although at a low level, indicating that scopoletin might also be considered a lignin monomer, at least in Arabidopsis (Hoengenaert et al., 2022).

3. Lignin monomers derived from beyond the monolignol biosynthetic pathway

Various investigations in recent years have discovered that several phenolic compounds derived from other polyphenolic biosynthetic pathways, namely the flavonoid, hydroxystilbene, or the ferulic acid amide pathways (Fig. 2B), can also participate in radical coupling reactions with monolignols and/or lignin oligomers to become incorporated into the lignin in several plants (del Río et al., 2020, 2021; del Río, Rencoret, et al., 2012b; del Río, Rencoret, Gutiérrez, Kim, & Ralph, 2017, 2018; Lan et al., 2015, 2016, 2016; Lan, Rencoret, del Río, & Ralph, 2019; Ralph et al., 2019; Rencoret et al., 2018, 2019, 2022; Rosado, Bausch, et al., 2021). These phenolic compounds may therefore be also considered as lignin monomers.

3.1 Flavonoids

Various phenolic compounds arising from different classes of flavonoids, namely the flavone tricin, as well as others recently discovered, such as the flavanones naringenin and dihydrotricin, and the chalconoid naringenin chalcone, have been found incorporated into the lignin polymer in several plants.

The discovery of the flavone tricin incorporated into the lignin polymer of various plants was an important milestone in lignin chemistry, as it was the first phenolic compound arising from outside the classical monolignol biosynthetic pathway that was established as an authentic lignin monomer participating in cross-coupling reactions with monolignols and being fully integrated into the lignin in several plants (del Río, Rencoret, et al., 2012b; Lan et al., 2015). Tricin was first described in the lignin from wheat straw (del Río, Rencoret, et al., 2012b), and its structure was fully established by 2D-NMR analyses that also indicated that it was incorporated into the lignin polymer through 4′–O–β ether bonds (Fig. 7). Subsequent studies indicated that tricin was widely present in the lignins of all grasses as well as in other monocots (del Río et al., 2015; Lan, Morreel, et al., 2016; Lan, Rencoret, et al., 2016; Rencoret et al., 2013; Rosado, Bausch, et al., 2021; Rosado, Rencoret, et al., 2021). Interestingly, tricin was also found in the lignins of the eudicotyledoneous plant alfalfa (*Medicago sativa*) (Lan, Rencoret, et al., 2016) and the related *M. truncatula* (Lui et al., 2020), a finding that revealed independent and convergent evolution of the enzymes leading to the biosynthesis of tricin in *Medicago* and in grasses (Lui et al., 2020; Ralph, 2020).

Tricin, as for the monolignols, is incorporated into the lignin via oxidative radical coupling reactions. Biomimetic radical coupling reactions of tricin with the three monolignols confirmed that tricin can cross-couple

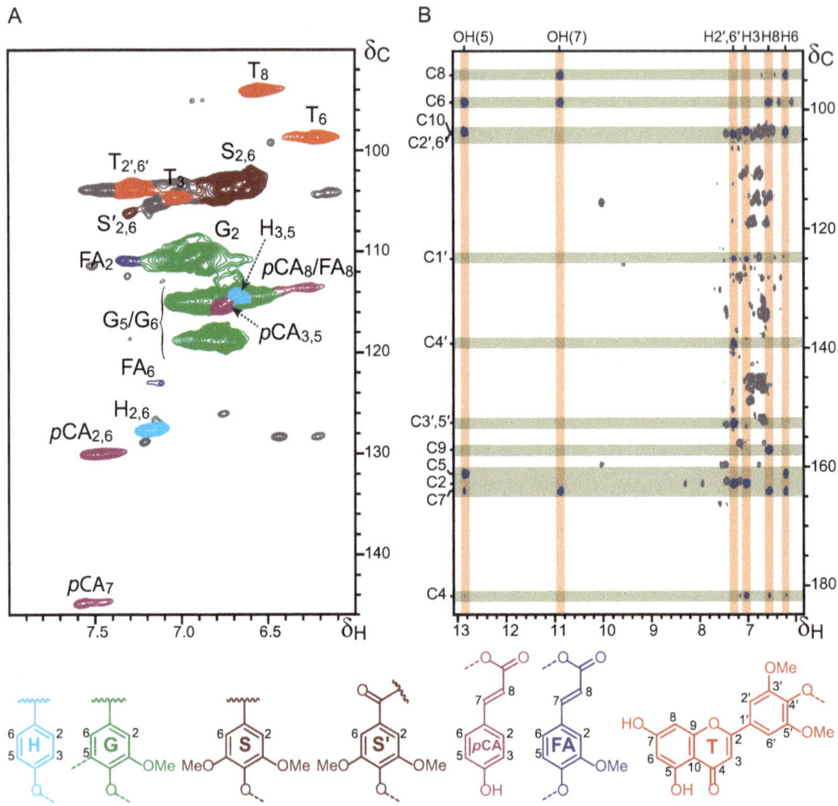

Fig. 7 (A) Aromatic region of the 2D HSQC NMR spectrum of the lignin isolated from wheat straw showing the main correlation signals for tricin and the different lignin units. (B) Partial HMBC spectrum of the lignin isolated from wheat straw showing the main correlations that allowed establishing the structure of tricin (T) in the lignin. Lignin units, H: *p*-hydroxyphenyl; G: guaiacyl; S: syringyl, S′: α-oxidized syringyl; *p*CA: *p*-coumarate; FA: ferulate; T: tricin. *Adapted from del Río, J. C., Rencoret, J., Prinsen, P., Martínez, A. T., Ralph, J., & Gutiérrez, A. (2012). Structural characterization of wheat straw lignin as revealed by analytical pyrolysis, 2D-NMR, and reductive cleavage methods.* Journal of Agricultural and Food Chemistry, 60, *5922–5935.*

with monolignols (or the growing lignin chain) exclusively through 4′–*O*–β-ether linkages (Lan et al., 2015). This, together with the fact that tricin cannot undergo radical dehydrodimerization, indicated that tricin can only appear at the starting end of the lignin chain, and it was therefore suggested that tricin may have a role as a nucleation site for lignification in grasses (Lan et al., 2015). Furthermore, due to the incorporation into the lignin through 4′–*O*–β linkages, tricin can be easily released from the lignin upon degradation by chemical degradative methods intended to cleave the β-ether bonds in lignin, such as thioacidolysis or Derivatization Followed by Reductive Cleavage (DFRC), which facilitated both the screening and the quantification of the amounts of tricin incorporated into the lignin (Lan, Rencoret, et al., 2016; Rosado, Rencoret, et al., 2021; Rosado, Bausch, et al., 2021).

The study of the phenolic metabolites present in the lignifying zone of maize cell walls identified a series of tricin-containing flavonolignan metabolites (Lan, Morreel, et al., 2016). These compounds included the dimeric products resulting from the coupling of tricin with the three monolignols, as well as with their γ-acylated ester conjugates (with acetates and *p*-coumarates), forming the respective tricin-4′–*O*–β-arylglyceryl ethers. Trimeric compounds produced from the additional coupling of the tricin-(4′–*O*–β)-monolignol dimeric products with a further 4–*O*–β or 5–β linkage to another monolignol, including acylated monolignols, were also identified. All of these tricin-containing dimeric and trimeric metabolites were found to be completely racemic, further supporting the combinatorial nature of the cross-coupling reactions between tricin and the monolignols. Therefore, it was clear that the dimeric compounds identified should not be termed flavonolignans, which should be optically active; these compounds should be considered as oligomers intended for the lignin polymer, and the more appropriate term "flavonolignol" or specifically "tricin-oligolignol" was coined for them (Lan, Morreel, et al., 2016).

The discovery of the flavone tricin incorporated into the lignin polymer boosted additional research to elucidate the enzymes involved in its biosynthesis as well as the interconnections with the monolignol biosynthetic pathway and provided additional insights into the mechanism of lignin formation, particularly in grasses (Lam, Liu, & Lo, 2015; Lam et al., 2017, 2021; del Río et al., 2020). The biosynthesis of flavones is controlled by chalcone synthase (CHS), which converts *p*-coumaroyl-CoA into naringenin chalcone, which is then converted to the flavanone naringenin by chalcone isomerase (CHI). The biosynthetic pathway leading from naringenin to tricin in grasses involves direct conversion of naringenin into apigenin by flavone synthase II (FNSII), and subsequent sequential and successive hydroxylations and *O*-methylations at the flavone B-ring by the corresponding flavonoid hydroxylases and *O*-methyltransferases, to produce the respective luteolin, chrysoeriol, selgin, and finally the flavone tricin (Lam et al., 2015, 2017). Disruption of the tricin biosynthetic pathway produced plants that incorporated flavonoids other than tricin into the lignin polymer. Hence, disruption of FNSII in rice resulted in a lignin that incorporated the intermediate naringenin, instead of tricin (Lam et al., 2017). Also, the lignin of culm tissues of a rice mutant lacking F3′H, the enzyme that hydroxylates the B-ring of apigenin to produce luteolin, incorporated the flavone apigenin (Lam et al., 2019). Expressing an apple CHS in poplar developing xylem led to the presence of naringenin in the lignin, suggesting that lignin-compatible flavonoid compounds can also be produced in lignifying tissues in poplar (Mahon et al., 2022).

More recently, flavonoids from different classes, such as the chalconoid naringenin chalcone, the flavanones naringenin and dihydrotricin, together with tricin, have been discovered to be incorporated into the lignin of papyrus (*Cyperus papyrus*) rind (Rencoret et al., 2022; Rosado, Bausch, et al., 2021). This is the first known case of a natural plant that incorporates such a variety of flavonoids from different classes, which makes this lignin unique. All of these flavonoids were released from the lignin by DFRC, indicating that at least some fraction of them was integrated into the lignin polymer as β-ether linked structures (Fig. 8). Dihydrotricin, as occurs with tricin, has the C-3′ and C-5′ positions at their B-rings occupied by methoxy groups, and can therefore only be incorporated into the lignin polymer through 4′–*O*–β bonds (Fig. 9). However, naringenin chalcone and naringenin have no substituents at these positions and can therefore form additional carbon-carbon linkages, including 3′– or 5′–β linkages producing phenylcoumaran structures upon incorporation into the lignin. Moreover, naringenin chalcone, with a double bond in the side-chain, can also form ether linkages through the 3-position of the side-chain, as shown in Fig. 9.

Several other flavonoids (taxifolin, quercetin, eriodictyol, apigenin, luteolin, and selgin) are known to participate in cross-coupling reactions with monolignols forming different flavonolignans (Begum, Sahai, & Ray, 2010; Chambers, Valentová, & Křen, 2015), which is an indication that these phenolic compounds may also be compatible with lignification. And, as shown above, apigenin has already been found incorporated into the lignin in rice mutants. Therefore, it cannot be ruled out that flavonoids other than those already discovered, are also incorporated into the lignins in normal plants and merely await identification.

3.2 Hydroxystilbenes

Hydroxystilbenes are another class of polyphenolic compounds derived from beyond the monolignol biosynthetic pathway that have also been reported to be incorporated into the lignin in some plants (del Río et al., 2017,

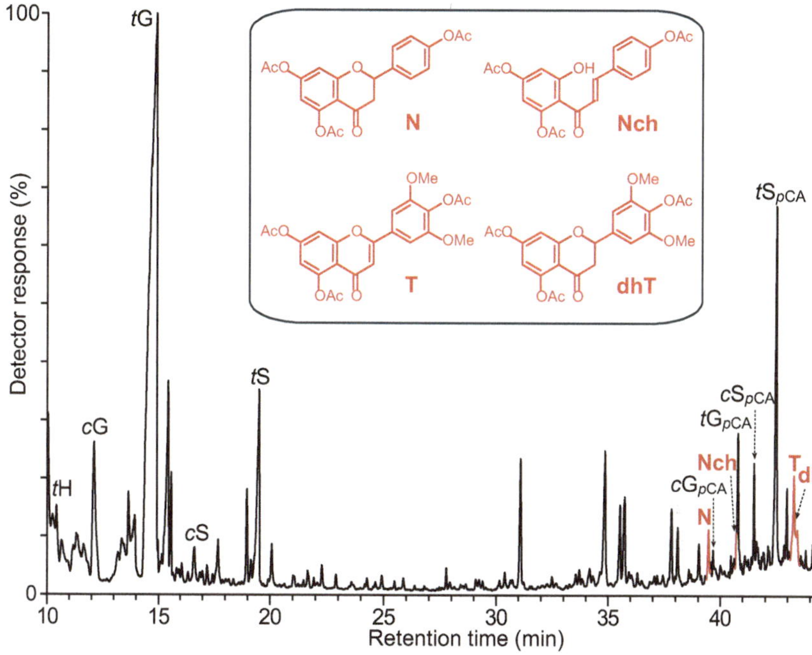

Fig. 8 Chromatogram of the DFRC degradation products released from the lignin of papyrus (*Cyperus papyrus*) rind, showing the presence of naringenin (**N**), naringenin chalcone (**Nch**), dihydrotricin (**dhT**), and tricin (**T**) (as their acetate derivatives). ***t*H**, ***c*G**, ***t*G**, ***c*S**, and ***t*S** are the *cis-* and *trans-p-*hydroxyphenyl, coniferyl, and sinapyl alcohols (as their acetate derivatives); ***c*G**$_{pCA}$, ***t*G**$_{pCA}$, ***c*S**$_{pCA}$, and ***t*S**$_{pCA}$ are the *cis-* and *trans-*coniferyl and sinapyl dihydro-*p*-coumarates (as their acetate derivatives). *Adapted from Rencoret, J., Rosado, M. J., Kim, H., Tymokhin, V., Gutiérrez, A., Bausch, F., et al. (2022). Flavonoids naringenin chalcone, naringenin, dihydrotricin, and tricin are lignin monomers in papyrus.* Plant Physiology, *188, 208–219.*

2020, 2021; Rencoret et al., 2018). Hydroxystilbenes can be oxidized to form radicals that are stabilized by resonance. These radicals can then participate in radical coupling reactions with other hydroxystilbenes as well as with monolignols and the growing lignin polymer to be incorporated into the lignin structure. Hydroxystilbenes are known to participate in cross-coupling reactions with monolignols producing the so-called stilbenolignans. For example, the hydroxystilbene piceatannol forms part of different stilbenolignans, such as aiphanol (Lee et al., 2001), and kompasinol A (Kobayashi et al., 1996), that are the β–O–4′, and β–8′ couplings, respectively, of sinapyl alcohol with piceatannol. The occurrence of all these stilbenolignans indicates the chemical compatibility of some hydroxystilbenes with lignification.

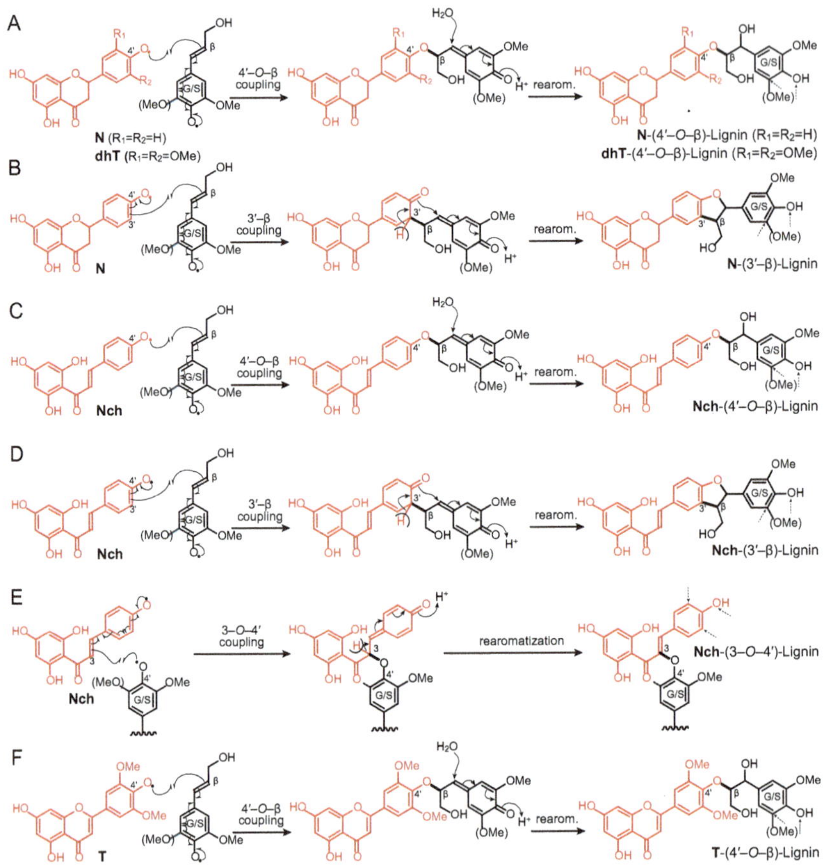

Fig. 9 Dehydrodimerization products arising from oxidative cross-coupling of the different flavonoids (naringenin, naringenin chalcone, dihydrotricin, and tricin) with monolignols. (A) 4′–*O*–β Coupling of naringenin (**N**, $R_1 = R_2 = H$) and dihydrotricin (**dhT**, $R_1 = R_2 = OMe$) with a monolignol producing aryl-alkyl ethers. (B) 3′–β Coupling of naringenin (**N**) and a monolignol producing phenylcoumaran structures. (C) 4′–*O*–β Coupling of naringenin chalcone (**Nch**) and a monolignol producing aryl-alkyl ethers. (D) 3′–β Coupling of naringenin chalcone (**Nch**) and a monolignol producing phenylcoumaran structures. (E) 3–*O*–4′ Coupling of naringenin chalcone (**Nch**) and the growing lignin chain producing alkyl-aryl ethers and conserving the double bond. (F) 4′–*O*–β Coupling of tricin (**T**) and a monolignol producing aryl-alkyl ethers. Sites of possible further radical coupling on the products are shown via small dashed arrows. *Adapted from Rencoret, J., Rosado, M. J., Kim, H., Tymokhin, V., Gutiérrez, A., Bausch, F., et al. (2022). Flavonoids naringenin chalcone, naringenin, dihydrotricin, and tricin are lignin monomers in papyrus.* Plant Physiology, *188, 208–219.*

Significant amounts of piceatannol (together with lesser quantities of resveratrol and isorhapontigenin) were found incorporated into the lignins of several palm fruit (macaúba, carnauba, coconut) endocarps. These compounds were released upon DFRC from these lignins (Fig. 10A), indicating that at least some fraction of them was incorporated into the lignins through β–O–4′ linkages, the ones cleaved by DFRC. Piceatannol can also couple and cross-couple with other piceatannols as well as with monolignols (and oligolignols) by a variety of linkages generating condensed structures that cannot be released by DFRC. The different modes of incorporation of piceatannol into the lignin was provided by 2D-NMR analyses (Fig. 10B) that allowed the identification of different linkages and structures involving piceatannol, including a benzodioxane structure (**Pb**) and a diphenylcoumaran structure (**Pc**) arising from homo-coupling of two piceatannol units, and a benzodioxane structure (**V**) produced by cross-coupling of piceatannol with a monolignol (or the growing lignin chain), as depicted in Fig. 10C.

The mechanisms of formation of the different structures involving piceatannol are depicted in Fig. 11. Structure **Pb** is formed by 8–*O*–4′ coupling of a piceatannol unit at its 8-position with the *O*–4′ position of another piceatannol unit followed by internal trapping of the quinone methide intermediate by the 3′-hydroxyl group and forming a benzodioxane ring (Fig. 11A). Similarly, structure **Pc** is formed by 8–10′ coupling of two piceatannol units followed by a subsequent 11′–*O*–7 bonding during rearomatization of the quinone methide intermediate and forming a phenylcoumaran structure (Fig. 11B). Finally, structure **V** is formed via β–*O*–4′ coupling of a monolignol (at its β-position) and the catechol part of piceatannol (at its *O*–4′ position) followed by internal trapping of the quinone methide intermediate by the 3′-hydroxyl group of piceatannol to form a benzodioxane ring (Fig. 11C). Biomimetic cross-coupling reactions between piceatannol and monolignols confirmed that benzodioxane structures **V** could be easily formed during the radical reaction (del Río et al., 2017). The identification of these benzodioxane structures **V** in the lignins from palm fruit endocarps conclusively demonstrates that piceatannol behaves as an authentic lignin monomer, participating in cross-coupling reactions with monolignols (or the growing lignin chain) during lignification and being integrated into the lignin polymer. A study of the thermodynamics of these coupling reactions using density functional theory (DFT) calculations indicated that the energetics of both

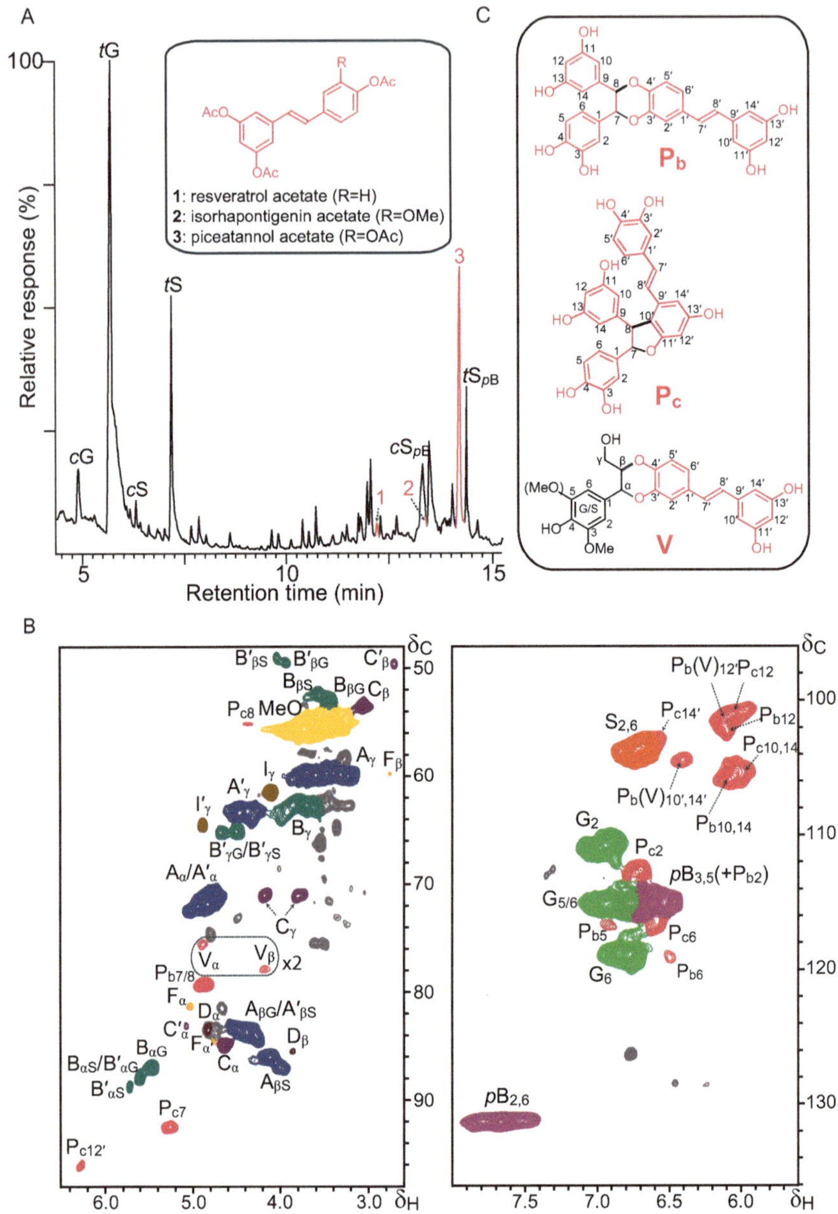

Fig. 10 (A) Chromatogram of the DFRC degradation products released from the lignin from macaúba (*Acrocomia aculeata*) fruit endocarp showing the presence of hydroxystilbenes (**1**, resveratrol, **2**, isorhapontigenin, **3**, piceatannol, as their acetate derivatives). *c***G**, *t***G**, *c***S**, and *t***S** are the normal *cis*- and *trans*-coniferyl (**G**) and sinapyl (**S**) alcohol monomers (as their acetate derivatives). ***c*S$_{pB}$** and ***t*S$_{pB}$** are the *cis*- and *trans*-sinapyl

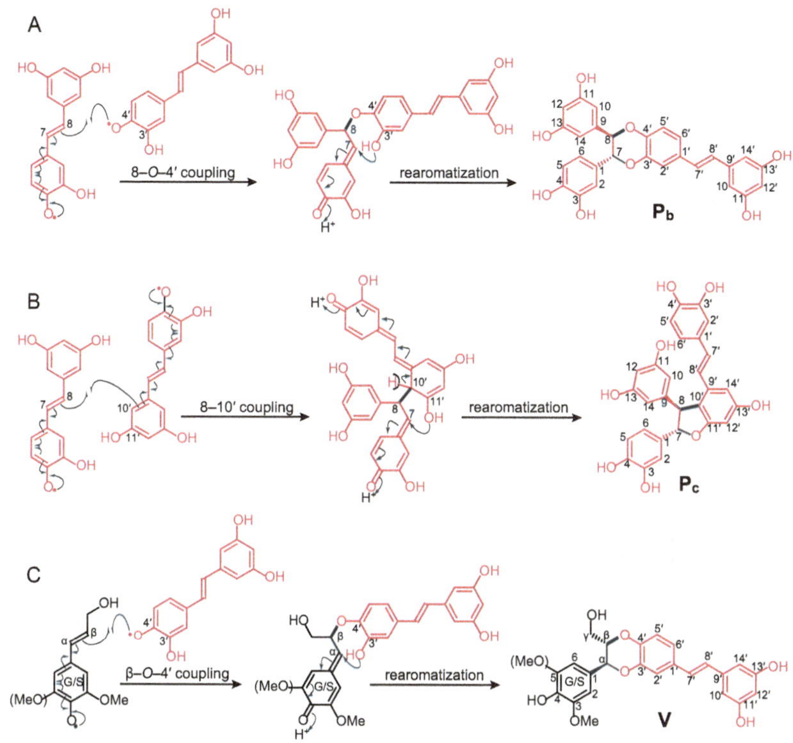

Fig. 11 Mechanism of radical coupling of piceatannol within the lignin polymer. (A) Piceatannol dehydrodimerization by 8–*O*–4′ coupling to give the benzodioxane structure **Pb**. (B) Piceatannol dehydrodimerization by 8–10′ coupling to give the diphenylcoumaran structure **Pc**. (C) Cross-coupling of monolignols with piceatannol via β–*O*–4′ coupling to give the benzodioxane structure **V**. *Adapted from del Río, J. C., Rencoret, J., Gutiérrez, A., Kim, H., & Ralph, J. (2017). Hydroxystilbenes are monomers in palm fruit endocarp lignins.* Plant Physiology, 174, *2072–2082.*

---

p-hydroxybenzoates (as their acetate derivatives). (B) Side-chain (δ_C/δ_H 48–98/2.6–6.5) and aromatic (δ_C/δ_H 96–155/5.6–8.0) regions of the 2D-NMR spectrum of the lignin isolated from macaúba fruit endocarp. **A**: β–*O*–4′ alkyl-aryl ethers, **A′**: β-ether structures with *p*-hydroxybenzoates acylating the γ-OH, **B**: phenylcoumarans, **B′**: phenylcoumarans with *p*-hydroxybenzoates acylating the γ-OH, **C**: resinols, **C′**: tetrahydrofuran structures formed by β–β′-coupling of monolignols acylated at the γ-OH, **D**: dibenzodioxocins, **F**: spirodienones, **I**: cinnamyl alcohol end-groups, **I′**: cinnamyl alcohol end-groups acylated at the γ-OH, ***J***: cinnamaldehyde end-groups, ***pB***: *p*-hydroxybenzoates, **G**: guaiacyl units, **S**: syringyl units, **P**: piceatannol units. (C) Structures of the main piceatannol homodimeric coupling (**Pb** and **Pc**) and piceatannol-lignin cross-coupling (**V**) products. *Adapted from del Río, J. C., Rencoret, J., Gutiérrez, A., Kim, H., & Ralph, J. (2017). Hydroxystilbenes are monomers in palm fruit endocarp lignins.* Plant Physiology, 174, *2072–2082.*

homo-coupling and cross-coupling reactions are comparable with those of the monolignol coupling, demonstrating the compatibility of piceatannol with the lignification process (Elder et al., 2019).

Hydroxystilbene glycosides, mostly isorhapontin (isorhapontigenin-*O*-glucoside) and, at lower levels, astringin (piceatannol-*O*-glucoside) and piceid (resveratrol-*O*-glucoside), have been reported to be incorporated into the lignin of Norway spruce bark (Neiva et al., 2020; Rencoret et al., 2019). The corresponding aglycones isorhapontigenin, piceatannol, and resveratrol, along with glucose, were released by DFRC (Fig. 12A), indicating that at least some fraction of the hydroxystilbene glucosides, in particular isorhapontin, was incorporated into the lignin as β-ether linked structures. Other types of linkages formed from the incorporation of the hydroxystilbene glucosides into the lignin of Norway spruce bark were identified by 2D-NMR (Fig. 12B), which revealed a series of structures involving hydroxystilbene glucosides that are similar to those formed from the incorporation of piceatannol into the lignins of palm fruit endocarps, including benzodioxane (**Pb**) and phenylcoumaran (**Pc**) structures involving coupling of two hydroxystilbene glucosides (mostly isorhapontin and astringin), as well as minor amounts of a benzodioxane structure (**V**) formed by the cross-coupling of astringin and coniferyl alcohol (Fig. 12C). Other phenylcoumaran structures arising from other types of linkages between two hydroxystilbene glucosides, such as 8–5′ and 8–12′, may also occur, but only the 8–10′ linkage could be unambiguously identified based on the occurrence of the characteristic signal $P_{c12'}$ in the HSQC spectrum (Rencoret et al., 2019). The mechanism of formation of these structures involving homo- and cross-coupling of hydroxystilbene glucosides and monolignols is similar to that described for piceatannol in Fig. 11. In addition, the hydroxystilbene glucoside isorhapontin contains a guaiacyl ring, and therefore it can also further cross-couple with coniferyl alcohol or lignin oligomers in different ways, including β–*O*–4′-alkyl aryl ethers, β–5′-phenylcoumarans, or 5–5′/4–*O*–β″-dibenzodioxocins (Rencoret et al., 2019). Further DFT studies indicated that there are no thermodynamic impediments to the incorporation of hydroxystilbene glucosides into the lignin polymer through these linkage types (Elder, Rencoret, del Río, Kim, & Ralph, 2021).

3.3 Ferulic acid amides

Several ferulic acid amides, such as feruloyltyramine and diferuloylputrescine, have been shown to be incorporated into the lignin structure in some plants through the radical coupling reactions that typify lignification.

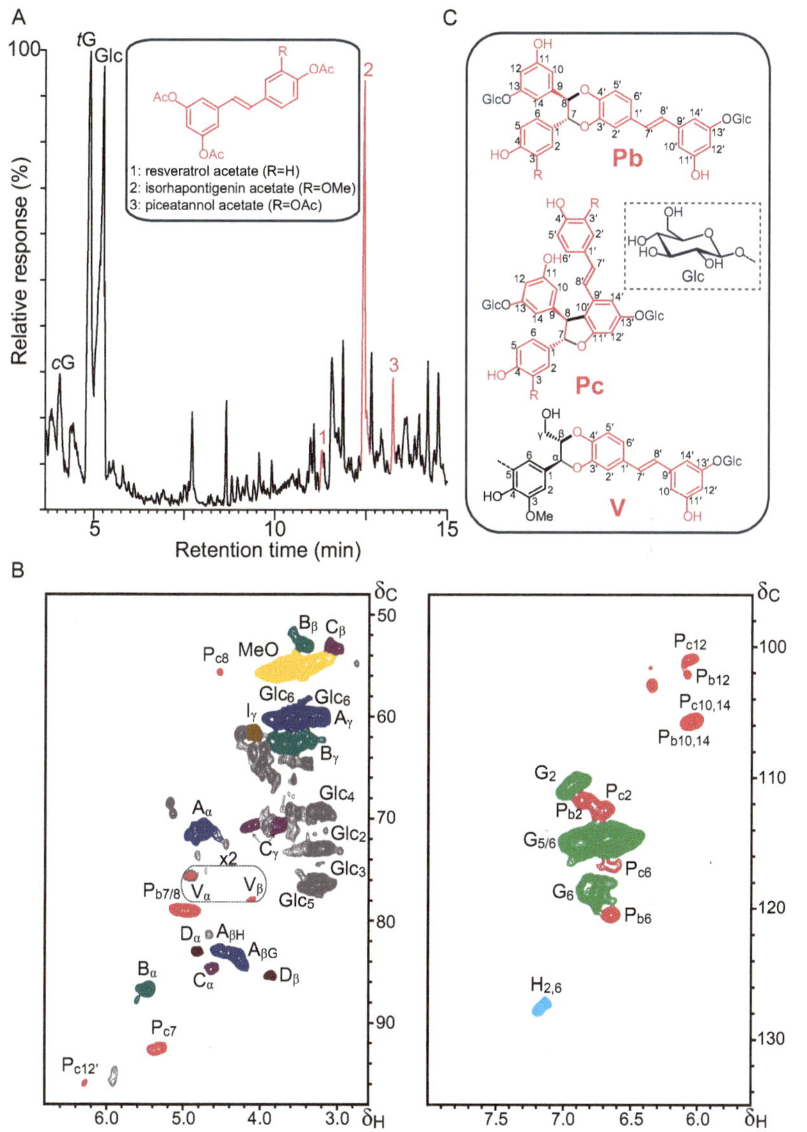

Fig. 12 (A) Chromatogram of the DFRC degradation products released from the lignin of Norway spruce bark showing the presence of hydroxystilbenes (**1**, resveratrol, **2**, isorhapontigenin, **3**, piceatannol), and glucose (Glc), as their acetate derivatives. *c***G** and *t***G**, are the *cis*- and *trans*-coniferyl alcohol monomers (as their acetate derivatives). (B) Side-chain (δ_C/δ_H 48–98/2.8–6.8) and aromatic (δ_C/δ_H 96–135/5.6–8.0) regions of the 2D-NMR spectrum of the lignin from Norway spruce bark showing the main signals from lignin and hydroxystilbene glucoside units. **P**: hydroxystilbene glucosides (isorhapontin, R=OMe; astringin, R=OH; piceid, R=H); **Pb**: 8–*O*–4′/3′–*O*–7 benzodioxane structures

(Continued)

Feruloyltyramine was the first ferulic acid amide described to be incorporated into the lignin in several plants from the Solanaceae. Feruloyltyramine was found in the lignin of tobacco (*Nicotiana tabacum*) plants (Kaur et al., 2012; Negrel & Jeandet, 1987; Ralph et al., 1998) and was shown to be covalently linked to the cell wall in potato (*Solanum tuberosum*) tubers (Negrel, Pollet, & Lapierre, 1996). More recently, feruloyltyramine was also found in the lignified pith of wild tobacco (*Nicotiana attenuata*) as a result of a defensive act after being attacked by the stem-boring insect *Trichobaris mucorea* (Joo et al., 2021). As in other ferulate conjugates, the ferulate moiety of feruloyltyramine can be oxidized by peroxidases and/or laccases forming radicals that are stabilized by resonance (Fig. 13) and that can participate in radical coupling reactions (with other ferulates, monolignols, or the growing lignin polymer) at their 4–O-, 5-, and 8- positions to become incorporated into the lignin. In principle, the phenolic group of the tyramine moiety can also be oxidized forming another radical that is delocalized over the tyramine aromatic ring system, providing additional sites for radical coupling (Fig. 13A). In potato tubers, the major fraction of the feruloyltyramine was attached to the cell wall via 8–O–4′-, and 4–O–β′- ether linkages, as well as via 8–5′ linkages forming a phenylcoumaran structure, involving the ferulate moiety, but a significant part (20% of all ether-linkages) was reportedly also attached to the cell wall through the phenolic group of the tyramine moiety (Bernards et al., 1995; Negrel et al., 1996); given the issues with the determinations, however, we suspect the etherification level of tyramine, as with *p*-coumarate, is considerably lower (Ralph, 2010). As feruloyltyramine has phenolic groups at both ends of the molecule, coupling with monolignols can therefore occur at both positions, and a compound arising from feruloyltyramine dicovalently bound to the cell wall through both phenolic

---

Fig. 12—Cont'd involving isorhapontin (R=OMe), astringin (R=OH) or piceid (R=H) units; **Pc**: 8–10′/11′–7 phenylcoumaran structures involving isorhapontin (R=OMe), astringin (R=OH) or piceid (R=H) units; **V**: β–*O*–4′/3′–*O*–α benzodioxane structure formed by cross-coupling of astringin and coniferyl alcohol; Glc: glucose; **A**: β–*O*–4′ alkyl-aryl ethers, **B**: phenylcoumarans, **C**: resinols, **D**: dibenzodioxocins, **I**: cinnamyl alcohol end-groups, **H**: *p*-hydroxyphenyl units, **G**: guaiacyl units. (**C**) Structures of the main hydroxystilbene glucosides homodimeric coupling products (**Pb**, and **Pc**) and hydroxystilbene glucosides-lignin cross-coupling products (**V**) present in the lignin of Norway spruce bark. *Adapted from Rencoret, J., Neiva, D., Marques, G., Gutiérrez, A., Kim, H., Gominho, J., et al. (2019). Hydroxystilbene glucosides are incorporated into Norway spruce bark lignin.* Plant Physiology, 180, *1310–1321.*

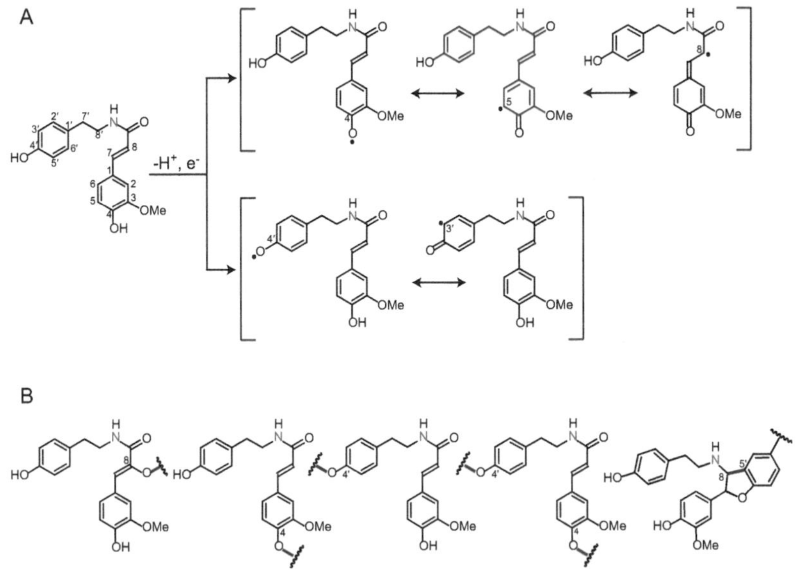

Fig. 13 (A) Oxidative radicalization of feruloyltyramine, and resonance forms at 4–*O*-, 5- and 8-positions of the ferulate moiety. The tyramine moiety may also be oxidized in the same manner producing a radical that is stabilized by resonance and can couple at its 4′–*O*-, and 3′- positions. (B) Different modes of incorporation of feruloyltyramine into the lignin through 8-, 4–*O*- and 4′–*O*- ether linkages and through 8–5′ linkages forming phenylcoumaran structures, as identified by thioacidolysis (Negrel et al., 1996) and NMR (Bernards, Lopez, Zajicek, & Lewis, 1995) analyses. In principle, tyramine units can react at 3′/5′ as well but the products have not been identified.

groups was released upon thioacidolysis of potato tubers (Negrel et al., 1996). The different modes of incorporation of feruloyltyramine into the lignin through 8-, 4–*O*- and 4′–*O*- ether linkages as well as through 8–5′ linkages forming phenylcoumaran structures, as identified by thioacidolysis (Negrel et al., 1996) and NMR (Bernards et al., 1995), are shown in Fig. 13B. In general terms, it is likely that the ferulate moiety of feruloyltyramine can be incorporated into the lignin through the same types of linkages as those depicted in Fig. 6 for ferulates attached to arabinoxylans.

The ferulic acid amide diferuloylputrescine was found incorporated into the lignin of maize kernels (del Río et al., 2018). As with feruloyltyramine and other ferulate conjugates, each ferulate moiety in the diferuloylputrescine molecule can be oxidized to form radicals that are stabilized by resonance at their 4–*O*-, 5-, and 8- positions (Fig. 14A). These radicals can then undergo coupling and cross-coupling reactions with another

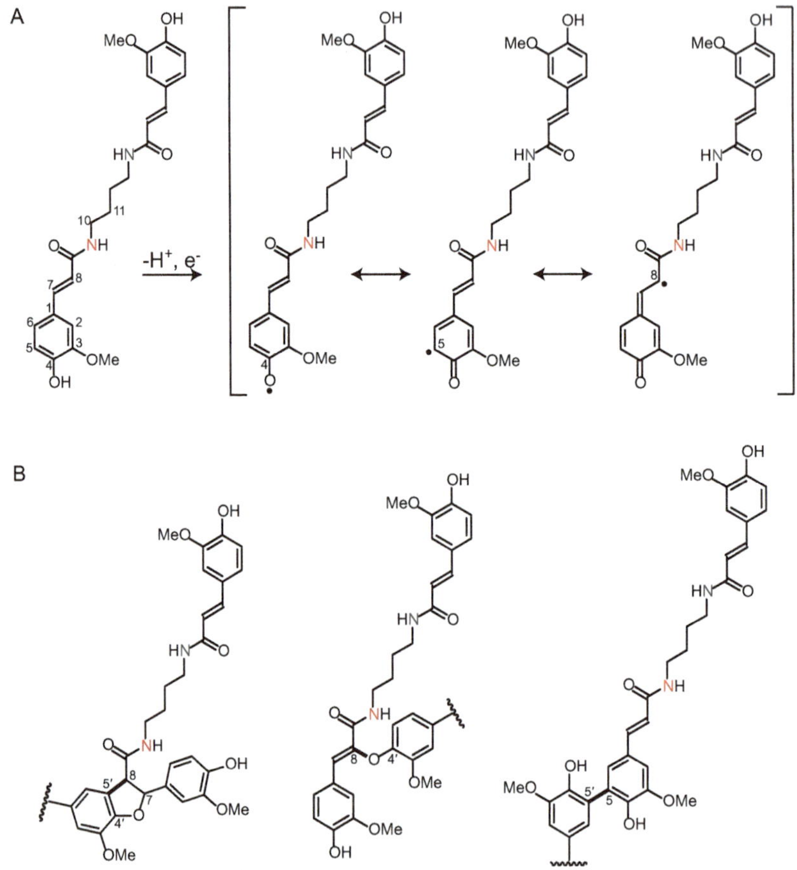

Fig. 14 (A) Oxidative radicalization of diferuloylputrescine, and various resonance forms of the radical of one ferulate moiety. The other ferulate moiety can also be oxidized in the same manner, allowing it to undergo independent radical coupling. (B) Modes of incorporation of diferuloylputrescine into the lignin polymer through cross-coupling with other ferulates, monolignols, or the growing lignin chain, through 8–5′, 8–*O*–4′, and 5–5′ coupling.

ferulate moiety, or with monolignols and the growing lignin chain, to become incorporated into the lignin polymer. The different modes of incorporation of diferuloylputrescine into the lignin of maize kernels were elucidated by NMR, and different types of linkages and structures were found. The analyses indicated that diferuloylputrescine was incorporated into the lignin polymer mostly through 8–5′ linkages involving one of the ferulate moieties and producing a phenylcoumaran structure (Fig. 14B). Other structures involving the incorporation of diferuloylputrescine into the lignin polymer through 8–*O*–4′, 5–5′, and other linkages were also evidenced by NMR (del Río et al., 2018). Also, other types of linkages similar to those produced during the incorporation of ferulates acylating polysaccharides, as depicted in Fig. 6, can also occur, including ethers through the 4–*O*- position of the ferulate moiety. As already indicated for feruloyltyramine, the particular structure of diferuloylputrescine, with two ferulates at both ends of the molecule, can allow lignification to proceed in both directions, forming covalent linkages at both ends of the molecule. Furthermore, since coupling is also possible at the 8- positions of the sidechain of the ferulate moiety, this implies that diferuloylputrescine can form branching points in the lignin polymer, thus producing a highly cross-linked polymeric network.

4. Conclusions and perspectives

Lignin has traditionally been considered to be derived from the oxidative polymerization of the three canonical monolignols. However, as shown here, it is clear that the lignin structure is much more complex than previously thought, and that many other phenolic compounds, including phenolic compounds derived from the classical monolignol biosynthetic pathway, as well as other phenolic compounds arising from different polyphenolic biosynthetic pathways (flavonoids, hydroxystilbenes, ferulic acid amides), can also be incorporated into the lignin polymer in several plants where they behave as authentic lignin monomers participating in radical coupling reactions during lignification, thus expanding and challenging the traditional definition of lignin. Moreover, from the examination of a large number of lignins during recent decades, it is obvious that most of these unconventional units are rather more common than previously thought, but they have only been poorly described. It is also likely that other unconventional lignin monomers may be discovered in the near future thus expanding the range of phenolic compounds that are incorporated into the lignin polymer. The occurrence of all of these unconventional phenolic compounds, and particularly the discovery of phenolic precursors arising from beyond the established monolignol biosynthetic pathway, reveals that the mechanism of lignification is remarkably flexible and demonstrates, as we reviewed many years ago (Ralph, 2010), that any phenolic compound that is transported to the cell wall can be oxidized and incorporated into the lignin polymer via radical coupling reactions, and subject exclusively to simple chemical compatibility. Therefore, it becomes evident that the definition of lignin should not be restricted only to a polymer made of the three canonical monolignols, but should also encompass a wide range of phenolic precursors that are also integrated into the lignin structure in many plants. The encouraging aspect of this added complexity is that there is still only a single reaction, radical coupling of phenols, known to be responsible for the polymerization events of lignification.

Acknowledgments

JCdR, JRe and AG were funded by the Spanish Project AGL2017-83036-R and PID2020-118968RB-100 (financed by Ministerio de Ciencia e Innovación, MCIN, Agencia Estatal de Investigación, AEI, and European Regional Development Fund, ERDF, "A way of making Europe"), the Regional Andalusian Government (Consejería de Transformación Económica, Industria, Conocimiento y Universidades, Junta de Andalucía, project P20-00017), and CSIC (project PIE 202040E185). JRa and HK were funded by the DOE Great Lakes Bioenergy Research Center (DOE Office of Science DE-SC0018409).

References

Ando, D., Lu, F., Kim, H., Eugene, A., Tobimatsu, Y., Vanholme, R., et al. (2021). Incorporation of catechyl monomers into lignins: Lignification from the non-phenolic end *via* Diels-Alder cycloaddition? *Green Chemistry*, *23*, 8995–9013.

Barros, J., Serrani-Yarce, J. C., Chen, F., Baxter, D., Venables, B. J., & Dixon, R. A. (2016). Role of bifunctional ammonia-lyase in grass cell wall biosynthesis. *Nature Plants*, *2*, 16050.

Begum, S. A., Sahai, M., & Ray, A. B. (2010). Non-conventional lignans: Coumarinolignans, flavonolignans, and stilbenolignans. *Fortschritte der Chemie organischer Naturstoffe/Progress in the Chemistry of Organic Natural Products*, *93*, 1–70.

Bernards, M. A., Lopez, M. L., Zajicek, J., & Lewis, N. G. (1995). Hydroxycinnamic acid-derived polymers constitute the polyaromatic domain of suberin. *Journal of Biological Chemistry*, *270*, 7382–7386.

Blaschek, L., Champagne, A., Dimotakis, C., Nuoendagula, C., Decou, R., Hishiyama, S., et al. (2020). Cellular and genetic regulation of coniferaldehyde incorporation in lignin of herbaceous and woody plants by quantitative Wiesner staining. *Frontiers in Plant Science*, *11*, 109.

Boerjan, W., Ralph, J., & Baucher, M. (2003). Lignin biosynthesis. *Annual Review of Plant Biology*, *54*, 519–546.

Bunzel, M., Ralph, J., Funk, C., & Steinhart, H. (2005). Structural elucidation of new ferulic acid-containing phenolic dimers and trimers isolated from maize bran. *Tetrahedron Letters*, *46*, 5845–5850.

Bunzel, M., Ralph, J., Lu, F., Hatfield, R. D., & Steinhart, H. (2004). Lignins and ferulate-coniferyl alcohol cross-coupling products in cereal grains. *Journal of Agricultural and Food Chemistry*, *52*, 6496–6502.

Campbell, M. M., & Sederoff, R. R. (1996). Variation in lignin content and composition. *Plant Physiology*, *110*, 3–13.

Chambers, S. C., Valentová, K., & Křen, V. (2015). Non-taxifolin derived flavonolignans: Phytochemistry and biology. *Current Pharmaceutical Design*, *21*, 5489–5500.

Chedgy, R. J., Köllner, T. G., & Constabel, C. P. (2015). Functional characterization of two acyltransferases from *Populus trichocarpa* capable of synthesizing benzyl benzoate and salicyl benzoate, potential intermediates in salicinoid phenolic glycoside biosynthesis. *Phytochemistry, 113*, 149–159.

Chen, C.-L. (1992). Determination of carbonyl groups. In Y. Lin, & W. W. Dence (Eds.), *Methods in lignin chemistry* (pp. 446–457). New York: Springer-Verlag.

Chen, F., Tobimatsu, Y., Havkin-Frenkel, D., Dixon, R. A., & Ralph, J. (2012). A polymer of caffeyl alcohol in plant seeds. *Proceedings of the National Academy of Sciences of the United States of America, 109*, 1772–1777.

Chen, F., Tobimatsu, Y., Jackson, L., Nakashima, J., Ralph, J., & Dixon, R. A. (2013). Novel seed coat lignins in the Cactaceae: Structure, distribution and implications for the evolution of lignin diversity. *The Plant Journal, 73*, 201–211.

del Río, J. C., Gutiérrez, A., & Martínez, A. T. (2004). Identifying acetylated lignin units in non-wood fibers using pyrolysis-gas chromatography/mass spectrometry. *Rapid Communications in Mass Spectrometry, 18*, 1181–1185.

del Río, J. C., Lino, A. G., Colodette, J. L., Lima, C. F., Gutiérrez, A., Martínez, A. T., et al. (2015). Differences in the chemical structures of the lignins from sugarcane bagasse and straw. *Biomass & Bioenergy, 81*, 322–338.

del Río, J. C., Marques, G., Rencoret, J., Martínez, A. T., & Gutiérrez, A. (2007). Occurrence of naturally acetylated lignin units. *Journal of Agricultural and Food Chemistry, 55*, 5461–5468.

del Río, J. C., Martínez, A. T., & Gutiérrez, A. (2007). Presence of 5-hydroxyguaiacyl units as native lignin constituents in plants as seen by Py-GC/MS. *Journal of Analytical and Applied Pyrolysis, 79*, 33–38.

del Río, J. C., Prinsen, P., Rencoret, J., Nieto, L., Jiménez-Barbero, J., Ralph, J., et al. (2012a). Structural characterization of the lignin in the cortex and pith of elephant grass (*Pennisetum purpureum*) stems. *Journal of Agricultural and Food Chemistry, 60*, 3619–3634.

del Río, J. C., Rencoret, J., Gutiérrez, A., Elder, T., Kim, H., & Ralph, J. (2020). Lignin monomers from beyond the canonical monolignol biosynthetic pathway: Another brick in the wall. *ACS Sustainable Chemistry & Engineering, 8*, 4997–5012.

del Río, J. C., Rencoret, J., Gutiérrez, A., Kim, H., & Ralph, J. (2017). Hydroxystilbenes are monomers in palm fruit endocarp lignins. *Plant Physiology, 174*, 2072–2082.

del Río, J. C., Rencoret, J., Gutiérrez, A., Kim, H., & Ralph, J. (2018). Structural characterization of lignin from maize (*Zea mays* L.) fibers: Evidences for diferuloylputrescine incorporated into the lignin polymer in maize kernels. *Journal of Agricultural and Food Chemistry, 66*, 4402–4413.

del Río, J. C., Rencoret, J., Gutiérrez, A., Lan, W., Kim, H., & Ralph, J. (2021). Lignin monomers derived from the flavonoid and hydroxystilbene biosynthetic pathways. In J. D. Reed, V. A. P. de Freitas, & S. Quideau (Eds.), *Recent advances in polyphenol research: Vol. 7* (pp. 177–206). Hoboken, N.J: John Wiley & Sons Ltd.

del Río, J. C., Rencoret, J., Marques, G., Gutiérrez, A., Ibarra, D., Santos, J. I., et al. (2008). Highly acylated (acetylated and/or *p*-coumaroylated) native lignins from diverse herbaceous plants. *Journal of Agricultural and Food Chemistry, 56*, 9525–9534.

del Río, J. C., Rencoret, J., Prinsen, P., Martínez, A. T., Ralph, J., & Gutiérrez, A. (2012b). Structural characterization of wheat straw lignin as revealed by analytical pyrolysis, 2D-NMR, and reductive cleavage methods. *Journal of Agricultural and Food Chemistry, 60*, 5922–5935.

Dixon, R. A., Chen, F., Guo, D., & Parvathi, K. (2001). The biosynthesis of monolignols: A "metabolic grid" or independent pathways to guaiacyl and syringyl units? *Phytochemistry, 57*, 1069–1084.

Dixon, R. A., & Reddy, M. S. S. (2003). Biosynthesis of monolignols. Genomic and reverse genetic approaches. *Phytochemical Reviews*, *2*, 289–306.
Donaldson, L. A. (2001). Lignification and lignin topochemistry—An ultrastructure view. *Phytochemistry*, *57*, 859–873.
Elder, T., del Río, J. C., Ralph, J., Rencoret, J., Kim, H., & Beckham, G. T. (2019). Radical coupling reactions of piceatannol and monolignols: A density functional theory study. *Phytochemistry*, *164*, 12–23.
Elder, T., Rencoret, J., del Río, J. C., Kim, H., & Ralph, J. (2021). Radical coupling reactions of hydroxystilbene glucosides and coniferyl alcohol: A density functional theory study. *Frontiers in Plant Science*, *12*, 642848.
Eugene, A., Lapierre, C., & Ralph, J. (2020). Improved analysis of arabinoxylan-bound hydroxycinnamate conjugates in grass cell walls. *Biotechnology for Biofuels*, *13*, 202.
Freudenberg, K., & Neish, A. C. (1968). *Constitution and biosynthesis of lignin*. Berlin-Heidelberg-New York: Springer-Verlag.
Fukagawa, N., Meshitsuka, G., & Ishizu, A. (1991). A two dimensional NMR study of birch milled wood lignin. *Journal of Wood Chemistry and Technology*, *11*, 373–396.
Grabber, J. H., Ralph, J., & Hatfield, R. D. (2000). Cross-linking of maize walls by ferulate dimerization and incorporation into lignin. *Journal of Agricultural and Food Chemistry*, *48*, 6106–6113.
Grabber, J. H., Ralph, J., & Hatfield, R. D. (2002). Model studies of ferulate-coniferyl alcohol cross-product formation in primary maize walls: Implications for lignification in grasses. *Journal of Agricultural and Food Chemistry*, *50*, 6008–6016.
Hatfield, R. D., Marita, J. M., Frost, K., Grabber, J. H., Lu, F., Kim, H., et al. (2009). Grass lignin acylation: *p*-Coumaroyl transferase activity and cell wall characteristics of C3 and C4 grasses. *Planta*, *229*, 1253–1267.
Hatfield, R. D., Ralph, J., & Grabber, J. H. (1999). Cell wall cross-linking by ferulates and diferulates in grasses. *Journal of the Science of Food and Agriculture*, *79*, 403–407.
Hoengenaert, L., Vanholme, R., Wouters, M., Kim, H., De Meester, B., Morreel, K., et al. (2022). Scopoletin is a natural lignin monomer in *Arabidopsis*, and its enhanced incorporation into lignin increases biomass processing efficiency. *Science Advances*. (in revision) Submitted for publication.
Joo, Y., Kim, H., Kang, M., Lee, G., Choung, S., Kaur, H., et al. (2021). Pith-specific lignification in *Nicotiana attenuata* as a defense against a stem-boring herbivore. *New Phytologist*, *232*, 332–344.
Kaal, J., Serrano, O., del Río, J. C., & Rencoret, J. (2018). Radically different lignin composition in *Posidonia* species may link to differences in organic carbon sequestration capacity. *Organic Geochemistry*, *124*, 247–256.
Karlen, S. D., Free, H. C. A., Padmakshan, D., Smith, B. G., Ralph, J., & Harris, P. J. (2018). Commelinid monocotyledon lignins are acylated by *p*-coumarate. *Plant Physiology*, *177*, 513–521.
Karlen, S. D., Smith, R. A., Kim, H., Padmakshan, D., Bartuce, A., Mobley, J. K., et al. (2017). Highly decorated lignins in leaf tissues of the Canary Island date palm *Phoenix canariensis*. *Plant Physiology*, *175*, 1058–1067.
Karlen, S. D., Zhang, C., Peck, M. L., Smith, R. A., Padmakshan, D., Helmich, K. E., et al. (2016). Monolignol ferulate conjugates are naturally incorporated into plant lignins. *Science Advances*, *2*, e1600393.
Kasahara, H., Jiao, Y., Bedgar, D. L., Kim, S. J., Patten, A. M., Xia, Z. Q., et al. (2006). *Pinus taeda* phenylpropenal double-bond reductase: Purification, cDNA cloning, heterologous expression in *Escherichia coli*, and subcellular localization in *P. taeda*. *Phytochemistry*, *67*, 1765–1780.

Kaur, H., Shaker, K., Heinzel, N., Ralph, J., Gális, I., & Baldwin, I. T. (2012). Environmental stresses of field growth allow cinnamyl alcohol dehydrogenase-deficient *Nicotiana attenuata* plants to compensate for their structural deficiencies. *Plant Physiology, 159*, 1545–1570.

Kim, H., Li, Q., Karlen, S. D., Smith, R. A., Shi, R., Liu, J., et al. (2020). Monolignol benzoates incorporate into the lignin of transgenic *Populus trichocarpa* depleted in C3H and C4H. *ACS Sustainable Chemistry & Engineering, 8*, 3644–3654.

Kim, H., Ralph, J., Lu, F., Pilate, G., Leplé, J. C., Pollet, B., et al. (2002). Identification of the structure and origin of thioacidolysis marker compounds for cinnamyl alcohol dehydrogenase deficiency in angiosperms. *Journal of Biological Chemistry, 277*, 47412–47419.

Kim, H., Ralph, J., Lu, F., Ralph, S. A., Boudet, A.-M., MacKay, J. J., et al. (2003). NMR analysis of lignins in CAD-deficient plants. Part 1. Incorporation of hydroxycinnamaldehydes and hydroxybenzaldehydes into lignins. *Organic & Biomolecular Chemistry, 1*, 168–281.

Kim, H., Ralph, J., Yahiaoui, N., Pean, M., & Boudet, A.-M. (2000). Cross-coupling of hydroxycinnamyl aldehydes into lignins. *Organic Letters, 2*, 2197–2200.

Kobayashi, M., Mahmud, T., Yoshioka, N., Hori, K., Shibuya, H., & Kitagawa, I. (1996). Indonesian medicinal plants. XVIII. Kompasinol A, a new stilbeno-phenylpropanoid from the bark of *Koompassia malaccensis* (Fabaceae). *Chemical and Pharmaceutical Bulletin, 44*, 2249–2253.

Kuroda, K.-I., Ozawa, T., & Ueno, T. (2001). Characterization of sago palm (*Metroxylon sagu* Rottb.) lignin by analytical pyrolysis. *Journal of Agricultural and Food Chemistry, 49*, 1840–1847.

Lam, P. Y., Liu, H., & Lo, C. (2015). Completion of tricin biosynthesis pathway in rice: Cytochrome P450 75B4 is a unique chrysoeriol 5′-hydroxylase. *Plant Physiology, 168*, 1527–1536.

Lam, P. Y., Lui, A. C. W., Wang, L., Liu, H., Umezawa, T., Tobimatsu, Y., et al. (2021). Tricin biosynthesis and bioengineering. *Frontiers in Plant Science, 12*, 733198.

Lam, P. Y., Lui, A. C. W., Yamamura, M., Wang, L., Takeda, Y., Suzuki, S., et al. (2019). Recruitment of specific flavonoid B-ring hydroxylases for two independent biosynthesis pathways of flavone derived metabolites in grasses. *New Phytologist, 223*, 204–219.

Lam, P. Y., Tobimatsu, Y., Takeda, Y., Suzuki, S., Yamamura, M., Umezawa, T., et al. (2017). Disrupting flavone synthase II alters lignin and improves biomass digestibility. *Plant Physiology, 174*, 972–985.

Lan, W., Lu, F., Regner, M., Zhu, Y., Rencoret, J., Ralph, S. A., et al. (2015). Tricin, a flavonoid monomer in monocot lignification. *Plant Physiology, 167*, 1284–1295.

Lan, W., Morreel, K., Lu, F., Rencoret, J., del Río, J. C., Vooren, W., et al. (2016). Maize tricin-oligolignol metabolites and their implications for monocot lignification. *Plant Physiology, 171*, 810–820.

Lan, W., Rencoret, J., del Río, J. C., & Ralph, J. (2019). Tricin in grass lignin: Biosynthesis, characterization, and quantitation. In F. Lu, & F. Yue (Eds.), *Lignin: Biosynthesis, functions and economic significance* (pp. 51–78). New York: Nova Science Publishers, Inc.

Lan, W., Rencoret, J., Lu, F., Karlen, S. D., Smith, B. G., Harris, P. J., et al. (2016). Tricin-lignins: Occurrence and quantitation of tricin in relation to phylogeny. *The Plant Journal, 88*, 1046–1057.

Landucci, L. L., Deka, G. C., & Roy, D. N. (1992). A ^{13}C NMR study of milled wood lignins from hybrid *Salix* clones. *Holzforschung, 46*, 505–511.

Lapierre, C., Pilate, G., Pollet, B., Mila, I., Leplé, J. C., Jouanin, L., et al. (2004). Signatures of cinnamyl alcohol dehydrogenase deficiency in poplar lignins. *Phytochemistry, 65*, 313–321.

Lapierre, C., Pollet, B., MacKay, J. J., & Sederoff, R. R. (2000). Lignin structure in a mutant pine deficient in cinnamyl alcohol dehydrogenase. *Journal of Agricultural and Food Chemistry*, *48*, 2326–2331.

Lapierre, C., Tollier, M. T., & Monties, B. (1988). A new type of constitutive unit in lignins from the corn *bm3* mutant. *Comptes Rendus de l'Académie des Sciences. Série III*, *307*, 723–728.

Lee, D., Cuendet, M., Vigo, J. S., Graham, J. G., Cabieses, F., Fong, H. H., et al. (2001). A novel cyclooxygenase-inhibitory stilbenolignan from the seeds of *Aiphanes aculeata*. *Organic Letters*, *3*, 2169–2171.

Li, L., Popko, J. L., Umezawa, T., & Chiang, V. L. (2000). 5-Hydroxyconiferyl aldehyde modulates enzymatic methylation for syringyl monolignol formation, a new view of monolignol biosynthesis in angiosperms. *Journal of Biological Chemistry*, *275*, 6537–6545.

Liu, C.-J. (2012). Deciphering the enigma of lignification: Precursor transport, oxidation and the topochemistry of lignin assembly. *Molecular Plant*, *5*, 304–317.

Liu, X., Van Acker, R., Voorend, W., Pallidis, A., Goeminne, G., Pollier, J., et al. (2020). Rewired phenolic metabolism and improved saccharification efficiency of a *Zea mays cinnamyl alcohol dehydrogenase 2 (Zmcad2)* mutant. *The Plant Journal*, *105*, 1240–1257.

Lourenço, A., Rencoret, J., Chematova, C., Gominho, J., Gutiérrez, A., del Río, J. C., et al. (2016). Lignin composition and structure differs between xylem, phloem and phellem in *Quercus suber* L. *Frontiers in Plant Science*, *7*, 1612.

Lu, F., Karlen, S. D., Regner, M., Kim, H., Ralph, S., Sun, R.-C., et al. (2015). Naturally *p*-hydroxybenzoylated lignins in palms. *Bioenergy Research*, *8*, 934–952.

Lu, F., & Ralph, J. (1999). Detection and determination of *p*-coumaroylated units in lignin. *Journal of Agricultural and Food Chemistry*, *47*, 1988–1992.

Lu, F., & Ralph, J. (2002). Preliminary evidence for sinapyl acetate as a lignin monomer in kenaf. *Chemical Communications*, *1*, 90–91.

Lu, F., & Ralph, J. (2005). Novel β–β structures in lignins incorporating acylated monolignols. *Appita*, 233–237.

Lu, F., & Ralph, J. (2008). Novel tetrahydrofuran structures derived from β–β-coupling reactions involving sinapyl acetates in kenaf lignins. *Organic & Biomolecular Chemistry*, *6*, 3681–3694.

Lu, F., Ralph, J., Morreel, K., Messens, E., & Boerjan, W. (2004). Preparation and relevance of a cross-coupling product between sinapyl alcohol and sinapyl *p*-hydroxybenzoate. *Organic & Biomolecular Chemistry*, *2*, 2888–2890.

Lüdemann, H.-D., & Nimz, H. (1974). ^{13}C Kernresonanz-spektren von Ligninen. Buchen- und Fichten-Björkman-lignin. *Die Makromolekulare Chemie*, *175*, 2409–2422.

Lui, A. C. W., Lam, P. Y., Chan, K. H., Wang, L., Tobimatsu, Y., & Lo, C. (2020). Convergent recruitment of 50-hydroxylase activities by CYP75B flavonoid B-ring hydroxylases for tricin biosynthesis in *Medicago* legumes. *New Phytologist*, *228*, 269–284.

Lundquist, K., & Stern, K. (1989). Analysis of lignins by ^{1}H NMR spectroscopy. *Nordic Pulp & Paper Research Journal*, *4*, 210–213.

Mahon, E. L., de Vries, L., Jang, S.-K., Middar, S., Kim, H., Unda, F., et al. (2022). Exogenous chalcone synthase expression in developing poplar xylem incorporates naringenin into lignins. *Plant Physiology*, *188*, 984–996.

Marita, J. M., Hatfield, R. D., Rancour, D. M., & Frost, K. E. (2014). Identification and suppression of the *p*-coumaroyl CoA:Hydroxycinnamyl alcohol transferase in *Zea mays* L. *The Plant Journal*, *78*, 850–864.

Marita, J. M., Ralph, J., Hatfield, R. D., Guo, D., Chen, F., & Dixon, R. A. (2003). Structural and compositional modifications in lignin of transgenic alfalfa down-regulated in caffeic acid 3-*O*-methyltransferase and caffeoyl coenzyme A 3-*O*-methyltransferase. *Phytochemistry*, *62*, 53–65.

Martin, A. F., Tobimatsu, Y., Kusumi, R., Matsumoto, N., Miyamoto, T., Lam, P. Y., et al. (2019). Altered lignocellulose chemical structure and molecular assembly in *CINNAMYL ALCOHOL DEHYDROGENASE*-deficient rice. *Scientific Reports*, *9*, 17153.

Martínez, A. T., Rencoret, J., Marques, G., Gutiérrez, A., Ibarra, D., Jiménez-Barbero, J., et al. (2008). Monolignol acylation and lignin structure in some non-woody plants: A 2D NMR study. *Phytochemistry*, *69*, 2831–2843.

Morreel, K., Ralph, J., Kim, H., Lu, F., Goeminne, G., Ralph, S. A., et al. (2004). Profiling of oligolignols reveals monolignol coupling conditions in lignifying poplar xylem. *Plant Physiology*, *136*, 3537–3549.

Negrel, J., & Jeandet, P. (1987). Metabolism of tyramine and feruloyltyramine in TMV inoculated leaves of *Nicotiana tabacum*. *Phytochemistry*, *26*, 2185–2190.

Negrel, J., Pollet, B., & Lapierre, C. (1996). Ether-linked ferulic acid amides in natural and wound periderms of potato tuber. *Phytochemistry*, *43*, 1195–1199.

Neiva, D., Rencoret, J., Marques, G., Gutiérrez, A., Gominho, J., Pereira, H., et al. (2020). Lignin from tree barks: Chemical structure and valorization. *ChemSusChem*, *13*, 4537–4547.

Parvathi, K., Chen, F., Guo, D. J., Blount, J. W., & Dixon, R. A. (2001). Substrate preferences of *O*-methyltransferases in alfalfa suggest new pathways for 3-*O*-methylation of monolignols. *The Plant Journal*, *25*, 193–202.

Petrik, D., Karlen, S. D., Cass, C., Padmakshan, D., Lu, F., Liu, S., et al. (2014). *p*-Coumaroyl-CoA:Monolignol transferase (PMT) acts specifically in the lignin biosynthetic pathway in *Brachypodium distachyon*. *The Plant Journal*, *77*, 713–726.

Ralph, J. (1996). An unusual lignin from kenaf. *Journal of Natural Products*, *59*, 341–342.

Ralph, J. (2010). Hydroxycinnamates in lignification. *Phytochemistry Reviews*, *9*, 65–83.

Ralph, J. (2020). Tricin and tricin-lignins in *Medicago* versus in monocots. *New Phytologist*, *228*, 11–14.

Ralph, J., Brunow, G., Harris, P. J., Dixon, R. A., Schatz, P. F., & Boerjan, W. (2008). Lignification: Are lignins biosynthesized via simple combinatorial chemistry or via proteinaceous control and template replication? In F. Daayf, A. El Hadrami, L. Adam, & G. M. Balance (Eds.), *Vol. 1. Recent advances in polyphenol research* (pp. 36–66). Oxford, UK: Wiley-Blackwell Publishing.

Ralph, J., Grabber, J. H., & Hatfield, R. D. (1995). Lignin-ferulate cross-links in grasses: Active incorporation of ferulate polysaccharide esters in ryegrass lignins. *Carbohydrate Research*, *275*, 167–178.

Ralph, J., Hatfield, R. D., Piquemal, J., Yahiaoui, N., Pean, M., Lapierre, C., et al. (1998). NMR characterization of altered lignins extracted from tobacco plants down-regulated for lignification enzymes cinnamyl-alcohol dehydrogenase and cinnamoyl-CoA reductase. *Proceedings of the National Academy of Sciences of the United States of America*, *95*, 12803–12808.

Ralph, J., Hatfield, R. D., Quideau, S., Helm, R. F., Grabber, J. H., & Jung, H.-J. G. (1994). Pathway of *p*-coumaric acid incorporation into maize lignin as revealed by NMR. *Journal of the American Chemical Society*, *116*, 9448–9456.

Ralph, J., Kim, H., Peng, J., & Lu, F. (1999). Arylpropane-1,3-diols in lignins from normal and CAD-deficient pines. *Organic Letters*, *1*, 323–326.

Ralph, J., Lapierre, C., & Boerjan, W. (2019). Lignin structure and its engineering. *Current Opinion in Biotechnology*, *56*, 240–249.

Ralph, J., Lapierre, C., Lu, F., Marita, J. M., Pilate, G., Van Doorsselaere, J., et al. (2001). NMR evidence for benzodioxane structures resulting from incorporation of 5-hydroxyconiferyl alcohol into lignins of *O*-methyltransferase-deficient poplars. *Journal of Agricultural and Food Chemistry*, *49*, 86–91.

Ralph, J., Lapierre, C., Marita, J. M., Kim, H., Lu, F., Hatfield, R. D., et al. (2001). Elucidation of new structures in lignins of CAD- and COMT-deficient plants by NMR. *Phytochemistry*, *57*, 993–1003.

Ralph, J., Lundquist, K., Brunow, G., Lu, F., Kim, H., Schatz, P. F., et al. (2004). Lignins: Natural polymers from oxidative coupling of 4-hydroxyphenylpropanoids. *Phytochemistry Reviews*, *3*, 29–60.

Ralph, J., MacKay, J. J., Hatfield, R. D., O'Malley, D. M., Whetten, R. W., & Sederoff, R. R. (1997). Abnormal lignin in a loblolly pine mutant. *Science*, *277*, 235–239.

Ralph, J., Marita, J. M., Ralph, S. A., Hatfield, R. D., Lu, F., Ede, R. M., et al. (1999). Solution-state NMR of lignins. In D. S. Argyropoulos, & T. Rials (Eds.), *Progress in lignocellulosics characterization* (pp. 55–108). Atlanta, GA: TAPPI Press.

Ralph, J., Quideau, S., Grabber, J. H., & Hatfield, R. D. (1994). Identification and synthesis of new ferulic acid dehydrodimers present in the grass cell walls. *Journal of the Chemical Society, Perkin Transactions*, *1*, 3485–3498.

Ralph, J., Schatz, P. F., Lu, F., Kim, H., Akiyama, T., & Nelsen, S. F. (2009). Quinone methides in lignification. In S. Rokita (Ed.), *Quinone methides* (pp. 385–420). Hoboken, NJ: Wiley-Blackwell.

Rencoret, J., Gutiérrez, A., Nieto, L., Jiménez-Barbero, J., Faulds, C. B., Kim, H., et al. (2011). Lignin composition and structure in young versus adult *Eucalyptus globulus* plants. *Plant Physiology*, *155*, 667–682.

Rencoret, J., Kim, H., Evaristo, A. B., Gutiérrez, A., Ralph, J., & del Río, J. C. (2018). Variability in lignin composition and structure in cell walls of different parts of macaúba (*Acrocomia aculeata*) palm fruit. *Journal of Agricultural and Food Chemistry*, *66*, 138–153.

Rencoret, J., Marques, G., Serrano, O., Kaal, J., Martínez, A. T., del Río, J. C., et al. (2020). Deciphering the unique structure and acylation pattern of *Posidonia oceanica* lignin. *ACS Sustainable Chemistry & Engineering*, *8*, 12521–12533.

Rencoret, J., Neiva, D., Marques, G., Gutiérrez, A., Kim, H., Gominho, J., et al. (2019). Hydroxystilbene glucosides are incorporated into Norway spruce bark lignin. *Plant Physiology*, *180*, 1310–1321.

Rencoret, J., Ralph, J., Marques, G., Gutiérrez, A., Martínez, A. T., & del Río, J. C. (2013). Structural characterization of lignin isolated from coconut (*Cocos nucifera*) coir fibers. *Journal of Agricultural and Food Chemistry*, *61*, 2434–2445.

Rencoret, J., Rosado, M. J., Kim, H., Tymokhin, V., Gutiérrez, A., Bausch, F., et al. (2022). Flavonoids naringenin chalcone, naringenin, dihydrotricin, and tricin are lignin monomers in papyrus. *Plant Physiology*, *188*, 208–2019.

Rosado, M. J., Bausch, F., Rencoret, J., Marques, G., Gutiérrez, A., Rosenau, T., et al. (2021). Differences in content, composition and structure of the lignins from the rind and pith of papyrus (*Cyperus papyrus* L.) culms. *Industrial Crops and Products*, *174*, 114226.

Rosado, M. J., Rencoret, J., Marques, G., Gutiérrez, A., & del Río, J. C. (2021). Structural characteristics of the guaiacyl-rich lignins from rice (*Oryza sativa* L.) husks and straw. *Frontiers in Plant Science*, *12*, 640475.

Sakakibara, A. (1980). A structural model of softwood lignin. *Wood Science and Technology*, *14*, 89–100.

Sakamoto, S., Kamimura, N., Tokue, Y., Nakata, M. T., Yamamoto, M., Hu, S., et al. (2020). Identification of enzymatic genes with the potential to reduce biomass recalcitrance through lignin manipulation in *Arabidopsis*. *Biotechnology for Biofuels*, *13*(97), 1–16.

Schatz, P. F., Ralph, J., Lu, F., Guzei, I. A., & Bunzel, M. (2006). Synthesis and identification of 2,5-bis-(4-hydroxy-3-methoxyphenyl)-tetrahydrofuran-3,4-dicarboxylic acid, an unanticipated ferulate 8–8-coupling product acylating cereal plant cell walls. *Organic and Biomolecular Chemistry*, *4*, 2801–2806.

Sun, R. C., Fang, J. M., & Tomkinson, J. (1999). Fractional isolation and structural characterization of lignins from oil palm trunk and empty fruit bunch fibres. *Journal of Wood Chemistry and Technology*, *19*, 335–356.

Terashima, N., Fukushima, K., He, L. F., & Takabe, K. (1993). Comprehensive model of the lignified plant cell wall. In H. G. Jung, D. R. Buxton, R. D. Hatfield, & J. Ralph (Eds.), *Forage cell wall structure and digestibility* (pp. 247–270). Madison: American Society of Agronomy, Crop Science Society of America, Soil Science Society of America.

Tobimatsu, Y., Chen, F., Nakashima, J., Escamilla-Treviño, L. L., Jackson, L., Dixon, R. A., et al. (2013). Coexistence but independent biosynthesis of catechyl and guaiacyl/syringyl lignin polymers in seed coats. *Plant Cell*, *25*, 2587–2600.

Tobimatsu, Y., & Schuetz, M. (2019). Lignin polymerization: How do plants manage the chemistry so well? *Current Opinion in Biotechnology*, *56*, 75–81.

Van Acker, R., Déjardin, A., Desmet, S., Hoengenaert, L., Vanholme, R., Morreel, K., et al. (2017). Different routes for conifer- and sinapaldehyde and higher saccharification upon deficiency in the dehydrogenase CAD1. *Plant Physiology*, *175*, 1018–1039.

Vanholme, R., De Meester, B., Ralph, J., & Boerjan, W. (2019). Lignin biosynthesis and its integration into metabolism. *Current Opinion in Biotechnology*, *56*, 230–239.

Vanholme, R., Demedts, B., Morreel, K., Ralph, J., & Boerjan, W. (2010). Lignin biosynthesis and structure. *Plant Physiology*, *152*, 895–905.

Vanholme, R., Morreel, K., Ralph, J., & Boerjan, W. (2008). Lignin engineering. *Current Opinion in Plant Biology*, *11*, 278–285.

Vanholme, R., Ralph, J., Akiyama, T., Lu, F., Rencoret, J., Kim, H., et al. (2010). Engineering traditional monolignols out of lignin by concomitant up-regulation of *F5H1* and down-regulation of *COMT* in Arabidopsis. *The Plant Journal*, *64*, 885–897.

Vanholme, R., Sundin, L., Seetso, K. C., Kim, H., Liu, X., De Meester, B., et al. (2019). COSY catalyzes *trans-cis* isomerization and lactonization in the biosynthesis of coumarins. *Nature Plants*, *5*, 1066–1075.

Vermerris, W., & Boon, J. J. (2001). Tissue specific patterns of lignification are disturbed in the *brown midrib2* mutant of maize (*Zea mays* L.). *Journal of Agricultural and Food Chemistry*, *49*, 721–728.

Wagner, A., Tobimatsu, Y., Phillips, L., Flint, H., Torr, K., Donaldson, L., et al. (2011). CCoAOMT suppression modifies lignin composition in *Pinus radiata*. *The Plant Journal*, *67*, 119–129.

Weng, J.-K., Mo, H., & Chapple, C. (2010). Over-expression of F5H in COMT-deficient Arabidopsis leads to enrichment of an unusual lignin and disruption of pollen wall formation. *The Plant Journal*, *64*, 898–911.

Wilkerson, C. G., Mansfield, S. D., Lu, F., Withers, S., Park, J., Karlen, S. D., et al. (2014). Monolignol ferulate transferase introduces chemically labile linkages into the lignin backbone. *Science*, *344*, 90–93.

Withers, S., Lu, F., Kim, H., Zhu, Y., Ralph, J., & Wilkerson, C. G. (2012). Identification of a grass-specific enzyme that acylates monolignols with *p*-coumarate. *Journal of Biological Chemistry*, *287*, 8347–8355.

Yamamoto, M., Tomiyama, H., Koyama, A., Okuizumi, H., Liu, S., Vanholme, R., et al. (2020). A century-old mystery unveiled: Sekizaisou is a natural lignin mutant. *Plant Physiology*, *182*, 1821–1828.

Yang, L., Lin, M., Zhang, H., Wang, C., Shi, L., Lan, W., et al. (2021). Ferulate-sinapyl alcohol cross-coupling reaction improves the understanding of grass cell wall lignification. *Industrial Crops and Products*, *168*, 113587.

Zhang, A., Lu, F., Sun, R., & Ralph, J. (2009). Ferulate-coniferyl alcohol cross-coupled products formed by radical coupling reactions. *Planta*, *229*, 1099–1108.

Zhao, Y., Yu, X., Lam, P.-Y., Zhang, K., Tobimatsu, Y., & Liu, C.-J. (2021). Monolignol acyltransferase for lignin *p*-hydroxybenzoylation in *Populus*. *Nature Plants*, 7, 1288–1300.

CHAPTER TWO

Lignin synthesis and bioengineering approaches toward lignin modification

Chang-Jun Liu[a,b,]* and Aymerick Eudes[a,c,]*

[a]Joint BioEnergy Institute, Emeryville, CA, United States
[b]Brookhaven National Laboratory, Biology Department, Upton, NY, United States
[c]Lawrence Berkeley National Laboratory, Biosciences Area, Environmental Genomics and Systems Biology, Berkeley, CA, United States
*Corresponding authors: e-mail address: cliu@bnl.gov; ageudes@lbl.gov

Contents

Abstract

Lignin is considered to represent the second most abundant terrestrial biopolymer next to cellulose. Although quite variable across vascular plants, its content can account for up to 30% of total biomass in certain woody species. Lignin plays important roles in plants, but often represents hurdles to the utilization of cellulosic biomass in different sectors such as pulp industry, forage digestibility, and bioenergy. Certain natural or mutagenesis-induced lignin mutants, which are notorious for exhibiting brown midribs in grasses or red-brown wood in trees, have long been recognized to produce biomass

Advances in Botanical Research, Volume 104
ISSN 0065-2296
https://doi.org/10.1016/bs.abr.2022.02.002

more amenable to animal digestion, biological conversion, and other agro-industrial processes. Advances in molecular biology techniques for DNA sequencing and manipulation have facilitated the identification of genes involved in lignin biosynthesis and enabled the modification of their expression. The first reports on plants modified in lignin via genetic engineering techniques trace back to the mid 90's. Since then, with the emergence of genome sequencing capabilities and the deployment of large-scale transcriptomic, proteomic, and metabolomic approaches, our understanding on lignin metabolism and regulation has expanded. Correspondingly, substantial efforts have been made in tailoring lignin biosynthesis. In this chapter, we cover several bioengineering strategies that leverage recent knowledge obtained on the mechanistic understanding of lignin formation and chemical composition. Modification of lignin content and/or monomeric composition can be achieved not only by tailoring gene expression, but also by exploiting enzyme post-translational modifications, altering enzyme cofactors and co-substrates, and rerouting lignin metabolic precursors.

1. Introduction

Lignin is an aromatic polymer deposited in secondary cell walls of vascular plants. Lignin interacts with cell wall polysaccharides (i.e., cellulose and hemicellulose) and provides strength, rigidity, and hydrophobicity to plant vascular tissues. Therefore, lignin plays various functions important for plant growth, development, defense against biotic and abiotic stresses, mechanical support, and water transport in xylem vessels. Lignin content varies between plant species—woody species typically contain more lignin than herbaceous species—and lignification tends to increase with plant age (Boerjan, Ralph, & Baucher, 2003). From an agro-industrial perspective, lignin has long been considered a negative factor that impacts ruminant performance in the forage sector, chemical pulping efficiencies in the pulp and paper industry, and saccharification yields for generation of advanced biofuels and bioproducts. As a result, tremendous research efforts have been carried out to elucidate the molecular mechanisms involved in lignin biosynthesis and to develop bioengineering strategies toward modifying lignin content and monomeric composition for crop improvement (Li, Weng, & Chapple, 2008). The aromatic amino acids phenylalanine and tyrosine synthesized in plastids from the shikimate pathway are the early precursors of the lignin monomers, namely monolignols (Fig. 1). The two enzymes phenylalanine ammonia-lyase (PAL) and cinnamate 4-hydroxylase (C4H) are needed to convert phenylalanine into the phenylpropanoid unit *p*-coumaric acid, whereas ammonia-lyase that has a substrate preference for tyrosine (tyrosine ammonia-lyase or TAL) enables the formation of *p*-coumaric acid

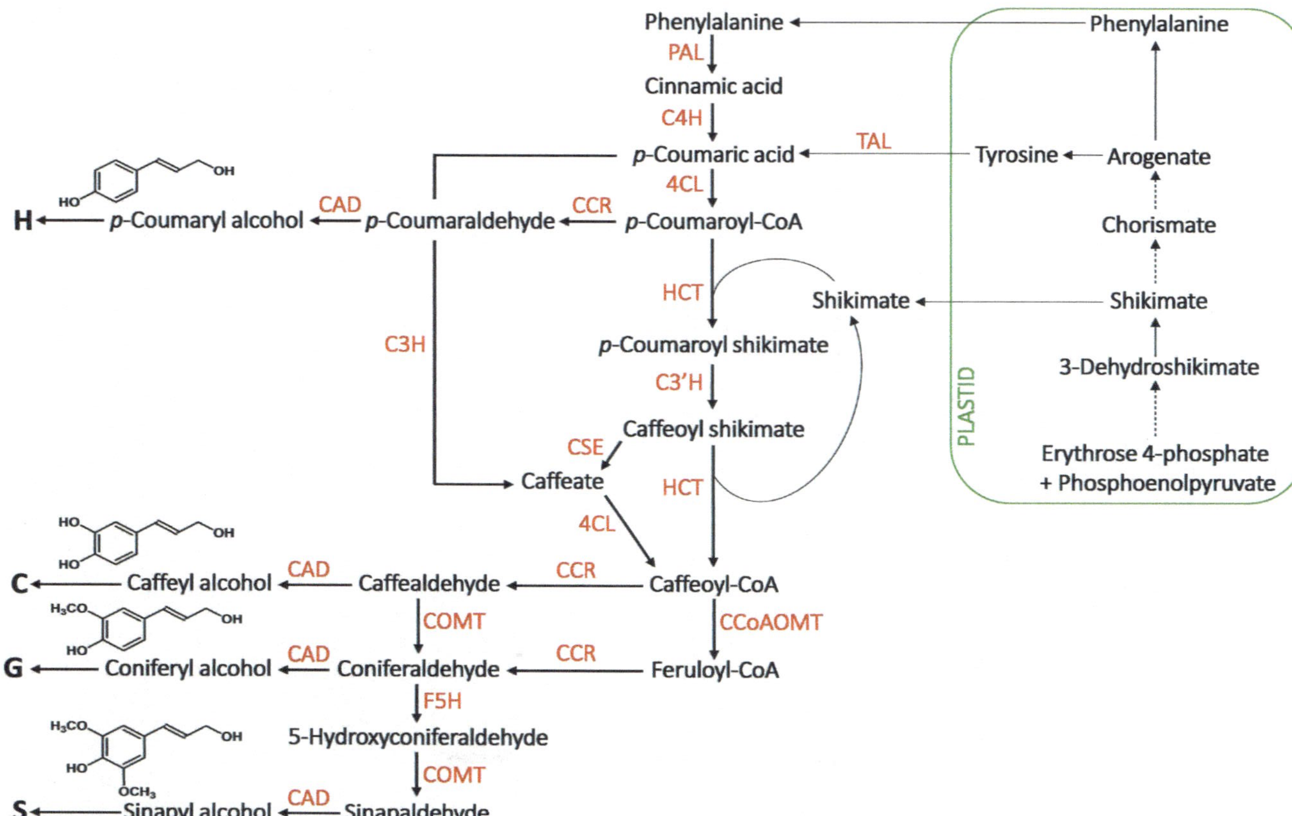

Fig. 1 See figure legend on next page.

directly from tyrosine. The two most prevalent monolignols in angiosperms are coniferyl alcohol and sinapyl alcohol and their incorporation into lignin gives rise to the so-called guaiacyl (G) and syringyl (S) lignin units. Small amount of hydroxyphenyl (H) units that originate from incorporation of *p*-coumaryl alcohol is also found in lignin, especially from grasses, whereas catechyl (C) units that result from the polymerization of caffeyl alcohol have been observed in the seed coats of certain plants (Renault, Werck-Reichhart, & Weng, 2019). Laccases and peroxidases are responsible for the oxidative polymerization of monolignols during lignin formation. Within the lignin pathway, different routes have been described for the synthesis of the precursor caffeoyl-CoA, which are distinct in the intermediate used for hydroxylation on the C3 position on the aromatic ring of the phenylpropanoid unit (Fig. 1). Recently, the simplest route involving *p*-coumarate 3-hydroxylase (C3H) for conversion of *p*-coumarate into caffeate followed by coenzyme A (CoA) ligation has been established genetically (Barros et al., 2019). Other routes involve the formation of a shikimate ester of *p*-coumarate prior to C3 hydroxylation catalyzed by *p*-coumaroyl-shikimate 3′-hydroxylase (C3'H) and followed by either ester hydrolysis or transesterification to release caffeate or caffeoyl-CoA, respectively (Fig. 1).

The modification of lignin via genetic engineering is a compelling and important aspect of crop improvement to potentially ameliorate the economics and the carbon footprint of several agro-industrial processes. For example, there is a negative correlation between lignin content and digestibility of forage materials in ruminant diets because lignin forms a physical barrier that reduces the access of rumen microbes to fermentable cell wall components. Genetic manipulation affecting lignin biosynthesis represents a solution to alter the antinutritive lignin component in legume forage crops and grasses (Halpin, 2019; Lei, Hannoufa, & Yu, 2017). In this regard, low-lignin alfalfa (branded as HarvXtra) obtained via downregulation of

Fig. 1 The lignin biosynthetic pathway. Tyrosine and phenylalanine are synthesized in plastids. Enzymes in red are: HCT, *p*-hydroxycinnamoyl-CoA shikimate hydroxycinnamoyl transferase; CAD, cinnamyl alcohol dehydrogenase; CCoAOMT, caffeoyl-CoA *O*-methyltransferase; CCR, cinnamoyl-CoA reductase; COMT, caffeate/5-hydroxyferulate *O*-methyltransferase; CSE, caffeoyl shikimate esterase; C3H, *p*-coumarate 3-hydroxylase; C3′H, *p*-coumaroyl-shikimate 3′-hydroxylase; C4H, cinnamate 4-hydroxylase; F5H, ferulate 5-hydroxylase; PAL, phenylalanine ammonia-lyase; TAL, tyrosine ammonia-lyase; 4CL, 4-coumarate-CoA ligase. Lignin units: H (hydroxyphenyl), C (catechyl), G (guaiacyl), and S (syringyl).

the gene encoding caffeoyl-CoA 3-*O*-methyltransferase (CCoAOMT) has been the first genetically engineered crop commercialized for the improved forage digestibility (Barros, Temple, & Dixon, 2019). Fine-tuned modification of lignin content and composition has also been contemplated to improve raw wood materials for pulp and paper production (Baucher, Halpin, Petit-Conil, & Boerjan, 2003). In mulberry trees, a mutation in a cinnamyl alcohol dehydrogenase (CAD) gene reduces lignin and leads to improved wood delignification efficiency and increased pulp yield under alkaline pulping conditions (Ikeda et al., 2021). Similarly, incorporation of chemically labile ester linkages in poplar lignin via overproduction of the rare monolignol coniferyl ferulate enhances wood chemical pulping efficiencies (Zhou et al., 2017). Finally, considering the negative impact of lignin on biomass deconstruction and enzymatic saccharification for the manufacturing of advanced bioproducts, engineering lignin in bioenergy crops such as woody perennials and grass feedstocks appears to be an essential path to developing a future sustainable bioeconomy (Baral et al., 2019; Chanoca, de Vries, & Boerjan, 2019; Mahon & Mansfield, 2019; Simmons, Loqué, & Ralph, 2010).

In this chapter, we illustrate various lignin engineering approaches that leverage recent knowledge on the genetics and biochemistry of lignin synthesis and metabolism. Opportunities to manipulate lignin physico-chemical properties and recalcitrance are also discussed considering the malleability of lignin's monomeric composition.

2. Gene editing approaches

Genome editing techniques include a set of methods to precisely modify the genome of organisms and enable a targeted intervention with or without the integration of a transgene. The molecular tools for genome editing include variants of site directed nucleases such as zinc-finger-nucleases, meganucleases, transcription activator like effector nucleases (TALENs), and clustered regularly interspaced short palindromic repeat (CRISPR)/CRISPR-associated proteins (CRISPR/Cas). Once introduced in a cell, the nuclease is capable of cutting the genome at a targeted site, which induces cellular DNA-repair mechanisms to fix the cut sites, but also potentially leads to insertions or deletions (indels) at the respective genomic site that result in loss-of-function edited gene sequences due to frame-shift mutations. Gene editing approaches using either CRISPR-Cas9 or

TALENs have been employed to generate genetic knock-out mutants to modify lignin in crops (Table 1). One of the advantages of gene editing is that higher-order knock-out mutants can be generated, which enables simultaneous disruption of closely related genes belonging to multigenic families and having redundant functions in a particular metabolic step. Plants derived by genome editing are potentially facing reduced regulatory barriers in certain countries (Menz, Modrzejewski, Hartung, Wilhelm, & Sprink, 2020).

2.1 CRISPR-Cas9

The CRISPR-Cas9 genome editing system allows precise gene targeting and mutations including gene indels, gene replacements, and single base pair conversions. The CRISPR-Cas9 machinery consists of a 20-nucleotide gene-specific guide RNA sequence (gRNA) and the Cas9 endonuclease. The gRNA binds to the specific target DNA sequence and ends with a 2–6-base pair DNA sequence called the protospacer adjacent motif (PAM). Adjacent to the 3′ end of the 20-nucleotide gRNA is an ~80-nucleotide long gRNA scaffold sequence that is necessary for Cas9 binding. Once the gRNA-Cas9 complex forms, Cas9 makes a double-strand cut exactly 3 base pairs before the PAM sequence (Jiang & Doudna, 2017). The break site is mainly repaired by non-homologous end joining, which is often error-prone.

The CRISPR-Cas9 technology has been used in various crop species to target genes encoding caffeate *O*-methyltransferase (COMT), C3'H, 4-coumarate-CoA ligase (4CL), ferulate 5-hydroxylase (F5H), cinnamoyl-CoA reductase (CCR), laccase, and caffeoyl shikimate esterase (CSE), as well as transcription factors that act as regulators of lignin biosynthesis (Table 1). In gray poplar (*Populus tremula* x *P. alba*), a double *cse1-cse2* mutant was obtained using a gRNA that targets both genes, resulting in 35% reduction of lignin, growth penalty, and large improvements of saccharification efficiencies. The approach validated the redundant function of *CSE1* and *CSE2* during lignin biosynthesis since single *cse1* and *cse2* knockouts have unaltered lignin, but also appeared to be more efficient than the *CSE*-targeted RNA downregulation strategy that results in residual *CSE* expression and 25% reduction of lignin content (de Vries, Brouckaert, et al., 2021). These results are in contrast with those obtained in a different poplar hybrid in which the editing of either *CSE1* or *CSE2* resulted in lignin reductions (up to 29%) and improvements of saccharification efficiency, but

Table 1 Gene editing approaches for modification of lignin biosynthesis.

Gene editing method	Target gene(s)	Engineered plant	Lignin modification (content and/or composition)	References
CRISPR-Cas9	*Pv4CL1*	Switchgrass (*Panicum virgatum*)	8%–30% reduction S/G increased	Park et al. (2017)
CRISPR-Cas9	*HvCOMT1*	Barley (*Hordeum vulgare*)	14% reduction S/G decreased	Lee et al. (2021)
CRISPR-Cas9	*OsC3'H1*	Rice (*Oryza sativa*)	Lethal phenotype	Takeda et al. (2018)
CRISPR-Cas9	*OsMYB108*	Rice (*O. sativa*)	Increased S/G increased *p*-coumarate and ferulate increased	Miyamoto et al. (2019)
CRISPR-Cas9	*OsWRKY36/ OsWRKY102*	Rice (*O. sativa*)	41% increase	Miyamoto et al. (2020)
CRISPR-Cas9	*OsCAld5H1*	Rice (*O. sativa*)	14% reduction S/G decreased	Takeda et al. (2019)
CRISPR-Cas9	*CCR2*	Gray poplar (*Populus tremula* x *P. alba*)	14% reduction S/G decreased	De Meester et al., 2020
CRISPR-Cas9	4CL1	Gray poplar (*P. tremula* x *P. alba*)	19%–23% reduction S/G decreased	Zhou, Jacobs, Xue, Harding, and Tsai (2015), Tsai et al. (2020)
CRISPR-Cas9	*CSE1/CSE2*	Gray poplar (*P. tremula* x *P. alba*)	35% reduction H units increased	De Vries et al. (2021)

Continued

Table 1 Gene editing approaches for modification of lignin biosynthesis.—cont'd

Gene editing method	Target gene(s)	Engineered plant	Lignin modification (content and/or composition)	References
CRISPR-Cas9	*PtoLAC14*	Chinese white poplar (*Populus tomentosa*)	7%–8% reduction S/G increased	Qin et al. (2020)
CRISPR-Cas9	*PtoMYB170*	Chinese white poplar (*P. tomentosa*)	8%–12% reduction	Xu et al. (2017)
CRISPR-Cas9	*PtrHSFB3-1* *PtrMYB092*	Black cottonwood (*Populus trichocarpa*)	17% reduction 27% reduction	Liu et al. (2021)
CRISPR-Cas9	*CSE1 or CSE2*	Poplar 84K (*Populus alba* × *P. glandulosa*)	Up to 16% reduction Up to 29% reduction in the field	Jang et al. (2021)
CRISPR-Cas9	*EgrCCR1*	Eucalyptus hairy roots (*Eucalyptus grandis*)	Reduced phloroglucinol-HCl staining	Dai et al. (2020)
CRISPR-Cas9	*CCR1*	*Arabidopsis thaliana*	Specific reduction in stem fibres H units increased	Yu, Liu, and Dixon (2021)
CRISPR-Cas9	*AtHCT*	*A. thaliana*	20%–30% reduction H units increased	Liang et al. (2019)
TALEN	*COMTa/COMTb*	Sugarcane (*Saccharum* spp. hybrids)	11%–32% reduction 11%–20% reduction in the field S/G decreased	Jung and Altpeter (2016); Kannan, Jung, Moxley, Lee, and Altpeter (2018)

exhibited normal growth performance compared to the control trees despite the occurrence of xylem collapsed vessels (Jang et al., 2021). Differences in the engineered poplar hybrid species and variations in the target sites used for CRISPR may explain the contrasting results obtained in these two studies. Using Arabidopsis as a model for the targeting of *HCT*, Liang et al. (2019) showed that tissue-specific expression of *Cas9* (i.e., restricted to tissues that develop secondary cell walls) could be an efficient strategy to reduce lignin without impacting plant growth. Indeed, it is often observed that disrupting genes involved in early steps of the lignin biosynthetic pathway results in plant growth penalty or even lethality (Takeda et al., 2018). To overcome this issue, Yu et al. (2021) demonstrated an engineering approach in Arabidopsis for the constitutive editing of a *CCR* gene and concomitant tissue-specific expression of codon-optimized *CCR* sequence that escapes CRISPR-Cas9-mediated editing using a promoter preferentially active in xylem vessels. These optimized *CCR*-edited Arabidopsis plants showed reduced lignin and higher saccharification yields without the detrimental effects usually observed in plants with *CCR* repressed in every plant tissue.

2.2 TALEN

TALENs are synthetic chimeric DNA-cutting enzymes built from the fusion of a DNA binding domain and the *Fok*I nuclease domain which can cut dimerized DNA (Miller et al., 2011). The DNA binding domain of a TALEN originates from the transcription activator-like (TAL) effectors from the bacteria *Xanthomonas* and consists of a repeated highly conserved 33–34 amino acid sequence with two divergent amino acids in position 12 and 13. These two variable amino acids influence the recognition of a particular base on a DNA sequence, which allows for the engineering of specific DNA-binding domains by joining multiple repeats containing the appropriate divergent residues. A pair of TALENs recognizing genomic targets in proximity can lead to the dimerization of the *Fok*I nuclease domain and induce a DNA double-strand break.

The first example of genome editing in sugarcane was obtained with TALENs designed to induce mutations in *COMT*. Such an approach is particularly relevant considering the highly diploid nature of sugarcane which contains multiple copies of functionally redundant *COMT* genes. Using a single TALEN pair, co-editing of 107 of the 109 unique *COMT* copies resulted in up to 30% reduction of lignin content and drastic improvements of saccharification efficiency (Jung & Altpeter, 2016; Kannan et al., 2018).

This strategy also appeared to be more effective than the *COMT* RNAi suppression approach previously used to reduce lignin in sugarcane.

3. Lignin engineering via post-transcriptional modifications

Post-transcriptional regulation of lignin biosynthetic genes plays important roles in modifying mRNA levels and mRNA stability. MicroRNAs (miRNAs) are a class of endogenous, noncoding, small RNAs of 20–24 nucleotides that regulate gene expression through cleaving mRNA targets or interfering with translation of the target mRNAs. For example, 36 miRNA genes have been identified in *Populus tomentosa* (Chinese white poplar) using an in silico approach (Quan et al., 2019). These are involved in many biological processes in plants and several members of the *miR397* and *miRNA528* families were shown to target the laccase (*LAC*) genes functioning in lignin synthesis (Abdel-Ghany & Pilon, 2008; Huang, Zhou, Gao, & Tang, 2021; Sun et al., 2018). In *Acacia mangium* (brown salwood), *miRNA166* was found to target class III homeodomain leucine zipper transcription factors to regulate the expression of *C4H*, *CAD*, and *CCoAOMT* genes (Ong & Wickneswari, 2012). Based on these observations, miRNAs have been employed as genetic tools to manipulate lignin and improve biomass in various plant species (Trumbo, Zhang, & Stewart, 2015) (Table 2). Overall, overexpression (OE) of miRNAs from the miR397 family triggers reductions of lignin content as a result of a downregulation of laccase genes. Conversely, silencing of miR397 using short tandem target mimics (STTM), endogenous target mimics (eTM), or CRISPR-Cas9 results in increased lignin content due to upregulation of laccase genes. In maize, manipulation of *Zm-miRNA528* expression enabled modification of lignin content via the targeting of *LAC3* and *LAC5* (Sun et al., 2018). In Arabidopsis roots, downregulation of *At-miR397b* increased *LAC2* expression and decreased lignin content while its overexpression had the opposite effects, which suggests an alternate regulatory mechanism for *miR397b-LAC2* in roots compared to other plant organs (Khandal et al., 2020). In sweet potato, overexpression of *miR828* led to the overexpression of several lignin-related genes and increased lignin content due to downregulation of transcriptional activators of lignin biosynthesis (Lin et al., 2012). miRNA such as *miR6443* in *P. tomentosa* and *miR172c* in brahmi were used to target *F5H* genes and modulate lignin in these plants (Fan et al., 2020; Jeena et al., 2021). Finally, by leveraging

Table 2 Examples of miRNA manipulations associated with lignin modifications.

Targeted miRNA	Gene(s) affected	Engineered plant	Lignin modification (content and/or composition)	References
mi-R393 (STTM)	Upregulation of *CCR1*, *CCRL6*, *OMT1*, *PER73*, and *PER9*	Poplar 84K (*Populus alba* × *glandulosa*)	Increased content	Chu, He, Shu, Wang, and Tang (2021)
At-miR397b (OE)	Downregulation of *AtLAC2*	*Arabidopsis thaliana*	Increased content in root	Khandal, Singh, and Chattopadhyay (2020)
At-miR397b (STTM)	Upregulation of *AtLAC2*		Decreased content in root	
At-miR397b (OE), *Sv-miR397b* (OE)	Downregulation of *AtLAC2*, *AtLAC4*, and *AtLAC17*	*A. thaliana*	Decreased content in stem and leaf	Nguyen, Brown, Pegler, Eamens, and Grof (2020)
At-miR397b (OE)	Downregulation of *AtLAC2*, *AtLAC4*, and *AtLAC17*	*A. thaliana*	Decreased content in stem	Wang et al. (2014)
Ptr-miR397a (OE)	Downregulation of 17 *PtrLAC genes*	Black cottonwood (*Populus trichocarpa*)	Decreased content	Lu et al. (2013)
Mh-miR397b (OE)	Downregulation of *MhLAC7*	Tea Crabapple (*Malus hupehensis*)	Decreased content	Yu, Gong, Cao, Hou, and Qu (2020)
Mh-miR397b (STTM)	Upregulation of *MhLAC7*		Increased content	
Os-miR397 (OE)	Downregulation of laccase gene(s)	Rice (*Oryza sativa*)	Decreased content	Swetha et al. (2018)
Pb-rmiR397a (OE)	Downregulation of *PbLAC1*, *PbLAC2*, *and PbLAC18*	Chinese pear (*Pyrus bretschneideri*)	Decreased content in fruit stone cells	Xue et al. (2019)

Continued

Table 2 Examples of miRNA manipulations associated with lignin modifications.—cont'd

Targeted miRNA	Gene(s) affected	Engineered plant	Lignin modification (content and/or composition)	References
Bm-miR172c-5p (OE)	Downregulation of *F5H*	Brahmi (*Bacopa monnieri*)	Decreased content, decreased S/G	Jeena, Joshi, and Shukla (2021)
Bm-miR172c-5p (eTM)	Upregulation of *F5H*		Increased content, increased S/G	
Zm-miR528 (OE)	Downregulation of *ZmLAC3 and ZmLAC5*	Maize (*Zea mays*)	Decreased content	Sun et al. (2018)
Zm-miR528 (STTM)	Upregulation of *ZmLAC3 and ZmLAC5*		Increased content	
Ib-miR828 (OE)	Downregulation of *IbMYB* and *IbTLD* Upregulation of *IbPAL02*, *IbPAL2-8*, *IbC4H*, *IbHCT*, and *IbCAD1*	Sweet potato (*Ipomoea batatas*)	Increased content	Lin, Lin, Lin, Chen, and Jeng (2012)
Pto-miR6443 (OE)	Downregulation of *PtoF5H2*	Chinese white poplar (*Populus tomentosa*)	Decreased S/G	Fan et al. (2020)
Pto-miR6443 (STTM)	Upregulation of *PtoF5H2*		Increased S/G	
C3'H–amiRNA and *F5H–amiRNA*	Downregulation of *C3'H* and *F5H*	Jute (*Corchorus olitorius*)	Decreased content	Shafrin, Das, Sanan-Mishra, and Khan (2015)

Abbreviations: eTM, endogenous target mimics; OE, overexpression; STTM, Short tandem target mimic.

the mode of action of miRNAs, artificial miRNAs (amiRNAs) have been developed to target specifically the genes encoding C3'H and F5H to reduce lignin in jute fibers (Shafrin et al., 2015).

Long non-coding RNAs (lncRNAs) are more than 200 nucleotides in length and have also been proposed to play a role in the regulation of lignin biosynthesis. lncRNA have been identified in the wood of *P. tomentosa* and 16 genes involved in wood formation processes including lignin biosynthesis appear to be potential targets (Chen, Quan, & Zhang, 2015). These lncRNAs can possibly bind to mRNAs of lignin biosynthetic genes to initiate their degradation and modulate lignin synthesis (Quan et al., 2019). Similarly, lncRNAs have been involved in fiber development, secondary cell wall formation, and lignin catabolic processes via in-silico analyses in cotton and moso bamboo (Wang et al., 2015; Wang, Hou, Wang, & Zhao, 2021). A possible mechanism is that miRNA could either bind to the mRNAs or IncRNAs to promote their degradations, causing an increase (binding to lncRNA) or decrease (binding to mRNA) of lignin, and thus enabling an RNA-based regulation mechanism for monolignol synthesis.

4. Lignin engineering via post-translational modifications

4.1 Post-translational modifications

Post-translational modifications (PTMs) are covalent processing events that change the properties of a protein by proteolytic cleavage and/or adding a modifying group, such as acetyl, phosphoryl, glycosyl and methyl, to one or more amino acids. PTMs have a significant impact on structure, function and dynamics of proteins. Therefore, they function as crucial molecular regulatory mechanisms to regulate diverse cellular processes, including phenylpropanoid-lignin metabolism. The emerging evidence shows that PTMs commonly occur on both lignin biosynthetic enzymes and the related regulatory proteins, constituting a layer of complex regulatory mechanism that can be harnessed for controlling lignin biosynthesis.

4.1.1 Phosphorylation

An integrated quantitative study of the transcriptome, proteome and phosphoproteome has discovered over 43,000 phosphorylation sites on 18,000 proteins from 30 tissues of Arabidopsis (Mergner et al., 2020). Among them, several lignin biosynthetic enzymes, including PAL, C4H, CCR, and

COMT have been found with potential phosphorylation sites (Samaras et al., 2020). Indeed, a phosphopeptide derived from PAL2 in *Phaseolus vulgaris* (French Bean) suspension-cultured cells was revealed in the earlier studies (Allwood, Davies, Gerrish, Ellis, & Bolwell, 1999; Cheng, Sheen, Gerrish, & Bolwell, 2001), and the calmodulin like domain protein kinase (CDPK) was suggested responsible for PAL2 phosphorylation (Allwood et al., 1999). Consistently, CDPK derived from French bean suspension cultures and from Arabidopsis were able to phosphorylate the recombinant PAL2 from hybrid poplar *Populus trichocarpa* × *P. deltoides* in vitro (Cheng et al., 2001). Presumably, the CDPK-mediated phosphorylation of PAL either marks the protein for turnover, thus reducing PAL activity or catalytic efficiency to maintain homeostasis, or mediates PAL subcellular translocation to membranes, which may contribute to the formation of metabolon in lignin biosynthesis (Allwood et al., 1999; Rasmussen & Dixon, 1999). The high degree sequence similarity of PAL peptides across plant species and their conserved phosphorylation sites suggest a potential common phosphoregulatory mechanism for this enzyme in phenylpropanoid biosynthesis (Sulis & Wang, 2020; Zhang & Liu, 2015).

A phosphoproteomics study on protein extracts from stem differentiating xylem (SDX) of *P. trichocarpa* (black cottonwood) revealed 4728 phosphopeptides of 1392 enriched proteins, including two phosphopeptides mapping to PtrAldOMT2/COMT2 (Wang et al., 2015). Monophosphorylation at the Ser123 or Ser125 residue was identified in this enzyme. The phosphorylation was mediated by kinases in the SDX protein extracts and could be reversed by phosphatase treatment. The phosphorylation strongly reduced enzymatic activity of the recombinant PtrAldOMT2; therefore, likely the dynamic phosphorylation and dephosphorylation mediate the on and off regulation of PtrAldOMT2 activity, and control S-lignin biosynthesis (Wang, Chuang, et al., 2015). Such postulated function of PtrAldOMT2 phosphorylation remains to be confirmed *in planta*. Nevertheless, the identified phosphorylation sites are highly conserved in a wide range of plant species, suggesting that this type of modification might be an evolutionally important means in functional regulation of the *O*-methyltransferases during monolignol biosynthesis, which might offer a more rapid and efficient regulatory mode compared to the transcriptional and translational controls (Sulis & Wang, 2020).

Phosphorylation modification occurs not only on monolignol biosynthetic enzymes but also on the regulators or transcription factors (TFs) controlling secondary cell wall formation and lignin biosynthesis. In Arabidopsis, AtNST1 is a master TF in the top level of a hierarchic multilayered regulatory network, which enhances secondary cell wall formation and lignin deposition in the stem fiber region. Three SNF1-related protein kinase 2s (i.e., SnRK2.2, SnRK2.3 and SnRK2.6), the key positive regulators of ABA signaling, were found to physically interact with and phosphorylate AtNST1 (Liu, Yu, Rao, Li, & Dixon, 2021). Point mutation in the SnRK2-mediated phosphorylation site of AtNST1 (S316A) abolished its regulatory activity, resulting in up to 75% reduction in the activation of the downstream TFs MYB46 and MYB83, although AtNST1 stability was not affected (Liu, Yu, et al., 2021). The S316A mutant of AtNST1 impaired ABA-induced lignin pathway gene expression. These results indicate that SnRK2-mediated phosphorylation of NST1 is a component of the transcriptional machinery regulating secondary cell wall thickening and lignification of fiber cells in response to ABA signal in Arabidopsis.

The members of the MYB family of TFs are known to act as either activators (e.g., MYB58, MYB63 and MYB85) or repressors (e.g., group 4 of R2R3-MYB members MYB4 and MYB32) to regulate lignin formation in Arabidopsis (Zhao & Dixon, 2011; Zhong, Lee, Zhou, McCarthy, & Ye, 2008). An AtMYB4 homolog in *Populus*, named as Lignin biosynthesis Transcription Factor 1 (LTF1), was found to bind directly to the promoters of *4CL1*, *CCoAOMT1* and *COMT2* and repress lignin biosynthesis during xylem development. Its transrepression activity is dynamically regulated by phosphorylation at the Thr146 and Thr178 residues by mitogen-activated protein kinase 6 (PdMAPK6 or PdMPK6), which accelerates the LTF1 protein degradation thus relieving the suppression on lignin biosynthetic genes and activating lignin deposition (Gui et al., 2019).

Conversely, phosphorylation of AtMYB75 (or Production of Anthocyanin Pigment 1, PAP1), a positive regulator of anthocyanin biosynthesis but a negative regulator of the lignin branch (Bhargava, Mansfield, Hall, Douglas, & Ellis, 2010; Borevitz, Xia, Blount, Dixon, & Lamb, 2000), was reported to increase its stability (Li et al., 2016). AtMYB75 was found to interact with and be phosphorylated by MAP kinase 4 (MPK4) at two canonical MPK target sites, Thr126 and Thr131. The phosphorylation is required for full function of AtMYB75 in regulation of the light-induced accumulation of anthocyanins (Li, Wang, et al., 2016).

Pinus taeda (loblolly pine) PtMYB4 is an ortholog of AtMYB56 and AtMYB83. This TF positively regulates lignin biosynthetic genes during the wood formation process (Patzlaff et al., 2003). In vitro assays demonstrated that *P. taeda* MAPK6 was able to phosphorylate PtMYB4 at its Ser236 residue. Phosphomimic mutation S236E that possesses constitutive phosphorylation of PtMYB4 did not interfere with its affinity for DNA binding but significantly reduced its transactivation ability (Morse, Whetten, Dubos, & Campbell, 2009), therefore potentially suppressing lignin biosynthesis.

Deciphering phosphorylation modification in lignin biosynthesis could be biotechnologically meaningful. Adopting phosphomimic or phosphonull mutants of the enzymes or regulators in lignin biosynthesis can be an effective strategy to tailor lignin or secondary cell wall formation. This is exemplified with a recent work carried by Gui and his colleagues, in which a phosphorylation-null version of LTF1 was created, which mitigates the phosphorylation-mediated protein degradation, thus conveying a persistent suppression of lignin biosynthesis. When the engineered LTF1 variant was introduced into poplar under the control of vessel- or fiber-specific promoters, the fiber-specific suppression of lignin biosynthesis resulted in the improvement of biomass quality with no growth defects (Gui et al., 2020).

4.1.2 Ubiquitination

The post-translational addition of ubiquitin (Ub) controls the half-life, interactions, and/or trafficking of numerous intracellular proteins. With adoption of the tandem repeated ubiquitin-affinity and nickel-chelate affinity chromatography, coupled with high sensitivity mass spectrometry, almost 950 Ub conjugates were identified from Arabidopsis young seedlings (Kim, Scalf, Smith, & Vierstra, 2013; Saracco et al., 2009). A number of identified ubiquitinated proteins connect to various aspects of metabolism, ranging from carbohydrate metabolism and glycolysis to secondary metabolism. Within the phenylpropanoid biosynthetic pathway, Arabidopsis PAL, C4H, HCT, CAD and COMT1 were detected with potential Ub modification, indicating that phenylpropanoid-monolignol throughput might be substantially regulated by ubiquitination (Kim et al., 2013). The (poly)ubiquitination of Arabidopsis PAL isozymes was then confirmed with the transiently expressed AtPALs in tobacco leaves probed with the anti-ubiquitin antibody (Zhang, Gou, & Liu, 2013). Ubiquitination of a target protein is achieved via an ATP-dependent conjugation cascade sequentially engaging ubiquitin activating enzyme (E1), ubiquitin conjugating enzymes (E2), and ubiquitin-protein ligase (E3) (Vierstra, 2009). Specific E3 ligases are responsible for the selectivity of ubiquitin-protein ligation. The Skp–Cullin–F-box (SCF) complex is the major type of E3 ligase that is composed of four structural components: Skp1, Cul1/Cdc53, Roc1/Rbx1/Hrt1, and an F-box protein (Vierstra, 2009). The F-box proteins in this complex are substrate recognition components whose specificity is determined by C-terminal protein–protein binding domains such as leucine-rich repeat, Kelch, and WD-40 repeats (Cardozo & Pagano, 2004). Arabidopsis has approximately 700 predicted F-box genes (Gagne, Downes, Shiu, Durski, & Vierstra, 2002), more than 100 of them encode kelch repeat domain-containing F-box proteins (KFBs) (Sun, Zhou, & Ma, 2007). Using AtPAL1 and AtPAL2 as baits to screen an Arabidopsis cDNA expression library in a yeast two-hybrid system (Y2H), four KFB proteins, namely KFB01, -20, -39 and -50, were captured as the common interactors of AtPALs. Pair-wise verification via Y2H, bimolecular fluorescence complementation (BiFC), and co-immunoprecipitation assays corroborated the physical interactions of four AtPAL isozymes with four KFB proteins, albeit their binding affinities are different (Zhang et al., 2013; Zhang, Gou, Guo, Yang, & Liu, 2015). Pair-wise co-expression of any four KFBs with AtPAL isozymes in tobacco leaves severely impaired the latter's stability and activity; whereas applying MG132, a specific inhibitor of the 26S proteasome, mitigated the PAL degradation, while the ubiquitinated PAL increased. Furthermore, up- or down-regulation of *KFB* expression in Arabidopsis reciprocally altered PAL's level and activity, and the accumulation of phenolic esters, flavonoids, anthocyanins, and the lignin deposited in the stem cell walls (Zhang et al., 2013; Zhang et al., 2015). Those in vitro and *in planta* evidences unambiguously demonstrate that PAL is ubiquitinated and degraded via 26S proteasome pathway and the identified KFBs mediate PAL's ubiquitination and selective degradation thus negatively regulating phenylpropanoid-lignin biosynthesis (Zhang & Liu, 2015). The identified PAL-regulating KFBs function redundantly. Up or down-regulation of four *KFBs* did not cause apparent growth effect, except for the alteration in the accumulation of extractable phenolics and polymer lignin (Zhang et al., 2013, 2015). Those PAL-regulating KFBs act as the critical regulatory components in the control of the allocation of photosynthetically fixed carbon into phenylpropanoid metabolism (Wang et al., 2021). Interestingly, disruption of glucosinolate biosynthesis in Arabidopsis *ref5* (*CYP83B1*) and *ref2* (*CYP83A1*) mutants strongly up-regulated the expression of *KFB39* and *KBF50*, which was accompanied with deficiency of phenolic esters; whereas disruption of those PAL-regulating KFBs partially restored phenolic ester synthesis in *ref5* and *ref2* mutants (Kim, Zhang, Pascuzzi, Liu, & Chapple, 2020). Those data suggest that the PAL-regulating KFBs are involved in the glucosinolate/phenylpropanoid metabolic crosstalk via controlling the degradation of PAL.

Apart from the KFB01, -20, -39, and -50, each of which shows physical interactions with four AtPAL isozymes (Zhang et al., 2013, 2015), an additional KFB homolog, denoted Small And Glossy Leaves 1 (SAGL1) was found to specifically associate with AtPAL1 and regulate its activity and stability, thus the lignification of vascular bundle in Arabidopsis (Yu, Kim, Yun, Suh, & Lee, 2019). This KFB member in fact also mediates proteasome-dependent degradation of ECERIFERUM3 (CER3), a biosynthetic enzyme involved in the production of very long chain alkanes, the major components of wax, thereby negatively regulating cuticular wax biosynthesis (Kim, Yu, Jung, Lee, & Suh, 2019). Disruption of SAGL1 resulted in severe growth retardation (Kim, Yu, et al., 2019; Yu et al., 2019). Therefore, SAGL1 appears to regulate multiple target proteins including CER3 and PAL1, and perhaps coordinate the processes of lipid barrier formation and vasculature lignification.

Besides mediating PAL ubiquitination and degradation, the KFB proteins are also involved in selective protein degradation of other phenylpropanoid-lignin biosynthetic enzymes. An auxin and ABA responsive KFB homolog in *Oryza sativa* (rice), the OsFBK1, was found to physically interact with rice cinnamoyl-CoA reductase 14 (OsCCR14), a homolog to AtCCR1/IRX4 responsible for lignin biosynthesis, and with a dirigent protein potentially involved in lignan formation (Borah & Khurana, 2018). The interactions led to their selective degradation by a 26S proteasome pathway, similar to the AtKFB-AtPAL pairs in Arabidopsis. The OsFBK1-OsCCR14 module controls lignification in roots and during anther development. In addition to the monolignol biosynthetic enzymes, chalcone synthase (CHS), the key branch-point enzyme leading to the formation of flavonoids including the non-canonical lignin monomer tricin, is also ubiquitinated and its ubiquitination and subsequent degradation are also controlled by a kelch domain containing F-box protein, the KFB^{CHS}, that distantly clusters with the PAL-regulating KFBs (Zhang, Abrahan, Colquhoun, & Liu, 2017).

In addition to F-box proteins, other adaptor proteins constituting E3 ubiquitin ligase complexes are also involved in regulating secondary cell wall formation and lignin deposition. In Arabidopsis, a WD40-repeat adaptor protein AtULCS1 that serves as a subunit of a cullin-RING E3 ligase, was shown to function in cell wall modification and lignin deposition in anther cell wall, and control anther dehiscence (Beris et al., 2016).

Ubiquitination components not only regulate phenylpropanoid biosynthetic enzymes but also modulate the activities of TFs. The ER-localized ubiquitin-conjugating enzyme (E2) PtoUBC34 in *P. tomentosa* interacted with subgroup 4 of R2R3-MYB TFs PtoMYB221 and PtoMYB156 that bind to the promoter regions of the cell wall biosynthetic genes *PdCESA8*, *PdGT47C*, and *PdCOMT2*. The interactions led to the translocation of those transcriptional repressors from nucleus to the ER membrane, which consequently relieved their repressor activity on secondary cell wall synthesis and lignin deposition (Zheng et al., 2019). Therefore, the ER-localized PtoUBC34 provides a potential mechanism for the temporal and spatial regulation of TFs in the compartment of ER prior to reaching the nucleus, thus finely tuning lignin deposition in specific cells under specific physiological conditions.

Taken together, the interactions of ubiquitination and protein degradation components with phenylpropanoid-lignin biosynthetic enzymes or regulators offer a fine-tuned regulatory mechanism underlying phenylpropanoid-lignin synthesis in response to plant growth and environmental signals. Continually exploring and understanding such higher level of regulatory mechanisms and the related components will facilitate the efficient and precise lignin manipulation.

4.2 Protein-protein interaction and multienzyme complex

A growing body of evidence indicates that many cellular reactions within metabolic pathways are catalyzed not by free-floating "soluble" enzymes, but via one or more membrane-associated multienzyme complexes. This type of macromolecular organization, also coined as "metabolon," has important implications for the overall efficiency, specificity, and regulation of metabolic pathways (Srere, 1987). Phenylpropanoid metabolism is composed of a sequence of enzyme-regulated reactions that direct the production of diverse end products, including monolignols and a variety of bioactive soluble phenolics (Fig. 1). The sequential enzymes in the general phenylpropanoid pathway and different branches have been long proposed to form multienzyme complexes, i.e., metabolon (Dixon, Chen, Guo, & Parvathi, 2001; Jørgensen et al., 2005). The first evidence of metabolon in this pathway was reported for the PAL-C4H couple. A partial association of the operationally soluble PAL within the purified microsomal fractions was detected from potato tuber and cucumber cotyledons (Czichi & Kindl, 1975, 1977). These microsomal fractions directly converted phenylalanine into *p*-coumarate, which was more efficient than converting the exogenously supplied cinnamate to *p*-coumarate, suggesting that PAL and membrane bound C4H might associate together to form a metabolic channeling. Subsequent studies differentiated the behavior of PAL1 and PAL2 in *Nicotiana tabacum* (tobacco) in localizing to the ER membrane (Rasmussen & Dixon, 1999). Both PALs were partially re-localized to the ER upon *C4H* overexpression but PAL1 showed stronger affinity for ER binding sites (Achnine, Blancaflor, Rasmussen, & Dixon, 2004). Co-localization of PALs with the ER-resident C4H in *Nicotiana benthamiana* (tobacco) was confirmed by double immunolabeling and fluorescence imaging (Achnine et al., 2004; Bassard, Mutterer, Duval, & Werck-Reichhart, 2012). Therefore, the entry point enzymes of the phenylpropanoid pathway PAL and C4H appear to partially associate with each other to control the metabolic flux toward the phenolic synthesis. Nevertheless, the channeling between *P. trichocarpa* PAL and C4H could not be confirmed in yeast system (Ro & Douglas, 2004) and the direct physical interaction of Arabidopsis PAL and C4H was also hardly detected in the split ubiquitin-based membrane Y2H system (Gou, Ran, Martin, & Liu, 2018), implicating the association of PAL and C4H is fairly loose and transient.

4CL catalyzes the formation of CoA thioesters of hydroxycinnamic acids (Fig. 1). The resulting *p*-coumaroyl-CoA represents a central metabolic precursor leading to the formation of a variety of phenolics. The followed actions catalyzed by HCT and C3'H convert *p*-coumaroyl CoA to caffeoyl CoA, eventually yielding G and S lignin monomers (Vanholme, Demedts, Morreel, Ralph, & Boerjan, 2010). Therefore, 4CL, HCT and C3'H are the sequential enzymes in the pathway and they catalyze the critical reactions to channel metabolic flux to monolignols. In *P. trichocarpa* SDX, Ptr4CL3 and Ptr4CL5 are the two 4CL isozymes specific for monolignol biosynthesis. They were found to interact with each other and form a heterotetrameric protein complex at 3:1 stoichiometry (Ptr4CL3 vs Ptr4CL5), where Ptr4CL5 was postulated to play a regulatory role (Chen et al., 2014). Mathematical modeling and quantifying the steady state flux distribution within monolignol pathway revealed that the presence of Ptr4CL3-Ptr4CL5 complex contributes to the maintenance of robustness and homeostasis of monolignol biosynthetic process, making the pathway resilient to mild perturbation (Naik et al., 2018). Although both 4CL and HCT were described as soluble proteins (Ehlting, Shin, & Douglas, 2001; Hoffmann, Maury, Martz, Geoffroy, & Legrand, 2003), similar to PAL, an early study also detected the 4CL activity in the membrane-associated enzyme aggregates (Hrazdina & Wagner, 1985), indicating the participation of 4CL in a membrane-associated protein complex. Similarly, microsomal fraction from *N. benthamiana* leaves exhibited significant HCT enzyme activity, implicating the partial association of HCT with the ER membrane (Bassard et al., 2012). Arabidopsis 4CL1- and HCT-GFP fusion proteins generated a fuzzy reticulate pattern in the parietal regions of the cell, further indicating a partial association of those soluble enzymes with the ER membrane (Bassard, Richert, et al., 2012). Co-expression with AtC3'H triggered a significant re-localization of both At4CL and AtHCT nearer to the ER. This suggests that three sequential enzymes physically associate together. Finally, the direct physical interaction of At4CL1 with AtC4H or AtC3'H was confirmed in a split ubiquitin membrane Y2H assay (Gou et al., 2018). Furthermore, the interactions of poplar xylem specific Ptr4CL3/5 with PtrHCT1/6 were also validated using BiFC and protein pull-down assays in *P. trichocarpa* SDX extracts (Lin et al., 2021). Disturbing PtrHCT1/6 substantially affected Ptr4CL activity, suggesting Ptr4CL and PtrHCT are in protein complexes modulating the metabolic flux of CoA ligation for monolignol biosynthesis.

CCR and CAD are the terminal enzymes of the monolignol pathway (Fig. 1). In *P. trichocarpa*, PtrCAD1 and PtrCCR2 were found to physically interact with each other and form heterodimers (Yan et al., 2019). Suppression of PtrCAD1 caused a reduction of CCR activity in SDX; conversely, suppression of PtrCCR2 reciprocally reduced CAD activity (Yan et al., 2019). So far there is no direct evidence if the PtrCAD1/PtrCCR2 complex is part of the supramolecular complex centered with monolignol cytochrome P450 enzymes (P450s). However, Y2H assay indicated that Arabidopsis CCR1 physically interacts with AtC4H (Gou et al., 2018), implicating that the terminal enzymes in monolignol biosynthesis might also associate with the large metabolon.

In the proposed supramolecular enzyme complex, the membrane proteins, primarily P450s, are envisaged to be the critical nucleation sites for organizing the complexes (Srere, 1987). Accordingly, the Arabidopsis P450s C4H, C3'H and F5H were found to be co-eluted with each other in immunoprecipitation assays and their spatial proximity or association in the ER network was also implicated with BiFC assay and/or fluorescence lifetime imaging (Bassard, Richert, et al., 2012; Gou et al., 2018). This led to the hypothesis of monolignol P450s forming homo- or hetero-dimer or oligomers in plant cells (Bassard, Richert, et al., 2012), although the direct physical interactions of those P450s were undetectable in split ubiquitin membrane Y2H system (Gou et al., 2018). The close proximity was also reported for the poplar homologs PtrC4H1, PtrC4H2 and PtrC3'H3 in xylem tissues, where they were proposed to form heterotrimers that drastically increase the overall catalytic efficiency and carry on the direct sequential *para*- and *meta*-hydroxylation of cinnamate yielding caffeic acid (Chen et al., 2011).

Monolignol P450s are not only spatially close to each other or associate with the soluble enzymes in the pathway, but also interact with different non-catalytic ancillary proteins/co-factors. These include their conventional redox partner NADPH-dependent cytochrome P450 reductase (CPR, or ATR in Arabidopsis) and the newly recognized obligatory electron donor protein, the cytochrome b_5 member AtCB5D in Arabidopsis, that specifically augments F5H-catalyzed reaction (Gou et al., 2019). Disruption of AtCB5D significantly suppressed S-lignin deposition with slight increase in G-lignin formation, which offers an additional genetic target for tailoring lignin monomeric composition via manipulating electron flow in addition to the carbon flux. Another set of non-catalytic partners of monolignol P450s are the membrane steroid binding proteins (MSBPs). MSBPs are a group of small, partially homologous proteins that share a similar non-covalent cytochrome b_5-like heme/steroid-binding domain. Arabidopsis MSBP1 and MSBP2 are prominently ER-localized and strongly interact with AtC4H, C3'H and F5H (Gou et al., 2018). Disruption of both MSBP genes impaired the stability and activity of the P450 enzymes and suppressed the syntheses of both the soluble phenolics and lignin, suggesting that they act as the ER structural protein necessary for monolignol P450s' spatial association and stability on the ER membrane (Gou et al., 2018).

In addition to the physical association of the sequential enzymes in phenylpropanoid-monolignol pathway, the pathway enzymes also interact with some partners involved in plant defense processes. When AtC4H or AtC3'H were used as baits, CYP71B15 was co-purified from protein extracts of Arabidopsis suspension-cultured cells that were induced for tracheary element formation (Bassard, Richert, et al., 2012). Conversely, CYP71B5 co-purified AtF5H from *Botrytis cinerea*-infected Arabidopsis leaves (Mucha et al., 2019). CYP71B15 is a key enzyme responsible for phytoalexin camalexin biosynthesis. Together with CYP79B2, CYP71A12/A13 and the electron donor protein ATR1, they form a multienzyme complex/metabolon at the ER surface for synthesizing antifungal camalexin (Mucha et al., 2019). The interactions of phenylpropanoid-monolignol biosynthetic P450s with the enzyme leading to phytoalexin camalexin formation implicates a metabolic crosstalk of two biological processes in plant disease resistance.

Arabidopsis caffeoyl shikimate esterase (CSE) was also identified as a lysophospholipase, named AtLYSOPL2, since it hydrolyzes lysophosphatidylcholine (lysoPC) by cleaving its carboxylic ester bond (de Torres Zabela, Fernandez-Delmond, Niittyla, Sanchez, & Grant, 2002). LyosPC is a signaling molecule reported to enhance pathogen susceptibility in plants, and associate with the accumulation of reactive oxygen species and ethylene (de Torres Zabela et al., 2002). AtCSE/AtLYSOPL2 was found to interact with Acyl-CoA binding Protein 2 (AtACBP2; AT4G27780) (Gao, Li, Xiao, & Chye, 2010a; Miao, Lung, Li, Li, & Chye, 2019). The interaction was interestingly verified to occur on the plasma membrane (Gao et al., 2010a). Both AtCSE/AtLYSOPL2 and AtACBP2 were reported to independently function in conferring tolerance to cadmium and oxidative stresses in the transgenic Arabidopsis (Gao et al., 2010a, 2010b). These data suggest that AtCSE/AtLYSOPL2 either acts as a dual function enzyme spatially involved in two different biological processes or it might serve as a metabolic connection point linking lignin biosynthesis with stress response signaling pathway.

Similarly, *O. sativa* (rice) CCR1 was found to interact with one of the Rac/Rop family of small GTPases, OsRac1, a defense effector important for the NADPH oxidase-mediated production of reactive oxygen species (Kawasaki et al., 2006). The interaction of OsCCR1 and OsRac1 was found necessary for the enzymatic activation of OsCCR1 in vitro. Constitutive expression of *OsRac1* resulted in the enhanced accumulation of lignin through increasing CCR activity and enhancing reactive oxygen species production. It is likely that physical interaction of OsCCR1 and OsRac1 offers a coordinated regulation mechanism by which OsRac1 controls lignin biosynthesis through regulation of both NADPH oxidase and OsCCR1 activities during defense responses (Kawasaki et al., 2006). The CCoAOMT and HCT homologs in *Zea mays* (maize) co-interact with the disease resistant R gene-encoded nucleotide binding Leu-rich-repeat (NLR) protein Rp1. They form a protein complex negatively modulating the activity of Rp1. In this model, HCT and CCoAOMT function primarily as decoys. A rust disease effector modifies CCoAOMT or HCT, which alters its physical interaction with Rp1, then triggers the switch between the inactive and active states of Rp1 to modulate defense response (Wang & Balint-Kurti, 2016).

Collectively, the in vitro and in vivo evidence from different species corroborate that phenylpropanoid-monolignol biosynthetic enzymes are organized as multienzyme complexes that coordinately modulate metabolic flux for monolignol synthesis. Meanwhile, the pathway enzymes also interact with the proteins/enzymes or even enzyme complexes involved in other metabolic processes or signaling cascades, which likely coordinates lignin biosynthesis with stress response processes to sustain plant growth and defense. Understanding those complicated protein-protein interactions and multienzyme organizations could be biotechnologically significant. Enzyme organization is a key design variable to be considered when engineering phenylpropanoid biosynthesis. This is exemplified with a previous exploration in using an unnatural fusion protein of *Arabidopsis thaliana* At4CL1 and *Vitis vinifera* stilbene synthase (VvSTS) to improve the biosynthesis of health-promoting resveratrol in yeast. A 15-fold increase in the production of resveratrol was achieved through artificial fusion protein, relative to the yeast co-expression of the individual At4CL1 and VvSTS (Zhang et al., 2006). On the other hand, given the prevalent occurrence of physical or functional associations of phenylpropanoid-monolignol biosynthetic enzymes in the pathway or with the proteins in other metabolic processes, caution must be paid when designing the genetic strategies to tailor the phenolics production via gene/genome editing, since simple elimination of one or more proteins may disturb the entire enzymatic complexes or impair other metabolic processes, causing undesired pleiotropic phenotypes on plant metabolism, growth and defense response.

4.3 Enzyme inhibition

Apart from transcriptional, translational and post-translational regulations, in vitro and in vivo experiments demonstrate that many pathway intermediates or metabolites substantially affect the metabolic flux through enzyme inhibition to fine tune the monolignol biosynthetic pathway. Early studies suggested that *trans*-cinnamic acid and caffeic acid are the competitive inhibitors of PAL, which thereby regulate the entry of phenylalanine into phenylpropanoid and monolignol biosynthesis (Blount et al., 2000; Lamb, 1979; Sato, Kiuchi, & Sankawa, 1982). Correspondingly, overexpression of French bean PAL along with co-suppression of C4H in tobacco significantly suppressed PAL activity, compared to the plants harboring PAL transgene alone, suggesting that PAL activity is feedback-regulated by the over-accumulated *trans*-cinnamic acid due to the suppression of C4H *in planta* (Blount et al., 2000). 4CL, the enzyme catalyzing the last step of the general phenylpropanoid pathway, not only exhibits substrate inhibition but also is regulated by downstream biosynthetic intermediates. Ptr4CL3 and Ptr4CL5, the two enzymes catalyzing the coenzyme A (CoA) ligation of *p*-coumaric acid to *p*-coumaroyl-CoA and caffeic acid to caffeoyl-CoA in poplar wood forming tissue, exhibited competitive and uncompetitive inhibition by substrates *p*-coumaric acid and caffeic acid, by which it modulates the CoA ligation flux (Chen et al., 2013). Moreover, the HCT and/or C3'H products *p*-coumaroyl shikimate and caffeoyl shikimate also showed strong inhibition to Ptr4CL3 and Ptr4CL5 activities (Lin et al., 2015). *P*-coumaroyl shikimate inhibits both *p*-coumaroyl-CoA and caffeoyl-CoA formation, while caffeoyl shikimate only inhibits the production of *p*-coumaroyl-CoA. *P*-coumaroyl shikimate is a stronger inhibitor compared to caffeoyl shikimate. With the inhibition of CoA ligation reaction by those intermediates, it increased flux through the direct 3-hydroxylation of *p*-coumaric acid to caffeic acid, catalysed by the PtrC4H1/C4H2/C3H3 complex, and subsequently to the formation of caffeoyl-CoA. Therefore, these shikimic acid esters play a regulatory role controlling the metabolic flux from *p*-coumaric acid to different paths.

In the monolignol pathway, HCT catalyzes the coupling of *p*-coumaroyl-CoA with acceptor shikimate to produce *p*-coumaroyl shikimate (Hoffmann et al., 2003, 2004) (Fig. 1). The enzymes from different land species spanning from bryophytes to angiosperms exhibit notable substrate promiscuity, recognizing a broad spectrum of acceptor molecules (Eudes et al., 2016). Besides the canonical substrate shikimic acid, the recombinant HCT from switchgrass showed considerable activity toward a set of 'non-canonical' substrates, such as 3-hydroxybenzoate, 2,5-dihydroxybenzoate, 2,3-dihydroxybenzoate, 3,4-dihydroxybenzoate (protocatechuate), 3-aminobenzoate, 3-hydroxyanthranilate, 5-hydroxyanthranilate, *ortho*-hydroxyphenol (catechol) and *para*-hydroxyphenol (hydroquinone). Interestingly, most of those non-canonical substrates can serve as competitive inhibitors of the enzyme suppressing the formation of *p*-coumaroyl shikimate. The in vitro activity of either switchgrass recombinant PvHCT2 or Arabidopsis stem protein extracts in the formation of *p*-coumaroyl shikimate was particularly inhibited protocatechuic acid (Eudes, Pereira, et al., 2016). The positioning and binding of protocatechuic acid to the PvHCT2 substrate binding site exhibited a similar pattern as that of shikimate in the crystal structure of PvHCT2, suggesting a mechanism of competitive inhibition (Eudes, Pereira, et al., 2016). This enzyme inhibition feature was then harnessed for manipulating lignin biosynthesis. A two-gene strategy was designed to divert the endogenous chorismic acid pool into protocatechuic acid in lignifying tissues of

Arabidopsis by heterologous co-expression of *E. coli* chorismate pyruvate-lyase (ubiC) and *Pseudomonas aeruginosa p*-hydroxybenzoate 3-mono-oxygenase (pobA), which resulted in up to 26-fold higher accumulation of protocatechuic acid in the ubiC-pobA transgenic lines compared with the wild type and the ubiC alone lines. Consequently, an approximate 30% reduction of lignin content was achieved in the ubiC-pobA lines, most likely due to the inhibition of HCT activity by the hyperaccumulated protocatechuic acid. Eventually this manipulation resulted in the improvement of stem biomass saccharification compared with the wild type (Eudes, Pereira, et al., 2016).

Monolignol biosynthesis not only connects to shikimate metabolism but also closely coordinates with one carbon metabolism that yields active methyl donor S-adenosylmethionine (AdoMet). Both CCoAOMT and COMT use AdoMet as a cofactor to methylate 3- and 5-hydroxyl groups of the aromatic intermediates respectively, ultimately leading to the biosynthesis of G and S lignin monomers (Fig. 2). In general, one carbon metabolism consists of tetrahydrofolate (THF) and methionine (Met) cycles. The former is responsible for the transfer of C1 units, while the latter is for the synthesis and recycling of AdoMet (Hanson & Roje, 2001). S-adenosylhomocysteine (SAH), a demethylated product of AdoMet, is a well-known competitive inhibitor of AdoMet dependent *O*-methyltransferases. It binds to the same cofactor binding pocket in OMTs (Liu et al., 2006). Both CCoAOMT and COMT enzymatic activities are strongly inhibited by SAH; moreover, the ratio of AdoMet:SAH but not the AdoMet concentration more critically determines the catalytic efficiency of the enzymes and the transmethylation reactions in monolignol biosynthesis (Wu et al., 2018). This was genetically validated via the examination of maize *brown midrib2* (*bm*2) mutant. It was found that downregulation of methylenetetrahydrofolate reductase (MTHFR), the enzyme converting 5,10-methylene-THF to 5-methyl-THF, in *bm*2 mutant affected SAH rather than AdoMet levels. Consequently, the *O*-methylation of lignin monomers was impaired, which resulted in a remarkable accumulation of novel phenolics and unusual C lignin that derives from the polymerization of caffeyl alcohol in the mutant. Although no impact on total lignin content and plant growth was observed, the altered lignin composition led to a significant improvement in the cell wall saccharification efficiency (Wu et al., 2018). The feature of AdoMet:SAH ratio on monolignol synthesis was further harnessed in engineering lignin synthesis in bioenergy crop switchgrass by down-regulation of cystathionine gamma-synthase (PvCGS), the first committed enzyme

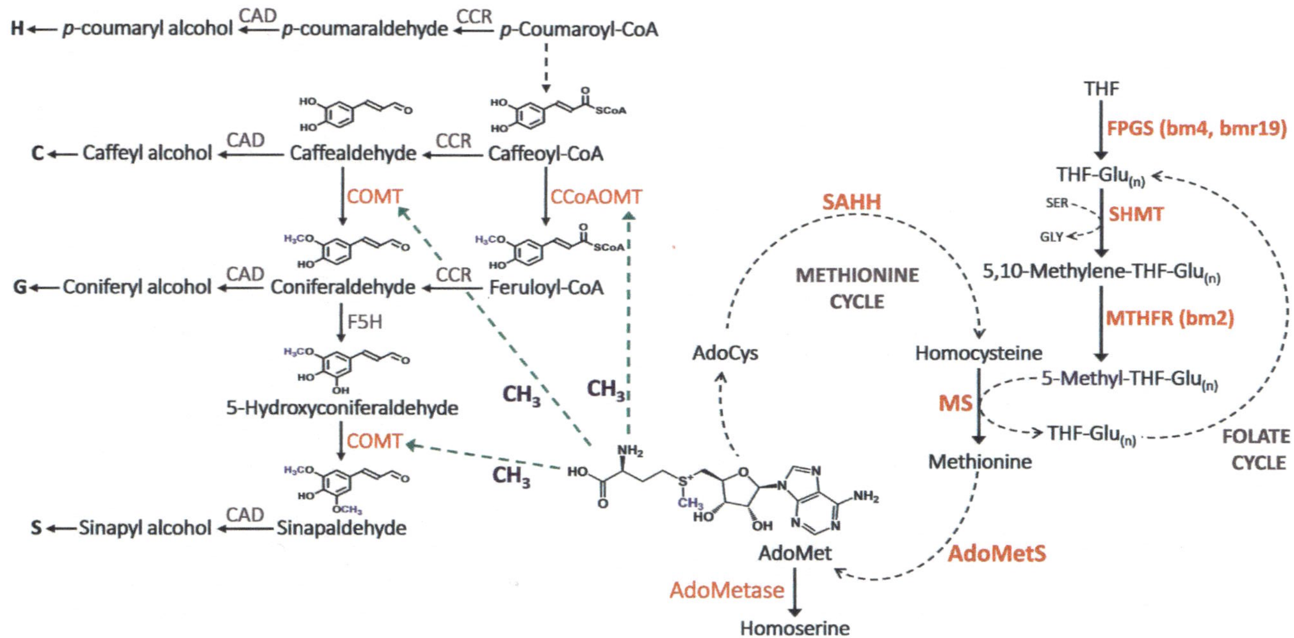

Fig. 2 Interconnection between the lignin biosynthetic pathway and the methionine and folate cycles. Enzymes are: AdoMetase, S-adenosylmethionine hydrolase; CAD, cinnamyl alcohol dehydrogenase; CCoAOMT, caffeoyl-CoA *O*-methyltransferase; CCR, cinnamoyl-CoA reductase; COMT, caffeate/5-hydroxyferulate *O*-methyltransferase; FPGS, folylpolyglutamate synthase; MS, methionine synthase; F5H, ferulate 5-hydroxylase; MTHFR, methylenetetrahydrofolate reductase; SAHH, S-adenosylhomocysteine hydrolase; SHMT, serine hydroxymethyltransferase. Other abbreviations are: AdoCys, S-adenosylhomocysteine; AdoMet, S-adenosylmethionine; GLY, glycine; SER, serine; THF, tetrahydrofolate; THF-Glu$_{(n)}$, tetrahydrofolate polyglutamate; 5,10-methylene-THF-Glu$_{(n)}$, 5,10-methylenetetrahydrofolate polyglutamate; 5-methyl-THF-Glu$_{(n)}$, 5-methyltetrahydrofolate polyglutamate.

for the biosynthesis of Met via homocysteine, and by suppression of S-adenosylhomocysteine hydrolase 1 (PvSAHH1), an enzyme responsible for SAH hydrolysis thus maintaining SAH levels in plants (Fig. 2). Suppression of PvCGS in switchgrass altered the accumulation of aspartate-derived amino acids and the aromatic amino acids, and reduced the content of SAH; thereby, enhancing lignin biosynthesis (Bai et al., 2018). In contrast, down-regulation of PvSAHH1 raised SAH levels, which impaired the biosynthesis of both G and S lignin and significantly increased saccharification efficiency of the cell walls (Bai et al., 2018). These results suggest that genes involved in Met and one carbon metabolism are valuable targets for lignin bioengineering, in addition to the conventional manipulation of monolignol biosynthetic genes and TFs. This opens another avenue for lignin engineering of bioenergy and feed crops.

5. Modification of enzymes cofactors and co-substrates

Several enzymes from the lignin biosynthetic pathway use cofactors and co-substrates such as ATP (4CL), NADPH (C4H, C3'H, C3H, CCR, CAD, and F5H), oxygen (C3'H, C4H, F5H, laccases), hydrogen peroxide (peroxidases, C3H), ascorbate (C3H), AdoMet (COMT and CCoAOMT), coenzyme A (4CL and HCT), and shikimate (HCT). Therefore, modifying intracellular pools of these cofactors and co-substrates has been contemplated for altering lignin biosynthesis. In this section, we highlight the recent discoveries specific to lignin biosynthesis concerning AdoMet and shikimate.

5.1 S-Adenosylmethionine

AdoMet is a key intermediate in one-carbon metabolism that serves as a universal methyl-group donor for the methylation of a large number of metabolites (Hanson & Roje, 2001). It is synthesized from methionine (Met) and ATP by the activity of AdoMet synthetase within the methionine cycle, which is itself connected to the folate cycle via Met synthase (Fig. 2). As previously mentioned, the lignin biosynthetic pathway contains two enzymes that consume AdoMet for transmethylation reactions: CCoAOMT is involved in the first methylation step required for the production of G units and COMT performs the second methylation step needed for the synthesis of S units (Fig. 2). The importance of AdoMet as a methyl donor for lignin biosynthesis has been illustrated by several mutant studies in Arabidopsis, sorghum (*Sorghum bicolor*), and maize. For example, mutation in one of the AdoMet synthetases (AdoMetS3) in

Arabidopsis results in concomitant reductions of AdoMet synthetase activity, AdoMet pools, and lignin content (Shen, Li, & Tarczynski, 2002). Moreover, it was shown that mutations in genes responsible for the synthesis of 5-methyltetrahydrofolate, which is used as methyl donor by Met synthase for the production of AdoMet via the Met cycle (Fig. 2), led to the reduction of lignin content in maize, sorghum, and Arabidopsis (Adeyanju et al., 2021; Li et al., 2015; Srivastava et al., 2015; Tang et al., 2014). In particular, these mutations affect MTHFR and folylpolyglutamate synthase (FPGS) in the folate cycle. In Arabidopsis, reductions in both pools of AdoMet and its precursor Met were measured in the Arabidopsis *fpgs* mutant (Srivastava et al., 2011, 2015). This mutant, which is deficient in a FPGS isoform preferentially expressed in vascular tissues, does not show any reductions in biomass yield (Srivastava et al., 2015). Both the *brown midrib4* (*bm4*) gene in maize and the *brown midrib19* (*bmr19*) gene in sorghum encode an FPGS. The corresponding *bm4* and *bmr19* mutants showed an increased lignin S/G; a phenotype also observed in the Arabidopsis *fpgs* mutant (Adeyanju et al., 2021; Barrière et al., 2004; Li et al., 2015; Srivastava et al., 2015). Similarly, a mutation in *brown midrib2* (*bm2*), which encodes a MTHFR, also resulted in a higher S/G in maize (Barrière et al., 2004; Tang et al., 2014). Further, it was recently shown in Arabidopsis that simultaneous downregulation of *FPGS1* and *CCoAOMT1* results in stronger lignin reduction and higher saccharification efficiency compared to the values observed in lines downregulated in either one of the two genes (Xie et al., 2019). These observations led researchers to silence *FPGS* in switchgrass in order to reduce biomass recalcitrance to enzymatic saccharification (Mazarei et al., 2020). Under field conditions, only the line showing the strongest reduction of *FPGS1* transcripts (decreased by ~85% compared to the wild-type control) had less lignin. However, lignin reduction was not accompanied with improvements of saccharification efficiency and the line had a significant biomass yield penalty. Interestingly, one engineered line with moderate reduction of *FPTGS1* transcripts had no yield drag and produced 18% more ethanol, although biomass did not contain less lignin nor released higher amounts of fermentable sugars after saccharification (Mazarei et al., 2020). Finally, overexpression of serine hydroxymethyltransferase (SHMT, Fig. 2) in poplar (*P. deltoides*) enabled reductions of lignin and cell wall recalcitrance (Zhang et al., 2019). The authors proposed a mechanism whereby SHMT has the capacity to regulate transcriptional master switches of secondary cell wall biosynthesis. In particular, down-regulation of transcription factors that act as activators of lignin biosynthetic genes was observed in SHMT overexpressing poplar lines. These results, in addition to the regulation of the activity of lignin-related *O*-methyltransferases by SAH pools (see Section 4.3), highlight another layer of complexity to the fine regulation and interconnection that exists between lignin biosynthesis and one-carbon metabolism.

In addition to being used as a substrate by AdoMet-utilizing methyltransferases, AdoMet also acts as a precursor in the synthesis of polyamines, nicotianamine, phytosiderophores, 5′-deoxyadenosyl radicals, and ethylene (Roje, 2006). Therefore, modification of the intracellular AdoMet pool may have pleiotropic effects besides altering lignin biosynthesis. The major demand for one-carbon units is allocated for lignin biosynthesis (Amthor, 2003), and reducing AdoMet pool specifically in the tissues that produce lignified cell walls may have limited impact on other metabolic pathways that have lower AdoMet demand (Srivastava et al., 2015). This approach was attempted by Eudes et al. (2016) in Arabidopsis via tissue-specific expression of an AdoMet hydrolase (AdoMetase) gene from *E. coli* phage using the promoter of a secondary cell wall synthase gene (*pAtCesa4*). Although plant height was slightly reduced, the lines expressing AdoMetase showed no reduction of biomass yields compared to the wild-type control. Reduction of the AdoMet pool in stems resulted in diminution of lignin content by ~30%, a decreased S/G ratio, and a ~2.5-fold increase in H lignin units, which led to the improvement in biomass saccharification efficiency.

One of the desired outcomes of reducing COMT and CCoAOMT activities is the potential accumulation of caffeoyl alcohol and formation of catechyl (C) lignin units. Considering a 'lignin-first' biorefinery approach, homopolymeric C-lignin represents an ideal lignin because its backbone contains only ether (C—O) interunit linkages so that it can be fully depolymerized, while its stability under acidic pretreatment conditions prevents condensation and formation of undesired new C—C bonds (Li et al., 2018). Moreover, the strength of the preponderant benzodioxane structures found in C-lignin as well as its linearity could offer a great advantage for diverse lignin-based biomaterials (Berstis, Elder, Crowley, & Beckham, 2016). Such benzodioxane structures are formed following the β-O-4 coupling of caffeyl alcohol monomers and upon rapid internal trapping of the quinone methide, which prevents benzyl ether and ester cross-linking of hemicellulosic alcohol or acid groups with C units (Vanholme et al., 2012). Therefore, accumulation of C-lignin (and benzodioxane-type lignins in general) in energy crops has also been proposed to reduce cell-wall crosslinkings between lignin and polysaccharides (Grabber et al., 2019). To elucidate the fine mechanism of C-lignin formation, transcriptomic studies have been conducted on plant species known to accumulate this type of lignin in seed coats. As expected, downregulation of *COMT* and/or *CCoAOMT* genes was observed in tissues that produce C-lignin in species like *Vanilla orchid*, *Cleome hassleriana*, and *Paphiopedilum armeniacum* (Fang et al., 2020; Rao et al., 2014; Zhuo et al., 2019). In addition, the requirement of a laccase with activity toward caffeyl alcohol has been recently demonstrated in *C. hassleriana* (Wang et al., 2020). In seeds coats from this plant, upregulation of the laccase gene *ChLAC8* was concomitant with the accumulation of C-lignin, while its downregulation reduced C-lignin content. Moreover, heterologous expression of *ChLAC8* in an Arabidopsis *comt* mutant background enabled formation of C-lignin upon exogenous feeding with caffeyl alcohol (Wang et al., 2020). These results may explain the accumulation of soluble caffeyl alcohol conjugates in the maize *bm2/mthfr* mutant, assuming that laccase activity toward caffeyl alcohol is too low for its efficient oxidative polymerization into lignin in maize midribs, albeit trace amount of C units was evidenced in this mutant (Wu et al., 2018).

5.2 Shikimate

Formation of coniferyl and sinapyl alcohols during lignin biosynthesis involves the formation of *p*-coumaroyl shikimate intermediate, catalyzed by the activity of HCT (Cardenas et al., 2021) (Fig. 2). HCT was shown to be a cytosolic enzyme partially associated with the ER membranes based on fluorescence imaging microscopy in leaves of both Arabidopsis and *N. benthamiana* (Bassard, Richert, et al., 2012). This implies that shikimate, which is produced in plastids via the shikimate pathway, needs to be exported to the cytosol to serve as substrate for HCT activity. How shikimate exits the chloroplast remains unknown and elucidation of the mechanism involved could enable the design of novel bioengineering approaches to inhibit HCT. Shikimate transporters have been identified from bacteria such as *E. coli* and *Corynebacterium glutamicum* (Kubota et al., 2015; Whipp, Camakaris, & Pittard, 1998), but proteins with similar function on the chloroplast envelope in plants or even cyanobacteria remain to be identified. One strategy that efficiently disturbs lignin synthesis, presumably via inhibition of HCT, is the expression of plastid-targeted 3-dehydroshikimate dehydratase (QsuB) from *C. glutamicum* (Eudes et al., 2015). Consistent with HCT inhibition, Arabidopsis plants transformed with the QsuB gene show up to 50% less lignin and ~6-fold increase in H units. Because the substrate of QsuB in plastids is also the direct precursor of shikimate (i.e., 3-dehydroshikimate), one of the consequences of this engineering approach is the reduction of the shikimate pool available for HCT in the cytosol. The QsuB low lignin trait has been successfully transferred to the bioenergy crop switchgrass (Hao et al., 2021). Recently, a gene stacking approach that enables a two-step conversion of protocatechuate—the product of QsuB—into the valuable polyester precursor 2-pyrone-4,6-dicarboxylic acid has been demonstrated in Arabidopsis, highlighting a potential for stacking the low lignin trait with a value-added coproduct trait (Lin et al., 2021). Finally, another approach for reducing the availability of shikimate could be the phosphorylation of its 3-hydroxyl group normally involved in the esterification of *p*-coumarate upon HCT activity, and expressing a shikimate kinase targeted to the cytosol has been proposed (Loqué & Eudes, 2021). Conclusively, a recent study from Hu and colleagues showed that heterologous expression of a shikimate kinase from *Helicobacter pylori* in hybrid aspen (*P. tremula* × *Populus tremuloides*) increased the amount of H lignin units, which is consistent with a reduction of HCT activity, and led to improvements of wood saccharification efficiency (Hu et al., 2022).

6. Modification of monolignols and metabolic precursors

Expression of enzymes known to use metabolic intermediates of the lignin biosynthetic pathway represents a strategy for potentially modifying lignin content and/or its monomeric composition. Several studies reported on the use of microbial enzymes as well as plant-derived enzymes that do not participate in the synthesis of the main monolignols to reroute lignin metabolic precursors. In many instances the product of newly introduced enzymes incorporates into lignin, thereby altering lignin structure and physico-chemical properties.

6.1 Expression of microbial enzymes

Expression of the *4-hydroxycinnamoyl-CoA hydratase/lyase* (*HCHL*) gene from *Pseudomonas fluorescens* in tobacco demonstrated the efficacy of using a bacterial enzyme to efficiently reroute intermediates of the lignin pathway (Mayer et al., 2001). Notably, HCHL cleaves *p*-coumaroyl-CoA and feruloyl-CoA into 4-hydroxybenzaldehyde and vanillin, which are readily converted inside the cells into glycosylated forms of 4-hydroxybenzoic acid and vanillic acid, respectively (Mayer et al., 2001). Strong constitutive *HCHL* expression in tobacco was detrimental to plant development and fertility, while reduction of lignin content and abnormal coloration of xylem tissues were also observed (Merali et al., 2007). In Arabidopsis, *HCHL* was expressed under the control of the promoter of a secondary cell wall cellulose synthase gene (*pCesa4*) to target specifically lignifying tissues and mitigate adverse effects. As a result, several HCHL Arabidopsis lines show no growth defects or biomass yield penalty. These lines showed improvements of saccharification efficiency despite having similar lignin content compared to wild-type controls, and analysis of lignin monomeric composition revealed the presence of various C6-C1 aromatics such as hydroxybenzoates, hydroxybenzaldehydes, and hydroxybenzyl alcohols that presumably derive from the two HCHL products 4-hydroxybenzaldehyde and vanillin. A reduction in the average molecular weight of lignin was also observed in transgenics, suggesting that incorporation of C6-C1 aromatics during the lignification process affects lignin's degree of polymerization and overall cell wall recalcitrance (Eudes et al., 2012; Kim et al., 2019). More generally, increasing non-canonical monomers only capable of single coupling reactions with conventional lignin monomers has been considered as an engineering strategy to increase polymer initiation events and the formation of shorter lignin chains (Mottiar, Vanholme, Boerjan, Ralph, & Mansfield, 2016). Consistent with this idea, *HCHL* expression in hybrid poplar (*P. deltoides* × *Populus euramericana*) using the promoter of the Arabidopsis *C4H* gene (*pAtC4H*) leads to improvements of saccharification despite no significant changes in lignin content. Nevertheless, the engineered HCHL trees show diminutions of height and fresh weight, possibly as a consequence of too broad *pAtC4H* activity in poplar (Zhang et al., 2020). Similar observations were made in Arabidopsis plants designed to express the *4-hydroxycinnamoyl-CoA hydratase/lyase* gene from *Rhodopseudomonas palustris* (*couA*) under the control of *pAtC4H*, and resulting lines also exhibited a higher lignin S/G ratio compared to the wildtype (Sakamoto et al., 2020). Another example of bacterial enzyme used to interfere with the lignin biosynthetic pathway is coniferaldehyde dehydrogenase CalB from *Pseudomonas* sp. HR199. Arabidopsis lines expressing *calB* under the control of *pAtC4H* do not have less lignin but exhibit an increased lignin S/G ratio and improvements of saccharification efficiency (Sakamoto et al., 2020). In these lines, a reduction of coniferaldehyde pools could be the cause of changes in lignin monomeric composition considering that coniferaldehyde is not only the precursor of coniferyl alcohol and G units, but also a potential inhibitor of CAD-catalyzed reduction of sinapaldehyde to sinapyl alcohol (see Section 4.2, Wang et al., 2014). Several studies showed that expression in plants of the chorismate pyruvate-lyase gene *ubiC* from *E. coli* results in the accumulation of 4-hydroxybenzoic acid glucose conjugates (4-hydroxybenzoic acid 4-*O*-glucoside and 4-hydroxybenzoic acid glucose ester) due to conversion of endogenous chorismate into 4-hydroxybenzoic acid (McQualter et al., 2005; Siebert et al., 1996). The potential bioenergy crop sorghum was recently transformed to express *ubiC* under the control of the rice ubiquitin gene promoter (*pRubi2*) and enable overproduction of 4-hydroxybenzoic acid *in planta* (Lin et al., unpublished). Two-dimensional nuclear magnetic resonance spectroscopy (2D NMR) analysis revealed the presence of 4-hydroxybenzoate ester groups on lignin isolated from *ubiC*-engineered sorghum (Wang et al., 2021). Although such ester end-groups are commonly found in lignins from willow, poplar, aspen, Neptune grass, and palms, they do not occur in wild-type sorghum, suggesting that a non-specific transferase is being recruited for transferring 4-hydroxybenzoate onto monolignols in modified 4-hydroxybenzoate-rich sorghum. This engineering approach potentially adds value to lignin since 4-hydroxybenzoate can be easily removed from lignin via hydrothermal treatment and further converted into deep eutectic solvents used for biomass pretreatment prior to saccharification, thereby enabling a closed-loop biorefinery (Wang, Meng, et al., 2021).

Another approach to modify lignin structure is via degradation of β-aryl ether bonds through expression of bacterial lignin-degrading enzymes. In Arabidopsis, expression of the gene encoding the Cα-dehydrogenase (LigD) from *Sphingobium* sp. strain SYK-6 altered the lignin structure, which was evidenced by an increased level of G-type α-keto-β–O–4 linkages (Tsuji et al., 2015). Heterologous expression of *LigD* in the cytoplasm resulted in the accumulation of neolignan-like compounds and α-oxidized products of two typical G-type β–O–4 dilignols. More likely, the α-oxidized dilignols incorporate lignin and contribute to the increased α-keto-β–O–4 units. These observations also indicate that monolignols not only couple in the apoplast, but also in the cytoplasm (Tsuji et al., 2015). In another study in Arabidopsis, co-expression of the genes encoding cell wall targeted LigD and glutathione S-transferases (LigF and LigG) from *Sphingomonas paucimobilis* was carried out to introduce postlignification modifications into the lignin structure. Successfully, partial disruption of β-aryl ether bonds in lignin was observed and the number of oxidized G and S units was increased but these units were no longer bound via α-keto-β–O–4 linkages. Successfully, biomass from these engineered plants showed improvements of biomass saccharification efficiency (Mnich et al., 2017).

6.2 Expression of plant enzymes

Lignification is initiated by the activity of laccases and/or peroxidases that oxidize monolignols and produce the corresponding phenolic radicals. These active monolignol radicals undergo cross-coupling reactions to incorporate the monomer units into oligomers or the growing lignin polymer. Considering that the *para*-hydroxyl of monolignol is crucial for the generation of phenolic radicals and the subsequent coupling, attempts were made to rationally design an enzyme with 4-*O*-methyltransferase activity against monolignols in order to prevent their oxidative radicalization. Using a 4-*O*-methyltransferase from *Clarkia breweri* known to be active on (iso)eugenol, Bhuiya and Liu (2010) generated a series of mutant variants with the novel capacity of 4-*O*-methylating coniferyl and sinapyl alcohols (the product 4-*O*-methylated coniferyl alcohol is shown in Fig. 3). One of these engineered monolignol 4-*O*-methyltransferase (MOMT4) was expressed in Arabidopsis using the promoter of a *PAL* gene from French bean, which resulted in transgenic lines with reduced lignin and improved saccharification yields (Zhang et al., 2012). Introducing *MOMT4* in gray poplar (*P. tremula* x *P. alba*) also resulted in lignin reduction, significant diminution of S/G ratio, and higher amount of simple sugars after saccharification compared to the control trees, which enabled higher yield of ethanol after yeast fermentation (Cai et al., 2016). Further, combining crystal structural information with combinatorial active site saturation mutagenesis, an evolved enzyme carrying nine mutated residues compared to the original enzyme was obtained to discriminately etherify the *para*-hydroxyl of coniferyl alcohol even in the presence of sinapyl alcohol (Cai, Bhuiya, Shanklin, & Liu, 2015). This enzyme could therefore be useful to increase lignin S/G ratio in crops considering that a higher S/G ratio is often associated with increased saccharification efficiencies after biomass pretreatment as well as improved delignification during pulping (Li, Pu, & Ragauskas, 2016).

Lignin from several plants can be acylated with various groups. As examples, acetate and *p*-coumarate decorate lignin in diverse herbaceous plants and grasses (del Río et al., 2008), 4-hydroxybenzoate is found in lignin from fruit bunches of palms (Lu et al., 2015), and benzoate, vanillate and ferulate groups are found in lignin from leaves of the Canary Island date palm

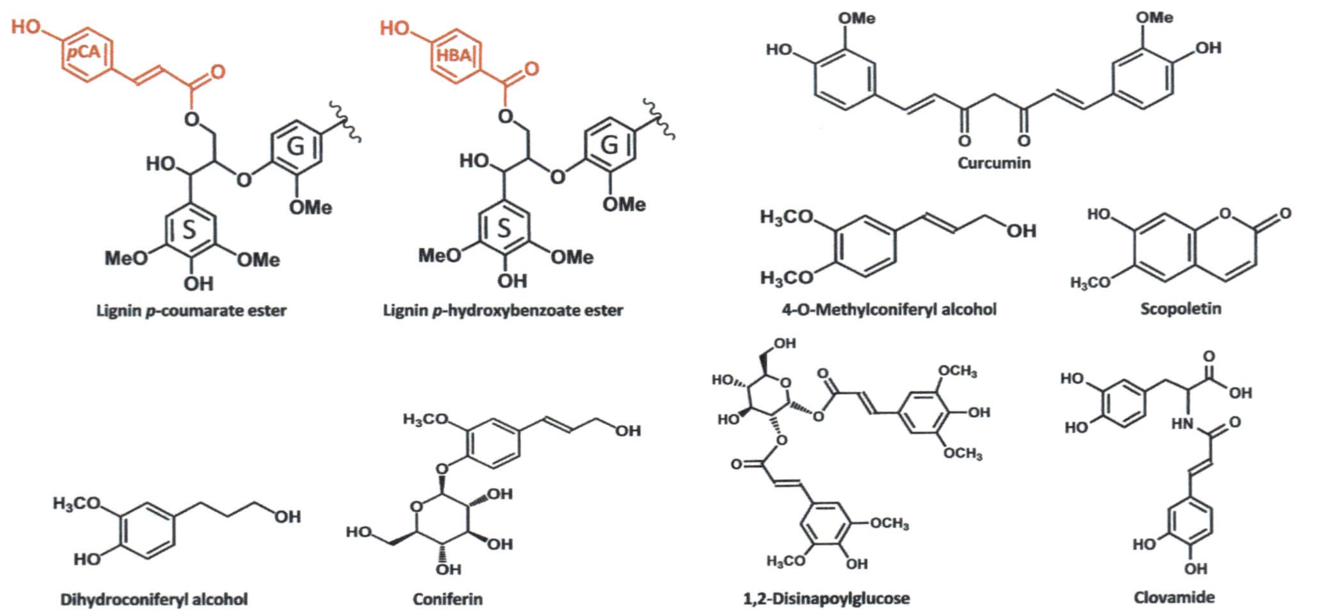

Fig. 3 Schematic of *p*-coumarate (*p*CA) and *p*-hydroxybenzoate (HBA) esters groups on lignin and structures of rare or potentially new monolignols.

(Karlen et al., 2017). These ester-linked groups are usually more frequent on lignin S units than G units (Fig. 3). Their formation involves the acyl-transferases that catalyze the formation of monolignol conjugates using CoA-activated acyl groups as donors and monolignols as acceptors. Acyltransferases responsible for the formation of monolignol *p*-coumarate conjugates have been identified in rice (Withers et al., 2012), maize (Marita, Hatfield, Rancour, & Frost, 2014) and Brachypodium (Petrik et al., 2014). These transferases have been exploited to engineer lignin in Arabidopsis, poplar and maize (Karlen et al., 2020; Lapierre et al., 2021; Sibout et al., 2016; Smith et al., 2015). One benefit of this approach is the enrichment of lignin with *p*-coumarate, which can potentially be easily cleaved and recovered from lignin streams generated in biorefineries. Lignin valorization and the upgrading of aromatics such as *p*-coumarate into more valuable bioproducts represents an option for achieving economically sustainable biorefineries (Becker & Wittmann, 2019). Another advantage of this strategy is that the increased *p*-coumaroylation of lignin has favorable effects on lignin structure: the engineered highly *p*-coumaroylated lignins in poplar and Arabidopsis show increased terminal units with free phenolic groups, presumably have reduced degree of polymerization, and present higher solubility under alkaline conditions, which improves subsequent enzymatic biomass saccharification (Lapierre et al., 2021; Sibout et al., 2016). At least one acyltransferase (PHBMT1 or PtACT54) from *P. trichocarpa* has been determined responsible for lignin 4-hydroxybenzoylation (de Vries et al., 2021; Zhao et al., 2021). PHBMT1 preferentially uses 4-hydroxybenzoyl-CoA as donor and sinapyl alcohol as acceptor in vitro. Disrupting *PHBMT1* via CRISPR-Cas9 gene editing almost completely depleted 4-hydroxybenzoate groups on lignin in stems, while overexpression of *PHBMT1* enhanced lignin 4-hydroxybenzoates. Interestingly, diminution of 4-hydroxybenzoate groups promoted lignin solubility under acidic condition, suggesting that 4-hydroxybenzoylation influences lignin physicochemical properties (Zhao et al., 2021). As previously mentioned for *p*-coumarate-rich lignin, augmenting the amount of 4-hydroxybenzoate groups on lignin represents an opportunity to add value to bioenergy crops given that 4-hydroxybenzoate is another aromatic with several industrial applications (Lin & Eudes, 2020). For an engineering purpose, it would be useful to identify the transferases that acylate efficiently coniferyl alcohol in order to expand lignin 4-hydroxybenzoylation to G units in crops. The seagrass *Posidonia oceanica* (Neptune grass) represents an appealing target for identifying such acyltransferase(s) considering that

4-hydroxybenzoate esters occur frequently on both G- and S-lignin units in this plant (Rencoret et al., 2020). Although they belong to the same family of BAHD acyltransferases, the enzymes involved in the feruloylation and *p*-coumaroylation of xylan in cell walls are distinct from those involved in lignin acylation (Mnich et al., 2020). Interestingly, it has been recently observed in sorghum that overexpression of such a xylan-specific *p*-coumaroyl transferase leads to the reductions of lignin content and modification of S/G ratio (Tian et al., 2021).

Hydroxycinnamoyl-CoAs (e.g., *p*-coumaroyl-CoA, feruloyl-CoA, caffeoyl-CoA) are not only metabolic intermediates in the lignin biosynthetic pathway but also the precursors to a wide array of secondary metabolites such as stilbenes (e.g., piceatannol), flavonoids (e.g., tricin), coumarins (e.g., scopoletin), polyketides (e.g., curcumin), hydroxycinnamic amides (e.g., feruloyltyramine), and hydroxycinnamic esters (e.g., coniferyl ferulate and sinapyl *p*-coumarate). Several of these metabolites were recently identified as part of lignin in some plant species, which has prompted the identification of their biosynthetic genes toward developing new lignins with altered structures and/or increased content of high-value products (del Río et al., 2020). Clovamide is a hydroxycinnamic amide found in cocoa and clover species that consists of a caffeate moiety linked via an amide bond to L-dopa (Fig. 3). Clovamide has been proposed as a good candidate for monolignol replacement due to its catechol groups that could promote the formation of benzodioxane structures during lignin formation and thereby prelude lignin-polysaccharides crosslinkings. Moreover, its amide bond linking the two aromatic moieties would be more cleavable than conventional ether and carbon-carbon bonds found in lignin, while its carboxyl group could increase lignin hydrophilicity. Using an artificial cell wall lignification system, clovamide was shown to incorporate into lignin when supplied with coniferyl and sinapyl alcohols, which increased enzymatic yields of glucose following alkaline or acid pretreatment, compared to the lignified cell wall controls obtained with only conventional monolignols and having a similar lignin content (Grabber et al., 2019). An enzyme from the BAHD acyltransferase family (HDT1) isolated from red clover was shown to catalyze the transfer of hydroxycinnamoyl-CoAs onto aromatic amino acids and L-dopa (Sullivan & Bonawitz, 2018), and its heterologous expression in the yeast *Saccharomyces cerevisiae* enabled the synthesis of clovamide (Bouchez et al., 2019). Considering the preference of HDT1 for tyrosine compared to L-dopa, it is plausible that clovamide formation *in planta* occurs via the coupling of caffeoyl-CoA with tyrosine, followed by hydroxylation of the C3 position on the tyrosine aromatic ring. Expression of *HDT1* in alfalfa resulted in foliar accumulation of *p*-coumaroyl-*L*-tyrosine and feruloyl-*L*-tyrosine, whereas transient co-expression in tobacco (*N. benthamiana*) of HDT1 with a tyrosine hydroxylase (CYP76AD6 from sugar beet) for the synthesis of *L*-dopa resulted in accumulation of clovamide (Sullivan & Knollenberg, 2021). As another example of potential novel monolignol, the polyketide curcumin (Fig. 3) was overproduced in Arabidopsis via rerouting of feruloyl-CoA upon heterologous co-expression of diketide-CoA synthase (*DCS*) and curcumin synthase (*CURS*) from turmeric (Oyarce et al., 2019). The incorporation of curcumin into lignin was evidenced and curcumin overproducing lines had improved saccharification yields after alkaline pretreatment despite having similar lignin content compared to the wildtype. These improvements are possibly due to the presence in lignin of alkali-labile β-diketone structures arising from the incorporation of curcumin. The coumarin scopoletin (Fig. 3) was also overproduced in Arabidopsis via overexpression of feruloyl-CoA 6′-hydroxylase (*F6'H1*) and was found to accumulate in the cell wall fraction. Incorporation of scopoletin into lignin was not determined but a lower amount of lignin monomers was released by thioacidolysis in these modified plants despite similar lignin content compared to the controls (Sakamoto et al., 2020).

Several metabolic engineering approaches have the potential to reduce lignin while concomitantly augmenting specific valuable chemicals *in planta* (Lin & Eudes, 2020). For example, eugenol is an important phenylpropene that derives from coniferyl alcohol. In flowers and leaves that produce eugenol, coniferyl alcohol is converted into coniferyl acetate by coniferyl alcohol: acetyl-CoA transferase (CAAT) (Dexter et al., 2007; Kim et al., 2014), which is in turn converted into eugenol by eugenol synthase. Expression in hybrid aspen of *CAAT* genes under the control of a *35S* promoter resulted in the production of eugenol and its glucosides (Koeduka et al., 2013; Lu et al., 2017). Preliminary data from field trials did not reveal any significant differences for lignin content in engineered trees (Lu et al., 2017), but optimized *CAAT* expression specifically in tissues that undergo lignification could represent a strategy to reduce lignin. Similarly, biosynthesis of the valuable aromatic compound 2-phenylethanol requires phenylalanine as precursor and therefore competes with the phenylpropanoid pathway. Overproduction of 2-phenylethanol in poplar and Arabidopsis has been reported via heterologous co-expression of phenylacetaldehyde synthase (PAAS) and phenylacetaldehyde reductase

(PAR) genes (Costa et al., 2013. Qi et al., 2015). Interestingly, in the case of Arabidopsis engineered lines, reduction of lignin content and improvement of biomass saccharification were observed in addition to 2-phenylethanol production (Qi et al., 2015).

As previously described for flavonoids and C6-C1 hydroxybenzenoids, dihydroconiferyl alcohol (Fig. 3) can only act as a chain-initiating monomer during lignin formation. Therefore, its incorporation could result in lignins with reduced degree of polymerization. Dihydroconiferyl alcohol was found in the lignin of a pine CAD mutant (Lapierre, Pollet, MacKay, & Sederoff, 2000; Ralph et al., 1997), possibly as a consequence of the activity of a C7-C8 double bond reductase (DBR) similar to those identified in pine and Arabidopsis (Kasahara et al., 2006; Youn et al., 2006). *In planta* overexpression of *DBRs* could represent a promising strategy to generate lignin with reduced average molecular weight and higher extractability.

Recently, the mechanisms for incorporation of monolignol glucosides in lignin have been proposed and the possible occurrence of phenyl glycoside-type lignin–carbohydrate linkages in cell wall samples from pine, eucalyptus, acacia, poplar, and moso bamboo has been observed (Miyagawa et al., 2020). The addition of carbohydrate moieties to lignin represents potentially an approach to increase lignin hydrophilicity, reduce hydrophobic interactions in the cell wall, and facilitate accessibility of degrading enzymes during biomass saccharification (Mottiar et al., 2016; Vanholme et al., 2012). Several UDP-glucosyltransferases (UGTs) have been implicated in the formation of monolignol glucosides. In Arabidopsis, a 90% reduction in both coniferyl and sinapyl alcohol 4-*O*-glucosides was observed in plants downregulated in three UGT genes (*UGT72E1, UGT72E2, UGT72E3*), while their overexpression resulted in increased levels of monolignol glucosides in leaves and/or roots (Lanot et al., 2006; Lanot, Hodge, Lim, Vaistij, & Bowles, 2008). Additionally, enhanced lignification was observed in the *ugt72b1* mutant (Lin et al., 2016), suggesting a regulating role for this UGT in the supply of monolignols implicated in lignification. Recently, using a combination of Raman microspectroscopy and histological ratiometric imaging techniques, Baldacci-Cresp et al. (2020) observed in the *ugt72e3* mutant an increase of lignin content in both interfascicular fibers and xylem from young stems. Labeling studies further indicated in this mutant an increased capacity of incorporating monolignols in lignin, suggesting a role for UGT72E3 in regulating the flow of monolignols to the cell wall. The expression profiles of homologous *UGT72* sequences found in poplar as well as the biochemical characterization of the corresponding enzymes highlighted several candidates involved in monolignol glycosylation and thereby potential engineering targets in this crop (Speeckaert et al., 2020). Engineering lignin with higher levels of monolignol glucosides would also require some knowledge on the export mechanisms involved in their transfer to the cell wall. Molecular dynamics simulations predict that uncharged aromatic compounds such as monolignols are likely transported passively across biological membranes (Vermaas et al., 2019). In this model, the translocation of monolignols involved in lignification is controlled by monolignols delivery, utilization rates, and membrane concentration gradients. However, the transport of monolignol glucosides appears to be an active process mediated by specific transporters that are located on both the plasma membrane and the vacuolar membrane (or tonoplast). In Arabidopsis, the uptake of coniferyl alcohol 4-*O*-glucoside (i.e., coniferin—Fig. 3) and sinapyl alcohol 4-*O*-glucoside (i.e., syringin) by vacuolar membrane vesicles is mediated by ATP-binding cassette-like transporters (Miao & Liu, 2010). In hybrid poplar (*P. sieboldii* × *Populus grandidentata*), Japanese cypress (*Chamaecyparis obtusa*), and bamboo (*Phyllostachys pubescens*), tonoplast and endomembrane vesicles prepared from tissues actively undergoing lignification transport coniferin and *p*-coumaryl alcohol 4-*O*-glucoside using a proton gradient formed by vacuolar-type H^+-ATPase (V-ATPase) depending on ATP. These results support the hypothesis that monolignol glucosides are loaded into secretory vesicles and delivered to the cell wall by exocytosis (Shimada et al., 2021; Tsuyama et al., 2013, 2019). A similar transport mechanism was shown in Norway spruce (*Picea abies*) using microsomal vesicles from developing xylem, but also in non-lignifying Bright Yellow-2 tobacco cells (*N. tabacum*) suggesting that vacuoles can represent storage compartments for monolignol glucosides (Väisänen et al., 2020). In addition, there is evidence that non-lignifying cells can synthesize monolignols and export lignin precursors to the cell walls of their lignifying neighbors to support their lignification, and potentially diverse and overlapping mechanisms of monolignol export may exist (Perkins, Smith, & Samuels, 2019). Understanding these mechanisms will be important for optimizing lignin engineering approaches aiming at incorporating non-conventional monolignols to alter lignin properties. For example, Arabidopsis lines engineered to overproduce 1,2-disinapoylglucose, a potential cleavable monolignol featuring two sinapates attached to a hydrophilic moiety by ester linkages (Fig. 3), accumulated preferentially the target compound in vacuolar compartments instead of lignin (Lee, Mo, Kim, & Chapple, 2017).

7. Conclusions and perspectives

Recent advances in the understanding of lignin biosynthesis and its fine regulation have enabled the manipulation of lignin in various model plants and crops. Engineering approaches not only include the alteration of the expression of genes that participate in lignin synthesis, but also the modification of the corresponding enzymes at the post-translational level or the conversion of cofactors and metabolic intermediates. Several strategies demonstrated in model plant species have been translated to crops. So far, low-lignin alfalfa obtained with traditional RNAi silencing technology for improved forage quality (branded as HarvXtra®) represents the only commercialized lignin-modified crop (Barros, Temple, & Dixon, 2019), while crops engineered in lignin via more recently developed techniques (e.g., CRISPR-Cas9, TALEN, etc.) have not yet reached the market. Research efforts are needed to facilitate the transformation of certain crops and accelerate the testing of engineering strategies (Altpeter et al., 2016). For example, genetic tools such as morphogenic genes have been successfully applied to transform "recalcitrant" sorghum varieties (Aregawi et al., 2021). Several lines of evidence suggest that manipulation of lignin, and more generally of the phenylpropanoid pathway, could represent a viable engineering method to protect plants against biotic and abiotic stresses (Cesarino, 2019; Yadav et al., 2020), which expands the scope of lignin engineering beyond biomass utilization applications. Considering the recent advances in lignin valorization (Rinaldi et al., 2016), increasing lignin content in bioenergy crops may represent a valuable option to promote the utilization of lignocellulosic biomass in biorefineries (Koshiba et al., 2017; Miyamoto et al., 2020). Lignin represents a substantial metabolic sink of the reduced carbon. The role of lignin on the turnover of plant residues in soils and its impact on soil carbon stabilization mechanisms remain debated topics (Hall, Huang, Timokhin, & Hammel, 2020). Pioneering studies revealed that detritus from plants with lignin modifications showed an increased decomposition rate in certain soils (Halpin et al., 2007). These observations suggest that lignin-engineered crops may influence the turnover of organic carbon in soils in the long term, and also highlight some research opportunities for modifying crops toward promoting soil carbon sequestration. Similarly, the effect of altered lignin biosynthesis on endophytic and rhizospheric microbiomes should be explored more thoroughly considering that recent work evidenced changes in microbial patterns in poplar engineered in phenylpropanoid and monolignol-related genes (Beckers et al., 2016; Behr et al., 2020).

Acknowledgments

This work was part of the DOE Joint BioEnergy Institute (http://www.jbei.org) supported by the U.S. Department of Energy, Office of Science, Office of Biological and Environmental Research, through contract DE-AC02-05CH11231 between Lawrence Berkeley National Laboratory and the U.S. Department of Energy. The authors are grateful to Dr. Marie Baucher and Dr. Marc Behr for critically reading the manuscript and providing suggestions.

References

Abdel-Ghany, S. E., & Pilon, M. (2008). MicroRNA-mediated systemic down-regulation of copper protein expression in response to low copper availability in Arabidopsis. *The Journal of Biological Chemistry*, *283*(23), 15932–15945.

Achnine, L., Blancaflor, E. B., Rasmussen, S., & Dixon, R. A. (2004). Colocalization of L-phenylalanine ammonia-lyase and cinnamate 4-hydroxylase for metabolic channeling in phenylpropanoid biosynthesis. *The Plant Cell*, *16*(11), 3098–3109.

Adeyanju, A. O., Sattler, S. E., Rich, P. J., Rivera-Burgos, L. A., Xu, X., & Ejeta, G. (2021). Sorghum brown midrib19 (bmr19) gene links lignin biosynthesis to folate metabolism. *Genes*, *12*(5).

Allwood, E. G., Davies, D. R., Gerrish, C., Ellis, B. E., & Bolwell, G. P. (1999). Phosphorylation of phenylalanine ammonia-lyase: Evidence for a novel protein kinase and identification of the phosphorylated residue. *FEBS Letters*, *457*(1), 47–52.

Altpeter, F., Springer, N. M., Bartley, L. E., Blechl, A. E., Brutnell, T. P., Citovsky, V., et al. (2016). Advancing crop transformation in the era of genome editing. *The Plant Cell*, *28*(7), 1510–1520.

Amthor, J. S. (2003). Efficiency of lignin biosynthesis: A quantitative analysis. *Annals of Botany*, *91*(6), 673–695.

Aregawi, K., Shen, J., Pierroz, G., Sharma, M. K., Dahlberg, J., Owiti, J., et al. (2021). Morphogene-assisted transformation of *Sorghum bicolor* allows more efficient genome editing. *Plant Biotechnology Journal*.

Bai, Z., Qi, T., Liu, Y., Wu, Z., Ma, L., Liu, W., et al. (2018). Alteration of S-adenosylhomocysteine levels affects lignin biosynthesis in switchgrass. *Plant Biotechnology Journal*, *16*(12), 2016–2026.

Baldacci-Cresp, F., Le Roy, J., Huss, B., Lion, C., Créach, A., Spriet, C., et al. (2020). UDP-GLYCOSYLTRANSFERASE 72E3 plays a role in lignification of secondary cell walls in Arabidopsis. *International Journal of Molecular Sciences*, *21*(17).

Baral, N. R., Sundstrom, E. R., Das, L., Gladden, J. M., Eudes, A., Mortimer, J., et al. (2019). Approaches for more efficient biological conversion of lignocellulosic feedstocks to biofuels and bioproducts. *ACS Sustainable Chemistry & Engineering*, 7(10), 9062–9079.

Barrière, Y., Ralph, J., Méchin, V., Guillaumie, S., Grabber, J. H., Argillier, O., et al. (2004). Genetic and molecular basis of grass cell wall biosynthesis and degradability. II. Lessons from brown-midrib mutants. *Comptes Rendus Biologies*, *327*(9–10), 847–860.

Barros, J., Escamilla-Trevino, L., Song, L., Rao, X., Serrani-Yarce, J. C., Palacios, M. D., et al. (2019). 4-Coumarate 3-hydroxylase in the lignin biosynthesis pathway is a cytosolic ascorbate peroxidase. *Nature Communications*, *10*(1), 1994.

Barros, J., Temple, S., & Dixon, R. A. (2019). Development and commercialization of reduced lignin alfalfa. *Current Opinion in Biotechnology*, *56*, 48–54.

Bassard, J.-E., Mutterer, J., Duval, F., & Werck-Reichhart, D. (2012). A novel method for monitoring the localization of cytochromes P450 and other endoplasmic reticulum membrane associated proteins: A tool for investigating the formation of metabolons. *The FEBS Journal*, *279*(9), 1576–1583.

Bassard, J.-E., Richert, L., Geerinck, J., Renault, H., Duval, F., Ullmann, P., et al. (2012). Protein-protein and protein-membrane associations in the lignin pathway. *The Plant Cell*, *24*(11), 4465–4482.

Baucher, M., Halpin, C., Petit-Conil, M., & Boerjan, W. (2003). Lignin: Genetic engineering and impact on pulping. *Critical Reviews in Biochemistry and Molecular Biology*, *38*(4), 305–350.

Becker, J., & Wittmann, C. (2019). A field of dreams: Lignin valorization into chemicals, materials, fuels, and health-care products. *Biotechnology Advances*, *37*(6), 107360.

Beckers, B., Op De Beeck, M., Weyens, N., Van Acker, R., Van Montagu, M., Boerjan, W., et al. (2016). Lignin engineering in field-grown poplar trees affects the endosphere bacterial microbiome. *Proceedings of the National Academy of Sciences of the United States of America*, *113*(8), 2312–2317.

Behr, M., Baldacci-Cresp, F., Kohler, A., Morreel, K., Goeminne, G., Van Acker, R., et al. (2020). Alterations in the phenylpropanoid pathway affect poplar ability for ectomycorrhizal colonisation and susceptibility to root-knot nematodes. *Mycorrhiza*, *30*(5), 555–566.

Beris, D., Kapolas, G., Livanos, P., Roussis, A., Milioni, D., & Haralampidis, K. (2016). RNAi-mediated silencing of the *Arabidopsis thaliana* ULCS1 gene, encoding a WDR protein, results in cell wall modification impairment and plant infertility. *Plant Science*, *245*, 71–83.

Berstis, L., Elder, T., Crowley, M., & Beckham, G. T. (2016). Radical nature of C-lignin. *ACS Sustainable Chemistry & Engineering*, *4*(10), 5327–5335.

Bhargava, A., Mansfield, S. D., Hall, H. C., Douglas, C. J., & Ellis, B. E. (2010). MYB75 functions in regulation of secondary cell wall formation in the Arabidopsis inflorescence stem. *Plant Physiology*, *154*(3), 1428–1438.

Bhuiya, M.-W., & Liu, C.-J. (2010). Engineering monolignol 4-O-methyltransferases to modulate lignin biosynthesis. *The Journal of Biological Chemistry*, *285*(1), 277–285.

Blount, J. W., Korth, K. L., Masoud, S. A., Rasmussen, S., Lamb, C., & Dixon, R. A. (2000). Altering expression of cinnamic acid 4-hydroxylase in transgenic plants provides evidence for a feedback loop at the entry point into the phenylpropanoid pathway. *Plant Physiology*, *122*(1), 107–116.

Boerjan, W., Ralph, J., & Baucher, M. (2003). Lignin biosynthesis. *Annual Review of Plant Biology*, *54*, 519–546.

Borah, P., & Khurana, J. P. (2018). The OsFBK1 E3 ligase subunit affects anther and root secondary cell wall thickenings by mediating turnover of a cinnamoyl-CoA reductase. *Plant Physiology*, *176*(3), 2148–2165.

Borevitz, J. O., Xia, Y., Blount, J., Dixon, R. A., & Lamb, C. (2000). Activation tagging identifies a conserved MYB regulator of phenylpropanoid biosynthesis. *The Plant Cell*, *12*(12), 2383–2394.

Bouchez, P., Teixeira Benites, V., Baidoo, E. E. K., Mortimer, J. C., Sullivan, M. L., Scheller, H. V., et al. (2019). Production of clovamide and its analogues in *Saccharomyces cerevisiae* and *Lactococcus lactis*. *Letters in Applied Microbiology*, *69*(3), 181–189.

Cai, Y., Bhuiya, M.-W., Shanklin, J., & Liu, C.-J. (2015). Engineering a monolignol 4-O-methyltransferase with high selectivity for the condensed lignin precursor coniferyl alcohol. *The Journal of Biological Chemistry*, *290*(44), 26715–26724.

Cai, Y., Zhang, K., Kim, H., Hou, G., Zhang, X., Yang, H., et al. (2016). Enhancing digestibility and ethanol yield of *Populus* wood via expression of an engineered monolignol 4-O-methyltransferase. *Nature Communications*, 7, 11989.

Cardenas, C. L., Costa, M. A., Laskar, D. D., Moinuddin, S. G. A., Lee, C., Davin, L. B., et al. (2021). RNAi modulation of chlorogenic acid and lignin deposition in *Nicotiana tabacum* and insufficient compensatory metabolic cross-talk. *Journal of Natural Products, 84*(3), 694–706.

Cardozo, T., & Pagano, M. (2004). The SCF ubiquitin ligase: Insights into a molecular machine. *Nature Reviews. Molecular Cell Biology, 5*(9), 739–751.

Cesarino, I. (2019). Structural features and regulation of lignin deposited upon biotic and abiotic stresses. *Current Opinion in Biotechnology, 56*, 209–214.

Chanoca, A., de Vries, L., & Boerjan, W. (2019). Lignin engineering in forest trees. *Frontiers in Plant Science, 10*, 912.

Chen, H.-C., Li, Q., Shuford, C. M., Liu, J., Muddiman, D. C., Sederoff, R. R., et al. (2011). Membrane protein complexes catalyze both 4- and 3-hydroxylation of cinnamic acid derivatives in monolignol biosynthesis. *Proceedings of the National Academy of Sciences of the United States of America, 108*(52), 21253–21258.

Chen, J., Quan, M., & Zhang, D. (2015). Genome-wide identification of novel long non-coding RNAs in *Populus tomentosa* tension wood, opposite wood and normal wood xylem by RNA-seq. *Planta, 241*(1), 125–143.

Chen, H.-C., Song, J., Wang, J. P., Lin, Y.-C., Ducoste, J., Shuford, C. M., et al. (2014). Systems biology of lignin biosynthesis in *Populus trichocarpa*: Heteromeric 4-coumaric acid:coenzyme A ligase protein complex formation, regulation, and numerical modeling. *The Plant Cell, 26*(3), 876–893.

Chen, H.-C., Song, J., Williams, C. M., Shuford, C. M., Liu, J., Wang, J. P., et al. (2013). Monolignol pathway 4-coumaric acid:coenzyme A ligases in *Populus trichocarpa*: Novel specificity, metabolic regulation, and simulation of coenzyme A ligation fluxes. *Plant Physiology, 161*(3), 1501–1516.

Cheng, S. H., Sheen, J., Gerrish, C., & Bolwell, G. P. (2001). Molecular identification of phenylalanine ammonia-lyase as a substrate of a specific constitutively active Arabidopsis CDPK expressed in maize protoplasts. *FEBS Letters, 503*(2–3), 185–188.

Chu, L., He, X., Shu, W., Wang, L., & Tang, F. (2021). Knockdown of miR393 promotes the growth and biomass production in poplar. *Frontiers in Plant Science, 12*, 714907.

Costa, M. A., Marques, J. V., Dalisay, D. S., Herman, B., Bedgar, D. L., Davin, L. B., et al. (2013). Transgenic hybrid poplar for sustainable and scalable production of the commodity/specialty chemical, 2-phenylegrassthanol. *PLoS One, 8*(12), e83169.

Czichi, U., & Kindl, H. (1975). Formation of *p*-coumaric acid and *o*-coumaric acid from L-phenylalanine by microsomal membrane fractions from potato: Evidence of membrane-bound enzyme complexes. *Planta, 125*(2), 115–125.

Czichi, U., & Kindl, H. (1977). Phenylalanine ammonia lyase and cinnamic acid hydroxylases as assembled consecutive enzymes on microsomal membranes of cucumber cotyledons: Cooperation and subcellular distribution. *Planta, 134*(2), 133–143.

Dai, Y., Hu, G., Dupas, A., Medina, L., Blandels, N., Clemente, H. S., et al. (2020). Implementing the CRISPR/Cas9 technology in Eucalyptus hairy roots using wood-related genes. *International Journal of Molecular Sciences, 21*(10).

De Meester, B., Madariaga Calderón, B., de Vries, L., Pollier, J., Goeminne, G., Van Doorsselaere, J., et al. (2020). Tailoring poplar lignin without yield penalty by combining a null and haploinsufficient CINNAMOYL-CoA REDUCTASE2 allele. *Nature Communications, 11*(1), 5020.

de Torres Zabela, M., Fernandez-Delmond, I., Niittyla, T., Sanchez, P., & Grant, M. (2002). Differential expression of genes encoding Arabidopsis phospholipases after challenge with virulent or avirulent *Pseudomonas* isolates. *Molecular Plant-Microbe Interactions, 15*(8), 808–816.

de Vries, L., Brouckaert, M., Chanoca, A., Kim, H., Regner, M. R., Timokhin, V. I., et al. (2021). CRISPR-Cas9 editing of CAFFEOYL SHIKIMATE ESTERASE 1 and 2 shows their importance and partial redundancy in lignification in *Populus tremula* x *P. alba*. *Plant Biotechnology Journal, 19*(11), 2221–2234.

de Vries, L., MacKay, H. A., Smith, R. A., Mottiar, Y., Karlen, S. D., Unda, F., et al. (2021). *p*HBMT1, a BAHD-family monolignol acyltransferase, mediates lignin acylation in poplar. *Plant Physiology*. https://doi.org/10.1093/plphys/kiab546.

del Río, J. C., Rencoret, J., Gutiérrez, A., Elder, T., Kim, H., & Ralph, J. (2020). Lignin monomers from beyond the canonical monolignol biosynthetic pathway: Another brick in the wall. *ACS Sustainable Chemistry & Engineering*, *8*(13), 4997–5012.

del Río, J. C., Rencoret, J., Marques, G., Gutiérrez, A., Ibarra, D., Santos, J. I., et al. (2008). Highly acylated (acetylated and/or *p*-coumaroylated) native lignins from diverse herbaceous plants. *Journal of Agricultural and Food Chemistry*, *56*(20), 9525–9534.

Dexter, R., Qualley, A., Kish, C. M., Ma, C. J., Koeduka, T., Nagegowda, D. A., et al. (2007). Characterization of a petunia acetyltransferase involved in the biosynthesis of the floral volatile isoeugenol. *The Plant Journal*, *49*(2), 265–275.

Dixon, R. A., Chen, F., Guo, D., & Parvathi, K. (2001). The biosynthesis of monolignols: A "metabolic grid", or independent pathways to guaiacyl and syringyl units? *Phytochemistry*, *57*(7), 1069–1084.

Ehlting, J., Shin, J. J., & Douglas, C. J. (2001). Identification of 4-coumarate:coenzyme A ligase (4CL) substrate recognition domains. *The Plant Journal*, *27*(5), 455–465.

Eudes, A., George, A., Mukerjee, P., Kim, J. S., Pollet, B., Benke, P. I., et al. (2012). Biosynthesis and incorporation of side-chain-truncated lignin monomers to reduce lignin polymerization and enhance saccharification. *Plant Biotechnology Journal*, *10*(5), 609–620.

Eudes, A., Pereira, J. H., Yogiswara, S., Wang, G., Teixeira Benites, V., Baidoo, E. E. K., et al. (2016). Exploiting the substrate promiscuity of hydroxycinnamoyl-CoA:shikimate hydroxycinnamoyl transferase to reduce lignin. *Plant & Cell Physiology*, *57*(3), 568–579.

Eudes, A., Sathitsuksanoh, N., Baidoo, E. E. K., George, A., Liang, Y., Yang, F., et al. (2015). Expression of a bacterial 3-dehydroshikimate dehydratase reduces lignin content and improves biomass saccharification efficiency. *Plant Biotechnology Journal*, *13*(9), 1241–1250.

Eudes, A., Zhao, N., Sathitsuksanoh, N., Baidoo, E. E. K., Lao, J., Wang, G., et al. (2016). Expression of S-adenosylmethionine hydrolase in tissues synthesizing secondary cell walls alters specific methylated cell wall fractions and improves biomass digestibility. *Frontiers in Bioengineering and Biotechnology*, *4*, 58.

Fan, D., Li, C., Fan, C., Hu, J., Li, J., Yao, S., et al. (2020). MicroRNA6443-mediated regulation of FERULATE 5-HYDROXYLASE gene alters lignin composition and enhances saccharification in *Populus tomentosa*. *The New Phytologist*, *226*(2), 410–425.

Fang, L., Xu, X., Li, J., Zheng, F., Li, M., Yan, J., et al. (2020). Transcriptome analysis provides insights into the non-methylated lignin synthesis in *Paphiopedilum armeniacum* seed. *BMC Genomics*, *21*(1), 524.

Gagne, J. M., Downes, B. P., Shiu, S.-H., Durski, A. M., & Vierstra, R. D. (2002). The F-box subunit of the SCF E3 complex is encoded by a diverse superfamily of genes in Arabidopsis. *Proceedings of the National Academy of Sciences of the United States of America*, *99*(17), 11519–11524.

Gao, W., Li, H.-Y., Xiao, S., & Chye, M.-L. (2010a). Protein interactors of acyl-CoA-binding protein ACBP2 mediate cadmium tolerance in Arabidopsis. *Plant Signaling & Behavior*, *5*(8), 1025–1027.

Gao, W., Li, H.-Y., Xiao, S., & Chye, M.-L. (2010b). Acyl-CoA-binding protein 2 binds lysophospholipase 2 and lysoPC to promote tolerance to cadmium-induced oxidative stress in transgenic Arabidopsis. *The Plant Journal*, *62*(6), 989–1003.

Gou, M., Ran, X., Martin, D. W., & Liu, C.-J. (2018). The scaffold proteins of lignin biosynthetic cytochrome P450 enzymes. *Nature Plants*, *4*(5), 299–310.

Gou, M., Yang, X., Zhao, Y., Ran, X., Song, Y., & Liu, C.-J. (2019). Cytochrome b5 is an obligate electron shuttle protein for syringyl lignin biosynthesis in Arabidopsis. *The Plant Cell*, *31*(6), 1344–1366.

Grabber, J. H., Davidson, C., Tobimatsu, Y., Kim, H., Lu, F., Zhu, Y., et al. (2019). Structural features of alternative lignin monomers associated with improved digestibility of artificially lignified maize cell walls. *Plant Science*, *287*, 110070.

Gui, J., Lam, P. Y., Tobimatsu, Y., Sun, J., Huang, C., Cao, S., et al. (2020). Fibre-specific regulation of lignin biosynthesis improves biomass quality in *Populus*. *The New Phytologist*, *226*(4), 1074–1087.

Gui, J., Luo, L., Zhong, Y., Sun, J., Umezawa, T., & Li, L. (2019). Phosphorylation of LTF1, an MYB transcription factor in *populus*, acts as a sensory switch regulating lignin biosynthesis in wood cells. *Molecular Plant*, *12*(10), 1325–1337.

Hall, S. J., Huang, W., Timokhin, V. I., & Hammel, K. E. (2020). Lignin lags, leads, or limits the decomposition of litter and soil organic carbon. *Ecology*, *101*(9), e03113.

Halpin, C. (2019). Lignin engineering to improve saccharification and digestibility in grasses. *Current Opinion in Biotechnology*, *56*, 223–229.

Halpin, C., Thain, S. C., Tilston, E. L., Guiney, E., Lapierre, C., & Hopkins, D. W. (2007). Ecological impacts of trees with modified lignin. *Tree Genetics & Genomes*, *3*(2), 101–110.

Hanson, A. D., & Roje, S. (2001). One-carbon metabolism in higher plants. *Annual Review of Plant Physiology and Plant Molecular Biology*, *52*, 119–137.

Hao, Z., Yogiswara, S., Wei, T., Benites, V. T., Sinha, A., Wang, G., et al. (2021). Expression of a bacterial 3-dehydroshikimate dehydratase (QsuB) reduces lignin and improves biomass saccharification efficiency in switchgrass (*Panicum virgatum* L.). *BMC Plant Biology*, *21*(1), 56.

Hoffmann, L., Besseau, S., Geoffroy, P., Ritzenthaler, C., Meyer, D., Lapierre, C., et al. (2004). Silencing of hydroxycinnamoyl-coenzyme A shikimate/quinate hydroxycinnamoyl-transferase affects phenylpropanoid biosynthesis. *The Plant Cell*, *16*(6), 1446–1465.

Hoffmann, L., Maury, S., Martz, F., Geoffroy, P., & Legrand, M. (2003). Purification, cloning, and properties of an acyltransferase controlling shikimate and quinate ester intermediates in phenylpropanoid metabolism. *The Journal of Biological Chemistry*, *278*(1), 95–103.

Hrazdina, G., & Wagner, G. J. (1985). Metabolic pathways as enzyme complexes: Evidence for the synthesis of phenylpropanoids and flavonoids on membrane associated enzyme complexes. *Archives of Biochemistry and Biophysics*, *237*(1), 88–100.

Hu, S., Kamimura, N., Sakamoto, S., Nagano, S., Takata, N., Liu, S., et al. (2022). Rerouting of the lignin biosynthetic pathway by inhibition of cytosolic shikimate recycling in transgenic hybrid aspen. *The Plant Journal*. https://doi.org/10.1111/tpj.15674.

Huang, S., Zhou, J., Gao, L., & Tang, Y. (2021). Plant miR397 and its functions. *Functional Plant Biology: FPB*, *48*(4), 361–370.

Ikeda, T., Takata, N., Sakamoto, S., Hu, S., Nuoendagula, Hishiyama, S., et al. (2021). Improved chemical pulping and saccharification of a natural mulberry mutant deficient in cinnamyl alcohol dehydrogenase. *Holzforschung*, *75*(10), 968–977.

Jang, H.-A., Bae, E.-K., Kim, M.-H., Park, S.-J., Choi, N.-Y., Pyo, S.-W., et al. (2021). CRISPR-knockout of CSE gene improves saccharification efficiency by reducing lignin content in hybrid poplar. *International Journal of Molecular Sciences*, *22*(18).

Jeena, G. S., Joshi, A., & Shukla, R. K. (2021). Bm-miR172c-5p regulates lignin biosynthesis and secondary xylem thickness by altering Ferulate 5 hydroxylase gene in *Bacopa monnieri*. *Plant & Cell Physiology*.

Jiang, F., & Doudna, J. A. (2017). CRISPR-Cas9 structures and mechanisms. *Annual Review of Biophysics*, *46*, 505–529.

Jørgensen, K., Rasmussen, A. V., Morant, M., Nielsen, A. H., Bjarnholt, N., Zagrobelny, M., et al. (2005). Metabolon formation and metabolic channeling in the biosynthesis of plant natural products. *Current Opinion in Plant Biology*, *8*(3), 280–291.

Jung, J. H., & Altpeter, F. (2016). TALEN mediated targeted mutagenesis of the caffeic acid O-methyltransferase in highly polyploid sugarcane improves cell wall composition for production of bioethanol. *Plant Molecular Biology*, *92*(1–2), 131–142.

Kannan, B., Jung, J. H., Moxley, G. W., Lee, S.-M., & Altpeter, F. (2018). TALEN-mediated targeted mutagenesis of more than 100 COMT copies/alleles in highly polyploid sugarcane improves saccharification efficiency without compromising biomass yield. *Plant Biotechnology Journal, 16*(4), 856–866.

Karlen, S. D., Fasahati, P., Mazaheri, M., Serate, J., Smith, R. A., Sirobhushanam, S., et al. (2020). Assessing the viability of recovery of hydroxycinnamic acids from lignocellulosic biorefinery alkaline pretreatment waste streams. *ChemSusChem, 13*(8), 2012–2024.

Karlen, S. D., Smith, R. A., Kim, H., Padmakshan, D., Bartuce, A., Mobley, J. K., et al. (2017). Highly decorated lignins in leaf tissues of the Canary Island date palm *Phoenix canariensis*. *Plant Physiology, 175*(3), 1058–1067.

Kasahara, H., Jiao, Y., Bedgar, D. L., Kim, S.-J., Patten, A. M., Xia, Z.-Q., et al. (2006). *Pinus taeda* phenylpropenal double-bond reductase: Purification, cDNA cloning, heterologous expression in *Escherichia coli*, and subcellular localization in *P. taeda*. *Phytochemistry, 67*(16), 1765–1780.

Kawasaki, T., Koita, H., Nakatsubo, T., Hasegawa, K., Wakabayashi, K., Takahashi, H., et al. (2006). Cinnamoyl-CoA reductase, a key enzyme in lignin biosynthesis, is an effector of small GTPase Rac in defense signaling in rice. *Proceedings of the National Academy of Sciences of the United States of America, 103*(1), 230–235.

Khandal, H., Singh, A. P., & Chattopadhyay, D. (2020). The MicroRNA397b-LACCASE2 module regulates root lignification under water and phosphate deficiency. *Plant Physiology, 182*(3), 1387–1403.

Kim, D.-Y., Scalf, M., Smith, L. M., & Vierstra, R. D. (2013). Advanced proteomic analyses yield a deep catalog of ubiquitylation targets in Arabidopsis. *The Plant Cell, 25*(5), 1523–1540.

Kim, S.-J., Vassão, D. G., Moinuddin, S. G. A., Bedgar, D. L., Davin, L. B., & Lewis, N. G. (2014). Allyl/propenyl phenol synthases from the creosote bush and engineering production of specialty/commodity chemicals, eugenol/isoeugenol, in *Escherichia coli*. *Archives of Biochemistry and Biophysics, 541*, 37–46.

Kim, K. H., Wang, Y., Takada, M., Eudes, A., Yoo, C. G., Kim, C. S., et al. (2019). Deep eutectic solvent pretreatment of transgenic biomass with increased C6C1 lignin monomers. *Frontiers in Plant Science, 10*, 1774.

Kim, H., Yu, S.-I., Jung, S. H., Lee, B.-H., & Suh, M. C. (2019). The F-box protein SAGL1 and ECERIFERUM3 regulate cuticular wax biosynthesis in response to changes in humidity in Arabidopsis. *The Plant Cell, 31*(9), 2223–2240.

Kim, J. I., Zhang, X., Pascuzzi, P. E., Liu, C.-J., & Chapple, C. (2020). Glucosinolate and phenylpropanoid biosynthesis are linked by proteasome-dependent degradation of PAL. *The New Phytologist, 225*(1), 154–168.

Koeduka, T., Suzuki, S., Iijima, Y., Ohnishi, T., Suzuki, H., Watanabe, B., et al. (2013). Enhancement of production of eugenol and its glycosides in transgenic aspen plants via genetic engineering. *Biochemical and Biophysical Research Communications, 436*(1), 73–78.

Koshiba, T., Yamamoto, N., Tobimatsu, Y., Yamamura, M., Suzuki, S., Hattori, T., et al. (2017). MYB-mediated upregulation of lignin biosynthesis in *Oryza sativa* towards biomass refinery. *Plant biotechnology (Tokyo, Japan), 34*(1), 7–15.

Kubota, T., Tanaka, Y., Takemoto, N., Hiraga, K., Yukawa, H., & Inui, M. (2015). Identification and expression analysis of a gene encoding a shikimate transporter of *Corynebacterium glutamicum*. *Microbiology, 161*(Pt. 2), 254–263.

Lamb, C. J. (1979). Regulation of enzyme levels in phenylpropanoid biosynthesis: Characterization of the modulation by light and pathway intermediates. *Archives of Biochemistry and Biophysics, 192*(1), 311–317.

Lanot, A., Hodge, D., Jackson, R. G., George, G. L., Elias, L., Lim, E.-K., et al. (2006). The glucosyltransferase UGT72E2 is responsible for monolignol 4-*O*-glucoside production in *Arabidopsis thaliana*. *The Plant Journal, 48*(2), 286–295.

Lanot, A., Hodge, D., Lim, E.-K., Vaistij, F. E., & Bowles, D. J. (2008). Redirection of flux through the phenylpropanoid pathway by increased glucosylation of soluble intermediates. *Planta, 228*(4), 609–616.

Lapierre, C., Pollet, B., MacKay, J. J., & Sederoff, R. R. (2000). Lignin structure in a mutant pine deficient in cinnamyl alcohol dehydrogenase. *Journal of Agricultural and Food Chemistry, 48*(6), 2326–2331.

Lapierre, C., Sibout, R., Laurans, F., Lesage-Descauses, M.-C., Déjardin, A., & Pilate, G. (2021). *p*-Coumaroylation of poplar lignins impacts lignin structure and improves wood saccharification. *Plant Physiology, 187*(3), 1374–1386.

Lee, S., Mo, H., Kim, J. I., & Chapple, C. (2017). Genetic engineering of Arabidopsis to overproduce disinapoyl esters, potential lignin modification molecules. *Biotechnology for Biofuels, 10*, 40.

Lee, J. H., Won, H. J., Hoang Nguyen Tran, P., Lee, S., Kim, H., & Jung, J. H. (2021). Improving lignocellulosic biofuel production by CRISPR/Cas9-mediated lignin modification in barley. *GCB Bioenergy, 13*(4), 742–752.

Lei, Y., Hannoufa, A., & Yu, P. (2017). The use of gene modification and advanced molecular structure analyses towards improving alfalfa forage. *International Journal of Molecular Sciences, 18*(2).

Li, L., Hill-Skinner, S., Liu, S., Beuchle, D., Tang, H. M., Yeh, C.-T., et al. (2015). The maize brown midrib4 (bm4) gene encodes a functional folylpolyglutamate synthase. *The Plant Journal, 81*(3), 493–504.

Li, M., Pu, Y., & Ragauskas, A. J. (2016). Current understanding of the correlation of lignin structure with biomass recalcitrance. *Frontiers in Chemistry, 4*, 45.

Li, Y., Shuai, L., Kim, H., Motagamwala, A. H., Mobley, J. K., Yue, F., et al. (2018). An "ideal lignin" facilitates full biomass utilization. *Science Advances, 4*(9), eaau2968.

Li, S., Wang, W., Gao, J., Yin, K., Wang, R., Wang, C., et al. (2016). MYB75 phosphorylation by MPK4 is required for light-induced anthocyanin accumulation in Arabidopsis. *The Plant Cell, 28*(11), 2866–2883.

Li, X., Weng, J.-K., & Chapple, C. (2008). Improvement of biomass through lignin modification. *The Plant Journal, 54*(4), 569–581.

Liang, Y., Eudes, A., Yogiswara, S., Jing, B., Benites, V. T., Yamanaka, R., et al. (2019). A screening method to identify efficient sgRNAs in Arabidopsis, used in conjunction with cell-specific lignin reduction. *Biotechnology for Biofuels, 12*, 130.

Lin, C.-Y., & Eudes, A. (2020). Strategies for the production of biochemicals in bioenergy crops. *Biotechnology for Biofuels, 13*, 71.

Lin, J.-S., Huang, X.-X., Li, Q., Cao, Y., Bao, Y., Meng, X.-F., et al. (2016). UDP-glycosyltransferase 72B1 catalyzes the glucose conjugation of monolignols and is essential for the normal cell wall lignification in *Arabidopsis thaliana*. *The Plant Journal, 88*(1), 26–42.

Lin, J.-S., Lin, C.-C., Lin, H.-H., Chen, Y.-C., & Jeng, S.-T. (2012). MicroR828 regulates lignin and H2O2 accumulation in sweet potato on wounding. *The New Phytologist, 196*(2), 427–440.

Lin, C.-Y., Sun, Y., Song, J., Chen, H.-C., Shi, R., Yang, C., et al. (2021). Enzyme complexes of Ptr4CL and PtrHCT modulate Co-enzyme A ligation of hydroxycinnamic acids for monolignol biosynthesis in *Populus trichocarpa*. *Frontiers in Plant Science, 12*, 727932.

Lin, C.-Y., Vuu, K. M., Amer, B., Shih, P. M., Baidoo, E. E. K., Scheller, H. V., et al. (2021). In-planta production of the biodegradable polyester precursor 2-pyrone-4,6-dicarboxylic acid (PDC): Stacking reduced biomass recalcitrance with value-added co-product. *Metabolic Engineering, 66*, 148–156.

Lin, C.-Y., Wang, J. P., Li, Q., Chen, H.-C., Liu, J., Loziuk, P., et al. (2015). 4-Coumaroyl and caffeoyl shikimic acids inhibit 4-coumaric acid:coenzyme A ligases and modulate metabolic flux for 3-hydroxylation in monolignol biosynthesis of *Populus trichocarpa*. *Molecular Plant, 8*(1), 176–187.

Liu, C.-J., Deavours, B. E., Richard, S. B., Ferrer, J.-L., Blount, J. W., Huhman, D., et al. (2006). Structural basis for dual functionality of isoflavonoid O-methyltransferases in the evolution of plant defense responses. *The Plant Cell*, *18*(12), 3656–3669.

Liu, B., Liu, J., Yu, J., Wang, Z., Sun, Y., Li, S., et al. (2021). Transcriptional reprogramming of xylem cell wall biosynthesis in tension wood. *Plant Physiology*, *186*(1), 250–269.

Liu, C., Yu, H., Rao, X., Li, L., & Dixon, R. A. (2021). Abscisic acid regulates secondary cell-wall formation and lignin deposition in *Arabidopsis thaliana* through phosphorylation of NST1. *Proceedings of the National Academy of Sciences of the United States of America*, *118*(5).

Loqué, D., Eudes, A. (2021), Tissue-specific reduction of lignin. U.S. Patent US10900049, B2.

Lu, F., Karlen, S. D., Regner, M., Kim, H., Ralph, S. A., Sun, R.-C., et al. (2015). Naturally *p*-hydroxybenzoylated lignins in palms. *Bioenergy Research*, *8*(3), 934–952.

Lu, S., Li, Q., Wei, H., Chang, M.-J., Tunlaya-Anukit, S., Kim, H., et al. (2013). Ptr-miR397a is a negative regulator of laccase genes affecting lignin content in *Populus trichocarpa*. *Proceedings of the National Academy of Sciences of the United States of America*, *110*(26), 10848–10853.

Lu, D., Yuan, X., Kim, S.-J., Marques, J. V., Chakravarthy, P. P., Moinuddin, S. G. A., et al. (2017). Eugenol specialty chemical production in transgenic poplar (*Populus tremula* × *P. alba*) field trials. *Plant Biotechnology Journal*, *15*(8), 970–981.

Mahon, E. L., & Mansfield, S. D. (2019). Tailor-made trees: Engineering lignin for ease of processing and tomorrow's bioeconomy. *Current Opinion in Biotechnology*, *56*, 147–155.

Marita, J. M., Hatfield, R. D., Rancour, D. M., & Frost, K. E. (2014). Identification and suppression of the *p*-coumaroyl CoA:hydroxycinnamyl alcohol transferase in *Zea mays* L. *The Plant Journal*, *78*(5), 850–864.

Mayer, M. J., Narbad, A., Parr, A. J., Parker, M. L., Walton, N. J., Mellon, F. A., et al. (2001). Rerouting the plant phenylpropanoid pathway by expression of a novel bacterial enoyl-CoA hydratase/lyase enzyme function. *The Plant Cell*, *13*(7), 1669–1682.

Mazarei, M., Baxter, H. L., Srivastava, A., Li, G., Xie, H., Dumitrache, A., et al. (2020). Silencing folylpolyglutamate synthetase1 (FPGS1) in switchgrass (*Panicum virgatum* L.) improves lignocellulosic biofuel production. *Frontiers in Plant Science*, *11*, 843.

McQualter, R. B., Chong, B. F., Meyer, K., Van Dyk, D. E., O'Shea, M. G., Walton, N. J., et al. (2005). Initial evaluation of sugarcane as a production platform for *p*-hydroxybenzoic acid. *Plant Biotechnology Journal*, *3*(1), 29–41.

Menz, J., Modrzejewski, D., Hartung, F., Wilhelm, R., & Sprink, T. (2020). Genome edited crops touch the market: A view on the global development and regulatory environment. *Frontiers in Plant Science*, *11*, 586027.

Merali, Z., Mayer, M. J., Parker, M. L., Michael, A. J., Smith, A. C., & Waldron, K. W. (2007). Metabolic diversion of the phenylpropanoid pathway causes cell wall and morphological changes in transgenic tobacco stems. *Planta*, *225*(5), 1165–1178.

Mergner, J., Frejno, M., List, M., Papacek, M., Chen, X., Chaudhary, A., et al. (2020). Mass-spectrometry-based draft of the Arabidopsis proteome. *Nature*, *579*(7799), 409–414.

Miao, Y.-C., & Liu, C.-J. (2010). ATP-binding cassette-like transporters are involved in the transport of lignin precursors across plasma and vacuolar membranes. *Proceedings of the National Academy of Sciences of the United States of America*, *107*(52), 22728–22733.

Miao, R., Lung, S.-C., Li, X., Li, X. D., & Chye, M.-L. (2019). Thermodynamic insights into an interaction between ACYL-CoA-BINDING PROTEIN2 and LYSOPHOSPHOLIPASE2 in Arabidopsis. *The Journal of Biological Chemistry*, *294*(16), 6214–6226.

Miller, J. C., Tan, S., Qiao, G., Barlow, K. A., Wang, J., Xia, D. F., et al. (2011). A TALE nuclease architecture for efficient genome editing. *Nature Biotechnology*, *29*(2), 143–148.

Miyagawa, Y., Tobimatsu, Y., Lam, P. Y., Mizukami, T., Sakurai, S., Kamitakahara, H., et al. (2020). Possible mechanisms for the generation of phenyl glycoside-type lignin-carbohydrate linkages in lignification with monolignol glucosides. *The Plant Journal*, *104*(1), 156–170.

Miyamoto, T., Takada, R., Tobimatsu, Y., Suzuki, S., Yamamura, M., Osakabe, K., et al. (2020). Double knockout of OsWRKY36 and OsWRKY102 boosts lignification with altering culm morphology of rice. *Plant Science*, *296*, 110466.

Miyamoto, T., Takada, R., Tobimatsu, Y., Takeda, Y., Suzuki, S., Yamamura, M., et al. (2019). OsMYB108 loss-of-function enriches *p*-coumaroylated and tricin lignin units in rice cell walls. *The Plant Journal*, *98*(6), 975–987.

Mnich, E., Bjarnholt, N., Eudes, A., Harholt, J., Holland, C., Jørgensen, B., et al. (2020). Phenolic cross-links: Building and de-constructing the plant cell wall. *Natural Product Reports*, *37*(7), 919–961.

Mnich, E., Vanholme, R., Oyarce, P., Liu, S., Lu, F., Goeminne, G., et al. (2017). Degradation of lignin β-aryl ether units in *Arabidopsis thaliana* expressing LigD, LigF and LigG from *Sphingomonas paucimobilis* SYK-6. *Plant Biotechnology Journal*, *15*(5), 581–593.

Morse, A. M., Whetten, R. W., Dubos, C., & Campbell, M. M. (2009). Post-translational modification of an R2R3-MYB transcription factor by a MAP kinase during xylem development. *The New Phytologist*, *183*(4), 1001–1013.

Mottiar, Y., Vanholme, R., Boerjan, W., Ralph, J., & Mansfield, S. D. (2016). Designer lignins: Harnessing the plasticity of lignification. *Current Opinion in Biotechnology*, *37*, 190–200.

Mucha, S., Heinzlmeir, S., Kriechbaumer, V., Strickland, B., Kirchhelle, C., Choudhary, M., et al. (2019). The formation of a camalexin-biosynthetic metabolon. *The Plant Cell*.

Naik, P., Wang, J. P., Sederoff, R., Chiang, V., Williams, C., & Ducoste, J. J. (2018). Assessing the impact of the 4CL enzyme complex on the robustness of monolignol biosynthesis using metabolic pathway analysis. *PLoS One*, *13*(3), e0193896.

Nguyen, D. Q., Brown, C. W., Pegler, J. L., Eamens, A. L., & Grof, C. P. L. (2020). Molecular manipulation of microRNA397 abundance influences the development and salt stress response of Arabidopsis thaliana. *International Journal of Molecular Sciences*, *21*(21).

Ong, S. S., & Wickneswari, R. (2012). Characterization of microRNAs expressed during secondary wall biosynthesis in *Acacia mangium*. *PLoS One*, 7(11), e49662.

Oyarce, P., De Meester, B., Fonseca, F., de Vries, L., Goeminne, G., Pallidis, A., et al. (2019). Introducing curcumin biosynthesis in Arabidopsis enhances lignocellulosic biomass processing. *Nature Plants*, *5*(2), 225–237.

Park, J.-J., Yoo, C. G., Flanagan, A., Pu, Y., Debnath, S., Ge, Y., et al. (2017). Defined tetra-allelic gene disruption of the 4-coumarate:coenzyme A ligase 1 (Pv4CL1) gene by CRISPR/Cas9 in switchgrass results in lignin reduction and improved sugar release. *Biotechnology for Biofuels*, *10*, 284.

Patzlaff, A., McInnis, S., Courtenay, A., Surman, C., Newman, L. J., Smith, C., et al. (2003). Characterisation of a pine MYB that regulates lignification. *The Plant Journal*, *36*(6), 743–754.

Perkins, M., Smith, R. A., & Samuels, L. (2019). The transport of monomers during lignification in plants: Anything goes but how? *Current Opinion in Biotechnology*, *56*, 69–74.

Petrik, D. L., Karlen, S. D., Cass, C. L., Padmakshan, D., Lu, F., Liu, S., et al. (2014). *p*-Coumaroyl-CoA:monolignol transferase (PMT) acts specifically in the lignin biosynthetic pathway in *Brachypodium distachyon*. *The Plant Journal*, 77(5), 713–726.

Qi, G., Wang, D., Yu, L., Tang, X., Chai, G., He, G., et al. (2015). Metabolic engineering of 2-phenylethanol pathway producing fragrance chemical and reducing lignin in Arabidopsis. *Plant Cell Reports, 34*(8), 1331–1342.

Qin, S., Fan, C., Li, X., Li, Y., Hu, J., Li, C., et al. (2020). LACCASE14 is required for the deposition of guaiacyl lignin and affects cell wall digestibility in poplar. *Biotechnology for Biofuels, 13*(1), 197.

Quan, M., Du, Q., Xiao, L., Lu, W., Wang, L., Xie, J., et al. (2019). Genetic architecture underlying the lignin biosynthesis pathway involves noncoding RNAs and transcription factors for growth and wood properties in *Populus*. *Plant Biotechnology Journal, 17*(1), 302–315.

Ralph, J., MacKay, J. J., Hatfield, R. D., O'Malley, D. M., Whetten, R. W., & Sederoff, R. R. (1997). Abnormal lignin in a loblolly pine mutant. *Science, 277*(5323), 235–239.

Rao, X., Krom, N., Tang, Y., Widiez, T., Havkin-Frenkel, D., Belanger, F. C., et al. (2014). A deep transcriptomic analysis of pod development in the vanilla orchid (*Vanilla planifolia*). *BMC Genomics, 15*, 964.

Rasmussen, S., & Dixon, R. A. (1999). Transgene-mediated and elicitor-induced perturbation of metabolic channeling at the entry point into the phenylpropanoid pathway. *The Plant Cell, 11*(8), 1537–1552.

Renault, H., Werck-Reichhart, D., & Weng, J.-K. (2019). Harnessing lignin evolution for biotechnological applications. *Current Opinion in Biotechnology, 56*, 105–111.

Rencoret, J., Marques, G., Serrano, O., Kaal, J., Martínez, A. T., del Río, J. C., et al. (2020). Deciphering the unique structure and acylation pattern of *Posidonia oceanica* lignin. *ACS Sustainable Chemistry & Engineering, 8*(33), 12521–12533.

Rinaldi, R., Jastrzebski, R., Clough, M. T., Ralph, J., Kennema, M., Bruijnincx, P. C. A., et al. (2016). Paving the way for lignin valorisation: Recent advances in bioengineering, biorefining and catalysis. *Angewandte Chemie (International Ed. in English), 55*(29), 8164–8215.

Ro, D.-K., & Douglas, C. J. (2004). Reconstitution of the entry point of plant phenylpropanoid metabolism in yeast (*Saccharomyces cerevisiae*): Implications for control of metabolic flux into the phenylpropanoid pathway. *The Journal of Biological Chemistry, 279*(4), 2600–2607.

Roje, S. (2006). S-adenosyl-L-methionine: Beyond the universal methyl group donor. *Phytochemistry, 67*(15), 1686–1698.

Sakamoto, S., Kamimura, N., Tokue, Y., Nakata, M. T., Yamamoto, M., Hu, S., et al. (2020). Identification of enzymatic genes with the potential to reduce biomass recalcitrance through lignin manipulation in Arabidopsis. *Biotechnology for Biofuels, 13*, 97.

Samaras, P., Schmidt, T., Frejno, M., Gessulat, S., Reinecke, M., Jarzab, A., et al. (2020). ProteomicsDB: A multi-omics and multi-organism resource for life science research. *Nucleic Acids Research, 48*(D1), D1153–D1163.

Saracco, S. A., Hansson, M., Scalf, M., Walker, J. M., Smith, L. M., & Vierstra, R. D. (2009). Tandem affinity purification and mass spectrometric analysis of ubiquitylated proteins in Arabidopsis. *The Plant Journal, 59*(2), 344–358.

Sato, T., Kiuchi, F., & Sankawa, U. (1982). Inhibition of phenylalanine ammonia-lyase by cinnamic acid derivatives and related compounds. *Phytochemistry, 21*(4), 845–850.

Shafrin, F., Das, S. S., Sanan-Mishra, N., & Khan, H. (2015). Artificial miRNA-mediated down-regulation of two monolignoid biosynthetic genes (C3H and F5H) cause reduction in lignin content in jute. *Plant Molecular Biology, 89*(4–5), 511–527.

Shen, B., Li, C., & Tarczynski, M. C. (2002). High free-methionine and decreased lignin content result from a mutation in the *Arabidopsis S* -adenosyl-L-methionine synthetase 3 gene. *The Plant Journal, 29*(3), 371–380.

Shimada, N., Munekata, N., Tsuyama, T., Matsushita, Y., Fukushima, K., Kijidani, Y., et al. (2021). Active transport of lignin precursors into membrane vesicles from lignifying tissues of bamboo. *Plants, 10*(11), 2237.

Sibout, R., Le Bris, P., Legée, F., Cézard, L., Renault, H., & Lapierre, C. (2016). Structural redesigning Arabidopsis lignins into alkali-soluble lignins through the expression of *p*-coumaroyl-CoA:monolignol transferase PMT. *Plant Physiology*, *170*(3), 1358–1366.

Siebert, M., Sommer, S., Li, S. M., Wang, Z. X., Severin, K., & Heide, L. (1996). Genetic engineering of plant secondary metabolism. Accumulation of 4-hydroxybenzoate glucosides as a result of the expression of the bacterial ubiC gene in tobacco. *Plant Physiology*, *112*(2), 811–819.

Simmons, B. A., Loqué, D., & Ralph, J. (2010). Advances in modifying lignin for enhanced biofuel production. *Current Opinion in Plant Biology*, *13*(3), 313–320.

Smith, R. A., Gonzales-Vigil, E., Karlen, S. D., Park, J.-Y., Lu, F., Wilkerson, C. G., et al. (2015). Engineering monolignol *p*-coumarate conjugates into poplar and Arabidopsis lignins. *Plant Physiology*, *169*(4), 2992–3001.

Speeckaert, N., Adamou, N. M., Hassane, H. A., Baldacci-Cresp, F., Mol, A., Goeminne, G., et al. (2020). Characterization of the UDP-glycosyltransferase UGT72 family in poplar and identification of genes involved in the glycosylation of monolignols. *International Journal of Molecular Sciences*, *21*(14).

Srere, P. A. (1987). Complexes of sequential metabolic enzymes. *Annual Review of Biochemistry*, *56*, 89–124.

Srivastava, A. C., Chen, F., Ray, T., Pattathil, S., Peña, M. J., Avci, U., et al. (2015). Loss of function of folylpolyglutamate synthetase 1 reduces lignin content and improves cell wall digestibility in Arabidopsis. *Biotechnology for Biofuels*, *8*, 224.

Srivastava, A. C., Ramos-Parra, P. A., Bedair, M., Robledo-Hernández, A. L., Tang, Y., Sumner, L. W., et al. (2011). The folylpolyglutamate synthetase plastidial isoform is required for postembryonic root development in Arabidopsis. *Plant Physiology*, *155*(3), 1237–1251.

Sulis, D. B., & Wang, J. P. (2020). Regulation of lignin biosynthesis by post-translational protein modifications. *Frontiers in Plant Science*, *11*, 914.

Sullivan, M. L., & Bonawitz, N. D. (2018). Spectrophotometric determination of reaction rates and kinetic parameters of a BAHD acyltransferase using DTNB (5,5′-dithio-bis-[2-nitrobenzoic acid]). *Plant Science*, *269*, 148–152.

Sullivan, M. L., & Knollenberg, B. J. (2021). Red clover HDT, a BAHD hydroxycinnamoyl-coenzyme A:L-3,4-dihydroxyphenylalanine (L-DOPA) hydroxycinnamoyl transferase that synthesizes clovamide and other N-hydroxycinnamoyl-aromatic amino acid amides. *Frontiers in Plant Science*, *12*, 727461.

Sun, Q., Liu, X., Yang, J., Liu, W., Du, Q., Wang, H., et al. (2018). MicroRNA528 affects lodging resistance of maize by regulating lignin biosynthesis under nitrogen-luxury conditions. *Molecular Plant*, *11*(6), 806–814.

Sun, Y., Zhou, X., & Ma, H. (2007). Genome-wide analysis of Kelch repeat-containing F-box family. *Journal of Integrative Plant Biology*, *49*(6), 940–952.

Swetha, C., Basu, D., Pachamuthu, K., Tirumalai, V., Nair, A., Prasad, M., et al. (2018). Major domestication-related phenotypes in indica rice are due to loss of miRNA-mediated laccase silencing. *The Plant Cell*, *30*(11), 2649–2662.

Takeda, Y., Suzuki, S., Tobimatsu, Y., Osakabe, K., Osakabe, Y., Ragamustari, S. K., et al. (2019). Lignin characterization of rice CONIFERALDEHYDE 5-HYDROXYLASE loss-of-function mutants generated with the CRISPR/Cas9 system. *The Plant Journal*, *97*(3), 543–554.

Takeda, Y., Tobimatsu, Y., Karlen, S. D., Koshiba, T., Suzuki, S., Yamamura, M., et al. (2018). Downregulation of *p*-COUMAROYL ESTER 3-HYDROXYLASE in rice leads to altered cell wall structures and improves biomass saccharification. *The Plant Journal*, *95*(5), 796–811.

Tang, H. M., Liu, S., Hill-Skinner, S., Wu, W., Reed, D., Yeh, C.-T., et al. (2014). The maize brown midrib2 (bm2) gene encodes a methylenetetrahydrofolate reductase that contributes to lignin accumulation. *The Plant Journal*, *77*(3), 380–392.

Tian, Y., Lin, C.-Y., Park, J.-H., Wu, C.-Y., Kakumanu, R., Pidatala, V. R., et al. (2021). Overexpression of the rice BAHD acyltransferase AT10 increases xylan-bound *p*-coumarate and reduces lignin in *Sorghum bicolor*. *Biotechnology for Biofuels, 14*(1), 217.
Trumbo, J. L., Zhang, B., & Stewart, C. N. (2015). Manipulating microRNAs for improved biomass and biofuels from plant feedstocks. *Plant Biotechnology Journal, 13*(3), 337–354.
Tsai, C.-J., Xu, P., Xue, L.-J., Hu, H., Nyamdari, B., Naran, R., et al. (2020). Compensatory guaiacyl lignin biosynthesis at the expense of syringyl lignin in 4CL1-knockout poplar. *Plant Physiology, 183*(1), 123–136.
Tsuji, Y., Vanholme, R., Tobimatsu, Y., Ishikawa, Y., Foster, C. E., Kamimura, N., et al. (2015). Introduction of chemically labile substructures into Arabidopsis lignin through the use of LigD, the Cα-dehydrogenase from *Sphingobium* sp. strain SYK-6. *Plant Biotechnology Journal, 13*(6), 821–832.
Tsuyama, T., Kawai, R., Shitan, N., Matoh, T., Sugiyama, J., Yoshinaga, A., et al. (2013). Proton-dependent coniferin transport, a common major transport event in differentiating xylem tissue of woody plants. *Plant Physiology, 162*(2), 918–926.
Tsuyama, T., Matsushita, Y., Fukushima, K., Takabe, K., Yazaki, K., & Kamei, I. (2019). Proton gradient-dependent transport of *p*-glucocoumaryl alcohol in differentiating xylem of woody plants. *Scientific Reports, 9*(1), 8900.
Väisänen, E., Takahashi, J., Obudulu, O., Bygdell, J., Karhunen, P., Blokhina, O., et al. (2020). Hunting monolignol transporters: Membrane proteomics and biochemical transport assays with membrane vesicles of Norway spruce. *Journal of Experimental Botany, 71*(20), 6379–6395.
Vanholme, R., Demedts, B., Morreel, K., Ralph, J., & Boerjan, W. (2010). Lignin biosynthesis and structure. *Plant Physiology, 153*(3), 895–905.
Vanholme, R., Morreel, K., Darrah, C., Oyarce, P., Grabber, J. H., Ralph, J., et al. (2012). Metabolic engineering of novel lignin in biomass crops. *The New Phytologist, 196*(4), 978–1000.
Vermaas, J. V., Dixon, R. A., Chen, F., Mansfield, S. D., Boerjan, W., Ralph, J., et al. (2019). Passive membrane transport of lignin-related compounds. *Proceedings of the National Academy of Sciences of the United States of America, 116*(46), 23117–23123.
Vierstra, R. D. (2009). The ubiquitin-26S proteasome system at the nexus of plant biology. *Nature Reviews. Molecular Cell Biology, 10*(6), 385–397.
Wang, G.-F., & Balint-Kurti, P. J. (2016). Maize homologs of CCoAOMT and HCT, two key enzymes in lignin biosynthesis, form complexes with the NLR Rp1 protein to modulate the defense response. *Plant Physiology, 171*(3), 2166–2177.
Wang, J. P., Chuang, L., Loziuk, P. L., Chen, H., Lin, Y.-C., Shi, R., et al. (2015). Phosphorylation is an on/off switch for 5-hydroxyconiferaldehyde O-methyltransferase activity in poplar monolignol biosynthesis. *Proceedings of the National Academy of Sciences of the United States of America, 112*(27), 8481–8486.
Wang, J., Hou, Y., Wang, Y., & Zhao, H. (2021). Integrative lncRNA landscape reveals lncRNA-coding gene networks in the secondary cell wall biosynthesis pathway of moso bamboo (*Phyllostachys edulis*). *BMC Genomics, 22*(1), 638.
Wang, Y., Meng, X., Tian, Y., Kim, K. H., Jia, L., Pu, Y., et al. (2021). Engineered sorghum bagasse enables a sustainable biorefinery with *p*-hydroxybenzoic acid-based deep eutectic solvent. *ChemSusChem*.
Wang, J. P., Naik, P. P., Chen, H.-C., Shi, R., Lin, C.-Y., Liu, J., et al. (2014). Complete proteomic-based enzyme reaction and inhibition kinetics reveal how monolignol biosynthetic enzyme families affect metabolic flux and lignin in *Populus trichocarpa*. *The Plant Cell, 26*(3), 894–914.
Wang, M., Yuan, D., Tu, L., Gao, W., He, Y., Hu, H., et al. (2015). Long noncoding RNAs and their proposed functions in fibre development of cotton (*Gossypium* spp.). *The New Phytologist, 207*(4), 1181–1197.

Wang, C.-Y., Zhang, S., Yu, Y., Luo, Y.-C., Liu, Q., Ju, C., et al. (2014). MiR397b regulates both lignin content and seed number in Arabidopsis via modulating a laccase involved in lignin biosynthesis. *Plant Biotechnology Journal, 12*(8), 1132–1142.

Wang, B., Zhao, X., Zhao, Y., Shanklin, J., Zhao, Q., & Liu, C.-J. (2021). Arabidopsis SnRK1 negatively regulates phenylpropanoid metabolism via Kelch domain-containing F-box proteins. *The New Phytologist, 229*(6), 3345–3359.

Wang, X., Zhuo, C., Xiao, X., Wang, X., Docampo-Palacios, M., Chen, F., et al. (2020). Substrate specificity of LACCASE8 facilitates polymerization of caffeyl alcohol for C-lignin biosynthesis in the seed coat of *Cleome hassleriana*. *The Plant Cell, 32*(12), 3825–3845.

Whipp, M. J., Camakaris, H., & Pittard, A. J. (1998). Cloning and analysis of the shiA gene, which encodes the shikimate transport system of *Escherichia coli* K-12. *Gene, 209*(1–2), 185–192.

Withers, S., Lu, F., Kim, H., Zhu, Y., Ralph, J., & Wilkerson, C. G. (2012). Identification of grass-specific enzyme that acylates monolignols with *p*-coumarate. *The Journal of Biological Chemistry, 287*(11), 8347–8355.

Wu, Z., Ren, H., Xiong, W., Roje, S., Liu, Y., Su, K., et al. (2018). Methylenetetrahydrofolate reductase modulates methyl metabolism and lignin monomer methylation in maize. *Journal of Experimental Botany, 69*(16), 3963–3973.

Xie, H., Engle, N. L., Venketachalam, S., Yoo, C. G., Barros, J., Lecoultre, M., et al. (2019). Combining loss of function of FOLYLPOLYGLUTAMATE SYNTHETASE1 and CAFFEOYL-COA 3-O-METHYLTRANSFERASE1 for lignin reduction and improved saccharification efficiency in *Arabidopsis thaliana*. *Biotechnology for Biofuels, 12*, 108.

Xu, C., Fu, X., Liu, R., Guo, L., Ran, L., Li, C., et al. (2017). PtoMYB170 positively regulates lignin deposition during wood formation in poplar and confers drought tolerance in transgenic Arabidopsis. *Tree Physiology, 37*(12), 1713–1726.

Xue, C., Yao, J.-L., Qin, M.-F., Zhang, M.-Y., Allan, A. C., Wang, D.-F., et al. (2019). PbrmiR397a regulates lignification during stone cell development in pear fruit. *Plant Biotechnology Journal, 17*(1), 103–117.

Yadav, V., Wang, Z., Wei, C., Amo, A., Ahmed, B., Yang, X., et al. (2020). Phenylpropanoid pathway engineering: An emerging approach towards plant defense. *Pathogens (Basel, Switzerland), 9*(4).

Yan, X., Liu, J., Kim, H., Liu, B., Huang, X., Yang, Z., et al. (2019). CAD1 and CCR2 protein complex formation in monolignol biosynthesis in *Populus trichocarpa*. *The New Phytologist, 222*(1), 244–260.

Youn, B., Kim, S.-J., Moinuddin, S. G. A., Lee, C., Bedgar, D. L., Harper, A. R., et al. (2006). Mechanistic and structural studies of apoform, binary, and ternary complexes of the Arabidopsis alkenal double bond reductase At5g16970. *The Journal of Biological Chemistry, 281*(52), 40076–40088.

Yu, X., Gong, H., Cao, L., Hou, Y., & Qu, S. (2020). MicroRNA397b negatively regulates resistance of *Malus hupehensis* to *Botryosphaeria dothidea* by modulating MhLAC7 involved in lignin biosynthesis. *Plant Science, 292*, 110390.

Yu, S.-I., Kim, H., Yun, D.-J., Suh, M. C., & Lee, B.-H. (2019). Post-translational and transcriptional regulation of phenylpropanoid biosynthesis pathway by Kelch repeat F-box protein SAGL1. *Plant Molecular Biology, 99*(1–2), 135–148.

Yu, H., Liu, C., & Dixon, R. A. (2021). A gene-editing/complementation strategy for tissue-specific lignin reduction while preserving biomass yield. *Biotechnology for Biofuels, 14*(1), 175.

Zhang, X., Abrahan, C., Colquhoun, T. A., & Liu, C.-J. (2017). A proteolytic regulator controlling chalcone synthase stability and flavonoid biosynthesis in Arabidopsis. *The Plant Cell, 29*(5), 1157–1174.

Zhang, K., Bhuiya, M.-W., Pazo, J. R., Miao, Y., Kim, H., Ralph, J., et al. (2012). An engineered monolignol 4-O-methyltransferase depresses lignin biosynthesis and confers novel metabolic capability in Arabidopsis. *The Plant Cell, 24*(7), 3135–3152.

Zhang, X., Gou, M., Guo, C., Yang, H., & Liu, C.-J. (2015). Down-regulation of Kelch domain-containing F-box protein in Arabidopsis enhances the production of (poly)phenols and tolerance to ultraviolet radiation. *Plant Physiology, 167*(2), 337–350.

Zhang, X., Gou, M., & Liu, C.-J. (2013). Arabidopsis Kelch repeat F-box proteins regulate phenylpropanoid biosynthesis via controlling the turnover of phenylalanine ammonia-lyase. *The Plant Cell, 25*(12), 4994–5010.

Zhang, J., Li, M., Bryan, A. C., Yoo, C. G., Rottmann, W., Winkeler, K. A., et al. (2019). Overexpression of a serine hydroxymethyltransferase increases biomass production and reduces recalcitrance in the bioenergy crop *Populus*. *Sustainable Energy & Fuels, 3*(1), 195–207.

Zhang, Y., Li, S.-Z., Li, J., Pan, X., Cahoon, R. E., Jaworski, J. G., et al. (2006). Using unnatural protein fusions to engineer resveratrol biosynthesis in yeast and mammalian cells. *Journal of the American Chemical Society, 128*(40), 13030–13031.

Zhang, X., & Liu, C.-J. (2015). Multifaceted regulations of gateway enzyme phenylalanine ammonia-lyase in the biosynthesis of phenylpropanoids. *Molecular Plant, 8*(1), 17–27.

Zhang, Y., Xu, H., Kong, Y., Hua, J., Tang, X., Zhuang, Y., et al. (2020). Wood forming tissue-specific expression of PdSuSy and HCHL increases holocellulose content and improves saccharification in *Populus*. *Journal of Forestry Research*.

Zhao, Q., & Dixon, R. A. (2011). Transcriptional networks for lignin biosynthesis: More complex than we thought? *Trends in Plant Science, 16*(4), 227–233.

Zhao, Y., Yu, X., Lam, P.-Y., Zhang, K., Tobimatsu, Y., & Liu, C.-J. (2021). Monolignol acyltransferase for lignin *p*-hydroxybenzoylation in *Populus*. *Nature Plants*, 7(9), 1288–1300.

Zheng, L., Chen, Y., Ding, D., Zhou, Y., Ding, L., Wei, J., et al. (2019). Endoplasmic reticulum-localized UBC34 interaction with lignin repressors MYB221 and MYB156 regulates the transactivity of the transcription factors in *Populus tomentosa*. *BMC Plant Biology, 19*(1), 97.

Zhong, R., Lee, C., Zhou, J., McCarthy, R. L., & Ye, Z.-H. (2008). A battery of transcription factors involved in the regulation of secondary cell wall biosynthesis in Arabidopsis. *The Plant Cell, 20*(10), 2763–2782.

Zhou, X., Jacobs, T. B., Xue, L.-J., Harding, S. A., & Tsai, C.-J. (2015). Exploiting SNPs for biallelic CRISPR mutations in the outcrossing woody perennial *Populus* reveals 4-coumarate:CoA ligase specificity and redundancy. *The New Phytologist, 208*(2), 298–301.

Zhou, S., Runge, T., Karlen, S. D., Ralph, J., Gonzales-Vigil, E., & Mansfield, S. D. (2017). Chemical pulping advantages of zip-lignin hybrid poplar. *ChemSusChem, 10*(18), 3565–3573.

Zhuo, C., Rao, X., Azad, R., Pandey, R., Xiao, X., Harkelroad, A., et al. (2019). Enzymatic basis for C-lignin monomer biosynthesis in the seed coat of *Cleome hassleriana*. *The Plant Journal, 99*(3), 506–520.

CHAPTER THREE

Glycobiology of the plant secondary cell wall dynamics

Marc Behr, Mondher El Jaziri, and Marie Baucher*
Laboratory of Plant Biotechnology, Department of Biology of the Organisms, Université libre de Bruxelles, Gosselies, Belgium
*Corresponding author: e-mail address: marie.baucher@ulb.be

Contents

Abstract

Primary cell walls are synthesized during plant cell division and elongation whereas secondary cell walls (SCWs) provide rigidity to plant specific tissues that have completed their growth. The cell wall matrix is rich in cellulose, hemicelluloses and pectin polysaccharides as well as various aromatic compounds and proteins. Lignin of the SCW is mainly polymerized after the deposition of the polysaccharide framework. This local environment may affect lignin structure and properties. Hence, the biosynthesis and the function of SCW depend on a large variety of sugar moieties as well as glycosylation reactions. Cellulose is polymerized at the plasma membrane by cellulose synthase complexes while hemicelluloses such as xylan are synthesized in the Golgi and exported to the thickening cell wall. Among other biological processes, cell wall glycoproteins such as arabinogalactan proteins are involved in SCW deposition and structure, as well as sensing of developmental and environmental signals. Finally, in addition to transcriptional regulation, the pools of monolignols and the polymerization of lignin are also modulated through glycosylation reaction. Overall, this chapter illustrates the roles of glucan polymerization and of the de/glycosylation of carbohydrates, proteins and monolignols in the building of SCW matrix, as well as in the regulation of monolignol

Advances in Botanical Research, Volume 104
ISSN 0065-2296
https://doi.org/10.1016/bs.abr.2022.02.003

homeostasis and lignification. We illustrate how the glycosylation of laccases and peroxidases modifies their enzymatic activities. We also point out the differential accumulation of monolignol glucosides in angiosperms and gymnosperms and sum up the main reasons for such a divergence. We explain how glycosylation reactions at the basis of polysaccharides biosynthesis and monolignols' fate partake in the dynamics of the SCW to reach the advanced stages of cellular differentiation characterizing the vascular cells, a specificity of the plant kingdom.

1. Introduction: Composition and occurrence of secondary cell walls

Secondary cell walls (SCWs) are formed in specific cell types involved in mechanical support and/or water conduction comprising vessels, fibers, tracheids and sclerenchyma as well as cells from the endosperm (Zhong, Cui, & Ye, 2019a). SCW is also required in anther endothecium and seed pod to release pollen grains and seeds, respectively (Zhong & Ye, 2015). SCWs are produced between the primary cell wall (PCW) and the plasma membrane (PM) after the cells reach their final size and shape. SCWs are made of three main components, polysaccharides, lignin and proteins, in proportions depending of the cell type (Kumar, Campbell, & Turner, 2016; Zhong et al., 2019a). The cellular program leading to biogenesis of SCWs is under strong transcriptional and post-transcriptional regulation to allow their precise deposition in targeted cell types (Behr, Guerriero, Grima-Pettenati, & Baucher, 2019; Kumar et al., 2016). Albeit traces of lignin oligomers have been detected in particular red (*Calliarthron cheilosporoides*) and green (*Coleochaete* sp.) algae as well as in a bryophyte (*Marchantia polymorpha*), lignified SCWs are only present in vascular plants (Behr, El Jaziri, & Baucher, 2021). The biosynthesis of SCW has been recently reviewed in detail (Meents, Watanabe, & Samuels, 2018; Zhang, Xie, Tuskan, Muchero, & Chen, 2018; Zhong et al., 2019a) and will be therefore summarized in this introduction. The wall polysaccharides form a huge network and display a very complex structure. Their biosynthesis results essentially from the activity of glycosyltransferases (GTs) (EC 2.4.x.y). These enzymes catalyze the transfer of a sugar moiety from activated donor molecules to specific acceptor molecules, forming glycosidic bonds (Bowles, Lim, Poppenberger, & Vaistij, 2006). Based on the available sequences of genomes, GTs are ubiquitous enzymes and 114 families have been identified so far (http://www.cazy.org/GlycosylTransferases.html). Donor sugar substrates are mostly activated in the form of nucleoside diphosphate sugars but nucleoside monophosphate sugars, lipid phosphates and unsubstituted phosphate are also utilized (Lairson, Henrissat, Davies, & Withers, 2008). SCW polysaccharides include mainly cellulose and hemicellulose polymers, as well as traces of pectins.

Cellulose is the most abundant component of SCWs, accounting for 40–80% of their biomass. This unbranched polymer is made of β-1, 4-linked D-glucopyranosyl units with a degree of polymerization of about 10.000 and even over 15.000 in cotton SCWs (Brett, 2000). The SCW is made of three different layers (S1, S2 and S3). In each of these layers, the microfibril angle is determined by the orientation of the cortical microtubule and by hemicellulose crosslinking. Microfibrils have a cross organization in S1, are oriented nearly parallel to the cell elongation axis within S2 and are oriented in a flat helix in S3 (Timell, 1967). The layers have also different thickness, S2 being thicker than S1 and S3. The spatial conformation of cellulose is a determinant feature of cell wall organization. Cellulosic chains are organized in microfibrils of various length, area and shape. This organization underlies the presence of hydrophilic (OH group of the sugar) and hydrophobic (CH group of the sugar) domains, providing different binding affinities for hemicelluloses, pectins or lignin, as well as different sites for enzymatic activity (Cosgrove, 2014).

Hemicelluloses are deposited simultaneously or subsequently with cellulose via secretory vesicles and may hydrogen bond with cellulose. Hemicelluloses account for 10–40% of the SCW and are mainly made of glucuronoxylans (xylans) and mannans depending on the taxon. Xylans are the most abundant hemicelluloses in angiosperms (representing 19–35% and 7–14% of wood in angiosperms and in gymnosperms, respectively; Timell, 1967) whereas conifer SCW contain mainly mannans (reviewed in Zhong et al., 2019a; Zhong, Cui, & Ye, 2019b). Hemicelluloses are synthesized in the Golgi by proteins such as GTs, acetyltransferases, methyltransferases, substrate transporters, as well as proteins involved in substrate synthesis (Hao & Mohnen, 2014; Kumar et al., 2016; Oikawa, Lund, Sakuragi, & Scheller, 2013; Rennie & Scheller, 2014; Wierzbicki, Maloney, Mizrachi, & Myburg, 2019). After secretory vesicles form, they are targeted to cortical microtubule arrays that line regions of the PM where the SCW is being actively deposited (Oda, Iida, Nagashima, Sugiyama, & Fukuda, 2015; Watanabe et al., 2015). The tethering of secretory vesicles to their target membranes prior to fusion can be facilitated in eukaryotes by the octomeric exocyst complex, which interacts with proteins on vesicles and their target membranes (reviewed in Synek, Sekeres, & Zarsky, 2014).

The presence of mixed linkage glucans was reported in rice and other cereals (Vega-Sánchez, Verhertbruggen, Scheller, & Ronald, 2013), but their regulation and their role in the SCW are not yet well established.

SCWs contain glycoproteins that play a role in their morphology and functional features. These proteins include arabinogalactan-rich proteins (AGPs), proline-rich proteins (PRPs), glycine-rich proteins (GRPs), and hydroxyproline-rich glycoproteins (HRGPs or extensins) (Showalter, 1993; Sommer-Knudsen, Bacic, & Clarke, 1998). AGPs are a class of hydroxyproline-rich glycoproteins that have highly complex and diverse structures and are potentially involved in signaling.

SCW's hemicelluloses, cellulose synthase complexes (CSCs) and glycoproteins are delivered through the *trans*-Golgi network via the cytoskeleton to the PM in a coordinated and dynamic manner to produce the spiral, annular or pitted patterns of the SCW with respect to the cell types. This process, essentially relying on GTs for the biosynthesis of polysaccharides, creates the scaffold of the SCW, into which the lignin polymer is later deposited (Ruel, Chevalier-Billosta, Guillemin, Berrio-Sierra, & Joseleau, 2006; Terashima & Fukushima, 1989). Cell corners are first lignified followed by the whole middle lamella and then outer to inner cell wall regions (Grünwald, Ruel, Kim, & Schmitt, 2002; Terashima, Nakashima, & Takabe, 1998). Lignin deposition is the last step of SCW biosynthesis. Lignin provides mechanical support to the cell walls and its hydrophobicity enables water transport through the vascular system. In addition, lignin forms a physico-chemical barrier against pathogens (Naoumkina et al., 2010).

Lignin, accounting for 19–33% of wood in gymnosperms and angiosperms, is a heteropolymer derived mainly from three monolignols differing in their degree of methoxylation, i.e., *p*-coumaryl alcohol, coniferyl alcohol and sinapyl alcohol (Boerjan, Ralph, & Baucher, 2003; Timell, 1967). During the lignification process, these hydroxycinnamyl alcohols are oxidized in the SCW by peroxidases and laccases. When incorporated within the lignin polymer, they form the *p*-hydroxyphenyl (H), guaiacyl (G) and syringyl (S) units, respectively. The proportion of these units differs among species, as well as within different cell types and cell wall layers (Behr et al., 2021). In gymnosperms, lignin is made principally of G units, whereas angiosperm lignin incorporates mainly G and S units. H units, which are minor component of lignin, are relatively more abundant in grasses. In addition, many other compounds can be incorporated within lignin polymer such as caffeyl alcohol, 5-hydroxyconiferyl alcohol, flavonoids (tricin, naringenin) or hydroxystilbenes (Behr et al., 2021).

Building blocks of lignin are oxidized in the cell wall by peroxidases using H_2O_2 and laccase using O_2 into radicals that couple spontaneously (Tobimatsu & Schuetz, 2019). The types of linkages in lignin are thought to be influenced by polysaccharides in proximity of monolignols during polymerization (Li et al., 2015). Xylan interacts with cellulose by forming a twofold helical screw that is induced by cellulose microfibrils (Busse-Wicher et al., 2014; Simmons et al., 2016). Hemicelluloses and lignin covalently cross-link in the cell wall (Barakat et al., 2007; Oinonen, Zhang, Lawoko, & Henriksson, 2015). The different substitutions impact the self-association of the xylan polymer with cellulose and lignin. Grass SCW is characterized by the presence of feruloylated and *p*-coumaroylated arabinosyl substitutions on the xylan backbone. Ferulate substitutions of xylan in grass have been demonstrated to undergo radical coupling similarly to lignification, forming xylan-xylan crosslinks (Ralph, Quideau, Grabber, & Hatfield, 1994) and to participate to the radical coupling reactions of lignification. Xylans of eudicots are not arabinosylated, while gymnosperms arabinosylated SCWs may not be feruloylated, consequently hemicellulose does not link to lignin through ferulate esters in these taxa. However, it has been described that hemicellulosic glycosyl residues may incorporate into lignin by rearomatization of lignification intermediates forming ether or ester bonds between hemicellulose and lignin (Terrett & Dupree, 2019).

In line with the focus of this book on lignin biosynthesis, this chapter highlights the glycobiology of the SCW as the main cellular compartment accumulating this important biopolymer, with a predilection for dicots. Since some of these aspects have not been investigated in SCWs, we will also bring examples of how the glycosylation partakes in the function of PCW, notably for cell wall proteins. In the first part, we will present the GTs required for the biosynthesis of cellulose and xylans, followed by the mechanisms of glycosylation of cell wall proteins and its significance during the formation of cell walls. Finally, the last part will cover the glycosylation of monolignols.

2. Glycosyltransferases and biosynthesis of secondary cell wall polysaccharides

2.1 Cellulose

Plant cellulose is synthesized by complexes of cellulose synthases (CESA), which are glycosyltransferase family 2 (GT2) proteins (Table 1) that synthesize β-1,4-glucan using UDP-α-D-Glc as substrate (Brown, Leijon, & Bulone, 2012; Cantarel et al., 2009). The invertase-mediated pathway is

Table 1 Secondary cell wall CAZy enzymes.

Cell wall constituent	Protein ID	GT family	References
Cellulose	CESA4/IRX5	GT2	Taylor, Howells, Huttly, Vickers, and Turner (2003); Zhang, Dominguez, et al., 2018; Zhang, Xie, et al., 2018
	CESA7/IRX3	GT2	
	CESA8/IRX1	GT2	
	KOR1/IRX2	GH9	Mansoori et al. (2014)
	CTL2	GH19	Sánchez-Rodríguez et al. (2012)
Hemicellulose			
Xylan backbone	IRX9	GT43	Peña et al. (2007)
	IRX14	GT43	Lee, Zhong, and Ye (2012)
	IRX10	GT47	Brown, Zhang, Stephens, Dupree, and Turner (2009)
XRES	FRA8/IRX7	GT47	Hao and Mohnen (2014)
	PARVUS/GATL1	GT8	Lee et al. (2007)
	F8H/IRX7L	GT47	Lee, Teng, Huang, Zhong, and Ye (2009)
	GAUT12/IRX8	GT8	Peña et al. (2007)
Xylan side chains	GUX1-2-3	GT8	Lee, Teng, Zhong, and Ye (2012)
	XAX1	GT61	Feijao et al. (2021)
Mannan	CSLA	GT2	Goubet et al. (2009)
	GalT	GT34	Edwards et al. (1999)

CESA, cellulose synthase; CSLA, cellulose synthase-like A; CTL chitinase-like; F8H, FRA 8 homolog; FRA, fragile fiber; GalT, galactosyltransferase; GATL, galacturonosyltransferase-like; GAUT, galacturonosyltransferase; GUX, Glucuronic acid substitution of xylan; IRX, irregular xylem; XAX, Xylosyl arabinosyl substitution of xylan.

the main provider of UDP-Glc for the biosynthesis of cellulose (Zhong et al., 2019a). In this pathway, invertase cleaves a sucrose into glucose and fructose. HEXOKINASE generates glucose-6-phosphate while fructose is 6-phosphorylated by FRUCTOKINASE before isomerisation in glucose-6-phosphate mediated by PHOSPHOGLUCOSE ISOMERASE. PHOSPHOGLUCOMUTASE converts glucose-6-phosphate to glucose-1-phosphate. Glucose-1-phosphate is the substrate of UDP-GLC

PYROPHOSPHORYLASE to yield UDP-Glc. UDP-Glc may also be synthesized from sucrose by SUCROSE SYNTHASE (SuSy), but this pathway does not seem to be predominant for cellulose deposition in vascular tissues (Zhong et al., 2019a). However an isoform of SuSy partakes in SCW deposition in cotton fibers (Brill et al., 2011) and overexpression of a cotton *SuSy* in poplar increases cellulose content in developing xylem (Coleman, Yan, & Mansfield, 2009), demonstrating that bot SuSy and invertase may provide UDP-Glc for cellulose biosynthesis.

In *Arabidopsis*, *CESA* genes involved in SCW biosynthesis of fibers and vessels are *CESA4/IRREGULAR XYLEM5* (*IRX5*), *CESA7/IRX3* and *CESA8/IRX1* (Taylor et al., 2003). The three CESAs have been demonstrated to interact with each other and the presence of the three proteins is necessary for the complex assembly and targeting to the PM (Taylor et al., 2003). SCW CSCs consist of a 25-nm-diameter rosette of CESA subunits in a hexagonal array with hexamer of a trimer of CESA4, CESA7 and CESA8 whose stoichiometry is variable (Zhang, Dominguez, et al., 2018; Zhang, Xie, et al., 2018). Plant CESAs contain multiple putative transmembrane regions (Slabaugh et al., 2016) that allow the protein complex to insert into a lipid bilayer forming a channel for cellulose translocation across the PM (Morgan, Strumillo, & Zimmer, 2013; Purushotham, Ho, & Zimmer, 2020). The CSCs are thought to be assembled in the Golgi apparatus and delivered to the PM via vesicle trafficking involving the actin cytoskeleton (Zhang et al., 2016).

STILLO1 and STILLO2 are GTs required for the correct assembly of the CSCs in the Golgi (Zhang et al., 2016). STILLO1 and STILLO2 might glycosylate some proteins associated with the CSC (Fig. 1), such as KORRIGAN/IRX2, whose *N*-glycosylation is important for its function (Rips et al., 2014; Zhang et al., 2016). Glucan chains are laid down outside the PM and assembled into cellulose microfibrils, held together by hydrogen bonds and van der Waals forces, and are thought to be composed of 18 chains (Nixon et al., 2016). These microfibrils, fold in a twofold helical screw conformation, aggregate into macrofibrils or higher-order bundles, forming crystalline cellulose. However, some microfibrils are less organized and are mentioned as amorphous cellulose. During SCW biosynthesis, dense clusters of CSCs move in the same direction, forming characteristics wall thickenings (Li et al., 2016).

Other proteins have been involved in cellulose biosynthesis during SCW formation. For instance, KORRIGAN (KOR), a membrane-bound

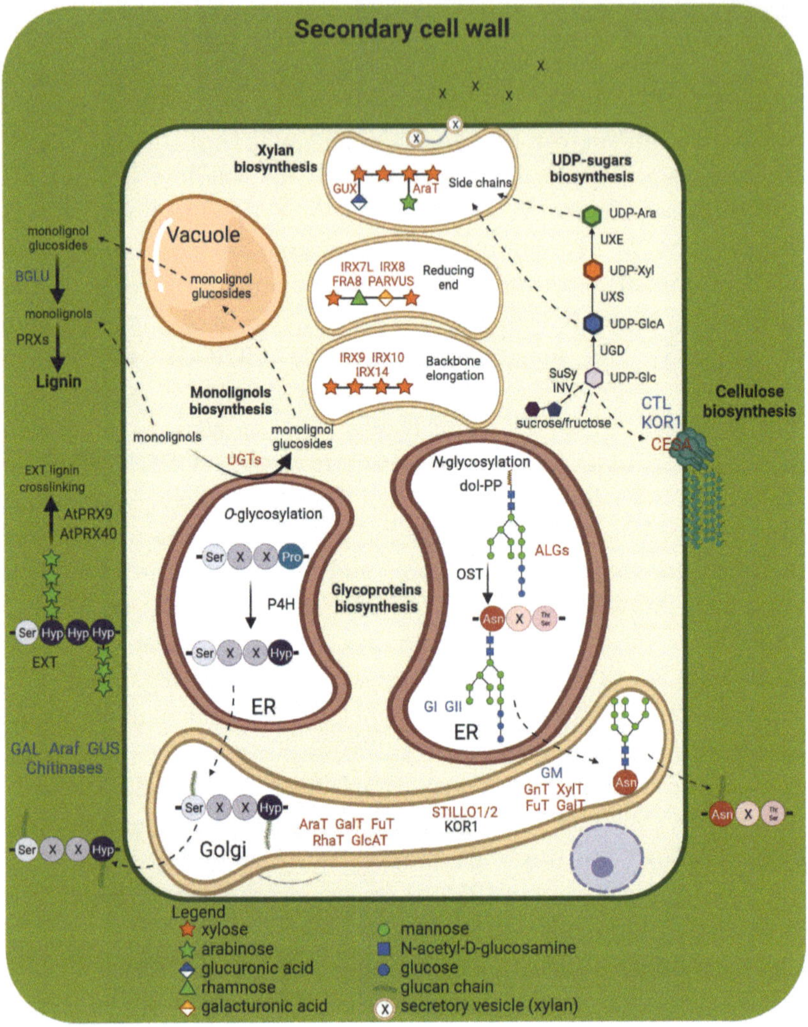

Fig. 1 Overview of the glycobiology of the SCW in a subcellular context. Xylan is synthesized in the Golgi with UDP-sugars produced by enzymes either localized in the Golgi, in the cytosol or membrane-bound. Membrane-bound SuSy associated with the cellulose synthase complex provides UDP-glucose for cellulose deposition. Monolignols biosynthesis takes place in the cytosol and the ER and some UGTs using UDP-glucose and monolignols as sugar acceptors are localized on the ER membrane (Speeckaert et al., 2020). Glycosyltransferases and glycosylhydrolases are shown in red and blue letters, respectively. In the Golgi, STILLO1 and STILLO2 may contribute to the glycosylation of KOR1, which is then targeted to the PM. Full and dotted arrows represent chemical reactions and translocations, respectively. ALG, asparagine-linked glycosylation; Araf, arabinofuranosidase; AraT, arabinosyltransferase; BGLU, β-glucosidase; CESA, cellulose synthase; CTL, chitinase-like; dol-PP, dolichol pyrophosphate; ER, endoplasmic reticulum; EXT, extensin; FuT, fucosyltransferase; GAL, β-galactosidase; GalT, galactosyltransferase; GlcAT, β-glucuronosyltransferase; GnT, *N*-acetylglucosaminyltransferase; G, α-glucosidase; GM, Golgi-α-mannosidase; GUS, β-glucuronidase; GUX, glucuronic acid substitution of xylan; INV, invertase; IRX, irregular xylem; OST, oligosaccharyltransferase complex; P4H, prolyl-4-hydroxylase; PRX, peroxidase; RhaT, rhamnosyltransferase; SuSy, sucrose synthase; UGD, UDP-glucose dehydrogenase; UGT, UDP-glycosyltransferase; UXE, UDP-xylose 4-epimerase; UXS, UDP-xylose synthase; XylT, xylosyltransferase.

endo-(1,4)-β-glucanase, interacts with SCW CESAs (Mansoori et al., 2014) and has been associated to the level of cellulose crystallinity in both gymnosperms and angiosperms (Maloney & Mansfield, 2010; Maloney, Samuels, & Mansfield, 2012). TRACHEARY ELEMENT DIFFERENTIATION-RELATED 6 (TED6) and TED7 are transmembrane proteins that may associate with SCWs' CSCs (Endo et al., 2009; Rejab, Nakano, Yoneda, Ohtani, & Demura, 2015; Vain et al., 2014). COBRA-LIKE4 is a lipid-anchored protein with a putative cellulose-binding domain and is possibly involved in producing highly crystalline cellulose in SCWs (Brown, Zeef, Ellis, Goodacre, & Turner, 2005; Li et al., 2003; Liu et al., 2013). CHITINASE-LIKE1 (CTL1)/POM-POM1 and CTL2 can bind to glucan polymers, but lack glucanase activity (Sánchez-Rodríguez et al., 2012; Zhang et al., 2004). CTL1 contributes to the velocity of PCW CSCs, while *CTL2* is expressed with SCW CSCs (Sánchez-Rodríguez et al., 2012).

2.2 Hemicelluloses

Xylans have a β-(1,4)-linked xylose backbone which is decorated with side chains of glucuronic acid (glucuronoxylan) in eudicots and non-grass monocots whereas conifer and grass xylans are also substituted with arabinose (Scheller & Ulvskov, 2010). Mannans have a β-(1,4)-linked mannose backbone and glucomannans refers to a chain of mannose and glucose. Mannans were detected in thickened SCWs of xylem elements and fibers of angiosperms, and more precisely in the S2 layer (Kaneda, Rensing, & Samuels, 2010). In conifers, side chains of galactose are formed, producing galactoglucomannans. Xylans and mannans are also variously acetylated and/or methylated (Rennie & Scheller, 2014).

Nucleotide sugars such as UDP-xylose, UDP-glucuronic acid and UDP-arabinose, deriving from UDP-glucose, are the sugar donors to GTs involved in the biosynthesis of xylan backbone and substitutions (Fig. 1). The conversion of UDP-glucuronic acid to UDP-xylose is mediated by UDP-XYLOSE SYNTHASES (UXS). Six *UXS* are present in the genome of *Arabidopsis thaliana*. Golgi-localized UXS1, UXS2 and UXS4 are weakly expressed in stem tissues undergoing SCW formation. In contrast, the cytosolic UXS activity of stem is significantly higher than in the microsome fraction. This cytosolic activity originates from *UXS3*, *UXS5* and *UXS6*, whose expressions are induced by transcription factors regulating SCW biogenesis. Simultaneous mutation/silencing of *UXS3*, *UXS5* and

UXS6 reduced the xylose content by 63%, while *uxs1/2/4* triple mutant displayed xylose level identical to the WT. Finally, impaired expression of *UXS3/5/6* does not reduce the chain length of the xylan backbone (Zhong et al., 2017).

Xylan biosynthesis is performed by a large set of enzymes mainly belonging to the family of GTs (Table 1) that are sequentially arranged within the Golgi (Fig. 1). Similarly to CESAs, some of the proteins involved in xylan biosynthesis must be arranged in a complex to show enzymatic activity (Zeng et al., 2016). The backbone and early side chains synthesis occur in protein complex at *cis*- and *medial*-cisternae while late side chains are synthesized in *trans*-cisternae (Hoffmann, King, Samuels, & McFarlane, 2021). Xylan is finally packed into secretory vesicles at the *trans*-Golgi network (TGN) (Meents, Motani, Mansfield, & Samuels, 2019). Glucuronoxylan is the most abundant form of xylan in dicots. A specific oligosaccharide, named xylan reducing-end sequence (XRES; Xyl-Rha-GalA-Xyl), is supposed to function as an initiator or terminator of xylan chain synthesis (Hao & Mohnen, 2014). Two glucuronoxylan biosynthesis models have been proposed based on this XRES assumption (Hao & Mohnen, 2014). In the first model, XRES acts as a primer and chain elongation occurs via the transfer of D-xylose from UDP-xylose onto the non-reducing end of XRES. The initial xylose residue is supposed to be anchored to an unknown moiety (lipid or protein such as AGPs). In the second model, XRES is de novo synthesized, without anchoring. The xylan backbone is elongated by the reducing-end and XRES is transferred at the end of the elongation process when the backbone has reached a certain length. These two mechanisms are not exclusive in different species, tissues or cell types (Hao & Mohnen, 2014). The enzymes involved in the XRES synthesis are partially characterized (e.g., FRAGILE FIBER 8 FRA8, PARVUS, IRX7L/F8H and IRX8; Hao & Mohnen, 2014; Lee et al., 2007; Wierzbicki et al., 2019). Backbone elongation is performed by another set of enzymes within the Golgi, including IRX9-IRX9L and IRX14-IRX14L, belonging to the GT43 family, and IRX10-IRX10L, belonging to the GT47 family (Jiang et al., 2016; Ratke et al., 2015; Zeng et al., 2016) (Table 1). Garden asparagus IRX9, IRX10 and IRX14A exclusively show xylan xylosyltransferase (XylT) activity and Golgi subcellular targeting when they are assembled in a xylan synthase complex (Zeng et al., 2016). A 93 average degree of polymerization of glucuronoxylan was determined in *Arabidopsis* WT, which was dropping to 28 in *irx9* and reaching 125 in *irx8* (Peña et al., 2007).

The GLUCURONIC ACID SUBSTITUTION OF XYLAN (GUX1 to GUX5 in *Arabidopsis*) enzymes (CAZy family GT8) are involved in adding the glucuronic acid side chains (Table 1). *GUX1* and *GUX2* are preferentially expressed in stem, while *GUX3* has no clear expression pattern, *GUX4* is exclusively expressed in roots and *GUX5* is highly expressed in leaf and flower (Lee, Teng, et al., 2012). The glucuronic acids can be methylated by glucuronoxylan methyltransferases (GXMTs) belonging to the domain of unknown function (DUF) 579 family (Lee, Teng, et al., 2012; Urbanowicz et al., 2012). Xylan acetylation is catalyzed by xylan *O*-acyltransferases belonging to the DUF231 family. The *REDUCED WALL ACETYLATION* genes (*RWA1* to *RWA4*) redundantly participate in the acetylation of xylan during SCW biosynthesis. They are regulated by the transcription factor SND1 (Lee, Teng, Zhong, & Ye, 2011). Xylan *O*-acetyltransferase 1 (XOAT1), also known as ESKIMO1, is able to acetylate xylose residues at the *O*-2 position (Grantham et al., 2017; Lunin et al., 2020). Acetylated polysaccharides have reduced solubility and thus are more easily deposited and cross-linked with cellulose hydrophilic surface (Grantham et al., 2017; Lee et al., 2011).

In grasses, the arabinose substitutions on the xylan backbone are catalyzed by arabinosyltransferases belonging to the GT61 family (Anders et al., 2012; Zhong, Cui, Phillips, & Ye, 2018). The arabinosyl side chains may be substituted with xylosyl residues or esterified by ferulic acid or *p*-coumaric acid (Scheller & Ulvskov, 2010). The rice GT61 XAX1 (XYLOSYL ARABINOSYL SUBSTITUTION OF XYLAN 1), has been described to transfer hydroxycinnamic acid arabinosyl substitution to the xylan backbone playing a role in the crosslinking between lignin and xylan (Feijao et al., 2021).

The biosynthesis of the 1,4-β-linked mannan and glucomannan backbone is catalyzed by mannan synthases, belonging to the GT2 CELLULOSE SYNTHASE-LIKE family A (CSLA) which are located in the Golgi (Davis, Brandizzi, Liepman, & Keegstra, 2010; Liepman & Cavalier, 2012). This backbone may be substituted by with α-galactose or with acetyl groups determining its properties. Mannan galactosyltransferases belonging to GT34 family (Edwards et al., 1999; Yu et al., 2018) and mannan *O*-acetyltransferases belonging to the DUF231 family have been identified (Zhong et al., 2019b; Zhong, Cui, & Ye, 2018).

3. Glycosylation of secondary cell wall proteins

Protein glycosylation fulfills a wide range of functions *in planta*, comprising solubilization, correct folding, protection against proteolysis and thermal denaturation, quality control in the endoplasmic reticulum (ER), subcellular targeting, protein-protein interaction and communication with the environment (Nguema-Ona et al., 2014; Schoberer & Strasser, 2018; Silva, Ferraz, Dupree, Showalter, & Coimbra, 2020).

An updated classification of cell wall proteins (CWP) in nine functional classes has been released in 2013 (Albenne, Canut, & Jamet, 2013). Proteins active on carbohydrates represent ca. 26% of CWP, followed by oxido-reductases (15%), proteases and CWP with interaction domains (11% each), signaling (6.6%), lipid metabolism (5.8%) and structural proteins (1.6%). This classification also reports proteins with unknown functions (12%) and proteins with miscellaneous functional domains (11%). There are now experimental evidence that most of these functional classes undergo glycosylation. Indeed, a glycoproteomic analysis performed in leaves of *Arabidopsis* highlighted glycosylation of proteases, oxido-reductases, proteins related to carbohydrates, lipid metabolism, signaling, as well as proteins with interaction domains and others with domain of unknown functions (Sultana, Florance, Johns, & Smirnoff, 2015). Glycosylation of these proteins' families were retrieved in elongating cotton fiber cells and other tissues from various species (Kumar et al., 2013), indicating a conserved glycosylation pattern of plant CWP with histologic specificities.

PCWs and SCWs display drastic structural and functional differences (Hoffmann et al., 2021). This differentiation is partially driven by CWP. For instance, expansins drive cell elongation by loosening PCW hotspots consisting of cellulose in close proximity with xyloglucan (Cosgrove, 2018), while several class III peroxidases produce the phenoxy radical form of monolignols that will polymerize into lignin during SCW biogenesis (Shigeto, Kiyonaga, Fujita, Kondo, & Tsutsumi, 2013). We will focus on glycoproteins located in SCWs, the genetic network underlying their glycosylation and their known functions *in vivo*.

3.1 Mechanisms of cell wall protein glycosylation

Glycosylation of CWP occurs in the ER and the Golgi apparatus (Hoffmann et al., 2021). There are two types of protein glycosylation (Fig. 1). Almost all CWP functional classes may be asparagine (*N*)-glycosylated (Song et al., 2013), with structural proteins appearing as an exception. The first steps of *N*-glycosylation being required for correct protein folding, they are performed by a conserved set of enzymes, which localize in the ER (Nguema-Ona et al., 2014). In polypeptide chains actively elongating in the lumen of the ER, a conserved glucan

($Glc_3Man_9GlcNAc_2$) pre-assembled in a lipid carrier dolichol pyrophosphate is *en bloc* attached to the Asn residue of the *N*-linked glycosylation sites (Asn-Xxx-Ser/Thr, Xxx not being Pro) (Schoberer & Strasser, 2018). The biosynthesis of this conserved glucan is guided by asparagine linked glycosylation proteins (ALGs). The glucan is transferred to targeted proteins by the oligosaccharyltransferase complex and is then digested by two α-glucosidases, resulting in a $Man_9GlcNAc_2$ motif. The proper folded glycoprotein exits the ER for stepwise maturation with α-MANNOSIDASES (GM), *N*-ACETYLGLUCOSAMINYLTRANSFERASES (GnT) and XYLOSYLTRANSFERASES (XylT) in the Golgi (Baker et al., 2016; Schoberer & Strasser, 2018). A Lewis A epitope (GalGlcNAcFuc) may optionally be obtained in the *trans*-Golgi through activity of β-1,3-GALACTOSYLTRANSFERASE (GalT) and α1,4-FUCOSYLTRANSFERASE (FuT) (Schoberer & Strasser, 2018; Strasser et al., 2007). This epitope, which is the last known step of *N*-glycan processing in plants, was detected in laccases involved in lignification, as well as in KORRIGAN (Beihammer, Maresch, Altmann, Van Damme, & Strasser, 2021). Many other glycans may arise from glucan precursors, as evidenced by mass spectrometry characterization of 35 different glycans detected on the peroxidase of windmill palm tree (Baker et al., 2016).

Serine (Ser), threonine (Thr) and hydroxyproline (Hyp) (*O*)-glycosylated CWP belong to the superfamily of HRGPs, whose main families are the highly glycosylated AGPs, the moderately glycosylated extensins and finally the PRPs with highly variable glycosylation degree (Hijazi, Velasquez, Jamet, Estevez, & Albenne, 2014). Since *O*-glycosylation occurs mainly on Hyp residues, conversion of proline to its hydroxylated counterpart is a prerequisite for the cascade of glycosylation reactions (Fig. 1). Proline hydroxylation is catalyzed in the ER through the action of several PROLYL-4-HYDROXYLASES (P4Hs). While the consequences of a defective proline hydroxylation at the SCW level are not yet described, several functional studies have highlighted the importance of this posttranslational modification during PCW assembly. In *Arabidopsis*, knockout T-DNA lines for *P4H2*, *P4H5* and *P4H13* show defective root hair tip growth and modifications in the cellulose fibrils angle while silencing *P4H1* in the green alga *Chlamydomonas reinhardtii* prevents the formation of the dense central cell wall layers probably by altering Hyp content (Keskiaho, Hieta, Sormunen, & Myllyharju, 2007; Marzol et al., 2018). Hyp and Ser residues are subsequently glycosylated in the Golgi by several GTs. The glycosylation pattern widely differs between the three main HRGP families, as well as within each family.

For instance, three types (I, II and III) of arabinogalactan polysaccharides (AG) are reported so far. A large number of functionally distinct GTs are required for the biosynthesis of AG, e.g., β-1,3-GalTs, β-1,6-GalTs, α-1,3- and α-1,5-ARABINOSYLTRANSFERASES, FuTs, RHAMN OSYLTRANSFERASES and GLUCURONOSYLTRANSFERASES (Knoch, Dilokpimol, & Geshi, 2014; Silva et al., 2020). AG are modified by the action of several GHs, such as α-ARABINOFURANOSIDASE (Araf, GH3), β-GALACTOSIDASE (GAL, GH35) and β-GLUCURONIDASE (GUS, GH79) mainly distributed in the cell wall (Silva et al., 2020). The activity of most of these enzymes requires the presence of manganese, since mutation in a Golgi Mn transporter leads to reduced mannosidase activity on $Man_9GlcNAc_2$ and lower abundance of KORRIGAN, probably due to its impaired glycosylation pattern (Yang et al., 2021).

3.2 Significance of glycosylation in secondary cell wall proteins

Within major SCW glycoproteins comprising HRGPs, laccases and peroxidases, glycan chains demonstrate their widespread biological properties. They contribute to the enzymatic activity of oxidoreductases, while in HRGPS they contribute to the cell wall network, as further explained below.

We will illustrate the role of SCW HRGPs glycosylation with two examples, namely AGPs and extensins. Covalent interactions between AGPs and plant cell walls have been demonstrated in *Arabidopsis*, where ARABINOXYLAN PECTIN ARABINOGALACTAN PROTEIN 1 (APAP1) was found to link with arabinoxylan and pectins (rhamnogalacturonan I/homogalacturonan region) (Tan et al., 2013). AGPs have been involved in SCW biosynthesis and stem mechanics (MacMillan et al., 2015; Wang et al., 2015). AGPs containing GlcNAc can be activated by enzymes such as endochitinases, producing various oligosaccharides (Hengel, Kammen, & Vries, 2002). Endochitinases have been extracted from powdered tobacco xylem, as well as from tobacco xylogenic cultures, suggesting their involvement in SCW biogenesis (Millar et al., 2009). Specific chitinases are highly expressed in bast fibers of flax and hemp, as well as in poplar tension wood, possibly regulating the formation of SCW G-layer in these cell types (Guerriero et al., 2017; Lafarguette et al., 2004; Mokshina, Gorshkova, & Deyholos, 2014). The accumulation of short (GlcNAc)n triggered by the heterologous expression of *N-ACETYLGLUCOSAMINYLTRANSFERASE* (*NodC*) of

Azorhizobium caulinodans alters internode length and leads to a general loss of cell wall strength in *Arabidopsis* (Vanholme et al., 2014). These oligosaccharides may compete in carbohydrate-carbohydrate and carbohydrate-protein interactions, leading to perturbation in the noncovalent interactions in the cell wall or at the cell wall-PM interface (Vanholme et al., 2014).

The protein backbone of AGPs is massively decorated by type II AG. Galactose is the first sugar added on the backbone. In *Arabidopsis*, this substitution is catalyzed by eight Hyp-*O*-galactosyltransferases whose functions are overlapping (Zhang, Held, Kaur, & Showalter, 2021). High-order CRISPR null mutants (up to five genes) exhibit lower AGP glycosylation, reduced growth of aerial and root parts, abnormal pollen and shorter siliques, as well as reduced seed set (Zhang et al., 2021). The monosaccharide composition of AGPs extracted from these mutants was characterized by lower arabinose and galactose content, while the proportion of rhamnose, glucose and xylose increased. These modifications potentially affect the associations between AGPs, PM and cell wall components, such as pectins and hemicelluloses. The same multiplexing approach evidenced a similar role for glucuronic acid AGPs substitution, with a triple β-glucuronosyltransferase null mutant displaying shorter root hair, abnormal pollen, shorter siliques, and reduced seed set, as well as delayed seed germination (Zhang, Held, & Showalter, 2020). While the importance of AGP glycosylation towards organization of the PCW and regulation of primary growth becomes well described, such functional studies are still lacking to decipher the involvement of AGPs in SCW structure and function. It should be worth investigating whether the AGPs substitution pattern and the associated GTs are conserved between PCW and SCW and how these side chains interact with other components of the SCW.

The FASCICLIN-LIKE AGP (FLA) are a class of CWPs whose functions are currently under deep investigations to unravel their functions in plant development. FLA are characterized by fasciclin domains and are thought to have structural and signaling functions mainly in the PCW (Seifert, 2021). A role of FLA in stem biomechanics via regulation of SCW architecture and/or composition has been highlighted in several species (Dahiya, Findlay, Roberts, & McCann, 2005; Liu et al., 2020; Ma et al., 2022; MacMillan et al., 2015; MacMillan, Mansfield, Stachurski, Evans, & Southerton, 2010; Wang et al., 2015, 2017).

Extensins are CWPs of prime importance in expanding cells such as pollen tubes, but they are also likely to play a role in the biomechanics of SCW. Ser-Hyp3–5 repeats is the signature motif of extensins. The contiguous Hyp residues account for >30 mol% of all amino acids and they are glycosylated with up to 4–5 units of L-arabinofuranose (L-Araf). These repeats alternate with hydrophobic regions containing a significant proportion of Tyr residues (Chen, Dong, Tan, Held, & Kieliszewski, 2015; Marzol et al., 2018). Both arabinosylation and Tyr residues are important parameters for extensin's function, notably in the SCW. *In vitro* data demonstrate that arabinosylation of extensins is required for efficient peroxidase-mediated extensin crosslinking (Chen et al., 2015). Extensins may indeed crosslink with other extensins, other CWPs but also with lignin (Fig. 1). This reaction is catalyzed by specific class III peroxidases, such as *Arabidopsis* PRX9 and PRX40 involved in pollen microspore cell wall integrity (Jacobowitz, Doyle, & Weng, 2019). Several extensins varying in their degree of deglycosylation were less prone to crosslinking than the native protein, with the completely dearabinosylated forms not being a substrate for a tomato extensin peroxidase (Chen et al., 2015). The fourth Ara of the side chain seems to play a prominent role in the crosslinking capacity of extensins (Chen et al., 2015). While such interactions are not yet well described, they appear to account significantly for the biological conversion of wood biomass to biofuel, as shown in poplar. Hot water pre-treatment of poplar biomass coupled to protease-mediated digestion of extensins increases the removal of Hyp (as a proxy for presence of extensins) as compared to either absence or single liquid hot water pretreatment, as well as glucose release (Fleming, Decker, & Bedinger, 2016).

Laccases involved in lignin polymerization generally show several *N*-glycosylation sites, as demonstrated in Norway spruce (Koutaniemi, Malmberg, Simola, Teeri, & Kärkönen, 2015) and loblolly pine (Sato, Wuli, Sederoff, & Whetten, 2001). Recent reporting of the crystal structure of *Zea mays* LACCASE 3 evidenced that all putative *N*-glycosylation sites were branched with a monomer or a dimer of *N*-acetyl-D-glucosamine (Xie, Liu, & Wang, 2020). Purification to homogeneity of a laccase secreted in cell suspension of sycamore indicated that carbohydrates (such as sidechains with xylose, fucose, galactose and mannose) account for 40% of the molecular weight of the glycoprotein (Driouich, Lainé, Vian, & Faye, 1992). Later on, these complex *N*-type glycans were demonstrated to show a carbohydrate sequence typical of Lewis A antigens, conferring localization at the cell surface (Fitchette-Lainé et al., 1997). This glycan motif may thus be required for addressing laccase to the apoplast. There are currently no data reporting the function of glycosylation of plant laccase regarding substrate specificity nor kinetic parameters. However, fungal laccases have been studied in much more details. Four deglycosylated laccases from *Steccherinum ochraceum* do not display drastically different catalytic properties as compared to their native forms, but they are slightly less thermostable (Glazunova et al., 2019). A comparable output was observed with a laccase from *Lentinus* sp. (Maestre-Reyna et al., 2015). The data available so far do not evidence a crucial role for glycosylation in the enzymatic properties of laccases, emphasizing the need to assess the significance of glycosylation of plant laccases.

Peroxidases involved in monolignol activation are glycosylated, as described in *Zinnia elegans* (Gabaldón, López-Serrano, Pedreño, & Barceló, 2005), Norway spruce (Koutaniemi et al., 2005), *Arabidopsis* and horseradish (Nielsen et al., 2001). Glycosylation increases the solubility of enzymes (Tams & Welinder, 1995). Several lines of evidence demonstrate that glycosylation of peroxidases is also important for their enzymatic activity. Comparative analysis of horseradish peroxidase HRP A2 and its ortholog in *Arabidopsis* ATP A2 (AtPRX53) highlighted the significance of protein glycosylation regarding specific activity and catalytic properties (Nielsen et al., 2001). HRP A2 and ATP A2 are 95% identical, differing by only 16 amino acid residues at the surface of the protein, thus unlikely to drive the activity. Recombinant ATP A2 was produced in *E. coli*, and thus was not glycosylated, in contrast with native HRP A2, isolated from horseradish and carrying seven *N*-linked glycan (major form Manα3(Manα6)(Xylβ2)Manβ4GlcNAcβ4(Fucα3)(GlcNAc)). Different activities therefore likely originate from the glycosylation pattern of the two proteins. Enzymatic turnover was systematically higher in HRP A2 than in ATP A2 when using ferulic, coumaric and sinapic acids. As compared to ATP A2, HRP A2 has a significantly higher reaction rate with ferulic acid, coniferyl alcohol and *p*-coumaric acid while no difference was observed for sinapic acid and ABTS. The authors propose a mechanism where the glycan chains might assist guidance, binding or release of the substrates and products. These glycans may also direct the diffusion of H_2O_2 in the distal cavity. Glycosylation of precise Asn residues might additionally contribute to differential affinity between pools of potential substrates. Even more interesting, two protein isoforms of the lignifying *Z. elegans* peroxidase only differing in their glycosylation patterns naturally occur *in planta* (Gabaldón et al., 2005). The two isoforms display a molecular mass shift of 1260 Da, with ZePrx33.44 (for $M_R = 33{,}440$ Da) and ZePrx34.70 being partially and fully glycosylated isoforms, respectively. Sinapyl alcohol was the preferred substrate of both enzymes, but ZePrx34.70 has a lower k_{cat}

for this substrate as compared to ZePrx33.44. By contrast, ZePrx34.70 displays significantly higher K_m values than ZePrx33.44 for *p*-coumaryl alcohol and coniferyl alcohol. The oxidation potential of the two enzymes was also significantly different when providing the cinnamaldehydes precursors of the three monolignols as substrate (Gabaldón et al., 2006). ZePrx33.44 was detected in tracheary elements, roots, old (6 days) hypocotyls and stems while ZePrx34.70 was evidenced in roots and young (3 days) hypocotyls. These different accumulation patterns, both in time and in space, as well as the variable affinity for the three monolignols, suggest that each isoform fulfill specific roles that may follow the peculiar pattern of lignification in stems and hypocotyl of *Z. elegans* (Gabaldón et al., 2005).

4. UDP-glycosyltransferases and glycosylation of monolignols

4.1 Monolignol (de)glucosylation

Monolignols, i.e., *p*-coumaryl alcohol, coniferyl alcohol and sinapyl alcohol can be glycosylated by URIDINE DIPHOSPHATE GLYCOSYLTRANSFERASES (UGT; E.C. 2.4.1.x). These enzymes use UDP-activated sugars as donor molecules giving rise to *p*-glucocoumaryl alcohol, coniferin and syringin, respectively (Liu, 2012; Wang, Chantreau, Sibout, & Hawkins, 2013). Monolignol glucosides are considered as either a storage or a transport form of monolignols (Wang et al., 2013). Glycosylation at the 4-O position of the aromatic ring of the monolignol renders them less reactive, increases their solubility and decreases their toxicity (Le Roy, Huss, Creach, Hawkins, & Neutelings, 2016; Väisänen et al., 2015).

The modulation between the free and conjugated forms allows regulating the storage, the stability and the solubility of compounds and UGTs are therefore involved in both developmental and defense processes. Together with glucosidases, UGTs rapidly shape the glycosylation status of such metabolites to allow plants to adapt to the environmental conditions. For instance, activity of the rice UGT GRAIN SIZE AND ABIOTIC STRESS TOLERANCE 1 (GSA1/UGT83A1) contributes to the down-regulation of the lignin pathway under abiotic stress and regulates grain size by glycosylating the three canonical monolignols and other substrates such as flavonoids (Dong et al., 2020). Several reports have shown that different recombinant GTs use hydroxycinnamic acids, cinnamaldehydes and cinnamyl alcohols as sugar acceptor (Table 2). The UGT72 sub-family, belonging to the group E of UGT (Ross, Li, Lim, & Bowles, 2001), has been

Table 2 Main UGTs using monolignols and precursors as substrate in recombinant assay.

Protein (*species*)	**Substrate**	**References**
UGT72B1 (*A. thaliana*)	Coniferaldehyde, *p*-coumaraldehyde, coniferyl alcohol, dihydroconiferyl alcohol, *p*-coumaryl alcohol	Lin et al. (2016)
UGT73B3 (*A. thaliana*)	Coniferaldehyde, sinapaldehyde	Lin et al. (2016)
UGT72E1 (*A. thaliana*)	Coniferaldehyde, sinapaldehyde	Lim, Jackson, and Bowles (2005)
UGT72E2 (*A. thaliana*)	Ferulic acid, sinapic acid, *p*-coumaric acid, caffeic acid, *o*-coumaric acid, *m*-coumaric acid, coniferaldehyde, sinapaldehyde, coniferyl alcohol, sinapyl alcohol	Lim et al. (2001, 2005)
UGT72E3 (*A. thaliana*)	Sinapic acid, caffeic acid, ferulic acid, coniferyl alcohol, sinapyl alcohol	Lim et al. (2001, 2005)
UGT72AM1 (*Camelia sinensis*)	Coniferaldehyde	Zhao et al. (2017)
UGT72AZ2 (*P. tremula* × *P. alba*)	Ferulic acid	Speeckaert et al. (2020)
UGT72B37 (*P. tremula* × *P. alba*)	*p*-Coumaraldehyde, coniferaldehyde, sinapaldehyde, coniferyl alcohol, sinapyl alcohol	Speeckaert et al. (2020)
UGT72B39 (*P. tremula* × *P. alba*)	Coniferyl alcohol	Speeckaert et al. (2020)
UGT84A1 (*A. thaliana*)	Cinnamic acid, *p*-coumaric acid, *o*-coumaric acid, caffeic acid, ferulic acid	Yang et al. (2018)
UGT84A2 (*A. thaliana*)	Cinnamic acid, *p*-coumaric acid, *o*-coumaric acid, caffeic acid, ferulic acid	
UGT84A3 (*A. thaliana*)	Cinnamic acid, *p*-coumaric acid, caffeic acid, ferulic acid	
UGT84A4 (*A. thaliana*)	Cinnamic acid, *p*-coumaric acid, caffeic acid, ferulic acid	
UGT84B1 (*A. thaliana*)	Cinnamic acid, *p*-coumaric acid, caffeic acid, ferulic acid	

Continued

Table 2 Main UGTs using monolignols and precursors as substrate in recombinant assay.—cont'd

Protein (*species*)	Substrate	References
UGT84A17 (*P. fremontii* × *angustifolia*)	Caffeic acid, *o*-coumaric acid, *p*-coumaric acid, sinapic acid, ferulic acid, cinnamic acid	Babst et al. (2014)
UGT71C1 (*A. thaliana*)	*p*-Coumaric acid, *o*-coumaric acid, caffeic acid	Yang et al. (2018)
UGT71C2 (*A. thaliana*)	*p*-Coumaric acid, *o*-coumaric acid, caffeic acid	
UGT73B3 (*A. thaliana*)	*p*-Coumaric acid, caffeic acid, ferulic acid	
UGT83A1 (*Oryza sativa*)	*p*-Coumaryl alcohol, sinapyl alcohol, coniferyl alcohol	Dong et al. (2020)

widely investigated regarding their role during monolignol glucosylation and regulation of the lignification process (Table 2). Indeed, the overexpression of several *UGT72s* in *Arabidopsis* and in poplar has led to an increase in glucosides of monolignol (or precursor) and/or lignin content (Lanot et al., 2006; Lanot, Hodge, Lim, Vaistij, & Bowles, 2008; Lin et al., 2016; Speeckaert et al., 2020). Besides, a reduction in the content of both coniferin and syringin was measured in some of the *Arabidopsis UGT72E* mutants and downregulated lines as compared to the WT plants (Lanot et al., 2006). On the opposite, surprisingly, the *Arabidopsis ugt72b1* mutant had about 3-fold more coniferin in young stems compared to the WT. These changes were attributed to a compensation due to an increased expression of both *UGT72B3* and *UGT72E2* in this mutant (Lin et al., 2016).

No impact on lignin was reported in transgenic *Arabidopsis* plant overexpressing *UGT72E1*, *UGT72E2* or *UGT72E3* nor in the single or triple *UGT72Es* down-regulated lines. The same observation applied to wood, bark and roots of poplar overexpressing *UGT72AZ1* and *UGT72AZ2* (Speeckaert et al., 2020). However, ectopic lignification with an arrested growth phenotype was reported for the *ugt72b1* mutant, indicating a major role of this gene in cell wall lignification in *Arabidopsis* (Lin et al., 2016). The RNA-Seq analysis of the *ugt72b1* mutant showed that genes involved in monolignol metabolism, monolignol transport, lignin polymerization as well as regulation of SCW formation were upregulated in the mutant.

The cleavage of the glucose, releasing the monolignol aglycone, is performed by β-glucosidases that are notably present in the xylem of gymnosperm and angiosperm (Dharmawardhana, Ellis, & Carlson, 1995; Tsuyama & Takabe, 2015). A β-glucosidase from *Pinus contorta* which localized in the cell walls of developing xylem, was shown to be specific for coniferin, syringin and coniferin synthetic analogs (Dharmawardhana et al., 1995; Dharmawardhana, Ellis, & Carlson, 1999). Proteins extracted from cell walls of poplar xylem tissue were able to hydrolyse coniferin as well as syringin to a less extent (Tsuyama & Takabe, 2015). In *Arabidopsis*, two β-glucosidases, BGLU45 and BGLU46 hydrolysing monolignol glucosides (Escamilla-Treviño et al., 2006), have been localized in the interfascicular fibers and in the protoxylem, respectively (Chapelle et al., 2012). An increase in coniferin content as well as other glycosylated phenylpropanoid compounds was measured in the *bglu45* and *bglu46* mutant lines. There was no differences in the content of coniferyl alcohol, sinapyl alcohol, coniferaldehyde, sinapaldehyde and syringin between the *bglu45* and *bglu46* mutants and the WT (Chapelle et al., 2012). Two different mutants of *bglu45* displayed slightly higher Klason lignin content as compared to their WT, but no differences were observed for *bglu46* and *bglu47* mutants or overexpressing lines. It should be noted that as far as we know, no double or triple knockout *bglu* mutants were produced, so the hypothesis of a compensation between these three proteins, or with other *BGLU* expressed in the stem, cannot be excluded. However, no compensation of gene expression was observed in *blgu45* and *bglu46* mutants. In short, while several β-glucosidases active on monolignol glucosides are present in vascular tissues of several species, there are yet no absolute evidence about their involvement in lignification in angiosperms.

4.2 Cellular transport of monolignols and glycosylated monolignols

The molecular mechanisms at the basis of monolignol and/or monolignol glucosides transport to the cell wall is not clearly established but several processes have been highlighted. Monolignol glucosides such as coniferin may be exported to the cell wall where they are hydrolysed by β-glucosidase liberating coniferyl alcohol that is then polymerized into lignin (Terashima, Ko, Matsushita, & Westermark, 2016). Monolignols have been shown to be transported either by active transport and/or passive diffusion to the cell wall where they are oxidized (Boija & Johansson, 2006; Perkins, Smith, & Samuels, 2019; Vermaas et al., 2019). A transport of monolignol glucoside via secretory vesicle and a delivery to the cell wall by exocytosis has also been evidenced (Tsuyama et al., 2019). Transport mechanisms may depend on the type of tissue/organs as well as cellular compartment.

On the one hand, a passive transport of monolignols and dilignols was evidenced by an *in vitro* partitioning assay made on immobilized liposomes and lipid bilayer disks (Boija, Lundquist, Edwards, & Johansson, 2007). Similarly, lignin precursors are transported at low level in the absence of ATP by both plasma and vacuolar membranes suggesting a passive transport (Miao & Liu, 2010). In accordance, *in silico* molecular dynamic simulation predicted that monolignols, but not monolignol glucosides, diffuse passively across PM (Vermaas et al., 2019).

On the other hand, an active transport of coniferaldehyde, coniferyl alcohol, sinapaldehyde and sinapyl alcohol was demonstrated using inverted PM vesicles from *Arabidopsis* leaves or poplar roots in the presence of ATP (Miao & Liu, 2010). However, there was no transport activity across PM vesicles for coniferin or syringin. Instead, monolignol glucosides but not monolignol aglycones were transported by vacuolar vesicles membrane prepared from *Arabidopsis* tissues in the presence of ATP (Miao & Liu, 2010).

The transport of *p*-glucocoumaryl alcohol across the tonoplast and endomembrane system has been suggested to be mediated by a proton gradient formed by vacuolar-type H^+-ATPase in differentiating xylem of both gymnosperm and angiosperms woody species (Tsuyama et al., 2013, 2019). Väisänen et al. (2020) showed that microsomal fractions prepared from developing xylem of Norway spruce has a H^+-gradient-dependent tonoplast localized transport of *p*-glucocoumaryl alcohol and coniferin. MFS or MATE tonoplast-located H^+/coniferin antiporters were proposed to be candidate transporters of coniferin and *p*-glucocoumaryl alcohol to the vacuole.

The transport of dilignols, such as a conjugate of sinapyl alcohol and *p*-coumaric acid produced in the cytoplasm by a transferase, towards the apoplast has been proposed in grass lignin (Hatfield et al., 2009; Hatfield, Ralph, & Grabber, 2008).

4.3 Roles of glycosylation of monolignols during lignification

The involvement of monolignol glucosides in lignification has been frequently proposed. However, whether their role is crucial as real intermediates in lignin biosynthesis or nonessential as storage form of monolignols in excess is still a matter of debate. Monolignol glucosides, mainly coniferin, have been found to accumulate in cambial tissue and differentiating xylem, mostly in conifers but also in angiosperms (Aoki et al., 2016; Terashima et al., 2016; Tsuyama & Takabe, 2014), but their role as lignin precursors is not clearly defined. By injecting radio-labeled glycosylated monolignols in stems of gymnosperms (reviewed by Terashima, 2007) and angiosperms (Rolando, Daubresse, Pollet, Jouanin, & Lapierre, 2004; Terashima et al., 2016; Tsuji & Fukushima, 2004), it was shown that they can be incorporated into the lignin polymer.

Overexpression of transcription factors *MYB58* and *MYB63* inducing the expression of monolignol biosynthesis genes in *Arabidopsis* (Zhou, Lee, Zhong, & Ye, 2009) results in dwarfed plantlets, together with strongly increased coniferin and syringin content in leaves, suggesting that these glucosides are synthesized when the pool of monolignols reaches critically high concentrations (Perkins et al., 2020). Co-overexpression of *MYB58* or *MYB63* with *LAC17* (involved in lignin polymerization) essentially alleviates the dwarfed phenotypes and restores WT coniferin and syringin levels, further suggesting that in angiosperms, glucosylation occurs when supply of monolignols by the biosynthesis pathway exceeds their consumption by sink compartments (e.g., for cell wall lignification). Accordingly, a strong increase in coniferin was observed in the core stem part of flax *lignified bast fiber 1* mutant, where a gene homolog to *Arabidopsis LAC11* involved in lignification was downregulated, possibly indicating lower capacity to polymerize monolignol into lignin (Chantreau et al., 2014). RNAi-mediated silencing of a laccase partaking in poplar lignification increases the coniferin content of developing xylem, further supporting monolignol glucosylation as a strategy to stabilize monolignols when lignin polymerization is impaired (Bryan et al., 2016).

High amounts of coniferin have been detected in the vacuoles of xylem of gymnosperms (Morikawa, Yoshinaga, Kamitakahara, Wada, & Takabe, 2010). In Japanese cypress, coniferin content peaks in the early stage of SCW formation and decreases when the cell wall formation proceeds, suggesting that coniferin is a precursor for lignification (Morikawa et al., 2010; Tsuyama & Takabe, 2014). In contrast, coniferyl alcohol, sinapyl alcohol and syringin were not detected in differentiating xylem of this gymnosperm. The vacuolar location of coniferin argues for a release (either through vesicles or after tonoplast disruption during programmed cell death) and hydrolysis by wall-bound β-glucosidases before incorporation of coniferyl alcohol into lignin. A similar analysis made in poplar showed that only a limited amount of coniferin was detected in differentiating xylem, whereas sinapyl alcohol was detected in the xylem, syringin in the phloem and coniferyl alcohol could not be detected in the analyzed samples (Tsuyama & Takabe, 2014). These differences are thought to be linked to a quicker consumption of coniferyl alcohol than sinapyl alcohol and may be related to the fact that vessels, containing mainly G units, encounter rapid cell death, whereas fibers live longer and accumulate S units in their lignin. The large difference between pools of monolignol glucosides between gymnosperms and evolved angiosperms may result from the specificities of their respective vascular systems (Terashima et al., 2016). Actually, gymnosperms rely on tracheids for both sap transport and plant support, while in angiosperms these two functions are ensured by vessels and fibers, respectively. Tracheids formed during spring and beginning of summer are referred to as earlywood and they are characterized by large diameters and thin cell walls (Rahman et al., 2016), in contrast with latewood. Yet, earlywood must resist to turgor pressure and support the stem, requiring quick supply of monolignols for lignification. The large pool of monolignol glucosides stored in the vacuole of cells of cambial zone and early differentiating xylem (Tsuyama & Takabe, 2014) aims at providing the building block for fast lignification of cell corners and compound middle lamella in newly formed tracheids (Terashima et al., 2016). In addition, wood of angiosperms contains higher proportion of parenchyma as compared to gymnosperms (>20% vs less than 10% for gymnosperms; Esau, 1953; Morris et al., 2016), and this cell type contributes to lignification of adjacent vessels and fibers, according to the neighbor hypothesis (Smith et al., 2013). Earlywood tracheids should therefore accumulate monolignols under their glycosylated forms to ease their storage in the vacuole before *post-mortem* hydrolysis and polymerization into lignin. The enzymatic activity of a pine UGT extract glucosylating coniferyl alcohol, sinapyl alcohol, coniferaldehyde and sinapaldehyde purified from cambium and developing xylem between May and September confirms this dynamic of monolignol glucosides in gymnosperms (Steeves, Förster, Pommer, & Savidge, 2001). This activity indeed increases from May to mid-June (corresponding to earlywood formation) and gradually decreases to reach a minimum corresponding to a near absence of activity at mid-September. Coniferin and syringin are subsequently hydrolysed through the action of β-glucosidases located in the differentiating xylem (Dharmawardhana et al., 1995). Altogether, studies performed on gymnosperms strongly suggest that the turnover of monolignol glucosides is a fundamental process to regulate the lignification of cell walls.

Acknowledgments

M. Behr is a post-doctoral researcher supported by Belgian Fonds de la Recherche Scientifique (FRS-FNRS) and M. Baucher is a senior research associate of the FRS-FNRS. This work was supported by the research project T.0068.18 (FRS-FNRS). Figure 1 was drawn using BioRender.

References

Albenne, C., Canut, H., & Jamet, E. (2013). Plant cell wall proteomics: The leadership of *Arabidopsis thaliana*. *Frontiers in Plant Science*, *4*, 111.

Anders, N., Wilkinson, M. D., Lovegrove, A., Freeman, J., Tryfona, T., Pellny, T. K., et al. (2012). Glycosyl transferases in family 61 mediate arabinofuranosyl transfer onto xylan in grasses. *Proceedings of the National Academy of Sciences*, *109*(3), 989–993.

Aoki, D., Hanaya, Y., Akita, T., Matsushita, Y., Yoshida, M., Kuroda, K., et al. (2016). Distribution of coniferin in freeze-fixed stem of *Ginkgo biloba* L. by cryo-TOF-SIMS/SEM. *Scientific Reports*, *6*(1), 31525.

Babst, B. A., Chen, H. Y., Wang, H. Q., Payyavula, R. S., Thomas, T. P., Harding, S. A., et al. (2014). Stress-responsive hydroxycinnamate glycosyltransferase modulates phenylpropanoid metabolism in *Populus*. *Journal of Experimental Botany*, *65*(15), 4191–4200.

Baker, M. R., Tabb, D. L., Ching, T., Zimmerman, L. J., Sakharov, I. Y., & Li, Q. X. (2016). Site-specific *N*-glycosylation characterization of windmill palm tree peroxidase using novel tools for analysis of plant glycopeptide mass spectrometry data. *Journal of Proteome Research*, *15*(6), 2026–2038.

Barakat, A., Winter, H., Rondeau-Mouro, C., Saake, B., Chabbert, B., & Cathala, B. (2007). Studies of xylan interactions and cross-linking to synthetic lignins formed by bulk and end-wise polymerization: A model study of lignin carbohydrate complex formation. *Planta*, *226*(1), 267.

Behr, M., El Jaziri, M., & Baucher, M. (2021). Lignin: An innovative, complex, and highly flexible plant material/component. In H. A. Santos, & P. Figueiredo (Eds.), *Lignin-based materials for biomedical applications - preparation, characterization and implementation* (pp. 35–60). Deans, Matthew: Elsevier.

Behr, M., Guerriero, G., Grima-Pettenati, J., & Baucher, M. (2019). A molecular blueprint of lignin repression. *Trends in Plant Science*, *24*(11), 1052–1064.

Beihammer, G., Maresch, D., Altmann, F., Van Damme, E. J. M., & Strasser, R. (2021). Lewis a glycans are present on proteins involved in cell wall biosynthesis and appear evolutionarily conserved among natural *Arabidopsis thaliana* accessions. *Frontiers in Plant Science*, *12*, 114.

Boerjan, W., Ralph, J., & Baucher, M. (2003). Lignin biosynthesis. *Annual Review of Plant Biology*, *54*(1), 519–546.

Boija, E., & Johansson, G. (2006). Interactions between model membranes and lignin-related compounds studied by immobilized liposome chromatography. *Biochimica et Biophysica Acta (BBA) - Biomembranes*, *1758*(5), 620–626.

Boija, E., Lundquist, A., Edwards, K., & Johansson, G. (2007). Evaluation of bilayer disks as plant cell membrane models in partition studies. *Analytical Biochemistry*, *364*(2), 145–152.

Bowles, D., Lim, E.-K., Poppenberger, B., & Vaistij, F. E. (2006). Glycosyltransferases of lipophilic small molecules. *Annual Review of Plant Biology*, *57*(1), 567–597.

Brett, C. T. (2000). Cellulose microfibrils in plants: Biosynthesis, deposition, and integration into the cell wall. *International Review of Cytology*, *199*, 161–199.

Brill, E., van Thournout, M., White, R. G., Llewellyn, D., Campbell, P. M., Engelen, S., et al. (2011). A novel isoform of sucrose synthase is targeted to the cell wall during secondary cell wall synthesis in cotton fiber. *Plant Physiology*, *157*(1), 40–54.

Brown, C., Leijon, F., & Bulone, V. (2012). Radiometric and spectrophotometric *in vitro* assays of glycosyltransferases involved in plant cell wall carbohydrate biosynthesis. *Nature Protocols*, 7(9), 1634–1650.

Brown, D. M., Zeef, L. A. H., Ellis, J., Goodacre, R., & Turner, S. R. (2005). Identification of novel genes in Arabidopsis involved in secondary cell wall formation using expression profiling and reverse genetics. *The Plant Cell*, *17*(8), 2281–2295.

Brown, D. M., Zhang, Z., Stephens, E., Dupree, P., & Turner, S. R. (2009). Characterization of IRX10 and IRX10-like reveals an essential role in glucuronoxylan biosynthesis in Arabidopsis. *The Plant Journal*, *57*(4), 732–746.

Bryan, A. C., Jawdy, S., Gunter, L., Gjersing, E., Sykes, R., Hinchee, M. A. W., et al. (2016). Knockdown of a laccase in *Populus deltoides* confers altered cell wall chemistry and increased sugar release. *Plant Biotechnology Journal*, *14*(10), 2010–2020.

Busse-Wicher, M., Gomes, T. C. F., Tryfona, T., Nikolovski, N., Stott, K., Grantham, N. J., et al. (2014). The pattern of xylan acetylation suggests xylan may interact with cellulose microfibrils as a twofold helical screw in the secondary plant cell wall of *Arabidopsis thaliana*. *The Plant Journal*, *79*(3), 492–506.

Cantarel, B. L., Coutinho, P. M., Rancurel, C., Bernard, T., Lombard, V., & Henrissat, B. (2009). The carbohydrate-active EnZymes database (CAZy): An expert resource for Glycogenomics. *Nucleic Acids Research*, *37*(suppl_1), D233–D238.

Chantreau, M., Portelette, A., Dauwe, R., Kiyoto, S., Crônier, D., Morreel, K., et al. (2014). Ectopic lignification in the flax *lignified bast fiber1* mutant stem is associated with tissue-specific modifications in gene expression and cell wall composition. *The Plant Cell*, *26*(11), 4462–4482.

Chapelle, A., Morreel, K., Vanholme, R., Le-Bris, P., Morin, H., Lapierre, C., et al. (2012). Impact of the absence of stem-specific ß-glucosidases on lignin and monolignols. *Plant Physiology*, *160*(3), 1204–1217.

Chen, Y., Dong, W., Tan, L., Held, M. A., & Kieliszewski, M. J. (2015). Arabinosylation plays a crucial role in extensin cross-linking *in vitro*. *Biochemistry Insights*, *8s2(8)*, BCI. S31353.

Coleman, H. D., Yan, J., & Mansfield, S. D. (2009). Sucrose synthase affects carbon partitioning to increase cellulose production and altered cell wall ultrastructure. *Proceedings of the National Academy of Sciences*, *106*(31), 13118–13123.

Cosgrove, D. J. (2014). Re-constructing our models of cellulose and primary cell wall assembly. *Current Opinion in Plant Biology*, *22*, 122–131.

Cosgrove, D. J. (2018). Diffuse growth of plant cell walls. *Plant Physiology*, *176*(1), 16–27.

Dahiya, P., Findlay, K., Roberts, K., & McCann, M. C. (2005). A fasciclin-domain containing gene, *ZeFLA11*, is expressed exclusively in xylem elements that have reticulate wall thickenings in the stem vascular system of *Zinnia elegans* cv envy. *Planta*, *223*(6), 1281–1291.

Davis, J., Brandizzi, F., Liepman, A. H., & Keegstra, K. (2010). Arabidopsis mannan synthase CSLA9 and glucan synthase CSLC4 have opposite orientations in the Golgi membrane. *The Plant Journal*, *64*(6), 1028–1037.

Dharmawardhana, D. P., Ellis, B. E., & Carlson, J. E. (1995). A [beta]-glucosidase from lodgepole pine xylem specific for the lignin precursor coniferin. *Plant Physiology*, *107*(2), 331–339.

Dharmawardhana, D. P., Ellis, B. E., & Carlson, J. E. (1999). cDNA cloning and heterologous expression of coniferin β-glucosidase. *Plant Molecular Biology*, *40*(2), 365–372.

Dong, N. Q., Sun, Y., Guo, T., Shi, C. L., Zhang, Y. M., Kan, Y., et al. (2020). UDP-glucosyltransferase regulates grain size and abiotic stress tolerance associated with metabolic flux redirection in rice. *Nature Communications*, *11*(1), 1–16.

Driouich, A., Lainé, A.-C., Vian, B., & Faye, L. (1992). Characterization and localization of laccase forms in stem and cell cultures of sycamore. *The Plant Journal*, *2*(1), 13–24.

Edwards, M. E., Dickson, C. A., Chengappa, S., Sidebottom, C., Gidley, M. J., & Reid, J. S. G. (1999). Molecular characterisation of a membrane-bound galactosyltransferase of plant cell wall matrix polysaccharide biosynthesis. *The Plant Journal, 19*(6), 691–697.

Endo, S., Pesquet, E., Yamaguchi, M., Tashiro, G., Sato, M., Toyooka, K., et al. (2009). Identifying new components participating in the secondary cell wall formation of vessel elements in *Zinnia* and *Arabidopsis*. *The Plant Cell, 21*(4), 1155–1165.

Esau, K. (1953). *Plant anatomy*. John Wiley.

Escamilla-Treviño, L. L., Chen, W., Card, M. L., Shih, M. C., Cheng, C. L., & Poulton, J. E. (2006). *Arabidopsis thaliana* β-glucosidases BGLU45 and BGLU46 hydrolyse monolignol glucosides. *Phytochemistry, 67*(15), 1651–1660.

Feijao, C., Morreel, K., Anders, N., Tryfona, T., Busse-Wicher, M., Kotake, T., et al. (2021). Hydroxycinnamic acid modified xylan side chains and their cross-linking products in rice cell walls are reduced in the *Xylosyl arabinosyl substitution of xylan 1* mutant. *The Plant Journal*. In press.

Fitchette-Lainé, A. C., Gomord, V., Cabanes, M., Michalski, J. C., Macary, M. S., Foucher, B., et al. (1997). *N*-glycans harboring the Lewis a epitope are expressed at the surface of plant cells. *The Plant Journal, 12*(6), 1411–1417.

Fleming, M. B., Decker, S. R., & Bedinger, P. A. (2016). Investigating the role of extensin proteins in poplar biomass recalcitrance. *BioResources, 11*(2), 4727–4744.

Gabaldón, C., López-Serrano, M., Pedreño, M. A., & Barceló, A. R. (2005). Cloning and molecular characterization of the basic peroxidase isoenzyme from *Zinnia elegans*, an enzyme involved in lignin biosynthesis. *Plant Physiology, 139*(3), 1138–1154.

Gabaldón, C., López-Serrano, M., Pomar, F., Merino, F., Cuello, J., Pedreño, M. A., et al. (2006). Characterization of the last step of lignin biosynthesis in *Zinnia elegans* suspension cell cultures. *FEBS Letters, 580*(18), 4311–4316.

Glazunova, O. A., Moiseenko, K. V., Kamenihina, I. A., Isaykina, T. U., Yaropolov, A. I., & Fedorova, T. V. (2019). Laccases with variable properties from different strains of *Steccherinum ochraceum*: Does glycosylation matter? *International Journal of Molecular Sciences, 20*(8), 2008.

Goubet, F., Barton, C. J., Mortimer, J. C., Yu, X., Zhang, Z., Miles, G. P., et al. (2009). Cell wall glucomannan in Arabidopsis is synthesised by CSLA glycosyltransferases, and influences the progression of embryogenesis. *The Plant Journal, 60*(3), 527–538.

Grantham, N. J., Wurman-Rodrich, J., Terrett, O. M., Lyczakowski, J. J., Stott, K., Iuga, D., et al. (2017). An even pattern of xylan substitution is critical for interaction with cellulose in plant cell walls. *Nature Plants, 3*(11), 859–865.

Grünwald, C., Ruel, K., Kim, Y. S., & Schmitt, U. (2002). On the cytochemistry of cell wall formation in poplar trees. *Plant Biology, 4*(01), 13–21.

Guerriero, G., Behr, M., Legay, S., Mangeot-Peter, L., Zorzan, S., Ghoniem, M., et al. (2017). Transcriptomic profiling of hemp bast fibres at different developmental stages. *Scientific Reports*, 7(1), 4961.

Hao, Z., & Mohnen, D. (2014). A review of xylan and lignin biosynthesis: Foundation for studying Arabidopsis irregular xylem mutants with pleiotropic phenotypes. *Critical Reviews in Biochemistry & Molecular Biology, 49*(3), 212–241.

Hatfield, R. D., Marita, J. M., Frost, K., Grabber, J., Ralph, J., Lu, F., et al. (2009). Grass lignin acylation: *P*-coumaroyl transferase activity and cell wall characteristics of C3 and C4 grasses. *Planta, 229*(6), 1253–1267.

Hatfield, R., Ralph, J., & Grabber, J. H. (2008). A potential role for sinapyl *p*-coumarate as a radical transfer mechanism in grass lignin formation. *Planta, 228*(6), 919–928.

Hengel, A. J. V., Kammen, A. V., & Vries, S. C. D. (2002). A relationship between seed development, arabinogalactan-proteins (AGPs) and the AGP mediated promotion of somatic embryogenesis. *Physiologia Plantarum, 114*(4), 637–644.

Hijazi, M., Velasquez, S. M., Jamet, E., Estevez, J. M., & Albenne, C. (2014). An update on post-translational modifications of hydroxyproline-rich glycoproteins: Toward a model highlighting their contribution to plant cell wall architecture. *Frontiers. Plant Science*, *5*(AUG), 395.

Hoffmann, N., King, S., Samuels, A. L., & McFarlane, H. E. (2021). Subcellular coordination of plant cell wall synthesis. *Developmental Cell*, *56*(7), 933–948.

Jacobowitz, J. R., Doyle, W. C., & Weng, J.-K. (2019). PRX9 and PRX40 are extensin peroxidases essential for maintaining tapetum and microspore cell wall integrity during Arabidopsis anther development. *The Plant Cell*, *31*(4), 848–861.

Jiang, N., Wiemels, R. E., Soya, A., Whitley, R., Held, M., & Faik, A. (2016). Composition, assembly, and trafficking of a wheat xylan synthase complex. *Plant Physiology*, *170*(4), 1999–2023.

Kaneda, M., Rensing, K., & Samuels, L. (2010). Secondary cell wall deposition in developing secondary xylem of poplar. *Journal of Integrative Plant Biology*, *52*(2), 234–243.

Keskiaho, K., Hieta, R., Sormunen, R., & Myllyharju, J. (2007). *Chlamydomonas reinhardtii* has multiple prolyl 4-hydroxylases, one of which is essential for proper cell wall assembly. *The Plant Cell*, *19*(1), 256–269.

Knoch, E., Dilokpimol, A., & Geshi, N. (2014). Arabinogalactan proteins: Focus on carbohydrate active enzymes. *Frontiers. Plant Science*, *5*(JUN), 198.

Koutaniemi, S., Malmberg, H. A., Simola, L. K., Teeri, T. H., & Kärkönen, A. (2015). Norway spruce (*Picea abies*) laccases: Characterization of a laccase in a lignin-forming tissue culture. *Journal of Integrative Plant Biology*, *57*(4), 341–348.

Koutaniemi, S., Toikka, M. M., Kärkönen, A., Mustonen, M., Lundell, T., Simola, L. K., et al. (2005). Characterization of basic *p*-coumaryl and coniferyl alcohol oxidizing peroxidases from a lignin-forming *Picea abies* suspension culture. *Plant Molecular Biology*, *58*(2), 141–157.

Kumar, M., Campbell, L., & Turner, S. (2016). Secondary cell walls: Biosynthesis and manipulation. *Journal of Experimental Botany*, *67*(2), 515–531.

Kumar, S., Kumar, K., Pandey, P., Rajamani, V., Padmalatha, K. V., Dhandapani, G., et al. (2013). Glycoproteome of elongating cotton fiber cells. *Molecular & Cellular Proteomics*, *12*(12), 3677–3689.

Lafarguette, F., Leplé, J. C., Déjardin, A., Laurans, F., Costa, G., Lesage-Descauses, M. C., et al. (2004). Poplar genes encoding fasciclin-like arabinogalactan proteins are highly expressed in tension wood. *New Phytologist*, *164*(1), 107–121.

Lairson, L. L., Henrissat, B., Davies, G. J., & Withers, S. G. (2008). Glycosyltransferases: Structures, functions, and mechanisms. *Annual Review of Biochemistry*, 77(1), 521–555.

Lanot, A., Hodge, D., Jackson, R. G., George, G. L., Elias, L., Lim, E.-K., et al. (2006). The glucosyltransferase UGT72E2 is responsible for monolignol 4- *O* -glucoside production in *Arabidopsis thaliana*. *The Plant Journal*, *48*(2), 286–295.

Lanot, A., Hodge, D., Lim, E. K., Vaistij, F. E., & Bowles, D. J. (2008). Redirection of flux through the phenylpropanoid pathway by increased glucosylation of soluble intermediates. *Planta*, *228*(4), 609–616.

Le Roy, J., Huss, B., Creach, A., Hawkins, S., & Neutelings, G. (2016). Glycosylation is a major regulator of phenylpropanoid availability and biological activity in plants. *Frontiers in Plant Science*, 7, 735.

Lee, C., Teng, Q., Huang, W., Zhong, R., & Ye, Z.-H. (2009). The F8H glycosyltransferase is a functional paralog of FRA8 involved in glucuronoxylan biosynthesis in *Arabidopsis*. *Plant and Cell Physiology*, *50*(4), 812–827.

Lee, C., Teng, Q., Zhong, R., & Ye, Z. H. (2011). The four Arabidopsis *REDUCED WALL ACETYLATION* genes are expressed in secondary wall-containing cells and required for the acetylation of xylan. *Plant and Cell Physiology*, *52*(8), 1289–1301.

Lee, C., Teng, Q., Zhong, R., & Ye, Z.-H. (2012). Arabidopsis GUX proteins are glucuronyltransferases responsible for the addition of glucuronic acid side chains onto xylan. *Plant and Cell Physiology*, *53*(7), 1204–1216.

Lee, C., Zhong, R., Richardson, E. A., Himmelsbach, D. S., McPhail, B. T., & Ye, Z.-H. (2007). The *PARVUS* gene is expressed in cells undergoing secondary wall thickening and is essential for glucuronoxylan biosynthesis. *Plant and Cell Physiology*, *48*(12), 1659–1672.

Lee, C., Zhong, R., & Ye, Z.-H. (2012). Arabidopsis family GT43 members are xylan xylosyltransferases required for the elongation of the xylan backbone. *Plant and Cell Physiology*, *53*(1), 135–143.

Li, S., Bashline, L., Zheng, Y., Xin, X., Huang, S., Kong, Z., et al. (2016). Cellulose synthase complexes act in a concerted fashion to synthesize highly aggregated cellulose in secondary cell walls of plants. *Proceedings of the National Academy of Sciences*, *113*(40), 11348–11353.

Li, Q., Koda, K., Yoshinaga, A., Takabe, K., Shimomura, M., Hirai, Y., et al. (2015). Dehydrogenative polymerization of coniferyl alcohol in artificial polysaccharides matrices: Effects of xylan on the polymerization. *Journal of Agricultural and Food Chemistry*, *63*(18), 4613–4620.

Li, Y., Qian, Q., Zhou, Y., Yan, M., Sun, L., Zhang, M., et al. (2003). *BRITTLE CULM1*, which encodes a COBRA-like protein, affects the mechanical properties of rice plants. *The Plant Cell*, *15*(9), 2020–2031.

Liepman, A. H., & Cavalier, D. M. (2012). The CELLULOSE SYNTHASE-LIKE A and CELLULOSE SYNTHASE-LIKE C families: Recent advances and future perspectives. *Frontiers in Plant Science*, *3*, 109.

Lim, E.-K., Jackson, R. G., & Bowles, D. J. (2005). Identification and characterisation of *Arabidopsis* glycosyltransferases capable of glucosylating coniferyl aldehyde and sinapyl aldehyde. *FEBS Letters*, *579*(13), 2802–2806.

Lim, E. K., Li, Y., Parr, A., Jackson, R., Ashford, D. A., & Bowles, D. J. (2001). Identification of glucosyltransferase genes involved in sinapate metabolism and lignin synthesis in Arabidopsis. *The Journal of Biological Chemistry*, *276*(6), 4344–4349.

Lin, J. S., Huang, X. X., Li, Q., Cao, Y., Bao, Y., Meng, X. F., et al. (2016). UDP-glycosyltransferase 72B1 catalyzes the glucose conjugation of monolignols and is essential for the normal cell wall lignification in *Arabidopsis thaliana*. *The Plant Journal*, *88*(1), 26–42.

Liu, C. J. (2012). Deciphering the enigma of lignification: Precursor transport, oxidation, and the topochemistry of lignin assembly. *Molecular Plant*, *5*(2), 304–317.

Liu, E., MacMillan, C. P., Shafee, T., Ma, Y., Ratcliffe, J., van de Meene, A., et al. (2020). Fasciclin-like arabinogalactan-protein 16 (FLA16) is required for stem development in *Arabidopsis*. *Frontiers in Plant Science*, *11*, 615392.

Liu, L., Shang-Guan, K., Zhang, B., Liu, X., Yan, M., Zhang, L., et al. (2013). Brittle Culm1, a COBRA-like protein, functions in cellulose assembly through binding cellulose microfibrils. *PLoS Genetics*, *9*(8), e1003704.

Lunin, V. V., Wang, H. T., Bharadwaj, V. S., Alahuhta, M., Peña, M. J., Yang, J. Y., et al. (2020). Molecular mechanism of polysaccharide acetylation by the arabidopsis xylan *O*-acetyltransferase XOAT1. *The Plant Cell*, *32*(7), 2367–2382.

Ma, Y., MacMillan, C. P., de Vries, L., Mansfield, S. D., Hao, P., Ratcliffe, J., et al. (2022). FLA11 and FLA12 glycoproteins fine-tune stem secondary wall properties in response to mechanical stresses. *New Phytologist.*, *233*(4), 1750–1767.

MacMillan, C. P., Mansfield, S. D., Stachurski, Z. H., Evans, R., & Southerton, S. G. (2010). Fasciclin-like arabinogalactan proteins: Specialization for stem biomechanics and cell wall architecture in Arabidopsis and Eucalyptus. *The Plant Journal*, *62*(4), 689–703.

MacMillan, C. P., Taylor, L., Bi, Y., Southerton, S. G., Evans, R., & Spokevicius, A. (2015). The fasciclin-like arabinogalactan protein family of *Eucalyptus grandis* contains members that impact wood biology and biomechanics. *New Phytologist, 206*(4), 1314–1327.

Maestre-Reyna, M., Liu, W.-C., Jeng, W.-Y., Lee, C.-C., Hsu, C.-A., Wen, T.-N., et al. (2015). Structural and functional roles of glycosylation in fungal laccase from *Lentinus* sp. *PLoS One, 10*(4), e0120601.

Maloney, V. J., & Mansfield, S. D. (2010). Characterization and varied expression of a membrane-bound endo-β-1,4-glucanase in hybrid poplar. *Plant Biotechnology Journal, 8*(3), 294–307.

Maloney, V. J., Samuels, A. L., & Mansfield, S. D. (2012). The endo-1,4-β-glucanase *Korrigan* exhibits functional conservation between gymnosperms and angiosperms and is required for proper cell wall formation in gymnosperms. *New Phytologist, 193*(4), 1076–1087.

Mansoori, N., Timmers, J., Desprez, T., Kamei, C. L. A., Dees, D. C. T., Vincken, J.-P., et al. (2014). KORRIGAN1 interacts specifically with integral components of the cellulose synthase machinery. *PLoS One, 9*(11), e112387.

Marzol, E., Borassi, C., Bringas, M., Sede, A., Rodríguez Garcia, D. R., Capece, L., et al. (2018). Filling the gaps to solve the extensin puzzle. *Molecular Plant, 11*(5), 645–658.

Meents, M. J., Motani, S., Mansfield, S. D., & Samuels, A. L. (2019). Organization of xylan production in the golgi during secondary cell wall biosynthesis. *Plant Physiology, 181*(2), 527–546.

Meents, M. J., Watanabe, Y., & Samuels, A. L. (2018). The cell biology of secondary cell wall biosynthesis. *Annals of Botany, 121*(6), 1107–1125.

Miao, Y. C., & Liu, C. J. (2010). ATP-binding cassette-like transporters are involved in the transport of lignin precursors across plasma and vacuolar membranes. *Proceedings of the National Academy of Sciences, 107*(52), 22728–22733.

Millar, D. J., Whitelegge, J. P., Bindschedler, L. V., Rayon, C., Boudet, A. M., Rossignol, M., et al. (2009). The cell wall and secretory proteome of a tobacco cell line synthesising secondary wall. *Proteomics, 9*(9), 2355–2372.

Mokshina, N., Gorshkova, T., & Deyholos, M. K. (2014). Chitinase-like and cellulose synthase gene expression in gelatinous-type cellulosic walls of flax (*Linum usitatissimum* L.) bast fibers. *PLoS One, 9*(6), e97949.

Morgan, J. L. W., Strumillo, J., & Zimmer, J. (2013). Crystallographic snapshot of cellulose synthesis and membrane translocation. *Nature, 493*(7431), 181–186.

Morikawa, Y., Yoshinaga, A., Kamitakahara, H., Wada, M., & Takabe, K. (2010). Cellular distribution of coniferin in differentiating xylem of *Chamaecyparis obtusa* as revealed by Raman microscopy. *Holzforschung, 64*(1), 61–67.

Morris, H., Plavcová, L., Cvecko, P., Fichtler, E., Gillingham, M. A. F., Martínez-Cabrera, H. I., et al. (2016). A global analysis of parenchyma tissue fractions in secondary xylem of seed plants. *New Phytologist, 209*(4), 1553–1565.

Naoumkina, M. A., Zhao, Q., Gallego-Giraldo, L., Dai, X., Zhao, P. X., & Dixon, R. A. (2010). Genome-wide analysis of phenylpropanoid defence pathways. *Molecular Plant Pathology, 11*(6), 829–846.

Nguema-Ona, E., Vicré-Gibouin, M., Gotté, M., Plancot, B., Lerouge, P., Bardor, M., et al. (2014). Cell wall *O*-glycoproteins and *N*-glycoproteins: Aspects of biosynthesis and function. *Frontiers in Plant Science, 5*, 499.

Nielsen, K. L., Indiani, C., Henriksen, A., Feis, A., Becucci, M., Gajhede, M., et al. (2001). Differential activity and structure of highly similar peroxidases. Spectroscopic, crystallographic, and enzymatic analyses of lignifying *Arabidopsis thaliana* peroxidase A2 and horseradish peroxidase A2. *Biochemistry, 40*(37), 11013–11021.

Nixon, B. T., Mansouri, K., Singh, A., Du, J., Davis, J. K., Lee, J.-G., et al. (2016). Comparative structural and computational analysis supports eighteen cellulose synthases in the plant cellulose synthesis complex. *Scientific Reports, 6*(1), 28696.

Oda, Y., Iida, Y., Nagashima, Y., Sugiyama, Y., & Fukuda, H. (2015). Novel coiled-coil proteins regulate exocyst association with cortical microtubules in xylem cells via the conserved oligomeric Golgi-complex 2 protein. *Plant and Cell Physiology*, *56*(2), 277–286.

Oikawa, A., Lund, C. H., Sakuragi, Y., & Scheller, H. V. (2013). Golgi-localized enzyme complexes for plant cell wall biosynthesis. *Trends in Plant Science*, *18*(1), 49–58.

Oinonen, P., Zhang, L., Lawoko, M., & Henriksson, G. (2015). On the formation of lignin polysaccharide networks in Norway spruce. *Phytochemistry*, *111*, 177–184.

Peña, M. J., Zhong, R., Zhou, G. K., Richardson, E. A., O'Neill, M. A., Darvill, A. G., et al. (2007). *Arabidopsis irregular xylem8* and *irregular xylem9*: Implications for the complexity of glucuronoxylan biosynthesis. *The Plant Cell*, *19*(2), 549–563.

Perkins, M. L., Schuetz, M., Unda, F., Smith, R. A., Sibout, R., Hoffmann, N. J., et al. (2020). Dwarfism of high-monolignol Arabidopsis plants is rescued by ectopic LACCASE overexpression. *Plant Direct*, *4*(9), e00265.

Perkins, M., Smith, R. A., & Samuels, L. (2019). The transport of monomers during lignification in plants: Anything goes but how? *Current Opinion in Biotechnology*, *56*, 69–74.

Purushotham, P., Ho, R., & Zimmer, J. (2020). Architecture of a catalytically active homotrimeric plant cellulose synthase complex. *Science*, *369*(6507), 1089–1094.

Rahman, M. H., Begum, S., Nakaba, S., Yamagishi, Y., Kudo, K., Nabeshima, E., et al. (2016). Relationship between the earlywood-to-latewood transition and changes in levels of stored starch around the cambium in locally heated stems of the evergreen conifer *Chamaecyparis pisifera*. *Trees*, *30*(5), 1619–1631.

Ralph, J., Quideau, S., Grabber, J. H., & Hatfield, R. D. (1994). Identification and synthesis of new ferulic acid dehydrodimers present in grass cell walls. *Journal of the Chemical Society, Perkin Transactions*, *1*(23), 3485–3498.

Ratke, C., Pawar, P. M.-A., Balasubramanian, V. K., Naumann, M., Duncranz, M. L., Derba-Maceluch, M., et al. (2015). *Populus GT43* family members group into distinct sets required for primary and secondary wall xylan biosynthesis and include useful promoters for wood modification. *Plant Biotechnology Journal*, *13*(1), 26–37.

Rejab, N. A., Nakano, Y., Yoneda, A., Ohtani, M., & Demura, T. (2015). Possible contribution of TED6 and TED7, secondary cell wall-related membrane proteins, to evolution of tracheary element in angiosperm lineage. *Plant Biotechnology*, *32*(4), 343–347.

Rennie, E. A., & Scheller, H. V. (2014). Xylan biosynthesis. *Current Opinion in Biotechnology*, *26*, 100–107.

Rips, S., Bentley, N., Jeong, I. S., Welch, J. L., von Schaewen, A., & Koiwa, H. (2014). Multiple *N*-glycans cooperate in the subcellular targeting and functioning of *Arabidopsis* KORRIGAN1. *The Plant Cell*, *26*(9), 3792–3808.

Rolando, C., Daubresse, N., Pollet, B., Jouanin, L., & Lapierre, C. (2004). Lignification in poplar plantlets fed with deuterium-labelled lignin precursors. *Comptes Rendus Biologies*, *327*(9–10), 799–807.

Ross, J., Li, Y., Lim, E.-K., & Bowles, D. J. (2001). Higher plant glycosyltransferases. *Genome Biology*, *2*(2), 1–6.

Ruel, K., Chevalier-Billosta, V., Guillemin, F., Berrio-Sierra, J., & Joseleau, J.-P. (2006). The wood cell wall at the ultrastructural scale - formation and topochemical organization. *Maderas. Ciencia y Tecnología*, *8*(2), 107–116.

Sánchez-Rodríguez, C., Bauer, S., Hématy, K., Saxe, F., Ibánez, A. B., Vodermaier, V., et al. (2012). CHITINASE-LIKE1/POM-POM1 and its homolog CTL2 are glucan-interacting proteins important for cellulose biosynthesis in Arabidopsis. *The Plant Cell*, *24*(2), 589–607.

Sato, Y., Wuli, B., Sederoff, R., & Whetten, R. (2001). Molecular cloning and expression of eight laccase cDNAs in loblolly pine (*Pinus taeda*). *Journal of Plant Research*, *114*(2), 147–155.

Scheller, H. V., & Ulvskov, P. (2010). Hemicelluloses. *Annual Review of Plant Biology, 61*(1), 263–289.

Schoberer, J., & Strasser, R. (2018). Plant glyco-biotechnology. *Seminars in Cell & Developmental Biology, 80*, 133–141.

Seifert, G. J. (2021). The FLA4-FEI pathway: A unique and mysterious signaling module related to cell wall structure and stress signaling. *Genes, 12*(2), 145.

Shigeto, J., Kiyonaga, Y., Fujita, K., Kondo, R., & Tsutsumi, Y. (2013). Putative cationic cell-wall-bound peroxidase homologues in Arabidopsis, AtPrx2, AtPrx25, and AtPrx71, are involved in lignification. *Journal of Agricultural and Food Chemistry, 61*(16), 3781–3788.

Showalter, A. M. (1993). Structure and function of plant cell wall proteins. *The Plant Cell, 5*(1), 9–23.

Silva, J., Ferraz, R., Dupree, P., Showalter, A. M., & Coimbra, S. (2020). Three decades of advances in arabinogalactan-protein biosynthesis. *Frontiers in Plant Science, 11*, 2014.

Simmons, T. J., Mortimer, J. C., Bernardinelli, O. D., Pöppler, A.-C., Brown, S. P., deAzevedo, E. R., et al. (2016). Folding of xylan onto cellulose fibrils in plant cell walls revealed by solid-state NMR. *Nature Communications*, 7(1), 1–9.

Slabaugh, E., Scavuzzo-Duggan, T., Chaves, A., Wilson, L., Wilson, C., Davis, J. K., et al. (2016). The valine and lysine residues in the conserved FxVTxK motif are important for the function of phylogenetically distant plant cellulose synthases. *Glycobiology, 26*(5), 509–519.

Smith, R. A., Schuetz, M., Roach, M., Mansfield, S. D., Ellis, B., & Samuels, L. (2013). Neighboring parenchyma cells contribute to Arabidopsis xylem lignification, while lignification of interfascicular fibers is cell autonomous. *The Plant Cell, 25*(10), 3988–3999.

Sommer-Knudsen, J., Bacic, A., & Clarke, A. E. (1998). Hydroxyproline-rich plant glycoproteins. *Phytochemistry, 47*(4), 483–497.

Song, W., Mentink, R. A., Henquet, M. G. L., Cordewener, J. H. G., Van Dijk, A. D. J., Bosch, D., et al. (2013). *N*-glycan occupancy of *Arabidopsis N*-glycoproteins. *Journal of Proteomics, 93*, 343–355.

Speeckaert, N., Adamou, N. M., Hassane, H. A., Baldacci-Cresp, F., Mol, A., Goeminne, G., et al. (2020). Characterization of the UDP-glycosyltransferase UGT72 family in poplar and identification of genes involved in the glycosylation of monolignols. *International Journal of Molecular Sciences, 21*(14), 5018.

Steeves, V., Förster, H., Pommer, U., & Savidge, R. (2001). Coniferyl alcohol metabolism in conifers — I. Glucosidic turnover of cinnamyl aldehydes by UDPG: Coniferyl alcohol glucosyltransferase from pine cambium. *Phytochemistry, 57*(7), 1085–1093.

Strasser, R., Bondili, J. S., Vavra, U., Schoberer, J., Svoboda, B., Glössl, J., et al. (2007). A unique β1,3-galactosyltransferase is indispensable for the biosynthesis of *N*-glycans containing Lewis a structures in *Arabidopsis thaliana. The Plant Cell, 19*(7), 2278–2292.

Sultana, N., Florance, H. V., Johns, A., & Smirnoff, N. (2015). Ascorbate deficiency influences the leaf cell wall glycoproteome in *Arabidopsis thaliana. Plant, Cell & Environment, 38*(2), 375–384.

Synek, L., Sekeres, J., & Zarsky, V. (2014). The exocyst at the interface between cytoskeleton and membranes in eukaryotic cells. *Frontiers in Plant Science, 4*, 543.

Tams, J. W., & Welinder, K. G. (1995). Mild chemical deglycosylation of horseradish peroxidase yields a fully active, homogeneous enzyme. *Analytical Biochemistry, 228*(1), 48–55.

Tan, L., Eberhard, S., Pattathil, S., Warder, C., Glushka, J., Yuan, C., et al. (2013). An *Arabidopsis* cell wall proteoglycan consists of pectin and arabinoxylan covalently linked to an arabinogalactan protein. *The Plant Cell, 25*(1), 270–287.

Taylor, N. G., Howells, R. M., Huttly, A. K., Vickers, K., & Turner, S. R. (2003). Interactions among three distinct CesA proteins essential for cellulose synthesis. *Proceedings of the National Academy of Sciences, 100*(3), 1450–1455.

Terashima, N. (2007). Non-destructive approaches to identify the ultrastructure of lignified ginkgo cell walls. *International Journal of Developmental Biology, 1*, 170–177.
Terashima, N., & Fukushima, K. (1989). Biogenesis and structure of macromolecular lignin in the cell wall of tree xylem as studied by microautoradiography. In *Plant Cell Wall polymers* (399th ed., pp. 160–168). American Chemical Society.
Terashima, N., Ko, C., Matsushita, Y., & Westermark, U. (2016). Monolignol glucosides as intermediate compounds in lignin biosynthesis. Revisiting the cell wall lignification and new 13C-tracer experiments with *Ginkgo biloba* and *Magnolia liliiflora. Holzforschung, 70*(9), 801–810.
Terashima, N., Nakashima, J., & Takabe, K. (1998). Proposed structure for protolignin in plant cell walls. In *ACS symposium series: Vol. 697* (pp. 180–193).
Terrett, O. M., & Dupree, P. (2019). Covalent interactions between lignin and hemicelluloses in plant secondary cell walls. *Current Opinion in Biotechnology, 56*, 97–104.
Timell, T. E. (1967). Recent progress in the chemistry of wood hemicelluloses. *Wood Science and Technology, 1*(1), 45–70.
Tobimatsu, Y., & Schuetz, M. (2019). Lignin polymerization: How do plants manage the chemistry so well? *Current Opinion in Biotechnology, 56*, 75–81.
Tsuji, Y., & Fukushima, K. (2004). Behavior of monolignol glucosides in angiosperms. *Journal of Agricultural and Food Chemistry, 52*(25), 7651–7659.
Tsuyama, T., Kawai, R., Shitan, N., Matoh, T., Sugiyama, J., Yoshinaga, A., et al. (2013). Proton-dependent coniferin transport, a common major transport event in differentiating xylem tissue of woody plants. *Plant Physiology, 162*(2), 918–926.
Tsuyama, T., Matsushita, Y., Fukushima, K., Takabe, K., Yazaki, K., & Kamei, I. (2019). Proton gradient-dependent transport of p-glucocoumaryl alcohol in differentiating xylem of woody plants. *Scientific Reports, 9*(1), 1–10.
Tsuyama, T., & Takabe, K. (2014). Distribution of lignin and lignin precursors in differentiating xylem of Japanese cypress and poplar. *Journal of Wood Science, 60*(5), 353–361.
Tsuyama, T., & Takabe, K. (2015). Coniferin β-glucosidase is ionically bound to cell wall in differentiating xylem of poplar. *Journal of Wood Science, 61*(4), 438–444.
Urbanowicz, B. R., Peña, M. J., Ratnaparkhe, S., Avci, U., Backe, J., Steet, H. F., et al. (2012). 4-O-methylation of glucuronic acid in *Arabidopsis* glucuronoxylan is catalyzed by a domain of unknown function family 579 protein. *Proceedings of the National Academy of Sciences of the United States of America, 109*(35), 14253–14258.
Vain, T., Crowell, E. F., Timpano, H., Biot, E., Desprez, T., Mansoori, N., et al. (2014). The cellulase KORRIGAN is part of the cellulose synthase complex. *Plant Physiology, 165*(4), 1521–1532.
Väisänen, E. E., Smeds, A. I., Fagerstedt, K. V., Teeri, T. H., Willför, S. M., & Kärkönen, A. (2015). Coniferyl alcohol hinders the growth of tobacco BY-2 cells and *Nicotiana benthamiana* seedlings. *Planta, 242*(3), 747–760.
Väisänen, E., Takahashi, J., Obudulu, O., Bygdell, J., Karhunen, P., Blokhina, O., et al. (2020). Hunting monolignol transporters: Membrane proteomics and biochemical transport assays with membrane vesicles of Norway spruce. *Journal of Experimental Botany, 71*(20), 6379–6395.
Vanholme, B., Vanholme, R., Turumtay, H., Goeminne, G., Cesarino, I., Goubet, F., et al. (2014). Accumulation of *N*-acetylglucosamine oligomers in the plant cell wall affects plant architecture in a dose-dependent and conditional manner. *Plant Physiology, 165*(1), 290–308.
Vega-Sánchez, M. E., Verhertbruggen, Y., Scheller, H. V., & Ronald, P. C. (2013). Abundance of mixed linkage glucan in mature tissues and secondary cell walls of grasses. *Plant Signaling and Behavior, 8*(2), e2314. https://doi.org/10.4161/Psb.23143.
Vermaas, J. V., Dixon, R. A., Chen, F., Mansfield, S. D., Boerjan, W., Ralph, J., et al. (2019). Passive membrane transport of lignin-related compounds. *Proceedings of the National Academy of Sciences of the United States of America, 116*(46), 23117–23123.

Wang, Y., Chantreau, M., Sibout, R., & Hawkins, S. (2013). Plant cell wall lignification and monolignol metabolism. *Frontiers in Plant Science, 4*(220).

Wang, H., Jiang, C., Wang, C., Yang, Y., Yang, L., Gao, X., et al. (2015). Antisense expression of the fasciclin-like arabinogalactan protein *FLA6* gene in *Populus* inhibits expression of its homologous genes and alters stem biomechanics and cell wall composition in transgenic trees. *Journal of Experimental Botany, 66*(5), 1291–1302.

Wang, H., Jin, Y., Wang, C., Li, B., Jiang, C., Sun, Z., et al. (2017). Fasciclin-like arabinogalactan proteins, PtFLAs, play important roles in GA-mediated tension wood formation in *Populus*. *Scientific Reports,* 7(1), 6182.

Watanabe, Y., Meents, M. J., McDonnell, L. M., Barkwill, S., Sampathkumar, A., Cartwright, H. N., et al. (2015). Visualization of cellulose synthases in *Arabidopsis* secondary cell walls. *Science, 350*(6257), 198–203.

Wierzbicki, M. P., Maloney, V., Mizrachi, E., & Myburg, A. A. (2019). Xylan in the middle: Understanding xylan biosynthesis and its metabolic dependencies toward improving wood fiber for industrial processing. *Frontiers in Plant Science, 10*, 176.

Xie, T., Liu, Z., & Wang, G. (2020). Structural basis for monolignol oxidation by a maize laccase. *Nature Plants, 6*(3), 231–237.

Yang, M., Fehl, C., Lees, K. V., Lim, E. K., Offen, W. A., Davies, G. J., et al. (2018). Functional and informatics analysis enables glycosyltransferase activity prediction. *Nature Chemical Biology, 14*(12), 1109–1117.

Yang, C.-H., Wang, C., Singh, S., Fan, N., Liu, S., Zhao, L., et al. (2021). Golgi-localised manganese transporter PML3 regulates *Arabidopsis* growth through modulating Golgi glycosylation and cell wall biosynthesis. *New Phytologist, 231*(6), 2200–2214.

Yu, L., Lyczakowski, J. J., Pereira, C. S., Kotake, T., Yu, X., Li, A., et al. (2018). The patterned structure of galactoglucomannan suggests it may bind to cellulose in seed mucilage. *Plant Physiology, 178*(3), 1011–1026.

Zeng, W., Lampugnani, E. R., Picard, K. L., Song, L., Wu, A.-M., Farion, I. M., et al. (2016). Asparagus IRX9, IRX10, and IRX14A are components of an active xylan backbone synthase complex that forms in the Golgi apparatus. *Plant Physiology, 171*(1), 93–109.

Zhang, X., Dominguez, P. G., Kumar, M., Bygdell, J., Miroshnichenko, S., Sundberg, B., et al. (2018). Cellulose synthase stoichiometry in aspen differs from Arabidopsis and Norway spruce. *Plant Physiology, 177*(3), 1096–1107.

Zhang, Y., Held, M. A., Kaur, D., & Showalter, A. M. (2021). CRISPR-Cas9 multiplex genome editing of the hydroxyproline- *O-* galactosyltransferase gene family alters arabinogalactan-protein glycosylation and function in *Arabidopsis*. *BMC Plant Biology, 21*(1), 1–16.

Zhang, Y., Held, M. A., & Showalter, A. M. (2020). Elucidating the roles of three β-glucuronosyltransferases (GLCATs) acting on arabinogalactan-proteins using a CRISPR-Cas9 multiplexing approach in Arabidopsis. *BMC Plant Biology, 20*(1), 1–20.

Zhang, D., Hrmova, M., Wan, C.-H., Wu, C., Balzen, J., Cai, W., et al. (2004). Members of a new group of chitinase-like genes are expressed preferentially in cotton cells with secondary walls. *Plant Molecular Biology, 54*(3), 353–372.

Zhang, Y., Nikolovski, N., Sorieul, M., Vellosillo, T., McFarlane, H. E., Dupree, R., et al. (2016). Golgi-localized STELLO proteins regulate the assembly and trafficking of cellulose synthase complexes in Arabidopsis. *Nature Communications,* 7(1), 11656.

Zhang, J., Xie, M., Tuskan, G. A., Muchero, W., & Chen, J. G. (2018). Recent advances in the transcriptional regulation of secondary cell wall biosynthesis in the woody plants. *Frontiers in Plant Science, 871*, 1535.

Zhao, X., Dai, X., Gao, L., Guo, L., Zhuang, J., Liu, Y., et al. (2017). Functional analysis of an uridine diphosphate glycosyltransferase involved in the biosynthesis of polyphenolic glucoside in tea plants (*Camellia sinensis*). *Journal of Agricultural and Food Chemistry, 65*(50), 10993–11001.

Zhong, R., Cui, D., Phillips, D. R., & Ye, Z.-H. (2018). A novel rice xylosyltransferase catalyzes the addition of 2-*O*-xylosyl side chains onto the xylan backbone. *Plant and Cell Physiology*, *59*(3), 554–565.

Zhong, R., Cui, D., & Ye, Z.-H. (2018). Members of the DUF231 family are *O*-acetyltransferases catalyzing 2-*O*- and 3-*O*-acetylation of mannan. *Plant and Cell Physiology*, *59*(11), 2339–2349.

Zhong, R., Cui, D., & Ye, Z.-H. (2019a). Secondary cell wall biosynthesis. *New Phytologist*, *221*(4), 1703–1723.

Zhong, R., Cui, D., & Ye, Z. (2019b). Evolutionary origin of O -acetyltransferases responsible for glucomannan acetylation in land plants. *New Phytologist*, *224*(1), 466–479.

Zhong, R., Teng, Q., Haghighat, M., Yuan, Y., Furey, S. T., Dasher, R. L., et al. (2017). Cytosol-localized UDP-xylose synthases provide the major source of UDP-xylose for the biosynthesis of xylan and xyloglucan. *Plant and Cell Physiology*, *58*(1), 156–174.

Zhong, R., & Ye, Z. H. (2015). Secondary cell walls: Biosynthesis, patterned deposition and transcriptional regulation. *Plant and Cell Physiology*, *56*(2), 195–214.

Zhou, J., Lee, C., Zhong, R., & Ye, Z. H. (2009). MYB58 and MYB63 are transcriptional activators of the lignin biosynthetic pathway during secondary cell wall formation in Arabidopsis. *The Plant Cell*, *21*(1), 248–266.

CHAPTER FOUR

Oxidative enzymes in lignification

Natalie Hoffmann[a], Eliana Gonzales-Vigil[a,b], Shawn D. Mansfield[c], and A. Lacey Samuels[d,*]

[a]Department of Cell & Systems Biology, University of Toronto, Toronto, ON, Canada
[b]Department of Biological Sciences, University of Toronto Scarborough, Toronto, ON, Canada
[c]Department of Wood Science, University of British Columbia, Vancouver, BC, Canada
[d]Department of Botany, University of British Columbia, Vancouver, BC, Canada
*Corresponding author: e-mail address: lsamuels@mail.ubc.ca

Contents

Abstract

The final stage of lignin polymerization in plant cell walls requires oxidation of lignin monomers by secreted enzymes, laccases and peroxidases. Identifying the role of oxidative enzymes during lignin deposition has been hindered by large gene families, functional redundancy, and involvement in diverse biological processes. In this chapter, we discuss how the location and composition of lignin in the secondary cell wall could be fine-tuned by localization and substrate specificity of specific oxidative enzymes. We use *Populus trichocarpa* (poplar) as a case study to demonstrate the initial steps for identifying putative lignin-associated oxidative enzymes that could be used for further experimental validation.

Advances in Botanical Research, Volume 104
ISSN 0065-2296
https://doi.org/10.1016/bs.abr.2022.02.004

Abbreviations

LAC laccase
PRX peroxidase

1. Introduction

Secondary cell walls are deposited in specific cell types following cessation of growth, and mainly consist of cellulose, hemicellulose, and the phenolic polymer lignin. Lignin is formed by oxidative coupling of lignin monomers, called monolignols, which are produced in the cytoplasm and exported to the cell wall. In the cell wall, two important classes of enzymes, laccases (LACs) and peroxidases (PRXs), catalyze the oxidation of monolignols into monolignol radicals that then spontaneously crosslink into the lignin polymer. In recent years, the roles that LACs and PRXs play in determining the distribution and timing of lignin deposition have been recognized as increasingly important. Common substrates of LAC and PRX are the major monolignols, *p*-coumaryl, coniferyl, and sinapyl alcohols, that are incorporated into lignin as *p*-hydroxyphenyl (H), guaiacyl (G) or syringyl (S) units, respectively (Ralph et al., 2004). The relative abundance of monolignols that are available to the oxidative enzymes during lignin deposition is an important factor in the relative H/G/S content of a cell wall during development. Studies examining the successive deposition of different types of lignin in cell walls of *Cleome hassleriana* have suggested that the final lignin composition is largely dependent on transcriptional regulation over biosynthetic genes producing the different monolignols (Tobimatsu et al., 2013; Zhuo et al., 2019).

Lignin composition in cell walls is remarkably diverse among species, cell types, regions of the cell wall, and developmental stages, and this may have functional importance for specialized lignified cells in structural support or conductance of water. Angiosperm lignin is composed primarily of G- and S-units, with secondary cell walls of xylem vessel elements being enriched in G-units and the secondary cell walls of fibers being rich in both G- and S-units (Chapple, Vogt, Ellis, & Somerville, 1992; Donaldson, 2001; Joseleau & Ruel, 1997), while gymnosperm xylem lignin is almost exclusively made up of G-units with trace amounts of H-units (Freudenberg, 1965). Specific plant taxa often incorporate non-canonical lignin units and produce a lignin polymer rich in ferulates, coniferylaldehydes, sinapylaldehydes, and acetate, *p*-coumarate, or *p*-hydroxybenzoate moieties (Goacher, Mottiar, & Mansfield, 2021; Hatfield, Rancour, & Marita, 2017; Jacquet, Pollet, Lapierre, Mhamdi, & Rolando, 1995; Ralph et al., 2004). The diversity of the resulting lignin polymers in cell walls indicates that plants not only have flexibility in the biosynthetic pathways producing different phenylpropanoid precursors (Mottiar, Vanholme, Boerjan, Ralph, & Mansfield, 2016), but also have polymerization machinery in the cell wall able to incorporate a wide range of potential subunits into the lignin polymer.

From a developmental standpoint, the timing and location of lignin deposition is highly regulated. Prior to secondary cell wall formation, lignin is initially deposited in high concentrations in the cell corners and middle lamella between adjacent cells (Fukushima & Terashima, 1991; Terashima & Fukushima, 1988) and is rich in G- and H-units (Joseleau & Ruel, 1997). Following deposition of cellulose and hemicellulose, lignin is polymerized within the secondary cell wall polysaccharide matrix. In addition to secondary cell walls, lignin may also be deposited in primary cell walls to act as a physical barrier to diffusion of molecules in endodermal tissues (Naseer et al., 2012), or in response to biotic (Hammerschmidt, 1984; Lee et al., 2019) and abiotic (Cabane, Afif, & Hawkins, 2012; Hawkins & Boudet, 1996) stresses.

2. Lignin deposition depends on oxidative activity of laccases (LACs) and peroxidases (PRXs)

The coupling of monolignols into the lignin polymer occurs in the cell wall following export from the cytoplasm. Early work by Erdtman (1933) proposed that lignin polymerization occurred through an oxidation (dehydrogenation) reaction involving radical intermediates (Erdtman, 1933). Further fundamental *in vitro* work by Freudenberg (1965) showed that following single-electron oxidation, monolignol radicals are estimated to have a half-life of approximately 45 s, as measured by electron spin resonance (Freudenberg, 1965). The high reactivity and short half-life of these radicals promotes bimolecular radical coupling, in which two radicals will covalently bond together. The lignin polymer thus increases in size *via* combination of two radicals: a monolignol radical and a radicalized growing end of a polymer (end-wise polymerization), or bulk polymerization of two radicalized polymers (Demont-Caulet, Lapierre, Jouanin, Baumberger, & Méchin, 2010; Ralph et al., 2004; Syrjänen & Brunow, 2000; Tobimatsu & Schuetz, 2019). Bond formation will quench both radicals, and depending on the structure, the resulting end groups may be radicalized again for further coupling. In addition to coupling with other monolignols or the lignin polymer, monolignol radicals can also react with other phenols or hydroxy groups of sugars in the cell wall (Freudenberg, 1965). For example, in monocots, lignin is covalently bound to the hemicellulose xylan *via* ferulate linkages, thus increasing the complexity of the secondary cell wall matrix (Kang et al., 2019).

The initial oxidation of monolignols or the lignin polymer occurs through the action of secreted oxidative enzymes, laccases and peroxidases. Laccases (LACs; EC 1.10.3.2) are glycosylated, multi-copper oxidoreductases that oxidize phenolic compounds through the reduction of molecular oxygen to water (Fig. 1A; Mayer & Harel, 1979). Class III secreted peroxidases (PRXs; EC 1.11.1.7) are glycosylated, heme-containing oxidoreductases that oxidize phenolic compounds using hydrogen peroxide (H_2O_2) as an electron acceptor (Fig. 1A; Shigeto & Tsutsumi, 2016). Protein *N*-glycosylation is important for the function of both LACs and PRXs, including secretion to the cell wall, protein stability, enzymatic activity, and substrate binding (Gabaldón, Gómez-Ros, Núñez-Flores, Esteban-Carrasco, & Ros Barceló, 2007; Lige, Ma, & van Huystee, 2001; Maestre-Reyna et al., 2015). Both classes of enzymes are encoded by large

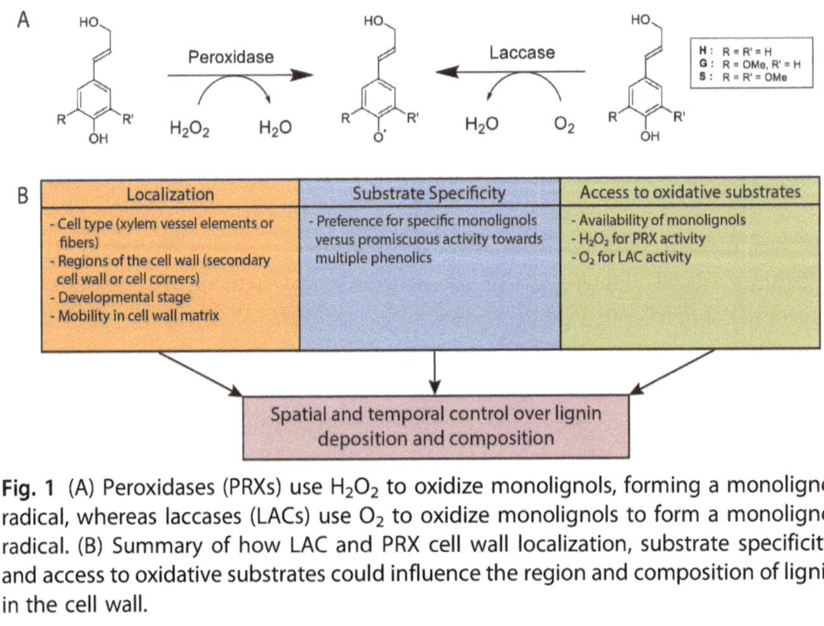

Fig. 1 (A) Peroxidases (PRXs) use H_2O_2 to oxidize monolignols, forming a monolignol radical, whereas laccases (LACs) use O_2 to oxidize monolignols to form a monolignol radical. (B) Summary of how LAC and PRX cell wall localization, substrate specificity, and access to oxidative substrates could influence the region and composition of lignin in the cell wall.

gene families; for example, there are 73 Class III PRXs in *Arabidopsis thaliana* (Arabidopsis) (Tognolli, Penel, Greppin, & Simon, 2002; Welinder et al., 2002), 138 in *Oryza sativa* (Passardi, Longet, Penel, & Dunand, 2004), 93 in *Populus trichocarpa* (Ren et al., 2014), 72 in *Cleome hassleriana* (Wang et al., 2020), and 374 in common wheat (*Triticum aestivum*) (Yan et al., 2019). LACs are also encoded by large gene families, with 17 LACs in Arabidopsis (Turlapati, Kim, Davin, & Lewis, 2011), 30 LACs in *Oryza sativa* (Swetha et al., 2018), 49 LACs in *Populus trichocarpa* (Lu et al., 2013), 95 LAC-like contigs in *Miscanthus sinensis* (He et al., 2019), 19 LACs in *Zea mays* (Xie, Liu, & Wang, 2020), and 29 LACs in *Brachypodium distachyon* (Le Bris et al., 2019). Due to the large number of genes, it is plausible that specific LAC and PRX isoenzymes play distinct roles in cell types, regions of the cell wall, or developmental stages. Identification of the roles of LACs and PRXs is further complicated by the fact that both LACs and PRXs have been shown to be involved in diverse cellular processes other than lignification, with PRXs implicated in response to abiotic and biotic stresses (Bindschedler et al., 2006; Kim, Ciani, & Schachtman, 2010), cell elongation (Liszkay, van der Zalm, & Schopfer, 2004), cell stiffening (Jacobowitz, Doyle, & Weng, 2019; Schopfer, 1996), auxin metabolism (Vatulescu et al., 2004), and seed coat mucilage extrusion (Kunieda et al., 2013). Therefore, identifying LACs and PRXs that are exclusively involved in lignification has been challenging.

Historically, both LACs and PRXs were thought to be involved in secondary cell wall lignification, due to the ability of each class of enzyme to produce a similar lignin-like dehydrogenation polymer (DHP) from monolignols *in vitro* (Freudenberg, 1959). Histochemical analyses by Harkin and Obst (1973) in angiosperm and gymnosperm wood cross-sections studied incorporation of the model phenolic compound syringaldazine, which was not oxidized unless exogenous H_2O_2 was added, leading to the hypothesis that PRXs were exclusively involved in lignification *in planta* and not LACs (Harkin & Obst, 1973). LAC involvement in lignification was later revived following evidence that extracted LACs from loblolly pine (*Pinus taeda* L.) and sycamore maple (*Acer pseudoplatanus* L.) suspension cultures could oxidize monolignols (Bao, O'Malley, Whetten, & Sederoff, 1993; Sterjiades, Dean, & Eriksson, 1992).

Arabidopsis has proven an important experimental model to demonstrate that the presence of lignin-associated oxidative enzymes with monolignol substrates in the cell wall are both sufficient and necessary for lignin deposition, as deletion mutants of LACs and PRXs show reduced lignin content

(Berthet et al., 2011; Rojas-Murcia et al., 2020; Zhao et al., 2013). Moreover, ectopic localization of a LAC to unlignified primary cell walls was sufficient to incorporate exogenously-supplied fluorescent monolignol analogs into the cell wall (Schuetz et al., 2014). LAC involvement in lignification of Arabidopsis was first demonstrated by identification of a mild irregular xylem phenotype in plants with a mutation in *LACCASE4 (AtLAC4/AtIRX12;* Brown, Zeef, Ellis, Goodacre, & Turner, 2005). Berthet et al. (2011) later demonstrated that simultaneous deletion of *AtLAC4* and *AtLAC17* decreased lignin content by 40% and led to an irregular xylem phenotype (Fig. 2; Berthet et al., 2011). Additionally, Arabidopsis triple *lac4-2/lac11/lac17* mutants are dwarfed and have decreased lignin content in stems and roots (Zhao et al., 2013), providing strong genetic evidence for *AtLAC4*, *AtLAC11*, and *AtLAC17* in lignification. An intriguing aspect of the *LAC* gene family is their use of microRNAs (miRNA) for post-transcriptional regulation. miRNA-based regulation of *LAC* expression appears to be a common feature in plants; for example, 7 of 17 Arabidopsis *LACs* (Abdel-Ghany & Pilon, 2008), 29 of 49 *Populus trichocarpa LACs* (Lu et al., 2013), 28 of 30 *Oryza sativa LACs* (Swetha et al., 2018), and 22 of 44 *Gossypium arboretum* (cotton) *LACs* (Balasubramanian, Rai, Thu, Hii, & Mendu, 2016) were predicted to have target sites for either *miR397*, *miR408*, or *miR857*. Interestingly, domesticated rice cultivars (*Oryza sativa*) show increased *LAC* expression and corresponding higher

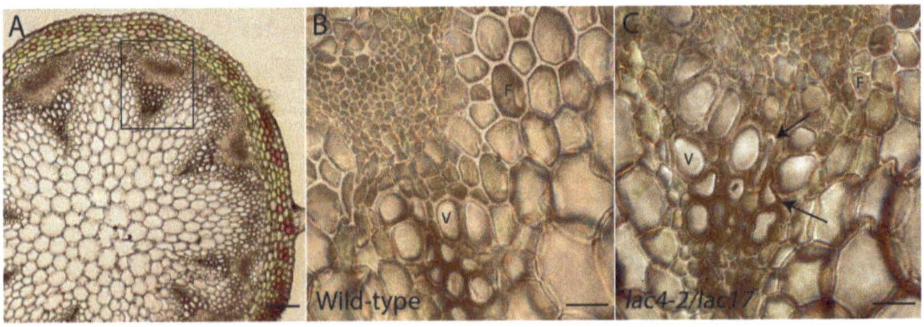

Fig. 2 (A) Unstained cross-section of a 5-week-old wild-type Columbia-0 Arabidopsis stem shows lignified vascular bundles (outlined in black box) connected by lignified interfascicular fibers. (B) Magnified view of a vascular bundle in the wild-type background shows large, expanded lignified xylem vessel elements (V) and lignified fibers (F). (C) Magnified view of a vascular bundle in the *lac4-2/lac17* mutant shows collapsed xylem vessel elements (arrows). Scale bars are 100 μm for A and 20 μm for B–C.

lignin content relative to wild species, thus creating shorter plants with more abundant seeds (Swetha et al., 2018). The increased *LAC* expression in cultivated lines appears to be due to reduced *miRNA397*-induced silencing of LAC genes (Swetha et al., 2018). These results suggest that miRNA regulation of *LAC* expression could have broad implications for plant growth, development, and evolution.

Arabidopsis has also been important in identifying PRXs implicated in lignification; however, unlike the *atlac* mutants, *atprx* mutants have mild phenotypes in lignified stem tissue. Lignin phenotypes have been reported for *AtPRX2* (Shigeto, Kiyonaga, Fujita, Kondo, & Tsutsumi, 2013), *AtPRX4* (Fernández-Pérez, Vivar, Pomar, Pedreño, & Novo-Uzal, 2015), *AtPRX17* (Cosio et al., 2017), *AtPRX25* (Shigeto et al., 2013), *AtPRX52* (Fernández-Pérez, Pomar, Pedreño, & Novo-Uzal, 2015b), *AtPRX64* (Lee, Rubio, Alassimone, & Geldner, 2013), *AtPRX71* (Shigeto et al., 2013, 2015), and *AtPRX72* (Fernández-Pérez, Pomar, Pedreño, & Novo-Uzal, 2015a; Herrero et al., 2013). These studies use a variety of lignin analytical tools and report large variation in lignin reduction in mutant plants, with *prx17, prx25,* and *prx72* mutant plants showing a 35%, 10%, and 58% reduction in stem lignin, respectively (Cosio et al., 2017; Shigeto et al., 2013; Fernández-Pérez, et al., 2015a). This contrasts with the comparably modest 3.5–11% reduction in stem Klason lignin consistently observed in single *atlac* mutants (Berthet et al., 2011). *atprx* mutants also show diverse phenotypic alterations not related to lignification, such as increased epidermal cell size in leaves for *prx71* mutants (Raggi et al., 2015), implying other biological functions in addition to secondary cell wall lignification.

Despite ample evidence for LAC and PRX involvement in lignification in diverse species, there are still unresolved questions that we are just beginning to address, including whether both classes of enzymes are required for lignification in all tissue types (Laitinen et al., 2017; Ros Barceló, 1995; Sterjiades, Dean, Gamble, Himmelsbach, & Eriksson, 1993; Wallace & Fry, 1999), whether LACs and PRXs act in different spatial or temporal locations within the cell wall (Chou et al., 2018; Hoffmann, Benske, Betz, Schuetz, & Samuels, 2020), and the degree to which enzyme activity or substrate specificity impacts lignin composition (Hiraide et al., 2021). In the following sections, we discuss accumulating evidence for how the spatial distribution and composition of lignin could be regulated by the localization, substrate specificity, and activity of specific LACs and PRXs (Fig. 1B).

3. Activity and localization of LACs and PRXs controls lignin deposition in different cell wall regions

3.1 Impact of LAC and PRX localization on lignin deposition in cell types and regions of the cell wall

The monolignols available to the oxidative enzymes in a secondary cell wall are determined by the activity of transcriptional networks in the surrounding cells (Wang et al., 2020; Hussey, 2022). Monolignols are exported from the cytoplasm to the cell wall by sink-driven diffusion (Perkins, Smith, & Samuels, 2019; Perkins et al., 2022). The monolignols in the cell wall are thought to be highly mobile and freely diffusible within the apoplast (Lion et al., 2017; Tobimatsu et al., 2013). From this random and rapid movement of monomers through the polysaccharide matrix of the cell wall, discrete differences in the lignin spatial composition between cell types and regions of the cell wall arise. Therefore, to control the location and composition of polymerizing lignin, cell wall-localized mechanisms are required. Localization studies have indicated that oxidative enzymes are secreted to highly specific cell types and regions of the cell wall, thus creating a model in which immobilized oxidative enzymes dictate lignin deposition spatially and temporally (Fig. 1B).

The localization of LACs and PRXs in the secondary cell wall can be studied by fusing a fluorescent protein to the enzyme, studying its distribution using confocal microscopy, and comparing it to the distribution of lignin (Chou et al., 2018). As a phenolic polymer, lignin naturally autofluoresces under ultraviolet light (Fukazawa, 1992). Historically, this property was used to identify the location as well as relative concentration of lignin in cell walls (Donaldson, 2001; Fergus, Procter, Scott, & Goring, 1969). More recently, imaging fluorescently-tagged enzymes in Arabidopsis has shown that multiple lignin-associated LACs and PRXs co-exist together in lignified cell walls (Hoffmann et al., 2020; Rojas-Murcia et al., 2020). Different lignified tissue types, such as water-conducting xylem vessel elements or supportive fiber cells in Arabidopsis stems, appear to have an overlapping but distinctive set of oxidative enzymes (Hoffmann et al., 2020). Within the lignified cell walls of a given cell type, LACs and PRXs show very specific localization patterns to different regions of the cell wall, with distinct enzymes being found within the pectin-rich middle lamella/cell corners or the cellulose- and xylan-rich thick secondary cell wall (Fig. 3A–B; Chou et al., 2018; Hoffmann et al., 2020). The delineated

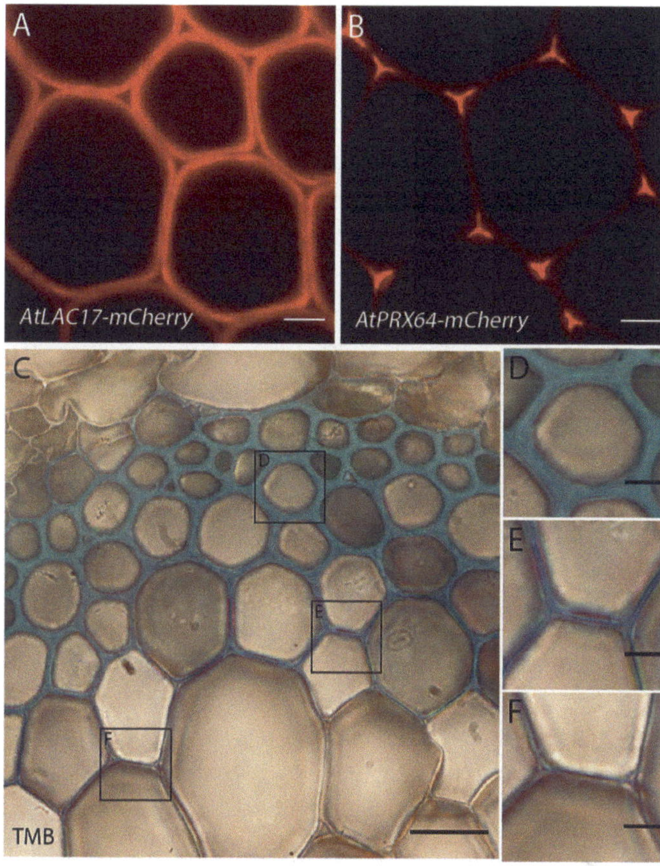

Fig. 3 (A) *prLAC17::LAC17-mCherry* localizes to the cellulose-rich secondary cell wall in interfascicular fibers of 5-week-old Arabidopsis stems. (B) *prPRX64::PRX64-mCherry* localizes to the cell corners and middle lamella between adjacent cells in Arabidopsis interfascicular fibers. (C) Histochemistry of interfascicular fibers in Arabidopsis stems using the substrate 3, 3′, 5, 5′-tetramethylbenzidine (TMB) reveals PRX activity (blue) in all regions of the cell wall. Boxes represent PRX activity in the thick secondary cell wall (D), early lignifying secondary cell wall (E) and cell corners and middle lamella (F). Scale bars for A–B and D–F are 5 μm, and 20 μm for C.

localization pattern of these enzymes not only to certain cell types but also to different cell wall domains is consistent with a specialized role within a certain cell wall region. Immunolocalization of CWPO-C, a lignin-associated PRX in *Populus alba*, detected labelling primarily in the cell corners and middle lamella of fibers and xylem (Sasaki, Baba, Nishida, Tsutsumi, & Kondo, 2006), suggesting that lignin-associated PRXs in woody species also show distinctive localizations to cell types and regions of the cell wall.

In addition to studying the final localization of LACs and PRXs in different cell types and regions of the cell wall, numerous studies have examined how LAC and PRX localization in the secondary cell wall correlates to lignin deposition during development. Early work using radiolabeled monolignol glucosides (Terashima & Fukushima, 1988) and more recent studies with fluorescent monolignol analogs (Lion et al., 2017) demonstrated that there are two distinct periods of coniferyl alcohol incorporation: first in the cell corners and then later in the cellulose-rich secondary cell wall. Interestingly, Wi, Singh, Lee, and Kim (2005) documented a substantial increase in PRX immunolabeling in the middle lamella following initiation of lignification in alfalfa (Wi et al., 2005), suggesting that oxidative enzyme secretion to cell wall domains is coordinated with induction of lignification. This was supported by a study which monitored the localization of fluorescently-tagged LACs and PRXs during stem development in Arabidopsis, with LACs and PRXs being secreted early to cell corners concomitantly with lignification in this region, and then later secretion of other LACs and PRXs to the secondary cell wall along with initiation of lignin deposition (Hoffmann et al., 2020).

As an alternative to fluorescently tagging selected enzymes, it is also possible to visualize overall LAC or PRX oxidative activity using colourimetric substrates or incorporation of fluorescently-tagged monolignol analogs into lignin. The colourimetric substrate 3, 3′, 5, 5′-tetramethylbenzidine (TMB) has been used to visualize PRX activity specifically in lignifying stems of *Zinnia elegans* (Ros-Barceló, 1998) and Arabidopsis (Hoffmann et al., 2020). In Arabidopsis, PRX oxidative activity is seen both within the cell corners/middle lamella as well as the secondary cell wall (Fig. 3C–F), implying a role for PRXs in lignification of both regions of the cell wall. High resolution studies in Japanese cypress (*Chamaecyparis obtusa*) used the substrate diaminobenzidine (DAB) to show LAC activity distributed throughout the middle lamella and the thick secondary cell wall, whereas PRX activity was generally restricted to the lignifying middle lamella in both normal and compression wood (Hiraide et al., 2021). In summary, studies examining oxidative enzyme localization, combined with visualization of enzymatic activity, support the hypothesis that lignin deposition is coordinated by controlled secretion of oxidative enzymes to distinct tissue types and regions of the cell wall.

As LACs and PRXs localize to specific cell types and regions of the cell wall, it is unclear what mechanisms underlie this localization pattern.

One possible mechanism is that the dense secondary cell wall matrix physically impedes diffusion of glycoproteins to different regions. Experimental evidence for the secondary cell wall impeding protein movement was provided by Fluorescence Recovery After Photobleaching (FRAP) experiments in Arabidopsis, where a fluorescently-tagged LAC, AtLAC4-mCherry, was mobile in the unlignified primary cell wall but was immobile in the lignified secondary cell wall (Chou et al., 2018). Therefore, the specific localization of putative lignin-associated LACs and PRXs may be due to either high mobility or the inability to diffuse to other regions of the cell wall following secretion. Alternatively, LACs and PRXs could directly bind to specific cell wall components to position them precisely in cell wall regions. In Arabidopsis, AtPRX36 was shown to bind to specific motifs of demethylesterified pectin in the primary cell wall of seed coats, providing a "platform" for restricted PRX activity in a given location (Francoz et al., 2019). Similarly, a motif of three basic amino acids on a PRX from zucchini facilitated attachment to Ca^{2+}-bound pectin (Carpin et al., 2001). The resolved crystal structure of ZmLAC3 from *Zea mays* showed an extended polyproline-II helix structure, which was hypothesized to play a role in protein-protein interactions within the cell wall (Xie et al., 2020). Together, these protein features may function to anchor or position the protein within particular cell wall domains, thus providing spatial control over where LACs and PRXs localize in the cell wall to facilitate lignin polymerization.

Besides controlling where lignin is deposited, spatial control over the localization of LACs and PRXs could also provide a mechanism to reduce or prevent lignin polymerization in a particular region. For example, the lack of oxidative enzymes, rather than the lack of available monolignol substrates, was hypothesized to be the cause for hypolignification of secondary cell walls in flax (*Linum usitatissimum* L.) bast fibers, as these cell wall layers could not incorporate exogenously-provided fluorescent monolignol analogs (Lion et al., 2017). Remarkably, the middle lamella of lignified tissues in eastern leatherwood (*Dirca palustris*) contains only trace amounts of lignin, unlike the heavily lignified middle lamella in other woody species (Mottiar, Gierlinger, Jeremic, Master, & Mansfield, 2020). It would be interesting to analyze whether the absence of lignin in this region is due to a lack of LACs and PRXs, potentially by quantifying fluorescent monolignol analog incorporation into different regions of the cell wall using tools such as the REPorter Ratiometrics Integrating Segmentation for Analyzing Lignification (REPRISAL) system (Morel et al., 2021).

3.2 Influence of LAC and PRX substrate specificity in fine-tuning lignin composition

In addition to spatial and temporal regulation over the location of lignin polymerization, plant cells can control the final lignin composition in the cell wall by incorporating different types of monolignols into specific cell types and cell wall domains at particular developmental stages (Pandey et al., 2016; Simon et al., 2018). It is unclear the extent to which lignin composition in different tissue types or regions of the cell wall is determined by the availability of monolignols (dependent on biosynthesis within the cell and export to the cell wall), by the substrate specificity of secreted oxidative enzymes, or by a combination of both. *In vitro*, the relative oxidation rate of different phenylpropanoids by a fungal LAC from *Trametes versicolor* was a determining factor for the composition of the final phenolic polymer (Kim & Sattely, 2021), suggesting that lignin compositions could be regulated not only by the types and amounts of monolignols produced by the cell but also the activity and substrate specificity of the oxidative enzymes in the cell wall (Fig. 1B).

Numerous studies have characterized the *in vitro* oxidative activity of LACs and PRXs on various phenolics, including LACs from *Populus tomentosa* (Qin et al., 2020), *Cleome hassleriana* (Wang et al., 2020), *Picea sitchensis* (Hiraide et al., 2021), *Miscanthus sinensis* (He et al., 2019), and *Zea mays* (Xie et al., 2020), as well as PRXs from Arabidopsis (Shigeto, Nagano, Fujita, & Tsutsumi, 2014), *Populus alba* L. (Sasaki et al., 2006), and *Picea abies* (Warinowski et al., 2016). Despite these results, assigning absolute substrate specificities remains challenging, as LACs and PRXs generally show broad substrate specificity toward numerous phenolics *in vitro* (Gumiero, Murphy, Metcalfe, Moody, & Raven, 2010). Recent molecular docking experiments on ChLAC8 from *Cleome hassleriana* (Wang et al., 2020), and elucidation of the crystal structure of ZmLAC3 from *Zea mays* (Xie et al., 2020) showed that the number and position of methoxy groups on the monolignol impacts binding within the enzyme active site, which could ultimately influence the ability of the LAC or PRX to oxidize a given monolignol. These studies provide additional evidence of preferential oxidation of certain monolignols over others.

Differential substrate specificity of oxidative enzymes is one hypothesis used to explain why LACs and PRXs are apparently non-redundant during lignification. This is based on early *in vitro* studies, in which PRXs from sycamore maple (*Acer pseudoplatanus*) showed greater oxidative activity on high-order phenolics compared with LACs (Sterjiades et al., 1993), suggesting that LACs may be responsible for oxidation of low molecular weight compounds while PRXs are required for oxidation of higher order polymers. Additional support for differential contribution of LACs and PRXs came from work in Norway spruce (*Picea abies* L. Karst.) cell culture studies, where cultures with inhibited PRX activity could not produce high-order lignin polymer in the media but rather showed accumulation of low-order phenolics and oligolignols (Laitinen et al., 2017). In contrast, other *in vitro* studies comparing the activity of horseradish PRX and isolated LACs from sycamore maple and *Populus euramericana* on six phenolic compounds showed that while the substrate specificity of the enzymes appeared similar, PRXs showed much higher oxidative activity compared to LACs (Wallace & Fry, 1999). While the relative contribution, substrate specificity, and oxidative activity of LACs and PRXs during lignification is still unresolved, it is likely that contribution to lignin polymerization is dependent on the particular enzyme isoform, and the species, tissue or cell type under analysis.

Although our increasing knowledge of the *in vitro* substrate specificities of these enzymes is important, these results should be interpreted in the context of the secondary cell wall. Important factors include the presence of a complex polysaccharide matrix (Lairez et al., 2005; Terashima et al., 1996), differential concentrations of oxidative enzymes, substrates, and products (Demont-Caulet et al., 2010; Grabber, Hatfield, & Ralph, 2003; Méchin, Baumberger, Pollet, & Lapierre, 2007), as well as pH (Fournand, Cathala, & Lapierre, 2003; Grabber et al., 2003; Terashima et al., 1996), all of which have been shown to impact the final composition of the produced lignin-like polymer. Furthermore, an additional caveat in interpreting these experiments is that despite possessing high oxidative activity toward a particular substrate *in vitro*, the LACs or PRXs embedded in a particular tissue type or cell wall domain may not have access to that substrate *in planta*, depending on monolignol biosynthesis. A striking illustration of contrasting *in vitro* and *in vivo* conditions is seen in a LAC from *Miscanthus sinensis*, MsLAC1, which showed *in vitro* preference for oxidizing sinapyl alcohol over coniferyl alcohol, but overexpression of *MsLAC1* in Arabidopsis led to a significant reduction in S:G ratio (He et al., 2019), indicating MsLAC1 *in planta* may preferentially oxidize coniferyl alcohol over sinapyl alcohol. Therefore, *in planta* characterization of oxidative activity and *in vitro* analyses provide complementary information, with benefits and limitations to each.

An elegant study using Japanese cypress (*Chamaecyparis obtusa*) highlighted the complementary roles of controlled monolignol production and oxidative enzyme substrate specificity to produce a specialized lignin composition in normal and compression wood (Hiraide et al., 2021). Interestingly, LACs isolated from normal wood showed higher oxidative activity for coniferyl alcohol, while LACs from compression wood showed higher activity toward *p*-coumaryl alcohol. These results therefore imply that the observed difference in lignin composition between G-rich lignin in normal wood and H-rich lignin in compression wood could be a combination of alternative biosynthesis of monolignols between tissues as well as selective oxidation in the cell wall. This study contributes to the emerging view that lignin composition is controlled by oxidative enzymes with a given substrate specificity that interact with available monolignols to provide the observed lignin heterogeneity in secondary cell walls.

3.3 LAC and PRX access to oxidative substrates could regulate lignin deposition spatially and temporally

Enzyme activity of lignin-associated LACs and PRXs, which are secreted precisely in space and time during development, can also be modulated by limiting access to oxidative substrates for lignification (Fig. 1B). As PRXs require H_2O_2 to oxidize monolignols, an additional mechanism to control PRX activity in the cell wall could be regulating apoplastic H_2O_2 production in cell walls (Cordoba-Pedregosa, Cordoba, Villalba, & Antonio Gonzalez-Reyes, 2003; Gabaldón et al., 2006; Lee et al., 2013; Ros Barceló, 2005, 1998). This is supported by the results of several studies which have observed accumulation of apoplastic reactive oxygen species (ROS) specifically in lignifying tissues of Arabidopsis (Fujita et al., 2020; Hoffmann et al., 2020), *Zinnia elegans* (Karlsson, Melzer, Prokhorenko, Johansson, & Wingsle, 2005), *Populus euramericana* (Czaninski, 1993), *Populus alba*, *Citrus aurantium*, and *Eucalyptus camaldulensis* (Ros-Barceló, Gómez-Ros, Ferrer, & Hernández, 2006). In addition, scavenging H_2O_2 in cell cultures of Norway spruce (Laitinen et al., 2017) or *Zinnia elegans* (Pesquet et al., 2013) inhibited production of extracellular polymeric lignin. Therefore, spatial restriction of H_2O_2 to specific cell types and locations during development may be a mechanism to regulate the activity of PRXs. The availability of H_2O_2 could also provide an additional explanation for non-redundancy of LACs and PRXs in lignified cell walls. For example, PRX activity could be high in tissues or regions of the cell wall where apoplastic H_2O_2 is produced in abundance, whereas in cell walls with low apoplastic H_2O_2 concentrations, LAC activity could predominate over PRX activity (Hoffmann et al., 2020). Therefore, despite similar localization patterns to regions of the cell wall, LAC and PRX oxidative activity may be non-redundant during lignification due to temporal and spatial differences in H_2O_2 availability.

Collectively, the current data have provided compelling evidence that LAC and PRX localization in cell wall domains, substrate specificity, and access to oxidative substrates all contribute to determining the location, amount, and composition of the resulting lignin polymer. This means that to understand lignification in a given secondary cell wall domain at a certain developmental stage, we must understand which oxidative enzymes are expressed there and determine the substrates that are available. Relatively few studies have taken this holistic, but rewarding, view (Hiraide et al., 2021; Wang et al., 2020).

3.4 Case study in *Populus trichocarpa*: Using *in silico* techniques to identify secondary cell wall-associated LACs and PRXs

Identifying and functionally characterizing the specific LACs and PRXs involved in lignification has been hindered by the occurrence of large gene families with overlapping expression profiles, broad substrate specificities, and functional redundancy (Cosio & Dunand, 2009; Duroux & Welinder, 2003; Shigeto & Tsutsumi, 2016; Turlapati et al., 2011). Consequently, to demonstrate LAC or PRX involvement in lignification, several lines of evidence must be provided. The first step in identifying candidate LACs and PRXs is to exploit available bioinformatic data for evidence of high expression in lignifying tissues and co-expression with other secondary cell wall genes. Use of co-expression data is based on the rationale that genes with similar expression patterns are likely involved in the same biological process (Persson, Wei, Milne, Page, & Somerville, 2005). There are also numerous *in silico* analyses available for PRXs using diverse online resources, including comparisons of protein structure, physicochemical properties, predicted subcellular localization, post-translational modifications, and substrate docking simulations (Lüthje & Ramanathan, 2020) that could also aid in predicting biological role. Following identification of candidate genes, functional characterization of a specific lignin-associated LAC or PRX generally involves experiments demonstrating the ability to oxidize monolignols *in vitro*, localization to lignified cell walls, and alterations in lignin content or structure following reduction of enzyme function (Cosio & Dunand, 2009; Sato et al., 2006). To demonstrate how *in silico* tools are useful to identify putative lignin-associated *LAC* and *PRX* genes, we used publicly available transcriptomic data from *Populus trichocarpa* (poplar; black cottonwood) to identify promising gene candidates for future analyses.

P. trichocarpa was the first tree species to have its genome sequenced and has rapidly become the model species for studying genes involved in wood formation due to its relatively small genome, fast growth, and capacity for genetic transformation (Geraldes et al., 2011; Mellerowicz, Baucher, Sundberg, & Boerjan, 2001; Tuskan et al., 2006). Despite the large amount of genomic and transcriptomic data available for analysis, in-depth functional characterization of lignin-associated LACs and PRXs has not been performed. Previous studies have demonstrated that the poplar genome contains at least 49 *LACs* and 93 *PRXs* (Lu et al., 2013; Ren et al., 2014). To identify LACs and PRXs putatively involved in lignification, we used publicly available transcriptomic data (https://phytozome-next.jgi.doe.gov) to identify *LACs* and *PRXs* that show increased expression in lignified stem tissue compared to young leaf tissue. We identified 31 *LACs* (Table 1) and 36 class III *PRXs* (Table 2) with more than two-fold increased expression in lignified stem tissue compared to young, poorly lignified leaf tissue. To further evaluate differentially expressed *LACs*, we filtered this list to only include highly-expressed *LACs* and *PRXs* with an expression value >10 fragments per kilobase of transcript per million mapped fragments (FPKM) in stem tissues. From this, we narrowed our candidate list to eleven *LACs* and nine *PRXs* (Tables 1 and 2).

Co-expression analyses can provide supporting evidence for oxidative enzymes' involvement in lignification; for example, a major lignin-associated LAC in Arabidopsis, *AtLAC4,* was first identified as playing a role in lignification due to tight co-expression with the secondary cell wall cellulose synthase gene *AtCESA7* (Brown et al., 2005). To further assess the potential involvement of *P. trichocarpa LACs* and *PRXs* in lignification, we utilized the AspWood co-expression database (available through PopGenie; https://popgenie.org) to test whether the candidate eleven *LACs* and nine *PRXs* from poplar are co-expressed with secondary cell wall-related genes, including secondary cell wall cellulose (poplar orthologs of Arabidopsis *AtCESA4*, *AtCESA7*, and *AtCESA8*) and monolignol biosynthetic genes. Since it is possible that some of these enzymes may be involved in modifying cell wall structure not relating to lignification, we also tested whether the candidate genes are co-expressed with primary cell wall genes, including primary cell wall *CESAs* (poplar homologs of Arabidopsis *AtCESA1*, *AtCESA3*, and *AtCESA6*) and pectin biosynthetic enzymes.

Table 1 List of *LAC* genes that show at least two-fold upregulation in poplar stem tissue relative to leaf tissue; selected candidate genes with stem expression >10 FPKM are bolded.

Gene	Young leaf expression	Stem expression	Fold change (stem/leaf)
Potri.008G073700	**0.09**	**33.84**	**398.15**
Potri.006G087100	**0.49**	**27.8**	**56.61**
Potri.009G042500	**0.3**	**19.64**	**66.13**
Potri.008G064000	**0.02**	**18.45**	**838.41**
Potri.001G401300	**3.73**	**17.1**	**4.58**
Potri.006G087500	**0.01**	**13.32**	**1024.62**
Potri.010G193100	**0.13**	**13.21**	**104.01**
Potri.011G120300	**1.8**	**11.47**	**6.39**
Potri.019G088500	**1.8**	**11.27**	**6.25**
Potri.006G097000	**0.49**	**11.04**	**22.39**
Potri.016G112000	**0.46**	**10.83**	**23.39**
Potri.009G034500	0.48	9.05	19.06
Potri.010G183500	0.49	8.13	16.65
Potri.010G183600	0.19	7.73	40.27
Potri.006G097100	0.07	5.03	70.8
Potri.008G073800	1.26	4.92	3.89
Potri.006G096900	0.25	4.62	18.56
Potri.009G156800	0.07	4.35	58.72
Potri.009G156600	0.04	3.41	89.82
Potri.001G054600	0.95	2.45	2.59
Potri.001G184300	0.05	2.37	43.91
Potri.001G341600	0.67	2.31	3.47
Potri.001G248700	0.16	1.8	11.15
Potri.004G156400	0.15	0.66	4.37
Potri.019G088800	0.13	0.43	3.38
Potri.019G088600	0.01	0.36	35.5
Potri.019G088900	0.04	0.33	7.56
Potri.006G094100	0.02	0.26	11.39
Potri.019G088700	0.01	0.17	24
Potri.013G152700	0.01	0.13	18.43
Potri.015G040400	0	0.03	6.5

Expression data (FPKM) was retrieved from Phytozome (https://phytozome-next.jgi.doe.gov).

Table 2 List of *PRX* genes that show at least two-fold upregulation in poplar stem tissue relative to leaf tissue; selected candidate genes with stem expression >10 FPKM are bolded.

Gene	Young leaf expression	Stem expression	Fold change (stem/leaf)
Potri.004G015300	**570.28**	**979.98**	**1.72**
Potri.005G108900	**13**	**159.13**	**12.25**
Potri.017G037900	**13.94**	**53.7**	**3.85**
Potri.003G214700	**1.51**	**37.01**	**24.54**
Potri.001G351000	**13.52**	**30.51**	**2.26**
Potri.003G214800	**3.15**	**21.33**	**6.78**
Potri.004G134800	**9.26**	**21.01**	**2.27**
Potri.005G072800	**0.33**	**16.37**	**49.16**
Potri.013G156500	**0.28**	**10.47**	**37.95**
Potri.004G006400	0.1	8.15	85.75
Potri.002G018000	0.02	6.4	355.61
Potri.T045500	0.16	5.91	36.95
Potri.007G096200	0.65	5.75	8.87
Potri.015G003500	0	5.09	1698
Potri.007G019300	1.12	4.81	4.31
Potri.006G129900	0.06	4.4	68.69
Potri.016G058200	0.18	3	16.86
Potri.017G038100	0.17	2.79	16.32
Potri.015G003600	0.03	2.76	89.06
Potri.005G112200	0.47	2.63	5.57
Potri.011G062300	0.08	2.24	29.12
Potri.006G069600	0.48	2.23	4.63
Potri.016G058800	0.21	2.17	10.31
Potri.004G052100	0.85	1.97	2.33
Potri.018G136900	0.05	0.96	18.75
Potri.014G143200	0.25	0.74	3

Table 2 List of *PRX* genes that show at least two-fold upregulation in poplar stem tissue relative to leaf tissue; selected candidate genes with stem expression >10 FPKM are bolded.—cont'd

Gene	Young leaf expression	Stem expression	Fold change (stem/leaf)
Potri.004G144600	0.1	0.68	6.58
Potri.017G038000	0.31	0.66	2.16
Potri.013G083600	0.13	0.37	2.94
Potri.T106200	0.04	0.31	8.71
Potri.018G131600	0.08	0.25	3.02
Potri.005G135300	0.01	0.16	23.14
Potri.003G214900	0.04	0.09	2.57
Potri.002G031200	0.03	0.08	2.53
Potri.012G042800	0.03	0.06	2.44
Potri.011G027300	0.03	0.06	2.1

Expression data (FPKM) was retrieved from Phytozome (https://phytozome-next.jgi.doe.gov).

Lists of cell wall biosynthetic genes for co-expression analyses were adapted from Bao et al. (2013) and Sundell et al. (2017).

Apart from Potri.019G088500, all poplar *LAC* candidate genes form a highly interconnected expression network with one another and with secondary cell wall *CESAs* (Fig. 4). Interestingly, some of the candidate *LAC* genes are co-expressed with both primary and secondary cell wall *CESAs*. Additionally, candidate *LACs* are co-expressed with monolignol biosynthetic genes but not with pectin biosynthetic genes (Table 3). These results indicate that expression of *LACs* is highly coordinated with the production of, and potentially the secretion of, secondary cell wall components in lignifying cells. In contrast to the close co-expression of *LACs* in lignifying secondary cell walls, candidate *PRXs* show varied co-expression patterns with primary cell wall *CESAs* (Fig. 4) and pectin biosynthetic genes (Table 4). Several candidates are also co-expressed with monolignol biosynthetic genes (Table 4), suggesting that they may also be involved in lignification. Due to the pectin-only co-expression pattern of *PRX* Potri.004G015300 and the lack of co-expression data for *PRX* Potri.017G037900, we removed these as *PRX* candidates. Overall, co-expression analysis indicates tight coordination of *LACs* with secondary cell wall biosynthetic genes in lignifying

Table 3 Summary of poplar *LAC* co-expression analyses from the AspWood database, using cellulose, monolignol, and pectin genes listed in Bao et al. (2013) and Sundell et al. (2017).

LACs	Co-expression
Potri.001G401300	Cellulose (*CESA8-A*)
Potri.006G087100	Cellulose (*CESA4, CESA7-A, CESA7-B, CESA8-A, CESA6-E*); monolignol (*PAL, F5'H, CAD*)
Potri.006G087500	Cellulose (*CESA3-A, CESA8-A*)
Potri.006G097000	Cellulose (*CESA4, CESA8-B*); monolignol (*CAD*)
Potri.008G064000	Monolignol (*PAL*)
Potri.008G073700	Cellulose (*CESA3-A, CESA8-A, CESA7-B*)
Potri.009G042500	Cellulose (*CESA4, CESA8-B*); monolignol (*PAL, HCT, C3'H, CSE, 4CL, CCoAOMT, CAD*)
Potri.010G193100	Monolignol (*PAL, C4'H, C3'H, CSE, 4CL, CCoAOMT, F5'H, COMT, CAD*)
Potri.011G120300	Cellulose (*CESA8-A*)
Potri.016G112000	Cellulose (*CESA4, CESA8-B*); monolignol (*CAD*)
Potri.019G088500	Not included in Sundell et al. (2017)

tissues, while *PRXs* have more broad expression patterns and diverse co-expression networks. The differences in expression patterns for *LACs* and *PRXs* across tissue types were also observed by Sundell et al. (2017), who hypothesized that one of the reasons why *LACs* and *PRXs* are non-redundant in lignifying tissues could be due to non-overlapping expression patterns of these enzymes (Sundell et al., 2017).

To provide additional evidence for the candidate poplar LACs and PRXs' involvement in lignification, we analyzed whether candidate genes are orthologs of well-characterized *LACs* and *PRXs* involved in lignification in Arabidopsis. Numerous putative lignin-related PRXs from Arabidopsis have been identified using homology-based searches, such as AtPRX52, which shows high amino acid similarity to the lignin peroxidase ZePRX in *Zinnia elegans* (Fernández-Pérez, Pomar, et al., 2015b), or AtPRX71, which is homologous to the lignin peroxidase CWPO-C in *Populus alba* (Sasaki et al., 2006; Shigeto et al., 2013). The Maximum Likelihood phylogenetic relationships for the *P. trichocarpa* 75 LACs and 191 class III PRXs and the Arabidopsis protein sequences are shown in Figs. 5 and 6,

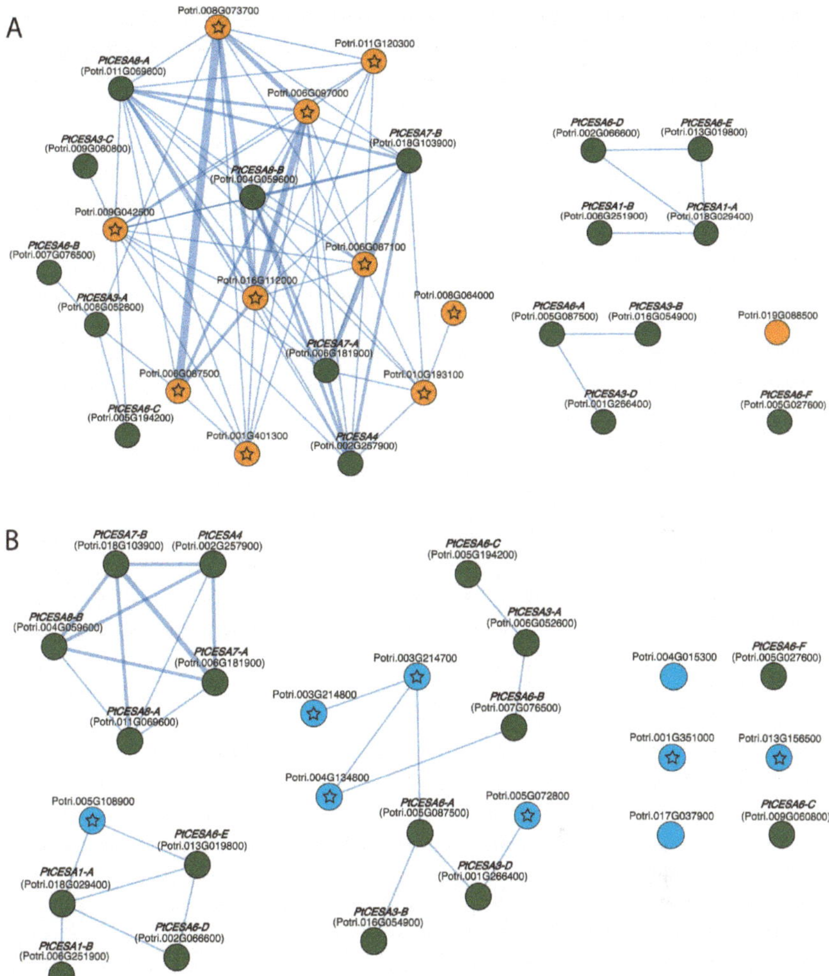

Fig. 4 (A) Co-expression network of selected LAC genes (orange circles) with poplar CESAs, with circles representing genes and lines connecting circles representing significant co-expression (Threshold $\geq$3). (B) Co-expression network of selected PRX genes (blue circles) with poplar CESAs, with circles representing genes and lines connecting circles representing significant co-expression (Threshold $\geq$ 3). Co-expression networks were generated using the AspWood database on Popgenie (https://popgenie.org).

respectively. The Arabidopsis and poplar LAC phylogenetic tree clusters into six clades with high bootstrap support, similar to previously reconstructed Arabidopsis and poplar LAC phylogenies (Fig. 5; Bryan et al., 2016; Lu et al., 2013; McCaig, Meagher, & Dean, 2005; Turlapati et al., 2011). To identify poplar LACs that are putatively involved in lignification, we looked

Table 4 Summary of poplar *PRX* co-expression analyses from the AspWood database, using cellulose, monolignol, and pectin genes listed in Bao et al. (2013) and Sundell et al. (2017).

PRXs	Co-expression
Potri.001G351000	Monolignol (*PAL, 4CL, CCR, F5'H, CAD*); pectin (*GALS2-B*)
Potri.003G214700	Cellulose (*CESA6-A*); monolignol (*4CL*)
Potri.003G214800	Monolignol (*4CL*)
Potri.004G015300	Pectin (*GAUT4-A*)
Potri.004G134800	Cellulose (*CESA6-B*); monolignol (*COMT, CAD*); pectin (*GAUT1-A*)
Potri.005G072800	Cellulose (*CESA3-D*); monolignol (*CCoAOMT*); pectin (*GAUT7-A, GAUT4-A*)
Potri.005G108900	Cellulose (*CESA1-A, CESA6-E*)
Potri.013G156500	Monolignol (*CCR*)
Potri.017G037900	Not included in Sundell et al. (2017)

for gene candidates that showed homology to well-characterized lignin-associated LACs in Arabidopsis. Of the 11 candidate poplar LACs, nine cluster with the lignin-associated AtLAC4 and AtLAC17 (Fig. 5), supporting a similar role for these genes in lignification. Unlike the clade-specific expansion of candidate poplar LACs, the candidate poplar PRXs showed homology to several diverse AtPRXs, including AtPRX3, AtPRX17, AtPRX31, AtPRX42, AtPRX47, AtPRX53, AtPRX64, and AtPRX67 (Fig. 6). Potri.005G108900 clusters with AtPRX64, which is associated with lignification of the Casparian strip in Arabidopsis roots (Lee et al., 2013) and localizes to lignified fiber cells in stem tissues (Chou et al., 2018) and therefore may represent an interesting candidate to further characterize. Based on the results of the transcriptomic, phylogenetic, and co-expression analyses, we propose 10 candidate LACs and seven candidate PRXs for involvement in cell wall biosynthesis, structure, or modification (stars; Figs. 4, 5, 6).

4. Assessment of candidate LACs and PRXs

Despite *in silico* support for involvement in lignification, further experimental work is required to test all candidate poplar LACs and PRXs for their role in lignifying tissues. The most robust test for implicating a gene

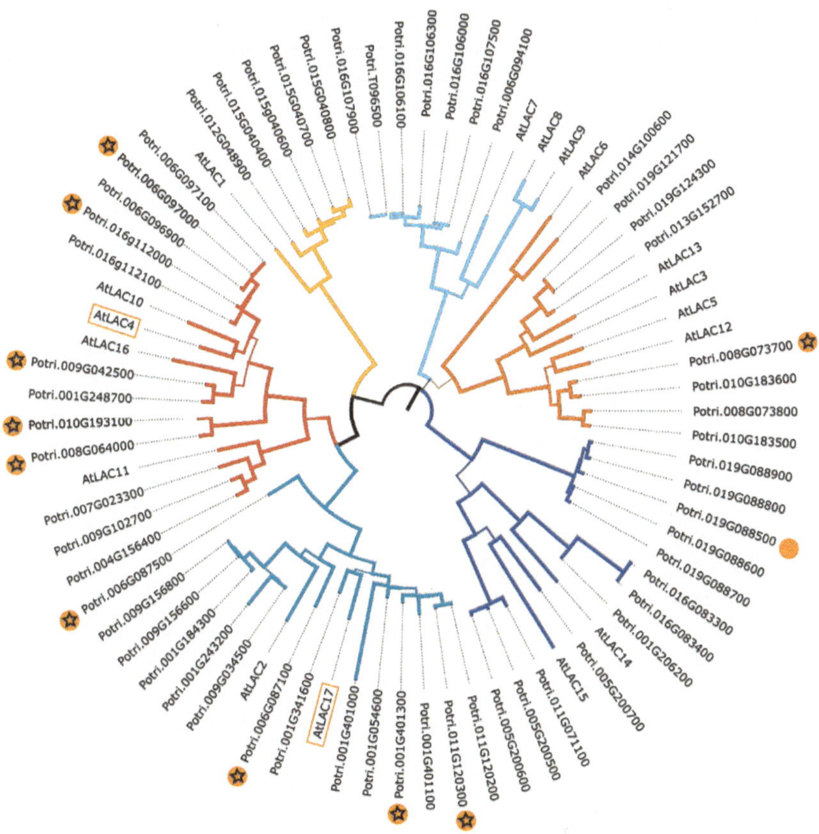

Fig. 5 Maximum-likelihood phylogeny of LACs from poplar and Arabidopsis; selected LACs are identified with an orange circle and final candidate genes are denoted with an orange circle with a star. Phylogenetic trees were reconstructed using MEGAX (Kumar, Stecher, Li, Knyaz, & Tamura, 2018) with 100 bootstrap replicates (bootstrap values are drawn proportional to the width of the branches). Amino acid sequences were aligned using GUIDANCE (http://guidance.tau.ac.il). Final trees were edited using FigTree (http://tree.bio.ed.ac.uk/software/figtree/). AtLAC4 and AtLAC17 are outlined in an orange box for reference.

in lignification is mutant analyses, where knockdown of expression of a *LAC* or *PRX* reduces lignin content or alters lignin composition. There is pre-existing experimental evidence for several of our candidate genes, including Potri.008G073700 (*PtLAC90*) and Potri.006G087100 (*PtLAC110*) in lignifying tissues, as well as Potri.010G1931000 (*PtLAC3*) in maintaining structure and integrity of fiber cell walls (Ranocha et al., 1999, 2002). This indicates that our *in silico* approach is suitable for identifying putative

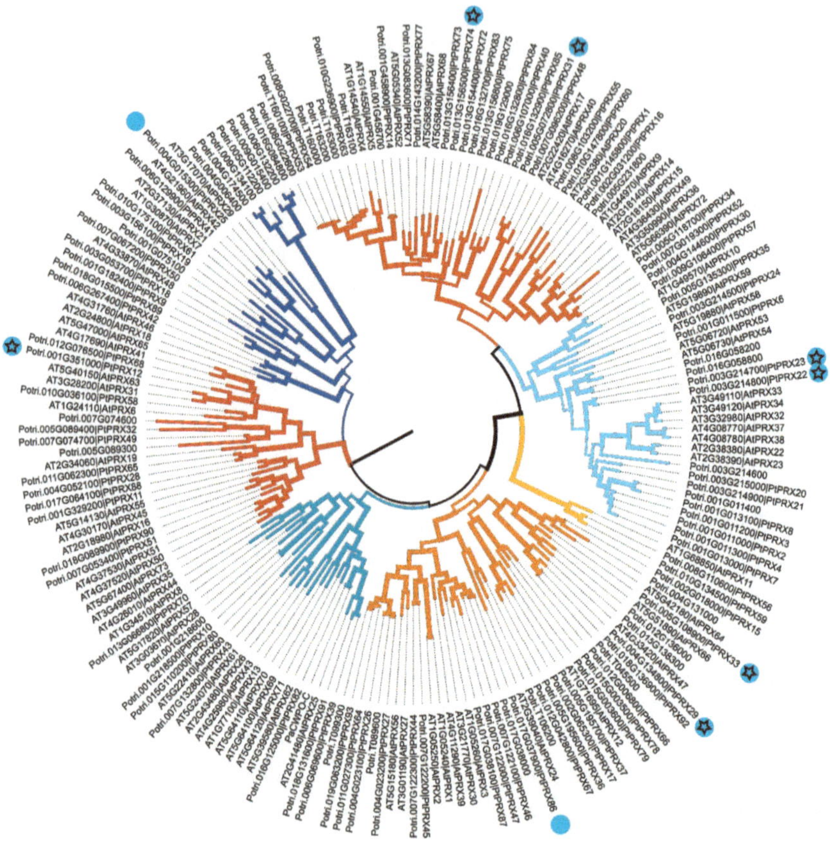

Fig. 6 Maximum-likelihood phylogeny of PRXs from poplar and Arabidopsis; selected PRXs are identified with a blue circle and final candidate genes are denoted with a blue circle with a star. Phylogenetic trees were reconstructed using MEGAX (Kumar et al., 2018) with 100 bootstrap replicates (bootstrap values are drawn proportional to the width of the branches). Amino acid sequences were aligned using GUIDANCE (http://guidance.tau.ac.il). Final trees were edited using FigTree (http:/tree.bio.ed.ac.uk/software/figtree/)

oxidative enzymes involved in lignification. Knockdown of *LAC2* from *Populus deltoides* (*PdLAC2*), which is the closest homolog to one of the candidate LACs in *P. trichocarpa* (Potri.008G06400), also led to a significant increase in S:G ratio and an altered soluble phenolic profile in transgenic plants (Bryan et al., 2016). Interestingly, the observed change in accumulated phenolics in *PdLAC2* mutants is similar to knockdown of the candidate gene Potri.010G1931000 (*PtLAC3*) in *P. trichocarpa* (Ranocha et al., 2002), implying that changing the polymerization of monolignols in the cell wall may feed back into intracellular metabolite pools. The combination of *in silico* analyses as well as the previous mutant results thus strongly implicate a lignin-associated role for several LACs in *P. trichocarpa*.

A caveat to our analyses is that the short-list includes only highly expressed genes in lignified stem tissue; however, it is possible that genes with low expression also may play a role in lignification. For example, although they were upregulated in lignified stem tissue relative to young leaf tissue, we did not include the LAC Potri.001G248700 (*PtLAC16*) or the PRX Potri.006G129900 (*PtrPO21*) in our candidate short-list due to overall low expression in stem tissues (<5 FPKM; Tables 1 and 2). Despite this, mutant analyses showed that knockout poplar transgenic lines of *PtLAC16* had slight but significant decreases to lignin content and cell wall thickness in xylem tissue (Liu et al., 2021), and knockdown lines of *PtrPO21* showed decreased height and reduced stem lignin content (Lin et al., 2016), implying that *LACs* and *PRXs* with low expression may also play a role in lignin polymerization.

Identification of lignin-associated genes in poplar will not only increase our understanding of lignification of secondary cell walls but is also relevant for the economic use of lignocellulosic biomass from trees. Currently, efficient processing of wood is hindered by the presence of lignin, which is largely resistant to enzymatic and chemical degradation. Previous studies in multiple species have demonstrated that down-regulation or mutation of LACs or PRXs can reduce lignin content or alter lignin composition (Berthet et al., 2011; Blee et al., 2003; Li, Kajita, Kawai, Katayama, & Morohoshi, 2003; Lu et al., 2013), which could aid in more efficient processing of wood. For example, overexpression of the *Populus tomentosa LAC14* (*PtoLAC14*) caused reduced growth, increased lignin content and decreased S:G ratio, while CRISPR/Cas9 knockout lines of *PtoLAC14* had increased growth, reduced cell wall thickness, and a higher S:G ratio (Qin et al., 2020), suggesting a role for *PtoLAC14* in oxidation of coniferyl alcohol. The closest homolog of *PtoLAC14* in *P. trichocarpa*, Potri.006 G096900, shows overall low expression in both xylem and leaf tissues but does show higher expression in xylem (Table 1), thus making it another potential lignin-associated gene in *P. trichocarpa*. Additionally, mutation of *BdLAC5* (Bradi1g74320) and *BdLAC8* (Bradi2g23370) in *Brachypodium distachyon* did not lead to any growth phenotypes but reduced stem lignin (20–30%) within the interfascicular fibers (Le Bris et al., 2019). The S:G ratio in double mutants was also significantly increased (Le Bris et al., 2019), suggesting that BdLAC5 and BdLAC8 show specificity for coniferyl alcohol over sinapyl alcohol. Importantly, saccharification of the double *bdlac5/bdlac8* mutant was substantially increased relative to wild-type plants (Le Bris et al., 2019). Therefore, genetic manipulation of LAC genes may be a useful way to improve biomass processing without imposing a large growth yield. BdLAC5 shows highest amino acid similarity to our poplar LAC candidate Potri.016G112000, while BdLAC8 shows highest similarity to another poplar candidate LAC, Potri.001G401300, suggesting that the poplar genes may play a similar role in lignification of xylem tissue. The selected gene candidates in poplar identified using publicly-available *in silico* resources therefore represent interesting targets to support genetic modification of lignified secondary cell walls.

5. Conclusions and future outlook

Lignin biosynthesis and deposition in the cell wall is a highly regulated and complex process. Emerging evidence indicates that secreted oxidative enzymes play a role in determining where and when lignin is deposited in the cell wall, as well as influencing the final polymer composition. This appears to be dependent on enzyme anchoring in specific regions of the cell wall, enzyme substrate specificity, and the availability of H_2O_2, in the case of PRXs. Due to the large number of oxidative enzymes and multiple biological roles these enzymes play, several diverse tools are required to identify putative lignin-associated LACs and PRXs. Use of bioinformatic or homology-based approaches to identify gene candidates should be followed by *in planta* characterization to confirm involvement in lignification, specifically testing whether altered enzyme function changes the amount or composition of lignin and other traits important for industrial treatment, such as saccharification. A more complete understanding of the specific LACs and PRXs involved in lignin polymerization in economically-important woody species will not only lead to a better understanding of the biological role of these oxidative enzymes, but could also guide the production of genetically engineered trees with decreased lignin content that are more amenable to industrial processing.

References

Abdel-Ghany, S. E., & Pilon, M. (2008). MicroRNA-mediated systemic down-regulation of copper protein expression in response to low copper availability in Arabidopsis. *Journal of Biological Chemistry*, *283*(23), 15932–15945. https://doi.org/10.1074/jbc.M801406200.

Balasubramanian, V. K., Rai, K. M., Thu, S. W., Hii, M. M., & Mendu, V. (2016). Genome-wide identification of multifunctional laccase gene family in cotton (*Gossypium* spp.); expression and biochemical analysis during fiber development. *Scientific Reports*, *6*(1), 34309. https://doi.org/10.1038/srep34309.

Bao, H., Li, E., Mansfield, S. D., Cronk, Q. C., El-Kassaby, Y. A., & Douglas, C. J. (2013). The developing xylem transcriptome and genome-wide analysis of alternative splicing in *Populus trichocarpa* (black cottonwood) populations. *BMC Genomics*, *14*(1), 359. https://doi.org/10.1186/1471-2164-14-359.

Bao, W., O'Malley, D. M., Whetten, R., & Sederoff, R. R. (1993). A laccase associated with lignification in loblolly pine xylem. *Science*, *260*(5108), 672–674. https://doi.org/10.1126/science.260.5108.672.

Berthet, S., Demont-Caulet, N., Pollet, B., Bidzinski, P., Cézard, L., Le Bris, P., et al. (2011). Disruption of *LACCASE4* and *17* results in tissue-specific alterations to lignification of *Arabidopsis thaliana* stems. *The Plant Cell*, *23*(3), 1124–1137. https://doi.org/10.1105/tpc.110.082792.

Bindschedler, L. V., Dewdney, J., Blee, K. A., Stone, J. M., Asai, T., Plotnikov, J., et al. (2006). Peroxidase-dependent apoplastic oxidative burst in Arabidopsis required for pathogen resistance. *The Plant Journal*, *47*(6), 851–863. https://doi.org/10.1111/j.1365-313X.2006.02837.x.

Blee, K. A., Choi, J. W., O'Connell, A. P., Schuch, W., Lewis, N. G., & Bolwell, G. P. (2003). A lignin-specific peroxidase in tobacco whose antisense suppression leads to vascular tissue modification. *Phytochemistry*, *64*(1), 163–176. https://doi.org/10.1016/S0031-9422(03)00212-7.

Brown, D. M., Zeef, L. A. H., Ellis, J., Goodacre, R., & Turner, S. R. (2005). Identification of novel genes in Arabidopsis involved in secondary cell wall formation using expression profiling and reverse genetics. *The Plant Cell*, *17*(8), 2281–2295. https://doi.org/10.1105/tpc.105.031542.

Bryan, A. C., Jawdy, S., Gunter, L., Gjersing, E., Sykes, R., Hinchee, M. A. W., et al. (2016). Knockdown of a laccase in *Populus deltoides* confers altered cell wall chemistry and increased sugar release. *Plant Biotechnology Journal*, *14*(10), 2010–2020. https://doi.org/10.1111/pbi.12560.

Cabane, M., Afif, D., & Hawkins, S. (2012). Lignins and abiotic stresses. In L. Jouanin, & C. Lapierre (Eds.), *Advances in Botanical Research* (pp. 219–262). Elsevier.

Carpin, S., Crèvecoeur, M., de Meyer, M., Simon, P., Greppin, H., & Penel, C. (2001). Identification of a Ca^{2+}-pectate binding site on an apoplastic peroxidase. *The Plant Cell*, *13*(3), 511–520. https://doi.org/10.1105/tpc.13.3.511.

Chapple, C. C., Vogt, T., Ellis, B. E., & Somerville, C. R. (1992). An Arabidopsis mutant defective in the general phenylpropanoid pathway. *The Plant Cell*, *4*(11), 1413–1424. https://doi.org/10.1105/tpc.4.11.1413.

Chou, E., Schuetz, M., Hoffmann, N., Watanabe, Y., Sibout, R., & Samuels, A. L. (2018). Distribution, mobility, and anchoring of lignin-related oxidative enzymes in Arabidopsis secondary cell walls. *Journal of Experimental Botany*. https://doi.org/10.1093/jxb/ery067.

Cordoba-Pedregosa, M., Cordoba, F., Villalba, J. M., & Antonio Gonzalez-Reyes, J. (2003). Differential distribution of ascorbic acid, peroxidase activity, and hydrogen peroxide along the root axis in *Allium cepa* L. and its possible relationship with cell growth and differentiation. *Protoplasma*, *221*(1–2), 57–65. https://doi.org/10.1007/s00709-002-0069-9.

Cosio, C., & Dunand, C. (2009). Specific functions of individual class III peroxidase genes. *Journal of Experimental Botany*, *60*(2), 391–408. https://doi.org/10.1093/jxb/ern318.

Cosio, C., Ranocha, P., Francoz, E., Burlat, V., Zheng, Y., Perry, S. E., et al. (2017). The class III peroxidase PRX 17 is a direct target of the MADS-box transcription factor AGAMOUS-LIKE15 (AGL15) and participates in lignified tissue formation. *New Phytologist*, *213*(1), 250–263. https://doi.org/10.1111/nph.14127.

Czaninski, Y. (1993). Cytochemical localization of hydrogen peroxide in lignifying cell walls. *Annals of Botany*, *72*(6), 547–550. https://doi.org/10.1006/anbo.1993.1143.

Demont-Caulet, N., Lapierre, C., Jouanin, L., Baumberger, S., & Méchin, V. (2010). Arabidopsis peroxidase-catalyzed copolymerization of coniferyl and sinapyl alcohols: Kinetics of an endwise process. *Phytochemistry*, *71*(14–15), 1673–1683. https://doi.org/10.1016/j.phytochem.2010.06.011.

Donaldson, L. A. (2001). Lignification and lignin topochemistry — An ultrastructural view. *Phytochemistry*, *57*(6), 859–873. https://doi.org/10.1016/S0031-9422(01)00049-8.

Duroux, L., & Welinder, K. G. (2003). The peroxidase gene family in plants: A phylogenetic overview. *Journal of Molecular Evolution*, *57*(4), 397–407. https://doi.org/10.1007/s00239-003-2489-3.

Erdtman, H. (1933). Dehydrierungen in der Coniferylreihe. II. Dehydrodi-isoeugenol. *Justus Liebig's Annalen Der. Chem*, *503*(1), 283–294. https://doi.org/10.1002/jlac.19335030116.

Fergus, B. J., Procter, A. R., Scott, J. A. N., & Goring, D. A. I. (1969). The distribution of lignin in sprucewood as determined by ultraviolet microscopy. *Wood Science and Technology*, *3*(2), 117–138. https://doi.org/10.1007/BF00639636.

Fernández-Pérez, F., Pomar, F., Pedreño, M. A., & Novo-Uzal, E. (2015a). Suppression of Arabidopsis peroxidase 72 alters cell wall and phenylpropanoid metabolism. *Plant Science*, *239*, 192–199. https://doi.org/10.1016/j.plantsci.2015.08.001.

Fernández-Pérez, F., Pomar, F., Pedreño, M. A., & Novo-Uzal, E. (2015b). The suppression of AtPrx52 affects fibers but not xylem lignification in Arabidopsis by altering the proportion of syringyl units. *Physiologia Plantarum*, *154*(3), 395–406. https://doi.org/10.1111/ppl.12310.

Fernández-Pérez, F., Vivar, T., Pomar, F., Pedreño, M. A., & Novo-Uzal, E. (2015). Peroxidase 4 is involved in syringyl lignin formation in *Arabidopsis thaliana*. *Journal of Plant Physiology*, *175*, 86–94. https://doi.org/10.1016/j.jplph.2014.11.006.

Fournand, D., Cathala, B., & Lapierre, C. (2003). Initial steps of the peroxidase-catalyzed polymerization of coniferyl alcohol and/or sinapyl aldehyde capillary zone electrophoresis study of pH effect. *Phytochemistry*, *62*(2), 139–146. https://doi.org/10.1016/S0031-9422(02)00573-3.

Francoz, E., Ranocha, P., Le Ru, A., Martinez, Y., Fourquaux, I., Jauneau, A., et al. (2019). Pectin demethylesterification generates platforms that anchor peroxidases to remodel plant cell wall domains. *Developmental Cell*, *48*(2), 261–276.e8. https://doi.org/10.1016/j.devcel.2018.11.016.

Freudenberg, K. (1959). Biosynthesis and constitution of lignin. *Nature*, *183*(4669), 1152–1155. https://doi.org/10.1038/1831152a0.

Freudenberg, K. (1965). Lignin: Its constitution and formation from *p*-hydroxycinnamyl alcohols. *Science*, *148*(3670), 595–600. https://doi.org/10.1126/science.148.3670.595.

Fujita, S., De Bellis, D., Edel, K. H., Köster, P., Andersen, T. G., Schmid-Siegert, E., et al. (2020). SCHENGEN receptor module drives localized ROS production and lignification in plant roots. *The EMBO Journal*, *39*(9), e103894. https://doi.org/10.15252/embj.2019103894.

Fukazawa, K. (1992). Ultraviolet microscopy. In S. Y. Lin, & C. W. Dence (Eds.), *Methods in lignin chemistry* (pp. 110–121). Springer Series in Wood Science. https://doi.org/10.1007/978-3-642-74065-7_8.

Fukushima, K., & Terashima, N. (1991). Heterogeneity in formation of lignin. XIV. Formation and structure of lignin in differentiating xylem of *Ginkgo biloba*. *Holzforschung*, *45*(2), 87–94. https://doi.org/10.1515/hfsg.1991.45.2.87.

Gabaldón, C., Gómez-Ros, L. V., Núñez-Flores, M. J., Esteban-Carrasco, A., & Ros Barceló, A. (2007). Post-translational modifications of the basic peroxidase isoenzyme from *Zinnea elegans*. *Plant Molecular Biology*, *65*, 43–61. https://doi.org/10.1007/s11103-007-9197-0.

Gabaldón, C., López-Serrano, M., Pomar, F., Merino, F., Cuello, J., Pedreño, M. A., et al. (2006). Characterization of the last step of lignin biosynthesis in *Zinnia elegans* suspension cell cultures. *FEBS Letters*, *580*(18), 4311–4316. https://doi.org/10.1016/j.febslet.2006.06.088.

Geraldes, A., Pang, J., Thiessen, N., Cezard, T., Moore, R., Zhao, Y., et al. (2011). SNP discovery in black cottonwood (*Populus trichocarpa*) by population transcriptome resequencing. *Molecular Ecology Resources*, *11*, 81–92. https://doi.org/10.1111/j.1755-0998.2010.02960.x.

Goacher, R. E., Mottiar, Y., & Mansfield, S. D. (2021). ToF-SIMS imaging reveals that *p*-hydroxybenzoate groups specifically decorate the lignin of fibers in the xylem of poplar and willow. *Holzforschung*, *75*(5), 452–462. https://doi.org/10.1515/hf-2020-0130.

Grabber, J. H., Hatfield, R. D., & Ralph, J. (2003). Apoplastic pH and monolignol addition rate effects on lignin formation and cell wall degradability in maize. *Journal of Agricultural and Food Chemistry*, *51*(17), 4984–4989. https://doi.org/10.1021/jf030027c.

Gumiero, A., Murphy, E. J., Metcalfe, C. L., Moody, P. C. E., & Raven, E. L. (2010). An analysis of substrate binding interactions in the heme peroxidase enzymes: A structural perspective. *Archives of Biochemistry and Biophysics*, *500*(1), 13–20. https://doi.org/10.1016/j.abb.2010.02.015.

Hammerschmidt, R. (1984). Rapid deposition of lignin in potato tuber tissue as a response to fungi non-pathogenic on potato. *Physiological Plant Pathology*, *24*(1), 33–42. https://doi.org/10.1016/0048-4059(84)90071-7.

Harkin, J. M., & Obst, J. R. (1973). Lignification in trees: Indication of exclusive peroxidase participation. *Science*, *180*(4083), 296–298. https://doi.org/10.1126/science.180.4083.296.

Hatfield, R. D., Rancour, D. M., & Marita, J. M. (2017). Grass cell walls: A story of cross-linking. *Frontiers in Plant Science*, 7. https://doi.org/10.3389/fpls.2016.02056.

Hawkins, S., & Boudet, A. (1996). Wound-induced lignin and suberin deposition in a woody angiosperm (*Eucalyptus gunnii* hook.): Histochemistry of early changes in young plants. *Protoplasma*, *191*(1–2), 96–104. https://doi.org/10.1007/BF01280829.

He, F., MacHemer-Noonan, K., Golfier, P., Unda, F., Dechert, J., Zhang, W., et al. (2019). The *in vivo* impact of MsLAC1, a *Miscanthus* laccase isoform, on lignification and lignin composition contrasts with its *in vitro* substrate preference. *BMC Plant Biology*, *19*(552). https://doi.org/10.1186/s12870-019-2174-3.

Herrero, J., Fernández-Pérez, F., Yebra, T., Novo-Uzal, E., Pomar, F., Pedreño, M.Á., et al. (2013). Bioinformatic and functional characterization of the basic peroxidase 72 from *Arabidopsis thaliana* involved in lignin biosynthesis. *Planta*, *237*(6), 1599–1612. https://doi.org/10.1007/s00425-013-1865-5.

Hiraide, H., Tobimatsu, Y., Yoshinaga, A., Lam, P. Y., Kobayashi, M., Matsushita, Y., et al. (2021). Localised laccase activity modulates distribution of lignin polymers in gymnosperm compression wood. *New Phytologist*, *230*(6), 2186–2199. https://doi.org/10.1111/nph.17264.

Hoffmann, N., Benske, A., Betz, H., Schuetz, M., & Samuels, A. L. (2020). Laccases and peroxidases co-localize in lignified secondary cell walls throughout stem development. *Plant Physiology*, *184*, 806–822. https://doi.org/10.1104/pp.20.00473.

Hussey, S. G. (2022). Transcriptional regulation of secondary cell wall formation and lignification. *Advances in Botanical Research*, *104*, 317–361. Elsevier.

Jacobowitz, J. R., Doyle, W. C., & Weng, J.-K. (2019). PRX9 and PRX40 are extensin peroxidases essential for maintaining tapetum and microspore cell wall integrity during Arabidopsis anther development. *The Plant Cell*, *31*(4), 848–861. https://doi.org/10.1105/tpc.18.00907.

Jacquet, G., Pollet, B., Lapierre, C., Mhamdi, F., & Rolando, C. (1995). New ether-linked ferulic acid-coniferyl alcohol dimers identified in grass straws. *Journal of Agricultural and Food Chemistry*, *43*(10), 2746–2751. https://doi.org/10.1021/jf00058a037.

Joseleau, J. P., & Ruel, K. (1997). Study of lignification by noninvasive techniques in growing maize internodes. *Plant Physiology*, *114*(3), 1123–1133. https://doi.org/10.1104/pp.114.3.1123.

Kang, X., Kirui, A., Dickwella Widanage, M. C., Mentink-Vigier, F., Cosgrove, D. J., & Wang, T. (2019). Lignin-polysaccharide interactions in plant secondary cell walls revealed by solid-state NMR. *Nature Communications*, *10*(1), 347. https://doi.org/10.1038/s41467-018-08252-0.

Karlsson, M., Melzer, M., Prokhorenko, I., Johansson, T., & Wingsle, G. (2005). Hydrogen peroxide and expression of hipI-superoxide dismutase are associated with the development of secondary cell walls in *Zinnia elegans*. *Journal of Experimental Botany*, *56*(418), 2085–2093. https://doi.org/10.1093/jxb/eri207.

Kim, M. J., Ciani, S., & Schachtman, D. P. (2010). A peroxidase contributes to ROS production during Arabidopsis root response to potassium deficiency. *Molecular Plant*, *3*(2), 420–427. https://doi.org/10.1093/mp/ssp121.

Kim, S. S., & Sattely, E. S. (2021). Dirigent proteins guide asymmetric heterocoupling for the synthesis of complex natural product analogues. *Journal of the American Chemical Society*. https://doi.org/10.1021/jacs.0c13164. jacs.0c13164.

Kumar, S., Stecher, G., Li, M., Knyaz, C., & Tamura, K. (2018). MEGA X: Molecular evolutionary genetics analysis across computing platforms. *Molecular Biology and Evolution*, *35*, 1547–1549. https://doi.org/10.1093/molbev/msy096.

Kunieda, T., Shimada, T., Kondo, M., Nishimura, M., Nishitani, K., & Hara-Nishimura, I. (2013). Spatiotemporal secretion of PEROXIDASE36 is required for seed coat mucilage extrusion in Arabidopsis. *The Plant Cell*, *25*(4), 1355–1367. https://doi.org/10.1105/tpc.113.110072.

Lairez, D., Cathala, B., Monties, B., Bedos-Belval, F., Duran, H., & Gorrichon, L. (2005). Aggregation during coniferyl alcohol polymerization in pectin solution: A biomimetic approach of the first steps of lignification. *Biomacromolecules*, *6*(2), 763–774. https://doi.org/10.1021/bm049390y.

Laitinen, T., Morreel, K., Delhomme, N., Gauthier, A., Schiffthaler, B., Nickolov, K., et al. (2017). A key role for apoplastic H_2O_2 in Norway spruce phenolic metabolism. *Plant Physiology*, *174*(3), 1449–1475. https://doi.org/10.1104/pp.17.00085.

Le Bris, P., Wang, Y., Barbereau, C., Antelme, S., Cézard, L., Legée, F., et al. (2019). Inactivation of *LACCASE8* and *LACCASE5* genes in *Brachypodium distachyon* leads to severe decrease in lignin content and high increase in saccharification yield without impacting plant integrity. *Biotechnology for Biofuels*, *12*(1), 181. https://doi.org/10.1186/s13068-019-1525-5.

Lee, M., Jeon, H. S., Kim, S. H., Chung, J. H., Roppolo, D., Lee, H., et al. (2019). Lignin-based barrier restricts pathogens to the infection site and confers resistance in plants. *The EMBO Journal*, *38*(23). https://doi.org/10.15252/embj.2019101948.

Lee, Y., Rubio, M. C., Alassimone, J., & Geldner, N. (2013). A mechanism for localized lignin deposition in the endodermis. *Cell*, *153*(2), 402–412. https://doi.org/10.1016/j.cell.2013.02.045.

Li, Y., Kajita, S., Kawai, S., Katayama, Y., & Morohoshi, N. (2003). Down-regulation of an anionic peroxidase in transgenic aspen and its effect on lignin characteristics. *Journal of Plant Research*, *116*(3), 175–182. https://doi.org/10.1007/s10265-003-0087-5.

Lige, B., Ma, S., & van Huystee, R. (2001). The effects of the site-directed removal of N-glycyslation from cationic peanut peroxidase and its function. *Archives of Biochemistry and Biophysics*, *386*(1), 17–24. https://doi.org/10.1006/abbi.2000.2187.

Lin, C., Li, Q., Tunlaya-Anukit, S., Shi, R., Sun, Y., Wang, J. P., et al. (2016). A cell wall-bound anionic peroxidase, PtrPO21, is involved in lignin polymerization in *Populus trichocarpa*. *Tree Genetics & Genomes*, *12*(2), 22. https://doi.org/10.1007/s11295-016-0978-y.

Lion, C., Simon, C., Huss, B., Blervacq, A.-S., Tirot, L., Toybou, D., et al. (2017). BLISS: A bioorthogonal dual-labeling strategy to unravel lignification dynamics in plants. *Cell Chemical Biology*, *24*(3), 326–338. https://doi.org/10.1016/j.chembiol.2017.02.009.

Liszkay, A., van der Zalm, E., & Schopfer, P. (2004). Production of reactive oxygen intermediates ($O2^{\bullet -}$, H_2O_2, and $^{\bullet}OH$) by maize roots and their role in wall loosening and elongation growth. *Plant Physiology*, *136*(2), 3114–3123. https://doi.org/10.1104/pp.104.044784.

Liu, Y., Cao, S., Liu, X., Li, Y., Wang, B., Sun, Y., et al. (2021). PtrLAC16 plays a key role in catalyzing lignin polymerization in the xylem cell wall of *Populus*. *International Journal of Biological Macromolecules*, *188*, 983–992. https://doi.org/10.1016/j.ijbiomac.2021.08.077.

Lu, S., Li, Q., Wei, H., Chang, M.-J., Tunlaya-Anukit, S., Kim, H., et al. (2013). *Ptr-miR397a* is a negative regulator of laccase genes affecting lignin content in Populus trichocarpa. *Proceedings of the National Academy of Sciences*, *110*(26), 10848–10853. https://doi.org/10.1073/pnas.1308936110.

Lüthje, S., & Ramanathan, K. (2020). *In silico* analysis of class III peroxidases: Hypothetical structure, ligand binding sites, posttranslational modifications, and interaction with substrates. In J. Jorrin-Novo, L. Valledor, M. Castillejo, & M. Rey (Eds.), *Plant proteomics. Methods in molecular biology* (2139th ed., pp. 325–339). Springer US.

Maestre-Reyna, M., Liu, W., Jeng, W., Lee, C., Hsu, C., Wen, T., et al. (2015). Structural and functional roles of glycosylation in fungal laccase from *Lentinus* sp. *PLoS One*, *10*(4), e0120601. https://doi.org/10.1371/journal/pone.0120601.

Mayer, A. M., & Harel, E. (1979). Polyphenol oxidases in plants. *Phytochemistry*, *18*(2), 193–215. https://doi.org/10.1016/0031-9422(79)80057-6.

McCaig, B. C., Meagher, R. B., & Dean, J. F. D. (2005). Gene structure and molecular analysis of the laccase-like multicopper oxidase (LMCO) gene family in *Arabidopsis thaliana*. *Planta*, *221*(5), 619–636. https://doi.org/10.1007/s00425-004-1472-6.

Méchin, V., Baumberger, S., Pollet, B., & Lapierre, C. (2007). Peroxidase activity can dictate the *in vitro* lignin dehydrogenative polymer structure. *Phytochemistry*, *68*(4), 571–579. https://doi.org/10.1016/j.phytochem.2006.11.024.

Mellerowicz, E. J., Baucher, M., Sundberg, B., & Boerjan, W. (2001). Unravelling cell wall formation in the woody dicot stem. In *Plant cell walls* (pp. 239–274). The Netherlands: Kluwer Academic Publishers.

Morel, O., Lion, C., Neutelings, G., Stefanov, J., Baldacci-Cresp, F., Simon, C., et al. (2021). REPRISAL: Mapping lignification dynamics using chemistry, data segmentation, and ratiometric analysis. *Plant Physiology*. https://doi.org/10.1093/plphys/kiab490.

Mottiar, Y., Gierlinger, N., Jeremic, D., Master, E. R., & Mansfield, S. D. (2020). Atypical lignification in eastern leatherwood (*Dirca palustris*). *New Phytologist*, *226*(3), 704–713. https://doi.org/10.1111/nph.16394.

Mottiar, Y., Vanholme, R., Boerjan, W., Ralph, J., & Mansfield, S. D. (2016). Designer lignins: Harnessing the plasticity of lignification. *Current Opinion in Biotechnology*, *37*, 190–200. https://doi.org/10.1016/j.copbio.2015.10.009.

Naseer, S., Lee, Y., Lapierre, C., Franke, R., Nawrath, C., & Geldner, N. (2012). Casparian strip diffusion barrier in Arabidopsis is made of a lignin polymer without suberin. *Proceedings of the National Academy of Sciences*, *109*(25), 10101–10106. https://doi.org/10.1073/pnas.1205726109.

Pandey, J. L., Kiemle, S. N., Richard, T. L., Zhu, Y., Cosgrove, D. J., & Anderson, C. T. (2016). Investigating biochemical and developmental dependencies of lignification with a click-compatible monolignol analog in *Arabidopsis thaliana* stems. *Frontiers in Plant Science*, 7. https://doi.org/10.3389/fpls.2016.01309.

Passardi, F., Longet, D., Penel, C., & Dunand, C. (2004). The class III peroxidase multigenic family in rice and its evolution in land plants. *Phytochemistry*, *65*(13), 1879–1893. https://doi.org/10.1016/j.phytochem.2004.06.023.

Perkins, M., Smith, R. A., & Samuels, A. L. (2019). The transport of monolignols during lignification in plants: Anything goes but how? *Current Opinion in Biotechnology*, *55*, 69–74. https://doi.org/10.1016/j.copbio.2018.09.011.

Perkins, M. L., Schuetz, M., Unda, F., Chen, K. T., Bally, M. B., Kulkarni, J. A., et al. (2022). Monolignol export by diffusion down a polymerization-induced concentration gradient. *The Plant Cell*. https://doi.org/10.1093/plcell/koac051. https://doi.org/10.1093/plcell/koac051.

Persson, S., Wei, H., Milne, J., Page, G. P., & Somerville, C. R. (2005). Identification of genes required for cellulose synthesis by regression analysis of public microarray data sets. *Proceedings of the National Academy of Sciences*, *102*(24), 8633–8638. https://doi.org/10.1073/pnas.0503392102.

Pesquet, E., Zhang, B., Gorzsás, A., Puhakainen, T., Serk, H., Escamez, S., et al. (2013). Non-cell-autonomous postmortem lignification of tracheary elements in *Zinnia elegans*. *The Plant Cell*, *25*(4), 1314–1328. https://doi.org/10.1105/tpc.113.110593.

Qin, S., Fan, C., Li, X., Li, Y., Hu, J., Li, C., et al. (2020). LACCASE14 is required for the deposition of guaiacyl lignin and affects cell wall digestibility in poplar. *Biotechnology for Biofuels*, *13*(197). https://doi.org/10.1186/s13068-020-01843-4.

Raggi, S., Ferrarini, A., Delledonne, M., Dunand, C., Ranocha, P., De Lorenzo, G., et al. (2015). The *Arabidopsis thaliana* class III peroxidase AtPRX71 negatively regulates growth under physiological conditions and in response to cell wall damage. *Plant Physiology*. https://doi.org/10.1104/pp.15.01464. pp.01464.2015.

Ralph, J., Lundquist, K., Brunow, G., Lu, F., Kim, H., Schatz, P. F., et al. (2004). Lignins: Natural polymers from oxidative coupling of 4-hydroxyphenyl- propanoids. *Phytochemistry Reviews*, *3*(1–2), 29–60. https://doi.org/10.1023/B:PHYT.0000047809.65444.a4.

Ranocha, P., Chabannes, M., Chamayou, S., Danoun, S., Jauneau, A., Boudet, A. M., et al. (2002). Laccase down-regulation causes alterations in phenolic metabolism and cell wall structure in poplar. *Plant Physiology*, *129*, 145–155. https://doi.org/10.1104/pp.010988.

Ranocha, P., McDougall, G., Hawkins, S., Sterjiades, R., Borderies, G., Stewart, D., et al. (1999). Biochemical characterization, molecular cloning and expression of laccases - a divergent gene family - in poplar. *European Journal of Biochemistry*, *259*(1–2), 485–495. https://doi.org/10.1046/j.1432-1327.1999.00061.x.

Ren, L.-L., Liu, Y.-J., Liu, H.-J., Qian, T.-T., Qi, L.-W., Wang, X.-R., et al. (2014). Subcellular relocalization and positive selection play key roles in the retention of duplicate genes of *Populus* class III peroxidase family. *The Plant Cell*, *26*(6), 2404–2419. https://doi.org/10.1105/tpc.114.124750.

Rojas-Murcia, N., Hématy, K., Lee, Y., Emonet, A., Ursache, R., Fujita, S., et al. (2020). High-order mutants reveal an essential requirement for peroxidases but not laccases in Casparian strip lignification. *Proceedings of the National Academy of Sciences*, *117*(46), 29166–29177. https://doi.org/10.1073/pnas.2012728117.

Ros Barceló, A. (1995). Peroxidase and not laccase is the enzyme responsible for cell wall lignification in the secondary thickening of xylem vessels in *Lupinus*. *Protoplasma*, *186*(1–2), 41–44. https://doi.org/10.1007/BF01276933.

Ros Barceló, A. (2005). Xylem parenchyma cells deliver the H_2O_2 necessary for lignification in differentiating xylem vessels. *Planta*, *220*(5), 747–756. https://doi.org/10.1007/s00425-004-1394-3.

Ros-Barceló, A. (1998). Hydrogen peroxide production is a general property of the lignifying xylem from vascular plants. *Annals of Botany*, *82*(1), 97–103.

Ros-Barceló, A., Gómez-Ros, L. V., Ferrer, M. A., & Hernández, J. A. (2006). The apoplastic antioxidant enzymatic system in the wood-forming tissues of trees. *Trees*, *20*(2), 145–156. https://doi.org/10.1007/s00468-005-0020-8.

Sasaki, S., Baba, K., Nishida, T., Tsutsumi, Y., & Kondo, R. (2006). The cationic cell-wall-peroxidase having oxidation ability for polymeric substrate participates in the late stage of lignification of *Populus alba* L. *Plant Molecular Biology*, *62*(6), 797–807. https://doi.org/10.1007/s11103-006-9057-3.

Sato, Y., Demura, T., Yamawaki, K., Inoue, Y., Sato, S., Sugiyama, M., et al. (2006). Isolation and characterization of a novel peroxidase gene *ZPO-C* whose expression and function are closely associated with lignification during tracheary element differentiation. *Plant and Cell Physiology*, *47*(4), 493–503. https://doi.org/10.1093/pcp/pcj016.

Schopfer, P. (1996). Hydrogen peroxide-mediated cell-wall stiffening *in vitro* in maize coleoptiles. *Planta*, *199*(1), 43–49. https://doi.org/10.1007/BF00196879.

Schuetz, M., Benske, A., Smith, R. A., Watanabe, Y., Tobimatsu, Y., Ralph, J., et al. (2014). Laccases direct lignification in the discrete secondary cell wall domains of protoxylem. *Plant Physiology*, *166*, 798–807. https://doi.org/10.1104/pp.114.245597.

Shigeto, J., Itoh, Y., Hirao, S., Ohira, K., Fujita, K., & Tsutsumi, Y. (2015). Simultaneously disrupting *AtPrx2*, *AtPrx25* and *AtPrx71* alters lignin content and structure in Arabidopsis stem. *Journal of Integrative Plant Biology*, *57*(4), 349–356. https://doi.org/10.1111/jipb.12334.

Shigeto, J., Kiyonaga, Y., Fujita, K., Kondo, R., & Tsutsumi, Y. (2013). Putative cationic cell-wall-bound peroxidase homologues in Arabidopsis, AtPrx2, AtPrx25, and AtPrx71, are involved in lignification. *Journal of Agricultural and Food Chemistry*, *61*(16), 3781–3788. https://doi.org/10.1021/jf400426g.

Shigeto, J., Nagano, M., Fujita, K., & Tsutsumi, Y. (2014). Catalytic profile of Arabidopsis peroxidases, AtPrx-2, 25 and 71, contributing to stem lignification. *PLoS One*, *9*(8), e105332. https://doi.org/10.1371/journal.pone.0105332.

Shigeto, J., & Tsutsumi, Y. (2016). Diverse functions and reactions of class III peroxidases. *New Phytologist*, *209*(4), 1395–1402. https://doi.org/10.1111/nph.13738.

Simon, C., Lion, C., Spriet, C., Baldacci-Cresp, F., Hawkins, S., & Biot, C. (2018). One, two, three: A bioorthogonal triple labelling strategy for studying the dynamics of plant cell wall formation *in vivo*. *Angewandte Chemie*, *130*(51), 16907–16913. https://doi.org/10.1002/ange.201808493.

Sterjiades, R., Dean, J. F. D., & Eriksson, K.-E. L. (1992). Laccase from sycamore maple (*Acer pseudoplatanus*) polymerizes monolignols. *Plant Physiology*, *99*(3), 1162–1168. https://doi.org/10.1104/pp.99.3.1162.

Sterjiades, R., Dean, J. D., Gamble, G., Himmelsbach, D., & Eriksson, K.-E. (1993). Extracellular laccases and peroxidases from sycamore maple (*Acer pseudoplatanus*) cell-suspension cultures. *Planta*, *190*(1). https://doi.org/10.1007/BF00195678.

Sundell, D., Street, N. R., Kumar, M., Mellerowicz, E. J., Kucukoglu, M., Johnsson, C., et al. (2017). AspWood: High-spatial-resolution transcriptome profiles reveal uncharacterized modularity of wood formation in *Populus tremula*. *The Plant Cell*, *29*(7), 1585–1604. https://doi.org/10.1105/tpc.17.00153.

Swetha, C., Basu, D., Pachamuthu, K., Tirumalai, V., Nair, A., Prasad, M., et al. (2018). Major domestication-related phenotypes in indica rice are due to loss of miRNA-dediated laccase silencing. *The Plant Cell*, *30*(11), 2649–2662. https://doi.org/10.1105/tpc.18.00472.

Syrjänen, K., & Brunow, G. (2000). Regioselectivity in lignin biosynthesis. The influence of dimerization and cross-coupling. *Journal of the chemical society, Perkin Transactions*, *1*(2), 183–187. https://doi.org/10.1039/a907919j.

Terashima, N., Atalla, R. H., Ralph, S. A., Landucci, L. L., Lapierre, C., & Monties, B. (1996). New preparations of lignin polymer models under conditions that approximate cell wall lignification. II. Structural characterization of the models by thioacidolysis. *Holzforschung*, *50*(1), 9–14. https://doi.org/10.1515/hfsg.1996.50.1.9.

Terashima, N., & Fukushima, K. (1988). Heterogeneity in formation of lignin. XI: An autoradiographic study of the heterogeneous formation and structure of pine lignin. *Wood Science and Technology*, *22*(3), 259–270. https://doi.org/10.1007/BF00386021.

Tobimatsu, Y., Chen, F., Nakashima, J., Escamilla-Trevino, L. L., Jackson, L., Dixon, R. A., et al. (2013). Coexistence but independent biosynthesis of catechyl and guaiacyl/syringyl lignin polymers in seed coats. *The Plant Cell*, *25*(7), 2587–2600. https://doi.org/10.1105/tpc.113.113142.

Tobimatsu, Y., & Schuetz, M. (2019). Lignin polymerization: How do plants manage the chemistry so well? *Current Opinion in Biotechnology*, *56*, 75–81. https://doi.org/10.1016/j.copbio.2018.10.001.

Tobimatsu, Y., Wagner, A., Donaldson, L., Mitra, P., Niculaes, C., Dima, O., et al. (2013). Visualization of plant cell wall lignification using fluorescence-tagged monolignols. *The Plant Journal*, *76*(3), 357–366. https://doi.org/10.1111/tpj.12299.

Tognolli, M., Penel, C., Greppin, H., & Simon, P. (2002). Analysis and expression of the class III peroxidase large gene family in *Arabidopsis thaliana*. *Gene*, *288*(1–2), 129–138. https://doi.org/10.1016/s0378-1119(02)00465-1.

Turlapati, P. V., Kim, K.-W., Davin, L. B., & Lewis, N. G. (2011). The laccase multigene family in *Arabidopsis thaliana*: Towards addressing the mystery of their gene function(s). *Planta*, *233*(3), 439–470. https://doi.org/10.1007/s00425-010-1298-3.

Tuskan, G. A., DiFazio, S., Jansson, S., Bohlmann, J., Grigoriev, I., Hellsten, U., et al. (2006). The genome of black cottonwood, *Populus trichocarpa* (Torr. & gray). *Science*, *313*(5793), 1596–1604. https://doi.org/10.1126/science.1128691.

Vatulescu, A. D., Fortunato, A. S., Sá, M. C., Amâncio, S., Ricardo, C. P. P., & Jackson, P. A. (2004). Cloning and characterisation of a basic IAA oxidase associated with root induction in *Vitis vinifera*. *Plant Physiology and Biochemistry*, *42*(7–8), 609–615. https://doi.org/10.1016/j.plaphy.2004.06.009.

Wallace, G., & Fry, S. C. (1999). Action of diverse peroxidases and laccases on six cell wall-related phenolic compounds. *Phytochemistry*, *52*(5), 769–773. https://doi.org/10.1016/S0031-9422(99)00342-8.

Wang, X., Zhuo, C., Xiao, X., Wang, X., Docampo-Palacios, M., Chen, F., et al. (2020). Substrate specificity of LACCASE8 facilitates polymerization of caffeyl alcohol for C-lignin biosynthesis in the seed coat of *Cleome hassleriana*. *The Plant Cell*, *32*(12), 3825–3845. https://doi.org/10.1105/tpc.20.00598.

Warinowski, T., Koutaniemi, S., Kärkönen, A., Sundberg, I., Toikka, M., Simola, L. K., et al. (2016). Peroxidases bound to the growing lignin polymer produce natural like extracellular lignin in a cell culture of Norway spruce. *Frontiers in Plant Science*, 7. https://doi.org/10.3389/fpls.2016.01523.

Welinder, K. G., Justesen, A. F., Kjærsgård, I. V. H., Jensen, R. B., Rasmussen, S. K., Jespersen, H. M., et al. (2002). Structural divesity and transcription of class III peroxidases from *Arabidopsis thaliana*. *European Journal of Biochemistry*, *269*(24), 6063–6081. https://doi.org/10.1046/j.1432-1033.2002.03311.x.

Wi, S. G., Singh, A. P., Lee, K. H., & Kim, Y. S. (2005). The pattern of distribution of pectin, peroxidase and lignin in the middle lamella of secondary xylem fibers in alfalfa (*Medicago sativa*). *Annals of Botany*, *95*(5), 863–868. https://doi.org/10.1093/aob/mci092.

Xie, T., Liu, Z., & Wang, G. (2020). Structural basis for monolignol oxidation by a maize laccase. *Nature Plants*, *6*(3), 231–237. https://doi.org/10.1038/s41477-020-0595-5.

Yan, J., Su, P., Li, W., Xiao, G., Zhao, Y., Ma, X., et al. (2019). Genome-wide and evolutionary analysis of the class III peroxidase gene family in wheat and *Aegilops tauschii* reveals that some members are involved in stress responses. *BMC Genomics*, *20*(1), 666. https://doi.org/10.1186/s12864-019-6006-5.

Zhao, Q., Nakashima, J., Chen, F., Yin, Y., Fu, C., Yun, J., et al. (2013). *LACCASE* is necessary and nonredundant with *PEROXIDASE* for lignin polymerization during vascular development in Arabidopsis. *The Plant Cell, 25*(10), 3976–3987. https://doi.org/10.1105/tpc.113.117770.

Zhuo, C., Rao, X., Azad, R., Pandey, R., Xiao, X., Harkelroad, A., et al. (2019). Enzymatic basis for C-lignin monomer biosynthesis in the seed coat of *Cleome hassleriana*. *The Plant Journal, 99*(3), 506–520. https://doi.org/10.1111/tpj.14340.

CHAPTER FIVE

Ferulic and coumaric acids in the cereal grain: Occurrence, biosynthesis, biological and technological functions

Anne-Laure Chateigner-Boutin and Luc Saulnier*
INRAE, UR1268 BIA, F-44316, Nantes, France
*Corresponding author: e-mail address: luc.saulnier@inrae.fr

Contents

Advances in Botanical Research, Volume 104
ISSN 0065-2296
https://doi.org/10.1016/bs.abr.2022.03.005

Abstract

Ferulic (FA) and *p*-coumaric (pCA) acids are phenolic compounds of considerable interest in plant science, human health and other industries. In the grain, FA and pCA are mainly bound to several cell wall polymers: arabinoxylan, lignin, cutin and suberin. They derive from the phenylpropanoid pathway which also produces lignin monomers and flavonoids. They are transferred to cell wall polymers by the action of acyl transferases of the BAHD family. In oxidative conditions they form radicals and can promote linkages between cell wall polymers. Their amount is variable between cereal species, between genotypes of a same species, in different tissues of the grain, during grain development and in response to pathogen attack. Biological functions were ascribed to bound FA in cell wall assembly, in tissue stiffening, in resistance to pathogens. Bound to cereal grain polymers FA has technological impacts in milling, bread-making and gelling. Free FA extracted from plant tissues has applications in bioconversion and has been associated to many health benefits.

1. Introduction

Cereals are usual name for plants of the *Poaceae* family which are cultivated for their edible grains. They have tremendous importance for human food, animal feed and other industries. Cereal derived food contributes to the dietary intake of energy and protein. They greatly impact human health providing dietary fibers, essential amino acids, minerals, vitamins and phytochemicals, but they also have adverse impacts such as allergy and intolerance for humans and digestive health issues for some feedstock animals. In addition, cereals are used for starch production and non-food applications such as biofuels (Shewry, 2009).

1.1 The cereal grain

The cereal fruit is a caryopsis, a dry fruit containing a unique seed. It is conventionally named grain. As a fruit it develops from the fertilized ovary and therefore contain maternal and filial tissues (Fig. 1). The embryo and endosperm are the products of the fertilization. The embryo, comprises an embryonic axis and a unique cotyledon, the scutellum. The endosperm is

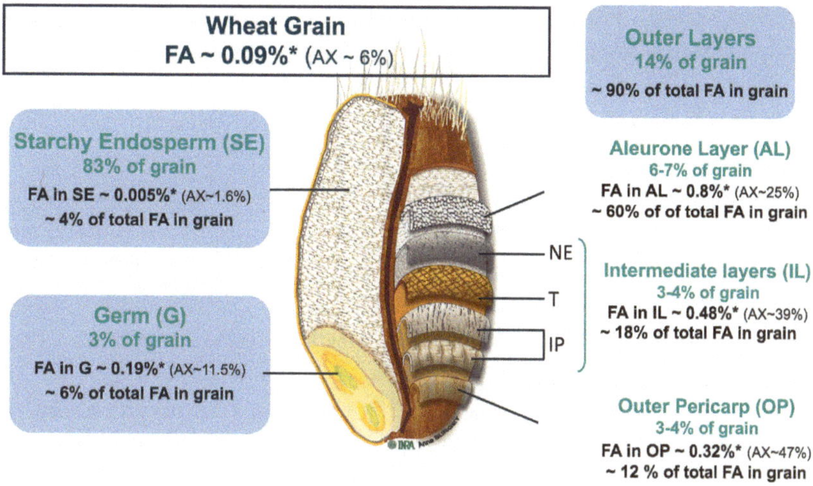

Fig. 1 Content and distribution of ferulic acid (FA) in wheat grain. Percentages (%) are based on FA released by mild alkaline hydrolysis. FA dimers are not included in the calculation nor FA ether-linked to lignin, which are as both as high as FA ester monomer in outer pericarp. * Content in g/100 g of grain/tissue. Value in parenthesis is the content (g/100 g) in Arabinoxylan (AX) of the fraction. NE, nucellar epidermis; T, testa; IP, inner pericarp. *Adapted from Surget and Barron (2005). Histologie du grain de blé.* Industrie des Céréales, 145, *3–7. Data from Barron, C., Surget, A., Rouau, X. (2007). Relative amounts of tissues in mature wheat (Triticum aestivum L.) grain and their carbohydrate and phenolic acid composition.* Journal of Cereal Science, 45*(1), 88–96. https://doi.org/10.1016/j.jcs.2006.07.004.*

divided into the starchy endosperm, a storage tissue in which starch and storage proteins are accumulated, and the aleurone layer a specialized and peripheral part of the endosperm that stays alive in mature dry grains and is implicated in the germination process. The seed is covered by a seed coat also called testa, and by pericarp tissues which originate from the carpel wall. In the cereal grain industry, the vocabulary employed is different in line with milling products. The embryo is called germ, the outer layers are called bran and comprise the pericarp tissues, testa, nucellar epidermis, aleurone and traces of endosperm, while the starchy endosperm gives the white flour (Evers & Millar, 2002).

1.2 Cereal grain composition and phenolic content

The biochemical composition of whole cereal grains varies depending on the species however the main compounds found in the grains are polysaccharides: starch and non-starch polysaccharides (cell wall polysaccharides, fructans), and proteins (storage proteins and functional proteins), and more minor compounds such as lipids (triacylglycerols, phospholipids and glycolipids (Marion & Saulnier, 2020) and lipids of the cuticles), minerals, terpenoids and phenolic compounds (Andersson et al., 2013; Marion & Saulnier, 2020; Shewry et al., 2013).

Phenolic compounds are molecules with a benzyl ring and among the existing phenolics those described in cereal grains are lignins (Chateigner-Boutin et al., 2018; Desvignes et al., 2006), flavonoids especially anthocyanins in colored grains (Zhu, 2018), tannins (Dykes, 2019), alkylresorcinols (Andersson, Kamal-Eldin, Fraś, Boros, & Åman, 2008; Andersson, Lampi, et al., 2008) and phenolic acids.

Phenolic acids are molecules with a benzyl ring, a carboxylic group and at least one hydroxyl group. These compounds are the focus of research and development because of their anti-oxidant properties and many articles are published to describe the phenolic acids content of diverse plants and food and report numerous health benefits (de Oliveira Silva & Batista, 2017). In cereal grain, they are found as free phenolic compounds, conjugated as ester of low molecular weight compounds or bound to cell wall polymers or cutin. Ferulic acid (FA) is the major phenolic compound found in cereal grain of which >90% is bound to cell wall polymers in wheat, rye or barley. Other phenolic compounds (vanillic, syringic, sinapic, caffeic acids) are found in much lower concentration (~10 times lower than FA in wheat) and mostly in conjugated form (Li, Shewry, & Ward, 2008).

This chapter will focus on ferulic acid (FA) and para-coumaric acid (pCA), the two main phenolic acids found in the cell walls of the cereal grain

1.3 Ferulic and *p*-coumaric acids

FA and pCA are phenolic acids and more precisely hydroxycinnamic acids (HCAs). In plants, HCAs are found linked to various compounds (Fig. 2). They are ester-linked to low molecular weight compounds such as glucose or organic acids forming for instance feruloyl malate (Mir Derikvand et al., 2008), feruloyl glucose (Yin et al., 2019), feruloyl sucrose (Chougui et al., 2013). They are ester-linked to phytosterols and an example is gamma-oryzanol, ester of ferulic acid with a mixture of phytosterols and triterpenoïds (Patel & Naik, 2004).

They are reported to be linked to proteins *in planta* (Fry, 2004; Obel, Porchia, & Scheller, 2003) and also in transformed plant products such as in bread dough (Piber & Koehler, 2005).

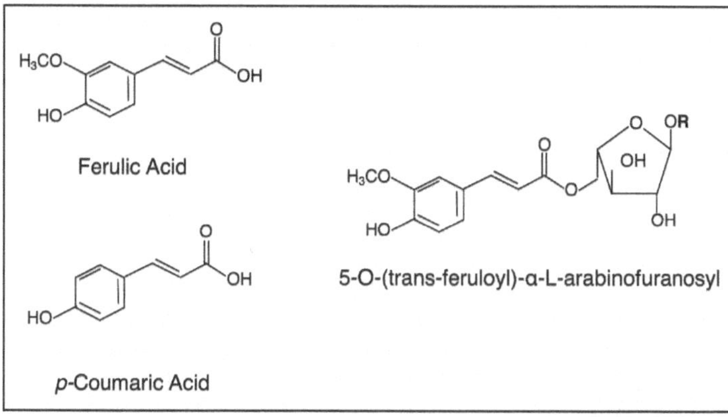

Fig. 2 Chemical structure of main phenolic acids in cereal grains and their ester linkage to arabinofuranose residue.

However, FA and pCA are mainly bound to polysaccharides (AX in Poaceae and pectin in eudicots), lignin and fatty acid containing polymers (suberin, cutin, wax and sporopollenin) (Domergue & Kosma, 2017).

The most abundant plant organ source of FA is the maize grain, especially the pericarp which accounts for 98% of the grain FA content (Harris & Trethewey, 2010). Acylation of plant cell wall with ferulic acid is particularly important in the grain of the Poaceae family.

2. Ferulic and *p*-coumaric acids in cereal grains: Soluble and bound hydroxycinnamic acids

The analysis of cereal grain compounds revealed the presence of FA and pCA, with FA being more abundant than pCA. In the HEALTHGRAIN study for instance, free, conjugated and bound phenolics were extracted and quantified from a panel of wheat lines (Li et al., 2008). Free and conjugated phenolic acids were extracted using 80% ethanol, and measured by HPLC either directly for free forms or after saponification of the EtOH extract for conjugated forms while bound phenolic acids were released from the EtOH residue by alkaline hydrolysis prior to HPLC analysis. The quantitative measurement of FA and pCA levels indicated that these compounds are predominantly found as bound to cell wall polymers (Fernandez-Orozco, Li, Harflett, Shewry, & Ward, 2010; Li et al., 2008).

2.1 Ferulic and *p*-coumaric acids are linked to arabinoxylans

In early reports, cell wall fractions from grasses that were treated with alkali resulted in the release of FA and pCA and in spectral changes diagnostics of ester bonds (Hartley, 1973). In addition, a fraction of soluble cell wall polysaccharides from wheat flour was found to be composed of xylose, arabinose and FA (Geissmann & Neukom, 1973). Since then many studies were conducted and confirmed that the presence of hydroxycinnamic acids linked to arabinoxylan is a common feature of grasses (Harris & Trethewey, 2010). In grasses, FA and pCA have only been detected esterified on the position O5 of the arabinofuranose (Ara) either as single side-chain of arabinoxylans (AX) in cereal grain endosperm or in more complex side chains in cereal bran heteroxylans (HX). In order to identify the location of FA/pCA on the AX or HX polymers, enzymatic or controlled acid hydrolysis of cell wall material and subsequent chromatographic fractionation were used to isolate oligosaccharides ester-linked to FA and pCA. Subsequently their structures were characterized by a combination of ^{1}H-,^{13}C NMR, mass spectrometry and methylation techniques. Using endoxylanase treatment feruloylated oligosaccharides including xylose residues derived from the beta (1,4)-linked xylose backbone of AX were isolated for instance: 2-O-[5-O-(trans-feruloyl)-α-L-arabinofuranosyl]-D-xylopyranose (FAraX) from wheat bran (Smith & Hartley, 1983), 4-O-[3-O-[5-O-(trans-feruloyl)-α-L-arabinofuranosyl]-β-D-xylopyranosyl]-D-xylose (FAraXX) from barley endosperm (Ahluwalia & Fry, 1986). Using controlled acid hydrolysis feruloylated oligosaccharides originated from the side chains of HX were isolated: 5-O-trans-feruloyl-L-arabinofuranose (FAra), β-D-xylopyranosyl- (1/2)-5-O-(trans-feruloyl) -L-arabinofuranose (XFAra) and α-L-galactopyranosyl- (1/2)-β-D-xylopyranosyl-(1/2)-5-O-trans-feruloyl-L-arabinofuranose (GXFAra) from maize bran (Allerdings, Ralph, Steinhart, & Bunzel, 2006; Saulnier, Vigouroux, & Thibault, 1995) and other cereal brans (Schendel, Meyer, & Bunzel, 2016), while combination of acid and enzymatic hydrolysis allowed the isolation of FAra, XFAra and GXFAra attached to xylan backbone fragments (Schendel, Becker, Tyl, & Bunzel, 2015). More recently, mild acidolysis in Dioxane/MetOH/2M aqueous HCl (60/30/10) have been proved as an efficient method to release arabinose linked to FA and pCA from grass AX (Lapierre, Voxeur, Karlen, Helm, & Ralph, 2018; Méchin et al., 2014). It is noteworthy that pCA linked to arabinose has been essentially identified in stem, leave or straw of grasses (Ishii, 1997; Lapierre et al., 2018) but has also been detected in maize bran (Lapierre, Voxeur, Boutet, & Ralph, 2019).

2.2 Oxidative coupling of ferulic acid

Ferulates acylating xylan polysaccharides in grasses are involved in radical coupling reactions, mediated by peroxidase *in vivo*. Dehydrodimers of FA

(DiFA) and oligomers of higher degree of polymerization possibly cross-linking cell wall polysaccharides have been identified in grasses and especially in cereal grain (Bunzel, 2010; Ralph, Quideau, Grabber, & Hatfield, 1994). The mechanism of formation and the structure of the main dimers and trimer is depicted on Fig. 3. Seven different DiFA have been identified in cereal grains, the main dimers found are in general 8-O-4′, 5-5′, 8-5′ (three forms) and 8-8′ (three forms) with traces of 4-O-5′ (Bunzel, 2010; Waterstraat & Bunzel, 2019). The different forms of 8-5′ and 8-8′ dimers are the consequence of the saponification reaction and only the cyclic form is occurring *in planta* (Bunzel, 2010). Next to the DiFA, dehydrotrimers (TriFA) were identified, albeit in lower quantity than dimers (Allerdings et al., 2005; Rouau et al., 2003). The first reported (5-5′/8′-O-4″) trimer of FA is major but eight other trimers have been identified in maize and wheat grains (Waterstraat & Bunzel, 2019). Finally, 4-O-8/5-5/8-O-4- tetramer has been identified in maize bran (Bunzel, Ralph, Brüning, & Steinhart, 2006), and more complex structure are probably not yet identified.

The structural evidence that dehydrodimers cross-link polysaccharide chains in the cell wall polysaccharides have been established after the isolation of 5-5′, 8-O-4′ and 8-8′c dehydrodimers attached to arabinose residues fragments following acidic hydrolysis of cell wall material (Allerdings et al., 2005; Bunzel, Allerdings, Ralph, & Steinhart, 2008; Saulnier et al., 1999). So far, evidence of TriFA attached to carbohydrate moiety has not been obtained. Although the most obvious hypothesis would support that DiFA couple two polysaccharides chain (and TriFA 3), molecular modeling has suggested that some cross linking might occur intra-molecularly, especially in the case of 5-5′ dimers (Hatfield & Ralph, 1999). In addition, it has been suggested that the formation of trimers might result of a back-crossing mechanism, e.g., coupling of a FA on an already formed intra-molecular dimer thus coupling two and not three AX chains (Ralph, Bunzel, et al., 2004). The presence of intra- and the proportion of intra-/inter-molecular FA bridges remain open questions, nevertheless the cross-linking of AX chains in cereal grain through oxidative coupling of FA is a structural feature that have tremendous impact on the cell wall properties and the degradation of cell walls by microbial enzymes (such as human microbiota).

2.3 UV photo-dimerization of ferulic and *p*-coumaric acids

The formation of phenolic cyclobutane dicarboxylic acids (truxillic and truxinic acids) under ultraviolet (UV) irradiation is another coupling

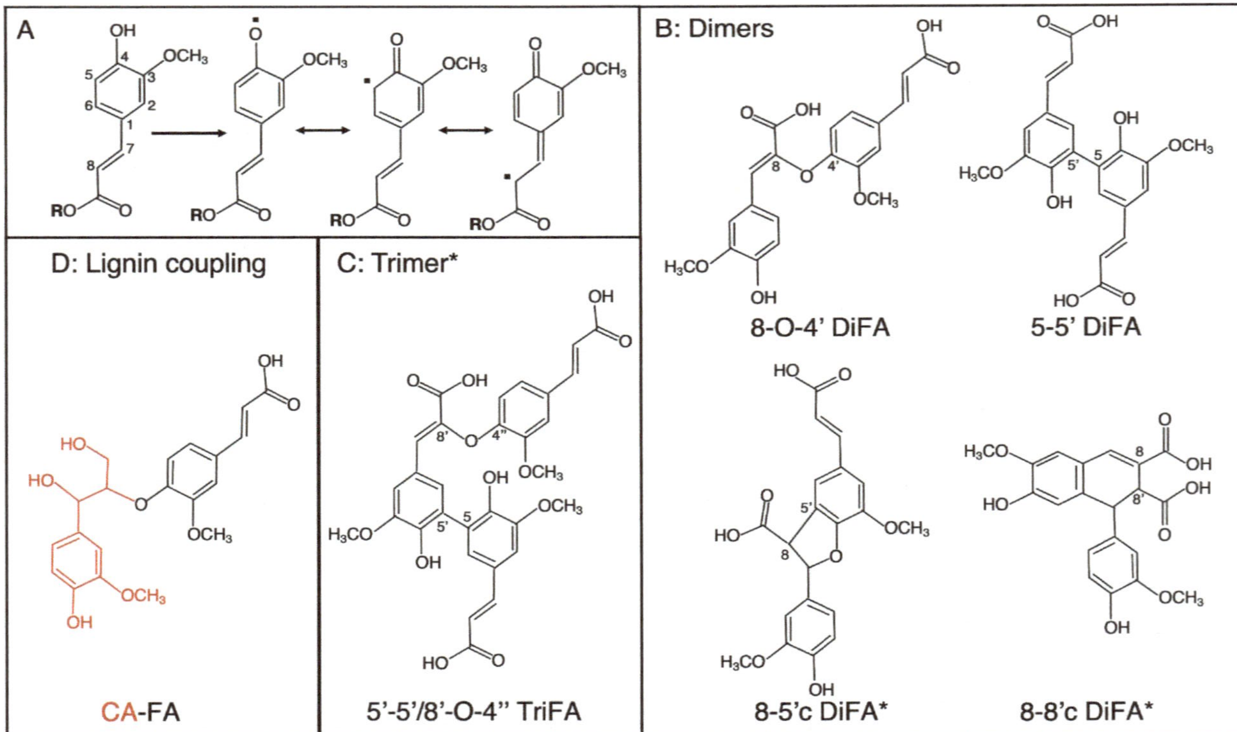

Fig. 3 Mechanism of oxidative coupling of ferulic acid and main oxidative products identified in cereal grains. (A) Mechanism of formation of phenoxy-radical of ferulic acid under oxidative conditions and resonance stabilization forms. (B) Main dimers of ferulic acid (DiFA) detected in cereal grains: * two other forms of 8-5′ and 8-8′ DiFA have been reported: see Bunzel (2010). (C) Main trimer of Ferulic acid (TriFA) detected in cereal grains: * eight other TriFA have been reported see Waterstraat and Bunzel (2019). (D) Ether linked monolignol Conyferyl Alcohol (CA) Ferulic acid (FA) dimer (Jacquet, Pollet, Lapierre, Mhamdi, & Rolando, 1995).

mechanism of AX in plant cell walls (Hartley, Morrison III, Himmelsbach, & Borneman, 1990; Morrison, Hartley, & Himmelsbach, 1992). However, such dimers are of minor importance in plant organs protected from direct light exposure such as cereal grains, and indeed only very minor amount of these dimers have been detected in maize bran (Waterstraat & Bunzel, 2019).

2.4 Ferulic and *p*-coumaric acids are linked to lignins

Grass lignins are essentially made of guaiacyl (G), and syringyl (S) units together with a low amount of *p*-hydroxyphenyl (H) units. Corresponding alcohol precursors: coniferyl alcohol, sinapyl alcohol and *p*-coumaryl alcohol are polymerized into lignin by radical coupling reactions. They are bound by different types of linkages: ether linkages (β-O-4, α-O-4) that are said labile because they are breakable by chemical treatments and condensed linkages such as C-C (5-5, β-1, β-5, β-β) and diaryl ether 4-O-5 that are resistant to chemical lysis (Dimmel, 2010). Additional components are naturally incorporated in lignins such as pCA and FA. Indeed, a distinguishing feature of grass lignins is the considerable incorporation of FA and pCA (Jung & Phillips, 2010). In grass straw, mild alkaline hydrolysis experiments lead to the isolation of ether-linked FA-coniferyl alcohol revealing linkages between FA and lignins (Jacquet et al., 1995; Lapierre et al., 2019). As a result, lignin is covalently bound to cell wall polysaccharides via FA cross-links (Fig. 4) and FA etherified to lignin is probably also esterified to HX (Iiyama, Lam, & Stone, 1994; Ralph, Bunzel, et al., 2004; Ralph, Lundquist, et al., 2004).

Most studies on grass lignins were conducted in vegetative tissues however some studies were carried out on grains where the outer layers are lignified (Wheat: (Beaugrand et al., 2004; Chateigner-Boutin et al., 2018; Desvignes et al., 2006); Brachypodium: (Ho-Yue-Kuang et al., 2016)). The lignified tissues vary between species since in wheat grains pericarp tissues contain lignin and in Brachypodium grains, pericarp, testa and palea are lignified (Chateigner-Boutin et al., 2018; Ho-Yue-Kuang et al., 2016). In wheat, treatment of outer tissue with alkali at high temperature, that is able to release ether-bound phenolic acids (Iiyama, Lam, & Stone, 1990), increased the phenolic acids released by twofold compared to mild alkaline treatment that only release ester bound ferulic acid (Table 1) (Barron et al., 2007) indicating interconnections of HX with lignin through ester-ether FA bonds. Although the amount of lignin in maize bran is very low, study of the phenolic fraction associated to HX extracted in mild alkaline

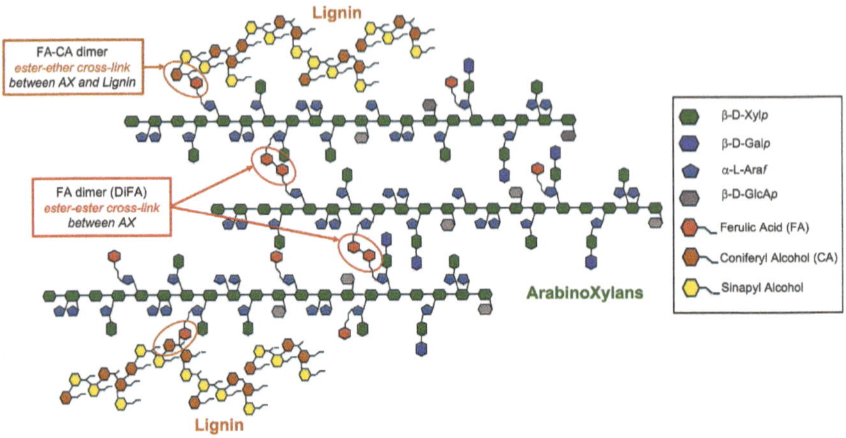

Fig. 4 Schematic view of interconnections between cell wall polymers involving ferulic acid in cereal grains. Ferulic acid (FA) dimers are linked through ester linkage to Arabinoxylans (AX). Ester linkages are broken in mild alkaline or acid conditions. Trimers of FA possibly connecting three AX chains are also detected in cereal grains. FA ester-linked to AX can be also coupled to Guaiacyl (G) unit of lignin through an ether bond (as evidenced by FA-CA diagnostic dimer). *Para*-coumaric acid (pCA) is present in very low amount in cereal grains and its possible connection to AX or more likely lignin is not indicated here. Xyl*p*, xylopyranose; Gal*p*, galactopyranose; Ara*f*, Arabinofuranose; GlcA*p*, glucuronic acid.

conditions revealed FA, 8-8′ dimers as main diferulate surviving alkaline extraction and typical lignin structures, mainly comprising S units interconnected through β-O-4, β-1 and β-β interunit (Lapierre, Pollet, Ralet, & Saulnier, 2001).

pCA is ester-linked to lignin S unit in grass more precisely at the γ-carbon of the side chain region of β-O-4 linked S unit and remains as a terminal residue with a free phenolic group (Ralph, Hatfield, et al., 1994; Ralph, Quideau, et al., 1994). pCA can also be found ester-linked to G units but to a lesser extent (Lu & Ralph, 1999). A minor fraction of pCA is also ester-linked to xylans, but unlike FA polysaccharide esters, pCA esters on lignin or polysaccharides do not extensively participate in oxidative coupling reactions (Grabber, 2005; Ralph, Bunzel, et al., 2004; Ralph, Lundquist, et al., 2004).

2.5 Ferulic and *p*-coumaric acids in aliphatic polymers

pCA and FA have been identified as constituents of plant lipidic polyesters cutin, suberin, and sporopollenin (Chateigner-Boutin et al., 2018; Kolattukudy, 1981; Li, Phyo, Jacobowitz, Hong, & Weng, 2019; Rautengarten et al., 2012). Cutins are the main components of plant cuticles;

Table 1 Bound hydroxycinnamate (HCA) contents in cereal grains.

Species	Line/cv/ panel	Tissue	pCA	FA_m	8-5′	8-O-4′ mg/100 g	5-5′	diFA tot	TriFA	Method/ tissue AIR or DF	References
Wheat *Triticum aestivum*	Camp Remy	Outer layers	15.4	514.8				97.7	23.9	Mild 2M NaOH 35 °C/tissue	Desvignes et al. (2006)
	Camp Remy	Outer layers	24.2	619.6				100.3	43.5	Severe 2M NaOH 170 °C/ tissue	Desvignes et al. (2006)
	Crousty	Whole grain	2	87				14	3	Mild 2M NaOH 35 °C/tissue	Barron et al. (2007)
		Starchy endosperm	nd	4				nd	nd		
		Aleurone	16	798				31	2		
		Nucellar layer	9	1083				46	2		
		Intermed. layers	6	496				59	6		
		Pericarp	11	450				259	85		
		Scutellum	1	332				30	3		
		Embryonic axis	nd	57				10	1		

Continued

Table 1 Bound hydroxycinnamate (HCA) contents in cereal grains.—cont'd

Species	Line/cv/ panel	Tissue	pCA	FA_m	8-5′	8-O-4′ mg/100 g	5-5′	diFA tot	TriFA	Method/ tissue AIR or DF	References
	Healthgrain panel 130 winter cv.	Whole grain flour	1.0 (0.3–1.9)	36.6 (16.2–72.1)				1.9 (0.8–5.1)		2M NaOH/ AIR	Li et al. (2008)
	120 spring cv.	Whole grain flour	0.4 (0.3–1.3)	37.5 (25.2–62.8)				2.5 (1.7–4.5)		2M NaOH/ AIR	Li et al. (2008)
	Not indicated	Grain dietary fiber (~equivalent to bran)			44.5	27.9	16.6	119.1	25.4	2M NaOH/ DF	Waterstraat and Bunzel (2019)
	Recital	Pericarp mature	5	272	121	51	41	213	63	2M NaOH 35 °C/AIR	Chateigner-Boutin et al. (2018)
		Pericarp 570° DAF	39	1805	118	50	46	214	21		
		Pericarp 450° DAF	49	1896	85	45	42	171	24		
		Pericarp 200° DAF	34	1380	90	60	45	195	28		
Durum wheat *T. durum*	Healthgrain Panel 10 cv	Whole grain flour	0.3 (0.3–0.4)	35.5 (24.1–70.1)				3.0 (1.7–4.5)		2M NaOH/ AIR	Li et al. (2008)
Spelt T. *spelta*	Healthgrain Panel 5 cv	Whole grain flour	0.5 (0.3–1.3)	33.4 (20.0–47.5)				1.9 (1.4–2.3)		2M NaOH/ AIR	Li et al. (2008)
Eikorn *T. monococcum*	Healthgrain Panel 5 lines	Whole grain flour	0.3 (0.28–0.29)	24.1 (16.3–37.2)				0.9 (0.7–1.3)		2M NaOH/ AIR	Li et al. (2008)

Emmer *T. urartu*	Healthgrain Panel 5 lines	Whole grain flour	0.3 (0.2–0.4)	44.9 (30.4–69.3)				2.4 (1.6–4.3)		2M NaOH/ AIR	Li et al. (2008)
Maize *Zea mays*	Mix of 5 hybrids	Whole grain	23.8	174.8						2 M NaOH at 90 °C/ AIR	Horvat et al. (2020)
	Not indicated	Bran?		2066						(esters)	Harris and Trethewey (2010)
	Panel 19 lines	Pericarp	238.3 (96.1–465.3)	3421 (1888.6–4992.3)						Mild 2M NaOH 35 °C/tissue	Chateigner-Boutin et al. (2016)
	Panel 19 lines	Endosperm	0.48 (0.26–0.87)	40.8 (25.6–74.5)							
	B73	Pericarp	62.6	4011.9	416.8	236.2	215.8	868.8	75.2		
	B73	Endosperm	0.81	37.5	5.54	2.85	1.72	10.11	nd		
	CO255	Pericarp	51.4	4992.3	275.5	208.3	159.5	643.3	106.7		
	CO255	Endosperm	0.48	74.5	2	1.26	0.74	4	nd		
	WF9	Pericarp	374.9	2422.6	408.6	289.6	209.5	907.7	223.9		
	WF9	Endosperm	0.29	53.4	3.56	1.9	1.18	6.64	nd		
	Everta	Grain dietary fiber (~equivalent to bran)			170.72	192.7	165.4	655.68	119.75	2M NaOH/ DF	Waterstraat and Bunzel (2019)
Rice *Oryza sativa*	Japonica (dehulled)	Whole grain	3.56	9.70	4.15	1.88	1.23	7.26	0.09	2M NaOH room temp/ AIR.	Feng et al. (2021)
		Bran	23.46	68.81	11.68	8.17	5.63	25.48	0.66		

Continued

Table 1 Bound hydroxycinnamate (HCA) contents in cereal grains.—cont'd

Species	Line/cv/ panel	Tissue	pCA	FA_m	8-5′	8-O-4′ mg/100 g	5-5′	diFA tot	TriFA	Method/ tissue AIR or DF	References
Triticale	Borwo	Whole grain	7.45	59.43				35.5		4M NaOH boiling (no ethanol extraction)/ tissue	Kaszuba, Kapusta, and Posadzka (2021)
Barley *Hordeum vulgare*	Hulled 8 lines	Whole grain flour	6.8 (3.9–11.0)	25.8 (19.1–36.5)						2M NaOH/ AIR	Andersson, Kamal-Eldin, et al. (2008), Andersson, Lampi, et al. (2008)
	Naked 2 lines		0.3	13.8 (10.4–17.3)							
Rye *Secale cereale*	17 lines	Whole grains	4.6 (3.8–6.5)	105.3 (94.1–117.4)	11.6 (9.3–14.0)	16.0 (12.8–20.0)	5.9 (4.6–6.9)	33.6 (27.8–40.9)		α-amylase +2M NaOH 25 °C/tissue	Andreasen, Christensen, Meyer, and Hansen (2000)
	Healthgrain 10 lines	Whole grain flour	1.9 (0.3–5.9)	31.9 (14.6–52.1)						2M NaOH/ AIR	Nyström et al. (2008)
Oat *Avena sativa*	Healthgrain 4 hulled lines	Whole grain flour	12.4 (6.7–28.7)	11.1 (4.5–26.0)						2M NaOH/ AIR	Shewry et al. (2008)
	Cacko naked	Whole grain flour	4.2	14.3							

Sorghum	UDG110	Whole grains	3.87	67.96			2M NaOH room temp/ AIR.	Salazar-López et al. (2017)
Sorghum bicolor		Bran	16.9	303.6			4 M NaOH room temp/ AIR	
Brachypodium distachyon	Bd21	Whole grains	98	183			1M NaOH room temp/ AIR	Ho-Yue-Kuang et al. (2016)

pCA: *p*-coumaric acid; FA_m: ferulic acid monomer; diFA: ferulic acid dehydrodimers (8-5′, 8-O-4′, 5-5′) other DiFA were reported in maize (8-8 3 forms) and 4-O-5 (Waterstraat & Bunzel, 2019). TriFA: ferulic acid dehydrotrimer: the main triFA is 5-5/8-O-4 but eight other TriFA were reported in maize (Waterstraat & Bunzel, 2019). Numbers followed by parentheses correspond to mean values from different lines (min-max values). “Method” depicts the method used to release HCAs (mild or strong alkaline treatments), a step or no step of alcohol extraction to remove soluble HCAs. The unit is mg/100g of tissue or of alcohol insoluble residue (AIR) or of dietary fiber (DF). See Waterstraat and Bunzel (2019) for the obtention of DF. nd, not detected.

they are composed of inter-esterified C16-C18 hydroxy fatty acids that also contain glycerol and FA and pCA. Cereal grains contain several cuticles, wheat for instance contain a thin cuticle at the surface of the grain, a thick cuticle covering the seed testa, and a thin cuticle between the testa and nucellar epidermis (Morrison, 1975). Few studies were carried out on cereal grain cuticles but in cutin isolated from wheat grains, FA and pCA were detected (Chateigner-Boutin et al., 2018). Suberin is reported to be deposited in a specific and localized structure called the pigmented strand of cereal grain studied in rice and wheat (Oparka & Gates, 1981; Zee & O'Brien, 1970). Suberin is a lipid-based polymer containing fatty acid with higher proportions of fatty alcohols, HCAs and glycerol, compared with cutin (Kolattukudy, 1981).

3. Ferulic and *p*-coumaric acids in cereal grains: Variability in content and distribution

Many articles report the phenolic content of plants among which cereal grains. In order to quantify and compare the FA and pCA content in cell walls of various cereal grains, it is important to consider (i) the plant material analyzed (the presence or absence of lignin in the tissue, a step of prior extraction of soluble compounds or not), (ii) the method used to release FA and pCA from cell walls (mild or strong alkaline hydrolysis) and the likeliness that only part of the FA and pCA is released from cell walls, (iii) the form of FA monitored (only FA or dimers and oligomers, ester-linked, ether-linked, etc.).

Table 1 summarizes a list of studies conducted in the grain of different cereal species and with the precautions presented above it tentatively compares the levels of FA and pCA. Most studies compare FA and pCA ester-linked to cell walls estimated by measuring HCAs released using mild alkaline hydrolysis. Some studies but not all of them include the monitoring of dimers and oligomers.

3.1 Variability in level of ferulic and *p*-coumaric acids in the whole grain of different grass species

In the whole grain of different species, there is a high variability of ester-linked FA and pCA (Table 1). Considering only FA "monomers," as previously described by a survey made by Harris and Trethewey in 2010, maize grains are effectively rich in FA. Whole grain studies are scarce but maize and Brachypodium grains are the richest in FA with 175–180 mg/100 g of cell walls (Table 1). The poorest whole grain in ester-linked FA are rice and oat with about 10 mg/100 g of cell walls (Feng et al., 2021; Shewry et al., 2008). The ester-linked pCA level also varies between species. It varies from less than 1 mg/100 g of cell walls in emmer (Li et al., 2008) to about 100 mg/100 g of cell walls for *Brachypodium distachyon* (Ho-Yue-Kuang et al., 2016). It is noteworthy that some species have much less pCA than FA ester-linked to their cell walls like wheat for instance with a ratio of about 1/40 while in other species such as oat, the grain cell walls contain almost equivalent levels of pCA and FA (Shewry et al., 2008).

3.2 Variability between grain tissues

The cereal grain is composed of several tissues. These tissues have different compositions and properties (Fig. 1).

In wheat grain, the FA and pCA level was monitored in dissected tissues and the studies revealed a considerable variability (Table 1). FA varies from 4 mg/100 g of endosperm to about 800 mg/100 g of isolated aleurone layer and 1000 mg/100 g of isolated nucellar epidermis. The most important part of dimers of FA are released from the pericarp tissue. The wheat endosperm tissue accounts for 80% of the wheat grain but only 4% of its FA level. In contrast, the aleurone layer accounts for about 6% of the grain and about 60% of its FA level. Ester-linked pCA was detected in different wheat grain tissues except in endosperm and in embryo axis and quantified. It varies from 1 mg/100 g in scutellum to 16 mg/100 g in aleurone.

A study conducted on maize also highlighted differences among tissues (pericarp and endosperm) and revealed the particularly high level of FA and pCA in the maize pericarp (Table 1). The FA content is extremely high in maize pericarp: 3421 mg/100 g (mean) and almost 100 times lower in starchy endosperm (mean 40.8 mg/100 g). The pCA content was generally 10–20 times lower than that of FA.

In all investigated cereal species, pericarp (or bran fractions) contain relatively more HCA than whole grains (Table 1).

The tissue composition and the relative mass of outer layers, endosperm and embryo vary from one species to another. This heterogeneity can account for the variability in the level of FA and pCA of whole grain among Poaceae species. For instance, a whole grain study can include the hull for hulled grain species such as Brachypodium (Ho-Yue-Kuang et al., 2016), barley (Andersson, Kamal-Eldin, et al., 2008; Andersson, Lampi, et al., 2008) and oat (Shewry et al., 2008). The hull, a structure consisting in two glume-like bracts named lemma and palea, is adhering to the pericarp in species such as barley. Hull-less cultivars of barley have been developed. Another difference is the number of layers of aleurone cells, the walls of which are thick and rich in FA in wheat. Modern wheat grains comprise only one aleurone cell layer whereas barley grains comprise two to four layers of aleurone cells (Evers & Millar, 2002; Hands, Kourmpetli, Sharples, Harris, & Drea, 2012).

3.3 Variability among lines or cultivars

Study among 19 lines of maize (Chateigner-Boutin et al., 2016) highlighted high variability between maize lines in the FA and pCA levels. FA level in pericarp ranges from 1.9% to 5.0% (1900–5000 mg/100 g), with a mean value of 3.5% (3421 mg/100 g) (Table 1).

pCA ranged from 51 to 374 mg/100 g and with a mean of 238.3 mg/100 g. The pericarp of B73 and CO255 contained notably low amounts of pCA (62 mg/100 g and 51 mg/100 g, respectively) compared with the other lines.

A panel of 175 wheat lines and related species, the HEALTHGRAIN panel, was analyzed to compare the composition and among other compounds their level of FA and pCA in whole grain flour (Li et al., 2008). Considerable variability for FA content were noticed: 16–70 mg/100 g even only between lines of bread wheat. In wheat, pCA varied with a six-time range from 0.3 to 1.9 mg/100 g.

Several other studies were conducted on different lines of a cereal grain species. For instance, rye, oat, sorghum and barley (Table 1). For barley, a considerable difference was noticed between hulled and naked species: hulled species contain more FA and much more pCA (10 times more) than naked species (Andersson, Kamal-Eldin, et al., 2008; Andersson, Lampi, et al., 2008) indicating a higher HCA, especially pCA, linked to a higher lignin content of these protective tissues related to leaves.

3.4 Variability due to the environment

In addition to the variability due to differences across lines, the environment, and possibly interactions between the genotype and environment possibly strongly influence the levels of grain compounds. Fernandez-Orozco et al. (2010) and Shewry et al. (2010) used the HEALTHGRAIN diversity panel cultivated at several locations with monitoring of soil and weather conditions to study the effect of the environment and interactions between genetic and environment on the phenolic acid compounds in wheat grain.

When analyzing the harvest of a same year at different locations, heterogeneity was observed among genotypes but generally the amounts of bound phenolics, of which FA is the more represented, were lowest when the wheat genotypes were grown in France and highest when grown in Hungary. A mean value of bound FA of 59mg/100g for 26 genotypes grown in Hungary was observed in 2007 while the mean value was 45mg/100g for the same genotypes grown in France for the same year. Correlations with temperatures and precipitations were studied and showed that the Hungarian site exhibited the more extremes temperature. The study also revealed that rainfall before heading may influence phenolic levels. In comparison to soluble phenolics, bound phenolic acids showed less variations across environments.

The climate change is predicted to have tremendous impact on crop cultures. Several works studied the effects of high temperature and/or drought in the phenolic composition of cereal grains. Different genotypes of sorghum were grown in optimal temperature (32 °C) and under heat stress (38 °C). An effect on the bound FA level was noticed but only for some genotypes and depending on the genotype it was either a decrease in the released FA or an increase (Wu et al., 2016). A study carried out on several genotypes of Durum wheat cultivated under three watering conditions and with or without heat stress revealed that there was no important changes in the measured FA content but significant increase in pCA level (whole level bound + soluble) with severe heat (Laddomada et al., 2021).

4. Biosynthesis of ferulic and *p*-coumaric acids, polymer acylation, cross-linkages

4.1 Deposition of ferulic and *p*-coumaric acids during cereal grain development

The evolution of FA and pCA levels in the tissues of the developing grain has been studied using biochemistry either on entire grains or dissected tissues (Barron et al., 2007; Chateigner-Boutin et al., 2018). Mild alkaline hydrolysis of pericarp cell walls from wheat grains harvested at different stages of development released increasing levels of FA from grain growth to maturation (Table 1). However, the level of released FA from pericarp samples harvested at late maturation and from dry mature samples dramatically decreased from 1896 to 272mg/100g while the level of diFA was stable and that of TriFA increased (Chateigner-Boutin et al., 2018). This is concomitant with an increase in level of recalcitrant lignin bonds in wheat pericarp which increases cross-linking of cell wall polymers and hinder FA release by NaOH. Therefore, the level of bound FA in the pericarp of mature wheat grain measured by mild alkaline hydrolysis is highly underestimated.

Other techniques allowed to determine precisely their localization in the grain tissues and to study developmental variations. For instance, the autofluorescence properties of FA and pCA upon UV excitation is well known and was used to localize HCAs in sections of mature wheat grains (Barron et al., 2007) and of developing grains (Beaugrand et al., 2004; Ghaffari et al., 2019). This method highlighted an increase in fluorescence across development and a high autofluorescence upon UV excitation of walls in the aleurone layer and in the pericarp.

In addition, antibodies have been raised against FA and pCA (Philippe, Tranquet, Utille, Saulnier, & Guillon, 2007; Tranquet, Saulnier, Utille, Ralph, & Guillon, 2009). The polyclonal anti-5-OFer-Ara was designed to target specifically feruloylated arabinose linked to AX (Philippe et al., 2007). Immunolabeling experiments, which consist in applying antibodies on sections of grains at different stages of development allowed to monitor the first appearance of feruloylated AX in the developing wheat grains (Philippe et al., 2007). Philippe et al. (2007) showed that feruloylated AX was not present in the cell walls of developing grains at very early stages but appeared at the end of the endosperm cellularization stage (when endosperm cells are formed) (Philippe et al., 2007). Immunolabeling experiments also allowed to localize feruloylated AX in cell walls (Philippe et al., 2007; Saulnier, Guillon, & Chateigner-Boutin, 2012). Signal first appeared in cell walls of the nucellar epidermis at 8–9 days after anthesis (DAA) and showed a polarity: the outer periclinal wall of the nucellar epidermis was more labeled than its inner wall and anticlinal walls. At the differentiation stage, cell walls of the nucellar epidermis also reacted positively with anti-5-OFer-Ara. At the mature stage, all cell walls of the grain were labeled (Philippe et al., 2007; Saulnier et al., 2012). Similarly, the use of anti-5-OFer-Ara revealed that in grains of *Brachypodium distachyon*, feruloylated AX also first occur in nucellar epidermis wall and then in other grain cell walls (Francin-allami et al., 2019). However, in the developing maize grain, no evidence of a strong signal corresponding to anti-5OFer-Ara in the walls of nucellar epidermis prior to signals in aleurone and pericarp was noticed (Chateigner-Boutin et al., 2016).

Antibodies were also raised against pCA (Tranquet et al., 2009). INRA-COU1 recognized a very small epitope restricted to the *p*-coumarate motif. The ester linkage was not essential but it strongly enhanced the binding. INRA-COU1 also bound to free pCA. The use of INRA-COU1 revealed a high signal in the cell walls of the wheat aleurone layer however, contrary to anti-5-OFer-Ara, INRA-COU1 did not produce labeling of cell walls in the outer layers (pericarp, testa and nucellar epidermis) of the developing wheat grain (Chateigner-Boutin et al., 2015; Tranquet et al., 2009).

4.2 Biosynthesis, polymer acylation, cross-linkages: Pathways and cellular machinery (Fig. 5)

FA and pCA are products of the phenylpropanoid pathway which also lead to the synthesis of flavonoids and lignin monomers. In contrast to eudicots, it was recently evidenced that Poaceae plants can use both Phenylalanine and Tyrosine to produce pCA which is subsequently transformed into many downstream metabolites (Barros et al., 2016).

The precursor of the acyl moiety of cell wall polymers has been the subject of a debate. It was suggested to be free FA or pCA, Feruloyl-glucose (Obel et al., 2003), or CoA esters (Meyer, Kohler, & Kauss, 1991; Yoshida-Shimokawa, Yoshida, Kakegawa, & Ischii, 2001). Subsequent identification of the nature of the reaction, of the transferases involved and the reconstruction of the reaction *in vitro* with recombinant proteins are all in favor of coenzyme A (CoA) esters to be the donors of FA and pCA.

p-coumaroylCoA is produced from pCA by the action of a 4-hydroxycinnamoyl CoA ligase (4CL). FeruloyCoA is produced from the conversion of *p*-coumaroylCoA by successive steps involving hydroxycinnamoyl CoA:shikimate hydroxycinnamoyl transferases (HCT), *p*-coumaroyl shikimate 3′-hydroxylase (C3′H) and caffeoyl CoA O-methyltransferase (CCoAOMT). Alternative routes have been described for example from pCA converted to caffeic acid to caffeoylCoA and feruloylCoA (Barros et al., 2016).

4.2.1 Acylation of grain CW polymers

An acyl-CoA transferase family carrying the PFAM PF02458 domain was proposed to be involved in the feruloylation of AX by Mitchell (Mitchell, Dupree, & Shewry, 2007). The PF02458-containing protein family is part of the BAHD family named according to the first letter of each of the first four biochemically characterized enzymes: benzyl alcohol O-acetyltransferase (BEAT), anthocyanin O-hydroxycinnamoyltransferase (AHCT), anthranilate N-hydroxycinnamoyl/benzoyltransferase (HCBT) and deacetylvindoline 4-O-acetyltransferase (DAT) (D'Auria, 2006).

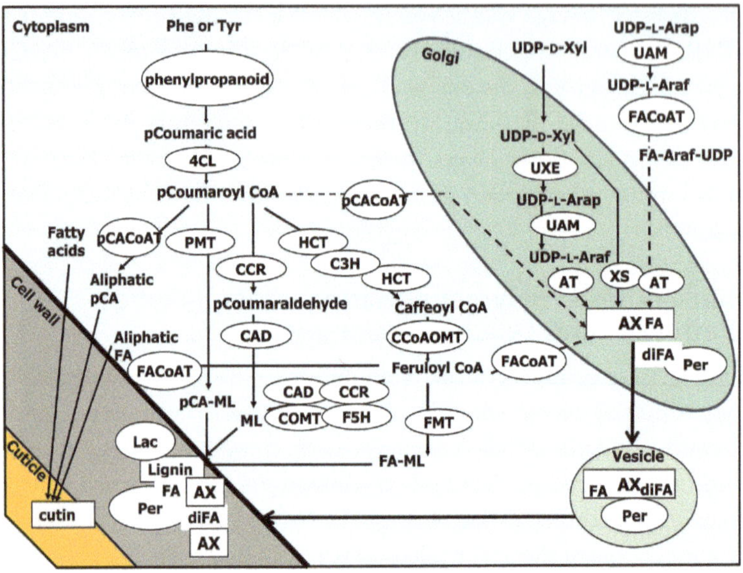

Fig. 5 Schematic representation of the biosynthesis of hydroxycinnamate-containing cell wall polymers in cereal grain. The cell depicted here is a combined model of cells in the various grain tissues at different stages of development since the composition of the grain cell wall varies both spatially and temporally. Only the polymers containing ferulic acid (FA) and *para*-coumaric acid (pCA) are represented (the cereal grain contain other cell wall polymers such as cellulose, mixed-linkage glucans, mannans, etc.). Biosynthetic pathway for the synthesis of precursors and the synthesis of the polymers as well as their transport to the wall are presented. The precursors and the different enzymes (circular boxes) involved are indicated. The nucleotide sugars are mostly synthesized in the cytosol but some are produced in the Golgi or in both subcellular locations. Enzymes such as UDP-D-xylose synthase and UDP-Arap mutase (UAM) were found as cytosolic and membrane bound in the Golgi. Cytosolic nucleotide sugars are transported to the Golgi by NDP-sugars/NMP antiporters. The phenylpropanoid pathway located in the cytosol produces monolignols (ML), FA and pCA. ML are thought to be transported to the apoplast via ABC transporters and/or by passive transport. They are polymerized *in muro* into lignin by peroxidases (Per) and laccases (Lac). The xylan backbone is synthesized by a xylan synthase complex (XS) and substituted by arabinose residues by arabinosyltransferases (AT). Others transferases (-T) add in minor amounts residues and side chains made of galactose and glucuronic acid (not shown). FA and pCA are linked to some arabinosyl residue. Not much is known about the assembly of cell wall components in wheat grain. FA is a cross-linking agent between arabinoxylan (AX) chains, or between AX and lignins. It forms di- (diFA) and oligo ferulic bonds via the action of peroxidases. FA and pCA are also found ester-linked to cutin in cuticles and the cereal grain contains several cuticles. FA and pCA are transferred to fatty acids in the cytosol. All cutin monomers are transported to the cuticle where they are polymerized. The Coenzyme A (CoA) esters of pCA and FA are the substrates for the acylation of monolignols, fatty acids and most probably for AX. The acceptor of FA

This family was revealed using a bioinformatics approach that identified gene families more expressed in AX rich plant species (Poaceae species) than in dicot species devoid of feruloylated AX. BAHD acyltransferases catalyze the acylation from a thioester of CoA to a wide range of acceptor molecules (D'Auria, 2006). Phylogeny identified several clades and subclades of BAHD, and clade A was retained as favorite for AX feruloylation because it contains genes co-expressed with AX biosynthesis genes (Molinari, Pellny, Freeman, Shewry, & Mitchell, 2013).

Several attempts were carried out to demonstrate the involvement of BAHD proteins in AX feruloylation and coumaroylation in transgenic plants, where the expression of one or several BAHD genes was either decreased or increased using RNA interference (RNAi) or overexpression (Bartley et al., 2013; de O. Buanafina, Fescemyer, Sharma, & Shearer, 2016; Piston et al., 2010). These assays resulted in moderate changes in bound FA and variations between generations.

BAHD genes are members of multigenic family and the first crystal clear evidence of a BAHD involved in AX acylation came from the suppression of SvBAHD01 (RNAi) from the species *Setaria viridis*, a monocot plant model amenable to genetic transformation (de Souza et al., 2018). A 60% reduction in bound FA was found in the stem of RNAi lines targeting SvBAHD01. Conversely an increase in pCA level was noticed in the same lines. Using mild acidolysis of Setaria alcohol insoluble residue the authors showed that the FA-Ara (linked to AX) was decreased while pCA-Ara increased in RNAi lines (de Souza et al., 2018).

Setaria viridis was also used to identify the BAHD protein involved in AX coumaroylation. Suppression of SvBAHD05 by RNAi lead to a decreased pCA levels on AX in leaves, stems and roots (Mota et al., 2020).

Several BAHD proteins were shown to acylate monolignols, the precursor of lignins (Withers et al., 2012). In this report, a rice recombinant BAHD protein co-expressed with enzymes of the lignin biosynthesis pathway was

and pCA for AX acylation is proposed to be either AX or UDP-Arabinose but this remain to be elucidated. UDP-D-Xyl, UDP-D-xylose; UXE, UDP-D-xylose epimerase; UDP-L-Arap, UDP-L-arabinopyranose; Araf, arabinofuranose; Phe, phenylalanine; Tyr, tyrosine; CCR, cinnamoyl CoA reductase; CCoAOMT, caffeoyl CoA O-methyltransferase; CAD, cinnamyl alcohol dehydrogenase; F5H, ferulate-5-hydroxylase; HCT, hydroxycinnamoyl CoA: shikimate hydroxycinnamoyl transferases; COMT, caffeic acid O-methyl transferase; PMT, pCA monolignol transferase; FMT, ferulate monolignol transferase; C3H, *p*-coumaroyl shikimate 3′-hydroxylase; 4CL, 4-hydroxycinnamoyl CoA ligase; pCACoAT, *p*-coumaroylCoA transferase; FACoAT, feruloylCoA transferase.

incubated with various CoA-esters and sinapyl or *p*-coumaryl alcohol, products identified by HPLC and kinetics parameters were monitored. sinapyl *p*-coumarate was the preferential product of the reactions. Transgenic plants confirmed the involvement of BAHD proteins in the acylation of monolignols: PMT *p*-coumaroyl CoA monolignol transferase (Petrik et al., 2014) and FMT feruloyl CoA monolignol transferase (Wilkerson et al., 2014).

The BAHD family includes also proteins involved in the feruloylation and coumaroylation of hydroxy fatty acids components of the cutin present in cuticles (Rautengarten et al., 2012). *In vitro* experiments were conducted with Arabidopsis BAHD DCF (deficient in cutin ferulate) which showed that this protein uses hydroxycinnamoyl-CoAs (preferentially feruloyl-CoA and sinapoyl-CoA) as acyl donor to transfer to γ-hydroxy fatty acids which will be incorporated in cutin polymers (Rautengarten et al., 2012). BAHDs are also involved in the acylation of other lipid containing polymers: suberin (Gou, Yu, & Liu, 2009).

BAHD proteins do not include any transmembrane domain nor targeting signals for Golgi. Fusion experiments with fluorescent reporters were carried out and revealed a cytosolic localization (Rautengarten et al., 2012). Xylan synthesis is known to occur in the Golgi apparatus (Meents, Motani, Mansfield, & Samuels, 2019). The location of AX feruloylation has been studied. Wheat seedling root segments incubated with $[^{14}C]$cinnamic acid and L-$[^{14}C]$arabinose were shown to produce $[^{14}C]$polysaccharides located both intracellularly and in the cell wall (Mastrangelo, Lenucci, Piro, & Dalessandro, 2009). A polyclonal antibody designed to target specifically feruloylated arabinose linked to AX incubated with wheat grain sections labeled cell walls, Golgi apparatus and secretory vesicles (Philippe et al., 2007).

de O. Buanafina (2009) and Molinari et al. (2013) proposed that AXs are synthesized in the Golgi apparatus and then feruloylated by feruloyl transferases using FA-CoA transported from cytosol. Alternatively, Molinari et al. (2013) and de Oliveira et al. (2015) proposed that UDP-Arabinose could be feruloylated to form FA-Ara-UDP in the cytosol which then is transported to the Golgi and incorporated in the AX chain by a glycosyltransferase. Molecular modeling analysis of the binding of FA and pCA to SvBAHD05 revealed that the presence of Araf residue around a conserved motif HLVFDG stabilizes the complex, which is in favor of the FA/pCA-Ara-UDP model (Mota et al., 2020).

Similarly, acylation of monolignols probably takes place in the cytosol and preacylated monomers are subsequently transported across the plasma membrane to the cell wall via a mechanism which is not totally elucidated. Both passive and transporter-assisted transports involving ATP-binding cassette (ABC) transporters were proposed (Vermaas et al., 2019; Perkins, Smit, & Samuels, 2019). Monolignols are then incorporated into lignins (Väisänen et al., 2020).

Acylation of fatty acid with FA and pCA is also proposed to take place in the cytosol (corresponding BAHD DCF was localized in the Golgi by GFP fusion experiment) (Rautengarten et al., 2012). Cutin monomers would be transported through the plasma membrane and cell wall to be polymerized in the forming cuticle. Alike monolignol transport, lipid transport across plasma membrane and cell wall is not completely elucidated (Wan, Wu, Li, An, & Tian, 2020). Trafficking of precursors across the plasma membrane has been shown to depend on ABC transporters and apoplastic Lipid Transfer Proteins have been proposed to play a role in transport of monomers across cell wall (Yeats & Rose, 2013).

4.2.2 Cross-linking of polymers

The FA oxidative coupling of AX chains, and FA oxidative coupling between AX and lignin involves oxidases. Phenol ring in monolignols and FA are oxidized and the radicals generate cross-coupling. The same enzyme families have also been implicated in lignin polymerization (Yi Chou et al., 2018). Plant class III peroxidases are iron-containing enzymes that use hydrogen peroxide to produce radicals while laccases are multi-copper-containing enzymes that uses O_2 to produce radicals. Laccase genes are mainly expressed in the lignified tissues (Blaschek & Pesquet, 2021).

Laccase and peroxidase were proposed to play non redundant roles in lignification (Zhao et al., 2013). Recently it was shown that a peroxidase and a laccase both involved in lignification were localized differently AtPRX 64 in middle lamella and cell corners of fibers while AtLAC4 was localized in secondary cell wall of vessels and fibers (Yi Chou et al., 2018). Peroxidases are known to oxidize many compounds. Specific peroxidases have been shown to act under particular developmental contexts such as the root Casparian strip (Lee, Rubio, Alassimone, & Geldner, 2013) or the seed testa (Jia et al., 2013).

FA dimerization localization was studied and shown to occur in the cell wall but also intracellularly in the endomembrane system (Fry, Willis, & Paterson, 2000). Since most peroxidases are targeted to cell walls by the secretory pathway it is likely that feruloylated AX could co-localize and be oxidized by peroxidase in Golgi and/or in vesicles. In maize suspension cells incubated with [^{14}C]cinnamate, radiolabeled [^{14}C]feruloylated AX were detected. Inhibition of endogenous peroxidase by H_2O_2 scavenger iodide lead to an increase in released [^{14}C]feruloylated AX suggesting the involvement of peroxidase in AX cross-linking (Lindsay & Fry, 2008).

4.2.3 In cereal grains

None of the above studies were conducted on cereal grains. It's reasonable to think that the same families are involved in FA, pCA synthesis, transfer to polymers and cross-linking in vegetative tissues and in cereal grains.

In cereal grains, proteomic and transcriptomic studies were carried out to identify cell wall associated genes and among them glycosyltransferases, BAHD, peroxidases were detected (Chateigner-Boutin et al., 2016, 2015; Mitchell et al., 2007; Pellny et al., 2012). RNAi attempts to down-regulate BAHD candidates in the endosperm of wheat grains failed, possibly because of gene redundancy and/or because the expression of the RNAi construct driven by an endosperm specific promoter (promoter of a storage protein) did not prevent an early accumulation of BAHD proteins (R. Mitchell, Personal Communication). Expression atlases are available for several cereal species and can help identifying genes involved in FA, pCA synthesis, transfer to polymers and cross-linking. For instance, in maize pericarp a tissue containing high level of FA and AX but poor in lignin, expressed genes of the phenylpropanoid pathway, lignin pathway and metabolism of AX were identified using the maize eFP browser. Among these genes, CCoAOMT genes, BAHD genes, type III peroxidase genes were highlighted and no laccase genes were found expressed while genes encoding peroxidases were highly expressed (Chateigner-Boutin et al., 2016).

5. Biological functions in cell wall architecture and in tissue protection

5.1 Cell wall architecture and cross-linking

FA cross-linking of AX chains is proposed to have a role in cell wall organization. FA-AX are secreted to the cell wall. According to Hatfield, Rancour, and Marita (2017), in the developing cell wall, secreted AX with high degree of arabinosyl substitution could move into the free space. As cell walls mature, arabinose substitutions are removed by arabinofuranosidases, and such a decrease in arabinose substitution was observed during the development of the endosperm in the wheat grain (Philippe, Robert, Barron, Saulnier, & Guillon, 2006).

The distribution of the arabinose side-chains on the backbone of the different types of xylans, and the distribution of FA ester on the arabinose side chains is still an unsolved question. GlucuronoArabinoXylans (GAX) encountered in cell walls of grass stems/leaves, ArabinoXylans (AX) in cereal grain endosperm and HeteroXylans (HX) in cereal grain pericarp, have very different arabinose to xylose, and FA to arabinose ratios. It has been hypothesized that for GAX present in highly lignified tissue of grasses, arabinose and FA are not randomly distributed along the xylose backbone favoring AX cross-linking through diferulic bridges and interactions through unsubstituted part of the xylan backbone, between the GAX polymers and with cellulose (Hatfield et al., 2017). This situation could not be encountered in cereal grain since AX and HX are much more substituted than GAX, and evidence of clusters of arabinose along the xylan backbone has never been demonstrated for AX or HX (Saulnier et al., 2012; Saulnier, Guillon, Sado, Chateigner-Boutin, & Rouau, 2013).

Apart from FA cross-linking AX chains, there is evidence that ester-linked FA to AX (FA-AX) act as nucleation sites for lignin (Ralph, Grabber, & Hatfield, 1995). A first step would be the binding of FA-AX to coniferyl alcohol that could oxidatively couple in the presence of appropriate oxidases with free monolignols in the cell wall to initiate lignin synthesis. It was proposed that numerous initiations might occur and could form small pools of lignin which would be responsible to the relatively high alkaline solubility of lignin in grasses especially in young tissues. The pools of lignin would subsequently connect as lignification increases (Hatfield et al., 2017).

pCA is only attached to lignin by ester-linkages (and mainly to S unit). It could act as a termination to lignin polymer and it could contribute to the fact that grass S rich lignin are more linear than dicot lignin (Hatfield et al., 2017). Adding pCA in dicot lignin by expressing Brachypodium *p*-coumaroyl CoA monolignol transferase in Arabidopsis increased lignin alkaline solubility and suggests that when *p*-coumaroylated monolignols participate in lignification, there is formation of alkali-releasable lignin domains (Sibout et al., 2016). A role for pCA in radical transfer during the lignification process was also proposed (Hatfield et al., 2017).

An hypothetical function in cell wall architecture was proposed for FA and pCA linked to suberin and cutin. Kolattukudy (1981) refers to an hypothesis stating that cell wall components could be attached to phenolics and aliphatic components. HCA would promote linkages between cutin/suberin and polysaccharides or anchor the cuticle/suberin to the cell wall.

The *A. thaliana* mutant defective in cuticular ridges (*dcr*), shows alterations in the cuticle organization: the interface between the cell wall and its cuticle is disrupted (Mazurek et al., 2017). In the latter article, no link was made with FA or pCA but FA and pCA could be involved in cutin-cell wall polymers interactions. In a recent work where they elucidated the composition of sporopollenin, the authors proposed that pCA would crosslink the sporopollenin components (Li et al., 2019).

5.2 Growth cessation and wall stiffening

In developing pericarp of wheat and maize, feruloylation monitored by immunolabeling increases (Chateigner-Boutin et al., 2016; Philippe et al., 2007). Mild alkaline released FA also increases across development up to a certain development stage (Chateigner-Boutin et al., 2018). It was proposed that in young tissues, the plant growth regulator gibberellin decreases apoplastic peroxidase activity to promote cell elongation (Fry, 1980).

A proposed consequence of increased FA and cross-linking in maturing tissue is wall stiffening which would constrain growth. In *Avena sativa* coleoptiles, the level of FA and diFA correlates with wall mechanical stiffness (Kamisaka, Takeda, Takahashi, & Shibata, 1990).

5.3 UV protection, resistance to pathogen: Cell wall integrity, mechanical reinforcement and inhibition of fungal growth

Phenolics are antioxidant and accordingly the presence of pCA and FA in the cuticle at the surface of aerial organs was proposed to be involved in protection against UV radiation. Cuticles were also proposed to provide protection against desiccation, and pathogen attack.

Pathogens have acquired mechanisms to attack plant cell wall structures. Several works suggest that FA is involved in plant protection against pathogens in vegetative tissues and in grains of cereals.

5.3.1 Fungi

Fungus contamination can spoil grain stocks and/or produce dangerous mycotoxin. Several studies reported roles of FA in relation to fungus contamination of grains. Free FA is reported to inhibit fungus growth. For instance, *in vitro* activity of FA on mycelial growth of *Fusarium graminearum, F. langsethiae and F. poae* were performed on milled cereals supplemented with increasing concentrations of FA. FA inhibited the growth of *F. graminearum* and *F. langsethiae* and reduced that of *F. poae* (Schöneberg et al., 2018).

However, it was reported that upon contamination, *Aspergillus niger* produces cell wall degrading enzymes among which a feruloyl esterase which hydrolyses ester linkages and releases FA from wheat and maize bran (Crepin, Faulds, & Connerton, 2004; Faulds, Kroon, Saulnier, Thibault, & Williamson, 1995). Transgenic expression of the ferulate esterase from *Aspergillus niger* in *Brachypodium* (vegetative tissues) lead to a decreased level of bound FA (a decreased release of FA and DiFA upon alkaline treatment) and increased susceptibility to *Bipolaris sorokiniana* (Reem et al., 2016).

Several works were conducted on maize grains to study the role of FA in maize resistance to pathogens since the maize grain outer layers are rich in FA with variability observed among genotypes (Assabgui, Reid, Hamilton, & Arnason, 1993; Bily et al., 2003). For instance, several lines were inoculated by *Fusarium gramianearum* and outer grain layers were separated from the endosperm and treated with 2N NaOH before analysis. Two genotypes with a very different genetic background exhibited high levels of DiFA and high levels of resistance (Bily et al., 2003). Moreover, a population of 75 individuals generated from a cross of a resistant and susceptible parent was analyzed. A negative correlation was found between levels of DiFA and disease ratings (Bily et al., 2003). It can be envisioned that DiFA would strengthen the physical integrity of the cell wall and hinder mycelium penetration.

5.3.2 Insects

Insects can damage plant vegetative tissues and also seeds and cereal grains either on the mother plant or during storage. Some studies investigated the role of grain cell wall and in particular phenolics in insect resistance. For instance, the role of maize pericarp cell wall components in resistance against weevil an insect that feed on stored grains was investigated in different maize lines. It was shown that resistant lines contained higher levels of DiFA than susceptible genotypes (Arnason, Baum, Gale, Lambert, et al., 1994; García-Lara et al., 2004). Moreover, mechanical hardness of the grain was measured and found negatively correlated with susceptibility to weevils but positively correlated to FA and diFA contents. Highly cross-linked and hard cell walls would be more resistant to the action of insect mandibles (Choong, 1996). Meta-QTL analysis targeting stem tissues confirmed the significant role of pCA, FA and diFA for maize resistance to insects (Badji et al., 2018).

6. Technological and nutritional impact of FA

FA as a cross linking agent of AX in cereal grains has clearly major effects on the properties of the cell walls and, consequently, on the processing of cereal grain into food products. The antioxidant properties of FA also confer specific nutritional properties on cereal grains with positive impact on human health. Besides cereal grains and especially outer tissues of the grain are sources of FA, feruloylated oligosaccharides (AXOS) and feruloylated AX than can be exploited in various applications. In this section we give a quick overview of these different aspects of FA in cereal grains.

6.1 Impact on grain processing and bread making

Mechanical properties of tissues are known to be determined by the composition and structure of their cell walls and by the interactions between the cell wall components (Burgert, 2006). In this respect, AX cross-linking by FA dehydrodimers and trimers and interconnection with lignin in grain outer layers affect the mechanical properties of these tissues and potentially impact their behavior during the milling process. Different studies using peroxidases/hydrogen peroxide systems, UV light or enhancing peroxidase activity *in situ* have shown the formation of covalent linkages involving phenolic moieties in cell walls or at tissue interfaces modifying bran extensibility and stiffness (Desvignes et al., 2006; Peyron, Abecassis, Autran, & Rouau, 2002; Peyron, Abecassis, Autran, & Rouau, 2001). The modifications in the mechanical properties of these tissues, result in an increased friability during grinding and a decrease in the size of generated bran particles but also impact endosperm/bran separability (Hemery, Rouau, Lullien-Pellerin, Barron, & Abecassis, 2007).

FA is likely to be involved in dough rheology, by interfering with the disulfide bond formation, due to its reductive capacity (Schefer, Oest, & Rohn, 2021). In bread-making the formation and the properties of the wheat gluten network are crucial to bread quality and relies to the formation and reorganization of disulfide bonds between gluten proteins components during the mixing of the dough. Although the effects of adding free FA on gluten network has been intensively studied, the exact mechanism is still not comprehensively understood (Saulnier et al., 2013; Schefer et al., 2021).

Gelation of feruloylated AX from wheat endosperm under oxidizing conditions has been recognized for a long time (Neukom & Markwalder, 1978). The occurrence of water extractable (WE)-AX gelation during bread-making has been hypothesized, with development of a secondary network superimposed to gluten, playing a role in dough rheology. However, the extensive setting up of such a network has not been demonstrated in normal dough (Saulnier et al., 2013). Next to the formation of AX networks, a reaction between AX-bound FA and proteins, likely via FA-tyrosine bridges, may also occur. Model studies have proved that low molecular weight adducts of tyrosine or tyrosine-rich peptides and FA can be produced in specific conditions (Figueroa-Espinoza, Morel, & Rouau, 1998). AX/casein adducts have been obtained (Boeriu et al., 2004), and diFA-tyrosine has been isolated from rye and wheat flour dough (Piber & Koehler, 2005). However, the amount of the diFA–tyrosine conjugates in wheat dough account for a very minor amount of the FA and tyrosine present in the flour and probably do not have a significant influence on dough functionality.

6.2 Nutritional impact of FA in cereal grain

Epidemiological studies have shown that high dietary fiber diets and whole grain consumption are associated with diminished risk of coronary heart disease, colon cancers, inflammatory bowel disease, and metabolic syndrome (Reynolds et al., 2019) and overall with a reduced risk of mortality (Huang, Xu, Lee, Cho, & Qi, 2015). Diverse mechanisms mediate the protective effects of whole wheat grain, some are related to the effects of cereal grain dietary fibers but synergistic action of other bioactive compounds also contributes to health protection (Fardet, 2010). Among these bioactive compounds, FA has received particular interest as it is the most abundant phenolic acid in cereals grains and also because contrary to other phenolic acids it is intrinsically linked to dietary fibers (AX); As pointed out in Table 1 FA is mainly present in the outermost layers of wheat that are recovered during the milling process in a fraction called bran. Among the outer layers, aleurone appears as very interesting as it concentrates most dietary fibers, bioactive compounds and FA in the grain (Brouns, Hemery, Price, & Anson, 2012; Rosa-Sibakov, Poutanen, & Micard, 2015). FA as other phenolic acids exhibit antioxidant activity which is the basis for promoting various health benefits, which for most of them are believed to be based on the neutralization of free radicals (e.g., Reactive Oxygen Species: ROS) present in the human organism. FA and other phenolic acids are further ascribed with anti-inflammatory properties, as well as providing antimicrobial activity (Schefer et al., 2021). Another interesting protective mechanism of FA is related with UV absorption (de Oliveira Silva & Batista, 2017). Many studies have focused on the effect of the administration of free FA in animals and humans and evidenced that FA can be absorbed along the entire gastrointestinal tract and metabolized mainly by the liver into a number of metabolites mainly FA-glucuronide and FA-sulfate (de Oliveira Silva & Batista, 2017; Zhao & Moghadasian, 2008). However, FA in cereal grains mainly exists in its bound form cross-linking cell wall polysaccharides together and with lignin, and therefore cannot be absorbed directly through the upper gastrointestinal tract. In the colon, esterases and xylanases are produced by the microbiota and therefore FA can be released and become available for absorption and further metabolism. However, the covalent linkages between polysaccharide and lignin network also restrict the degradation of cell walls by microbial enzymes (Rose, Patterson, & Hamaker, 2010) and the bioavailability of FA is therefore dependent on cell wall polysaccharide structure and interconnection and may also be impacted by the cell wall integrity. In this respect the reduction of aleurone particle size enhances its antioxidant capacity and subsequent treatment with microbial xylanase and esterase acts synergistically to release bio-accessible FA (Rosa, Aura, et al., 2013; Rosa, Barron, Gaiani, Dufour, & Micard, 2013; Rosa, Dufour, Lullien-Pellerin, & Micard, 2013). Regarding dimers, that are in high amount in grain outer layers and could be released by microbial enzymes in the colon, some studies indicate that they possess a higher radical scavenging activity than FA (Amić, Marković, Dimitrić Marković, Milenković, & Stepanić, 2020; Schefer et al., 2021). *In vitro* assay of antioxidant capacity of feruloylated-AXOS and cross-linked feruloylated-AXOS indicated a lowered capacity compared to free FA but still high compared to water soluble vitamin E analog which makes feruloylated-AXOS interesting water-soluble antioxidant (Snelders, Dornez, Delcour, & Courtin, 2013).

6.3 Cereal grain as sources of ferulic acid and feruloylated polymers

FA and other phenolic acids are interesting chemicals that can be used in a variety of applications or further modified to high value-added compounds (Tinikul, Chenprakhon, Maenpuen, & Chaiyen, 2018). The transformation of FA into vanillin using biological process has received considerable attention as the product labeling as "natural vanilla flavor" can increase vanillin price hugely. Vanillin is an important aromatic flavor compound used in foods, beverages, perfume and pharmaceuticals produced on a large scale in the industry through chemical synthesis, but that cannot be considered as natural. The biotransformation of FA into vanillin can be achieved following different pathways (Kumar & Pruthi, 2014) and production of "natural" vanillin using microorganism has been implemented at the industrial scale. A variety a phenolic acid and especially FA and pCA can be derived from the chemical, physicochemical, biochemical, and biological depolymerization of different types of lignin (Fache, Boutevin, & Caillol, 2016). FA is in relatively high amount in cereal by-products such as cereal brans (wheat, maize). The combination of physicochemical and enzymatic treatments can release efficiently FA from maize bran (Saulnier, Marot, Elgorriaga, Bonnin, & Thibault, 2001) or wheat bran (Dupoiron et al., 2017; Ferri et al., 2020) but the separation of FA from these agricultural feedstocks and downstream processes demands energy and cost and the development of efficient separation and recovery process is a key issue that is not yet solved at industrial scale.

As pointed above, gelation of water-soluble feruloylated AX occurring in wheat flour has been reported since many decades. In mild alkaline conditions (pH 11–12 temperature 40–50 °C), HXs from cereal brans can be extracted while preserving some of the FA-arabinose ester linkages. Feruloylated xylan polymers extracted in mild alkaline conditions can form gels under oxidative conditions similarly to feruloylated AX from the endosperm of wheat grains (Fig. 6) (Carvajal-Millan et al., 2007; Ng, Greenshields, & Waldron, 1997). The gelation primarily results from the

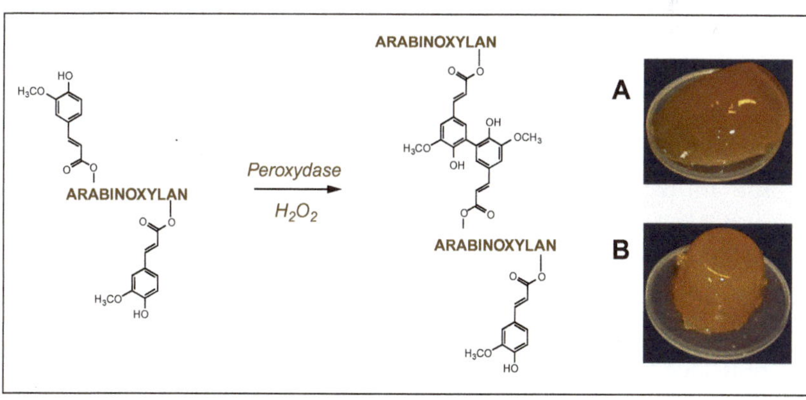

Fig. 6 Oxidative gelation of arabinoxylans (AX) extracted from maize bran under mild alkaline conditions. (A) Soft gel obtained with 3% AX solution (AX containing 0.1% of ferulic acid (FA)). (B) Rigid gel obtained with 3% AX solution (AX containing 0.3% of ferulic acid (FA)).

formation of three-dimensional networks of AX chains anchored by DiFA. The 8-5′, 8-O-4′, 5-5′, and 8-8′ DiFA have been detected in AX gels, with the 8-5′ and 8-O-4′ being the most abundant (Bunzel, Ralph, Marita, Hatfield, & Steinhart, 2001; Dervilly-Pinel, Rimsten, Saulnier, Andersson, & Åman, 2001). Apart of it is possible role on dough rheology during bread-making, AX gels have a potential interest for controlled drug release (Carvajal-Millan, Guilbert, Doublier, & Micard, 2006; Martínez-López et al., 2019), e.g., in colon as the AX gel could entrap drugs and protect them during the transit in the gastro-intestinal tract and delay their delivery in the colon where AX gels could be degraded by microbial enzymes. Another use of AX gels is related to their high water-binding properties and soft texture that can be used for specific wound dressing (Miraftab, Qiao, Kennedy, Anand, & Groocock, 2003).

7. Conclusion

FA and pCA are components of cereal grains mainly found associated to cell wall polymers. A considerable variability in the level of FA and pCA has been uncovered among the grain of different cereal species even between the three most cultivated cereals: maize, wheat and rice, and between different lines of a same species. It is difficult to find clear explanations for this variability apart from variations in anatomy, tissue compositions and lignin content. Maize pericarp is the tissue with the highest level of FA, while more lignified tissues such as stems are richer in pCA.

FA plays multiple important roles *in planta* in cell wall assembly and mechanical properties. Among the many remaining open questions is the role of FA and pCA in the interactions between the different polymers of the cell wall such as cutin, polysaccharides and lignin. A role of FA in plant response to pathogens is suggested by several studies. The existing intra-specific variability in FA level could be used to select genotypes with grains less susceptible to pathogens. In this respect, further studies are needed to investigate variations in FA content with environmental conditions and the possible impact on the properties of the grain cell walls and tissues and their resistance to pathogens.

Numerous reports are published these days to study the phenolic compounds of diverse plant sources among which FA and pCA, in relation to benefits for human health. Many applications could be envisioned to exploit these sources of phenolics. However cereal grains, and in the last decades not only flour but whole grains, are mainly used for human consumption therefore other applications could compete with food supply. For instance, extraction would be technically feasible from maize and wheat bran (usually used in animal feed), or from by-products of alkaline cooking of maize (tortillas-making) rich in soluble FA. However, life cycle costs and life cycle assessments must be conducted to prove the sustainability of such procedures. Moreover, in food applications, it should be more reasonable to promote whole grain products and use biological process (sprouting, malting, etc.) of grain/bran to enhance FA bioavailability rather than supplementing food with extracted FA for health benefits.

References

Ahluwalia, B., & Fry, S. C. (1986). Barley endosperm cell walls contain a feruloylated arabinoxylan and a non-feruloylated β-glucan. *Journal of Cereal Science*, *4*(3), 287–295. https://doi.org/10.1016/S0733-5210(86)80031-5.

Allerdings, E., Ralph, J., Schatz, P. F., Gniechwitz, D., Steinhart, H., & Bunzel, M. (2005). Isolation and structural identification of diarabinosyl 8-O-4-dehydrodiferulate from maize bran insoluble fibre. *Phytochemistry*, *66*(1), 113–124.

Allerdings, E., Ralph, J., Steinhart, H., & Bunzel, M. (2006). Isolation and structural identification of complex feruloylated heteroxylan side-chains from maize bran. *Phytochemistry*, *67*(12), 1276–1286. https://doi.org/10.1016/j.phytochem.2006.04.018.

Amić, A., Marković, Z., Dimitrić Marković, J. M., Milenković, D., & Stepanić, V. (2020). Antioxidative potential of ferulic acid phenoxyl radical. *Phytochemistry*, *170*(October 2019). https://doi.org/10.1016/j.phytochem.2019.112218.

Andersson, A. A. M., Andersson, R., Piironen, V., Lampi, A.-M., Nyström, L., Boros, D., et al. (2013). Contents of dietary fibre components and their relation to associated bioactive components in whole grain wheat samples from the HEALTHGRAIN diversity screen. *Food Chemistry*, *136*(3–4), 1243–1248. https://doi.org/10.1016/j.foodchem.2012.09.074.

Andersson, A. A. M., Kamal-Eldin, A., Fraś, A., Boros, D., & Åman, P. (2008). Alkylresorcinols in wheat varieties in the HEALTHGRAIN diversity screen. *Journal of Agricultural and Food Chemistry*, *56*(21), 9722–9725. https://doi.org/10.1021/jf8011344.

Andersson, A. A. M., Lampi, A.-M., Nyström, L., Piironen, V., Li, L., Ward, J. L., et al. (2008). Phytochemical and dietary fiber components in barley varieties in the HEALTHGRAIN diversity screen. *Journal of Agricultural and Food Chemistry*, *56*(21), 9767. https://doi.org/10.1021/jf802037f.

Andreasen, M. F., Christensen, L. P., Meyer, A. S., & Hansen, A. (2000). Content of phenolic acids and ferulic acid dehydrodimers in 17 rye (Secale cereale L.) varieties. *Journal of Agricultural and Food Chemistry*, *48*(7), 2837–2842. https://doi.org/10.1021/jf991266w.

Arnason, J. T., Baum, B., Gale, J., Lambert, J. D. H., et al. (1994). Variation in resistance of Mexican landraces of maize to maize weevil Sitophilus zeamais, in relation to taxonomic and biochemical parameters. *Euphytica*, *74*(3), 227–236.

Assabgui, R. A., Reid, L. M., Hamilton, R. I., & Arnason, J. T. (1993). Correlation of kernel (E)-ferulic acid content of maize with resistance to Fusarium graminearum. *Phytopathology*, *83*, 949–953.

Badji, A., Otim, M., Machida, L., Odong, T., Kwemoi, D. B., Okii, D., et al. (2018). Maize combined insect resistance genomic regions and their co-localization with cell wall constituents revealed by tissue-specific QTL meta-analyses. *Frontiers in Plant Science*, *9*, 895. https://doi.org/10.3389/fpls.2018.00895.

Barron, C., Surget, A., & Rouau, X. (2007). Relative amounts of tissues in mature wheat (Triticum aestivum L.) grain and their carbohydrate and phenolic acid composition. *Journal of Cereal Science*, *45*(1), 88–96. https://doi.org/10.1016/j.jcs.2006.07.004.

Barros, J., Serrani-Yarce, J. C., Chen, F., Baxter, D., Venables, B. J., & Dixon, R. A. (2016). Role of bifunctional ammonia-lyase in grass cell wall biosynthesis. *Nature Plants*, *2*, 16050.

Bartley, L. E., Peck, M. L., Kim, S.-R., Ebert, B., Manisseri, C., Chiniquy, D. M., et al. (2013). Overexpression of a BAHD acyltransferase, OsAt10, alters rice cell wall hydroxycinnamic acid content and saccharification. *Plant Physiology*, *161*(4), 1615–1633. https://doi.org/10.1104/pp.112.208694.

Beaugrand, J., Crônier, D., Thiebeau, P., Schreiber, L., Debeire, P., & Chabbert, B. (2004). Structure, chemical composition, and xylanase degradation of external layers isolated from developing wheat grain. *Journal of Agricultural and Food Chemistry*, *52*, 7108–7117. https://doi.org/10.1021/jf049529w.

Bily, A. C., Reid, L. M., Taylor, J. H., Johnston, D., Malouin, C., Burt, A. J., et al. (2003). Dehydrodimers of ferulic acid in maize grain pericarp and aleurone: Resistance factors to fusarium graminearum. *Phytopathology*, *93*(6), 712–719. https://doi.org/10.1094/PHYTO.2003.93.6.712.

Blaschek, L., & Pesquet, E. (2021). Phenoloxidases in plants—How structural diversity enables functional specificity. *Frontiers in Plant Science*, *12*(October), 1–24. https://doi.org/10.3389/fpls.2021.754601.

Boeriu, C. G., Oudgenoeg, G., Spekking, W. T. J., Berendsen, L. B. J. M., Vancon, L., Boumans, H., et al. (2004). Horseradish peroxidase-catalyzed cross-linking of feruloylated arabinoxylans with β-casein. *Journal of Agricultural and Food Chemistry*, *52*(21), 6633–6639. https://doi.org/10.1021/jf049622k.

Brouns, F., Hemery, Y., Price, R., & Anson, N. M. (2012). Wheat aleurone: Separation, composition, health aspects, and potential food use. *Critical Reviews in Food Science and Nutrition*, *52*(6), 553–568. https://doi.org/10.1080/10408398.2011.589540.

Bunzel, M. (2010). Chemistry and occurrence of hydroxycinnamate oligomers. *Phytochemistry Reviews*, *9*(1), 47–64. https://doi.org/10.1007/s11101-009-9139-3.

Bunzel, M., Allerdings, E., Ralph, J., & Steinhart, H. (2008). Cross-linking of arabinoxylans via 8-8-coupled diferulates as demonstrated by isolation and identification of diarabinosyl 8-8(cyclic)-dehydrodiferulate from maize bran. *Journal of Cereal Science*, *47*(1), 29–40.

Bunzel, M., Ralph, J., Brüning, P., & Steinhart, H. (2006). Structural identification of dehydrotriferulic and dehydrotetraferulic acids isolated from insoluble maize bran fiber. *Journal of Agricultural and Food Chemistry*, *54*(17), 6409–6418. https://doi.org/10.1021/jf061196a.

Bunzel, M., Ralph, J., Marita, J. M., Hatfield, R. D., & Steinhart, H. (2001). Diferulates as structural components in soluble and insoluble cereal dietary fibre. *Journal of the Science of Food and Agriculture*, *81*(7), 653–660.

Burgert, I. (2006). Exploring the micromechanical design of plant cell walls. *American Journal of Botany*, *93*(10), 1391–1401. https://doi.org/10.3732/ajb.93.10.1391.

Carvajal-Millan, E., Guilbert, S., Doublier, J.-L., & Micard, V. (2006). Arabinoxylan/protein gels: Structural, rheological and controlled release properties. *Food Hydrocolloids*, *20*(1), 53–61. https://doi.org/10.1016/j.foodhyd.2005.02.011.

Carvajal-Millan, E., Rascón-Chu, A., Márquez-Escalante, J. A., Micard, V., de León, N. P., & Gardea, A. (2007). Maize bran gum: Extraction, characterization and functional properties. *Carbohydrate Polymers*, *69*(2), 280–285. https://doi.org/10.1016/j.carbpol.2006.10.006.

Chateigner-Boutin, A.-L., Lapierre, C., Alvarado, C., Yoshinaga, A., Barron, C., Bouchet, B., et al. (2018). Ferulate and lignin cross-links increase in cell walls of wheat grain outer layers during late development. *Plant Science*, *276*, 199–207. https://doi.org/10.1016/j.plantsci.2018.08.022.

Chateigner-Boutin, A.-L., Ordaz-Ortiz, J. J., Alvarado, C., Bouchet, B., Durand, S., Verhertbruggen, Y., et al. (2016). Developing pericarp of maize: A model to study arabinoxylan synthesis and feruloylation. *Frontiers in Plant Science*, 7(September2016), 1476. https://doi.org/10.3389/fpls.2016.01476.

Chateigner-Boutin, A. L., Suliman, M., Bouchet, B., Alvarado, C., Lollier, V., Rogniaux, H., et al. (2015). Endomembrane proteomics reveals putative enzymes involved in cell wall metabolism in wheat grain outer layers. *Journal of Experimental Botany*, *66*(9), 2649–2658. https://doi.org/10.1093/jxb/erv075.

Choong, M. F. (1996). What makes a leaf tough and how this affects the pattern of Castanopsis fissa leaf consumption by caterpillars. *Functional Ecology*, *10*(5), 668–674.

Chougui, N., Tamendjari, A., Hamidj, W., Hallal, S., Barras, A., Richard, T., et al. (2013). Oil composition and characterisation of phenolic compounds of Opuntia ficus-indica seeds. *Food Chemistry*, *139*, 796–803. https://doi.org/10.1016/j.foodchem.2013.01.054.

Crepin, V. F., Faulds, C. B., & Connerton, I. F. (2004). Functional classification of the microbial feruloyl esterases. *Applied Microbiology and Biotechnology*, *63*(6), 647–652. https://doi.org/10.1007/s00253-003-1476-3.

D'Auria, J. C. (2006). Acyltransferases in plants: A good time to be BAHD. *Current Opinion in Plant Biology*, *9*(3), 331–340. https://doi.org/10.1016/j.pbi.2006.03.016.

de O. Buanafina, M. M. (2009). Feruloylation in grasses: Current and future perspectives. *Molecular Plant*, *2*(5), 861–872. https://doi.org/10.1093/mp/ssp067.

de O. Buanafina, M. M., Fescemyer, H. W., Sharma, M., & Shearer, E. A. (2016). Functional testing of a PF02458 homologue of putative rice arabinoxylan feruloyl transferase genes in Brachypodium distachyon. *Planta*, *243*(3), 659–674. https://doi.org/10.1007/s00425-015-2430-1.

de Oliveira, D. M., Finger-Teixeira, A., Mota, T. R., Salvador, V. H., Moreira-Vilar, F. C., Molinari, H. B., et al. (2015). Ferulic acid: A key component in grass lignocellulose recalcitrance to hydrolysis. *Plant Biotechnology Journal*, *13*(9), 1224–1232. https://doi.org/10.1111/pbi.12292.

de Oliveira Silva, E., & Batista, R. (2017). Ferulic acid and naturally occurring compounds bearing a feruloyl moiety: A review on their structures, occurrence, and potential health benefits. *Comprehensive Reviews in Food Science and Food Safety*, *16*(4), 580–616. https://doi.org/10.1111/1541-4337.12266.

de Souza, W. R., Martins, P. K., Freeman, J., Pellny, T. K., Michaelson, L. V., Sampaio, B. L., et al. (2018). Suppression of a single BAHD gene in Setaria viridis causes large, stable decreases in cell wall feruloylation and increases biomass digestibility. *New Phytologist*, *218*, 81–93. https://doi.org/10.1111/nph.14970.

Dervilly-Pinel, G., Rimsten, L., Saulnier, L., Andersson, R., & Åman, P. (2001). Water-extractable arabinoxylan from pearled flours of wheat, barley, rye and triticale. Evidence for the Presence of Ferulic Acid Dimers and their Involvement in Gel Formation. *Journal of Cereal Science*, *34*(2), 207–214. https://doi.org/10.1006/jcrs.2001.0392.

Desvignes, C., Olivé, C., Lapierre, C., Rouau, X., Pollet, B., & Lullien-Pellerin, V. (2006). Effects of calcium chloride treatments on wheat grain peroxidase activity and outer layer mechanical properties. *Journal of the Science of Food and Agriculture*, *86*(11), 1596–1603. https://doi.org/10.1002/jsfa.2499.

Dimmel, D. R. (2010). Overview. In C. H. Heitner, D. R. Dimmel, & J. A. Schmidt (Eds.), *Lignin and lignans. Advances in chemistry* (pp. 1–10). Florida: CRC Press.

Domergue, F., & Kosma, D. K. (2017). Occurrence and biosynthesis of alkyl hydroxycinnamates in plant lipid barriers. *Plants*, *6*(3), 25. https://doi.org/10.3390/plants6030025.

Dupoiron, S., Lameloise, M. L., Pommet, M., Bennaceur, O., Lewandowski, R., Allais, F., et al. (2017). A novel and integrative process: From enzymatic fractionation of wheat bran with a hemicellulasic cocktail to the recovery of ferulic acid by weak anion exchange resin. *Industrial Crops and Products, 105*(May), 148–155. https://doi.org/10.1016/j.indcrop.2017.05.004.

Dykes, L. (2019). Tannin analysis in Sorghum grains. *Methods in Molecular Biology, 1931*, 109–120. https://doi.org/10.1007/978-1-4939-9039-9_8.

Evers, T., & Millar, S. (2002). Cereal grain structure and development: Some implications for quality. *Journal of Cereal Science, 36*(3), 261–284. https://doi.org/10.1006/jcrs.2002.0435.

Fache, M., Boutevin, B., & Caillol, S. (2016). Vanillin production from lignin and its use as a renewable chemical. *ACS Sustainable Chemistry & Engineering, 4*(1), 35–46. https://doi.org/10.1021/acssuschemeng.5b01344.

Fardet, A. (2010). New hypotheses for the health-protective mechanisms of whole-grain cereals: What is beyond fibre? *Nutrition Research Reviews, 23*(01), 65–134. https://doi.org/10.1017/S0954422410000041.

Faulds, C. B., Kroon, P. A., Saulnier, L., Thibault, J.-F., & Williamson, G. (1995). Release of ferulic acid from maize bran and derived oligosaccharides by Aspergillus niger esterases. *Carbohydrate Polymers, 27*(3), 187–190. https://doi.org/10.1016/0144-8617(95)00073-G.

Feng, Z., Dong, L., Zhang, R., Chi, J., Liu, L., Zhang, M., et al. (2021). Structural elucidation, distribution and antioxidant activity of bound phenolics from whole grain brown rice. *Food Chemistry, 358*, 129872. https://doi.org/10.1016/j.foodchem.2021.129872.

Fernandez-Orozco, R., Li, L., Harflett, C., Shewry, P. R., & Ward, J. L. (2010). Effects of environment and genotype on phenolic acids in wheat in the HEALTHGRAIN diversity screen. *Journal of Agricultural and Food Chemistry, 58*, 9341–9352. https://doi.org/10.1021/jf102017s.

Ferri, M., Happel, A., Zanaroli, G., Bertolini, M., Chiesa, S., Commisso, M., et al. (2020). Advances in combined enzymatic extraction of ferulic acid from wheat bran. *New Biotechnology, 56*(November 2019), 38–45. https://doi.org/10.1016/j.nbt.2019.10.010.

Figueroa-Espinoza, M. C., Morel, M. H., & Rouau, X. (1998). Effect of lysine, tyrosine, cysteine, and glutathione on the oxidative cross-linking of feruloylated arabinoxylans by a fungal laccase. *Journal of Agricultural and Food Chemistry, 46*(7), 2583–2589.

Francin-allami, M., Alvarado, C., Daniel, S., Geairon, A., Saulnier, L., & Guillon, F. (2019). Plant Science Spatial and temporal distribution of cell wall polysaccharides during grain development of Brachypodium distachyon. *Plant Science, 280*(December 2018), 367–382. https://doi.org/10.1016/j.plantsci.2018.12.018.

Fry, S. C. (1980). Gibberellin-controlled pectinic acid and protein secretion in growing cells. *Phytochemistry, 19*, 735–740.

Fry, S. C. (2004). Oxidative coupling of tyrosine and ferulic acid residues: Intra- and extra-protoplasmic occurrence, predominance of trimers and larger products, and possible role in inter-polymeric cross-linking. *Phytochemistry Reviews, 3*(1–2), 97–111. https://doi.org/10.1023/B:PHYT.0000047808.74647.43.

Fry, S. C., Willis, S. C., & Paterson, A. E. J. (2000). Intraprotoplasmic and wall-localised formation of arabinoxylan- bound diferulates and larger ferulate coupling-products in maize cell-suspension cultures. *Planta, 211*(5), 679–692.

García-Lara, S., Bergvinson, D. J., Burt, A. J., Ramputh, A. I., Díaz-Pontones, D. M., & Arnason, J. T. (2004). The role of pericarp cell wall components in maize weevil resistance. *Crop Science, 44*(5), 1546–1552. https://doi.org/10.2135/cropsci2004.1546.

Geissmann, T., & Neukom, H. (1973). On the composition of the water soluble wheat flour pentosans and their oxidative gelation. *Lebensmittel-Wissenschaft + Technologie. Food Science and Technology*, 59–62.

Ghaffari, M., Chateigner-Boutin, A.-L., Guillon, F., Devaux, M.-F., Abdollahi, H., & Duponchel, L. (2019). Multi-excitation hyperspectral autofluorescence imaging for the exploration of biological samples. *Analytica Chimica Acta, 1062*, 47–59. https://doi.org/10.1016/j.aca.2019.03.003.

Gou, J. Y., Yu, X. H., & Liu, C. J. (2009). A hydroxycinnamoyltransferase responsible for synthesizing suberin aromatics in Arabidopsis. *Proceedings of the National Academy of Sciences of the United States of America, 106*(44), 18855–18860. https://doi.org/10.1073/pnas.0905555106.

Grabber, J. H. (2005). How do lignin composition, structure, and cross-linking affect degradability? A review of cell wall model studies. *Crop Science, 45*(3), 820. https://doi.org/10.2135/cropsci2004.0191.

Hands, P., Kourmpetli, S., Sharples, D., Harris, R. G., & Drea, S. (2012). Analysis of grain characters in temperate grasses reveals distinctive patterns of endosperm organization associated with grain shape. *Journal of Experimental Botany, 63*(17), 6253–6266. https://doi.org/10.1093/jxb/ers281.

Harris, P., & Trethewey, J. K. (2010). The distribution of ester-linked ferulic acid in the cell walls of angiosperms. *Phytochemistry Reviews, 9*(1), 19–33. https://doi.org/10.1007/s11101-009-9146-4.

Hartley, R. D. (1973). Carbohydrate esters of ferulic acid as components of cell walls of Lolium multiflorum. *Phytochemistry, 12*, 661–665.

Hartley, R. D., Morrison, W. H., III, Himmelsbach, D. S., & Borneman, W. S. (1990). Cross-linking of cell wall phenolic arabinoxylans in graminaceous plants. *Phytochemistry, 29*(12), 3705–3709. https://doi.org/10.1016/0031-9422(90)85317-9.

Hatfield, R. D., & Ralph, J. (1999). Modelling the feasibility of intramolecular dehydrodiferulate formation in grass walls. *Journal of the Science of Food and Agriculture, 79*(3), 425–427. https://doi.org/10.1002/(SICI)1097-0010(19990301)79:3<425::AID-JSFA282>3.0.CO;2-U.

Hatfield, R. D., Rancour, D. M., & Marita, J. M. (2017). Grass cell walls: A story of cross-linking. *Frontiers in Plant Science*, 7(January). https://doi.org/10.3389/fpls.2016.02056.

Hemery, Y., Rouau, X., Lullien-Pellerin, V., Barron, C., & Abecassis, J. (2007). Dry processes to develop wheat fractions and products with enhanced nutritional quality. *Journal of Cereal Science, 46*(3), 327–347. https://doi.org/10.1016/j.jcs.2007.09.008.

Horvat, D., Šimić, G., Drezner, G., Lalić, A., Ledenčan, T., Tucak, M., et al. (2020). Phenolic acid profiles and antioxidant activity of major cereal crops. *Antioxidants, 9*, 527. https://doi.org/10.3390/antiox9060527.

Ho-Yue-Kuang, S., Alvarado, C., Antelme, S., Bouchet, B., Cézard, L., Le Bris, P., et al. (2016). Mutation in Brachypodium caffeic acid O-methyltransferase 6 alters stem and grain lignins and improves straw saccharification without deteriorating grain quality. *Journal of Experimental Botany, 67*(1), 227–237. https://doi.org/10.1093/jxb/erv446.

Huang, T., Xu, M., Lee, A., Cho, S., & Qi, L. (2015). Consumption of whole grains and cereal fiber and total and cause-specific mortality: Prospective analysis of 367,442 individuals. *BMC Medicine, 13*(1), 59. https://doi.org/10.1186/s12916-015-0294-7.

Iiyama, K., Lam, T. B. T., & Stone, B. A. (1990). Phenolic-acid bridges between polysaccharides and lignin in wheat internodes. *Phytochemistry, 29*(3), 733–737.

Iiyama, K., Lam, T. B. T., & Stone, B. A. (1994). Covalent cross-links in the cell-wall. *Plant Physiology, 104*(2), 315–320.

Ishii, T. (1997). Structure and functions of feruloylated polysaccharides. *Plant Science, 127*(2), 111–127. https://doi.org/10.1016/S0168-9452(97)00130-1.

Jacquet, G., Pollet, B., Lapierre, C., Mhamdi, F., & Rolando, C. (1995). New ether-linked ferulic acid-coniferyl alcohol dimers identified in grass straws. *Journal of Agricultural and Food Chemistry, 43*(10), 2746–2751. https://doi.org/10.1021/jf00058a037.

Jia, L., Xu, W., Li, W., Ye, N., Liu, R., Shi, L., et al. (2013). Class III peroxidases are activated in proanthocyanidin-deficient Arabidopsis thaliana seeds. *Annals of Botany, 111*(5), 839–847. https://doi.org/10.1093/aob/mct045.

Jung, H. G., & Phillips, R. L. (2010). Putative seedling ferulate ester (sfe) maize mutant: Morphology, biomass yield, and stover cell wall composition and rumen degradability. *Crop Science, 50*(1), 403–418. https://doi.org/10.2135/cropsci2009.04.0191.

Kamisaka, S., Takeda, S., Takahashi, K., & Shibata, K. (1990). Diferulic and ferulic acid in the cell wall of Avena coleoptiles—Their relationships to mechanical properties of the cell wall. *Physiologia Plantarum, 78*, 1–7.

Kaszuba, J., Kapusta, I., & Posadzka, Z. (2021). Content of phenolic acids in the grain of selected polish triticale cultivars and its products. *Molecules, 26*(3), 562. https://doi.org/10.3390/molecules26030562.

Kolattukudy, P. E. (1981). Structure, biosynthesis, and biodegradation of cutin and suberin. *Annual Review of Plant Physiology, 32*(1), 539–567. https://doi.org/10.1146/annurev.pp.32.060181.002543.

Kumar, N., & Pruthi, V. (2014). Potential applications of ferulic acid from natural sources. *Biotechnology Reports, 4*(1), 86–93. https://doi.org/10.1016/j.btre.2014.09.002.

Laddomada, B., Blanco, A., Mita, G., D'Amico, L., Singh, R. P., Ammar, K., et al. (2021). Drought and heat stress impacts on phenolic acids accumulation in durum wheat cultivars. *Food, 10*(9), 2142. https://doi.org/10.3390/foods10092142.

Lapierre, C., Pollet, B., Ralet, M.-C., & Saulnier, L. (2001). The phenolic fraction of maize bran: Evidence for lignin-heteroxylan association. *Phytochemistry, 57*(5), 765–772. https://doi.org/10.1016/S0031-9422(01)00104-2.

Lapierre, C., Voxeur, A., Boutet, S., & Ralph, J. (2019). Arabinose conjugates diagnostic of ferulate-ferulate and ferulate-monolignol cross-coupling are released by mild acidolysis of grass cell walls. *Journal of Agricultural and Food Chemistry, 67*(46), 12962–12971. https://doi.org/10.1021/acs.jafc.9b05840.

Lapierre, C., Voxeur, A., Karlen, S. D., Helm, R. F., & Ralph, J. (2018). Evaluation of feruloylated and p-coumaroylated arabinosyl units in grass arabinoxylans by acidolysis in dioxane/methanol. *Journal of Agricultural and Food Chemistry, 66*(21), 5418–5424. https://doi.org/10.1021/acs.jafc.8b01618.

Lee, Y., Rubio, M. C., Alassimone, J., & Geldner, N. (2013). A mechanism for localized lignin deposition in the endodermis. *Cell, 153*(2), 402–412. https://doi.org/10.1016/j.cell.2013.02.045.

Li, F. S., Phyo, P., Jacobowitz, J., Hong, M., & Weng, J. K. (2019). The molecular structure of plant sporopollenin. *Nature Plants, 5*(1), 41–46. https://doi.org/10.1038/s41477-018-0330-7.

Li, L., Shewry, P. R., & Ward, J. L. (2008). Phenolic acids in wheat varieties in the healthgrain diversity screen. *Journal of Agricultural and Food Chemistry, 56*(21), 9732–9739. https://doi.org/10.1021/jf801069s.

Lindsay, S., & Fry, S. (2008). Control of diferulate formation in dicotyledonous and gramineous cell-suspension cultures. *Planta, 227*(2), 439–452. https://doi.org/10.1007/s00425-007-0630-z.

Lu, F. C., & Ralph, J. (1999). Detection and determination of p-coumaroylated units in lignins. *Journal of Agricultural and Food Chemistry, 47*(5), 1988–1992.

Marion, D., & Saulnier, L. (2020). Minor components and wheat quality: Perspectives on climate changes. *Journal of Cereal Science, 94*(May), 103001. https://doi.org/10.1016/j.jcs.2020.103001.

Martínez-López, A. L., Carvajal-Millan, E., Sotelo-Cruz, N., Micard, V., Rascón-Chu, A., López-Franco, Y. L., et al. (2019). Enzymatically cross-linked arabinoxylan microspheres as oral insulin delivery system. *International Journal of Biological Macromolecules, 126*, 952–959. https://doi.org/10.1016/j.ijbiomac.2018.12.192.

Mastrangelo, L. I., Lenucci, M. S., Piro, G., & Dalessandro, G. (2009). Evidence for intra- and extra-protoplasmic feruloylation and cross-linking in wheat seedling roots. *Planta, 229*(2), 343–355. https://doi.org/10.1007/s00425-008-0834-x.

Mazurek, S., Garroum, I., Daraspe, J., De Bellis, D., Olsson, V., Mucciolo, A., et al. (2017). Connecting the molecular structure of cutin to ultrastructure and physical properties of the cuticle in petals of Arabidopsis. *Plant Physiology, 173*(2), 1146–1163. https://doi.org/10.1104/pp.16.01637.

Méchin, V., Laluc, A., Legée, F., Cézard, L., Denoue, D., Barrière, Y., et al. (2014). Impact of the brown-midrib bm 5 mutation on maize lignins. *Journal of Agricultural and Food Chemistry*, *62*(22), 5102–5107. https://doi.org/10.1021/jf5019998.

Meents, M. J., Motani, S., Mansfield, S. D., & Samuels, A. L. (2019). Organization of xylan production in the Golgi during secondary cell wall biosynthesis. *Plant Physiology*, *181*(2), 527–546. https://doi.org/10.1104/pp.19.00715.

Meyer, K., Kohler, A., & Kauss, H. (1991). Biosynthesis of ferulic acid esters of plant cell wall polysaccharides in endomembranes from parsley cells. *FEBS Letters*, *290*, 209–212.

Mir Derikvand, M., Sierra, J. B., Ruel, K., Pollet, B., Do, C. T., Thévenin, J., et al. (2008). Redirection of the phenylpropanoid pathway to feruloyl malate in Arabidopsis mutants deficient for cinnamoyl-CoA reductase 1. *Planta*, *227*(5), 943–956. https://doi.org/10.1007/s00425-007-0669-x.

Miraftab, M., Qiao, Q., Kennedy, J. F., Anand, S. C., & Groocock, M. R. (2003). Fibres for wound dressings based on mixed carbohydrate polymer fibres. *Carbohydrate Polymers*, *53*(3), 225–231. Retrieved from file://localhost/Users/saulnier/Documents/BiblioLuc/2003/Miraftab + 03.pdf.

Mitchell, R. A. C., Dupree, P., & Shewry, P. R. (2007). A novel bioinformatics approach identifies candidate genes for the synthesis and feruloylation of arabinoxylan. *Plant Physiology*, *144*(1), 43–53. https://doi.org/10.1104/pp.106.094995.

Molinari, H. B. C., Pellny, T. K., Freeman, J., Shewry, P. R., & Mitchell, R. A. C. (2013). Grass cell wall feruloylation: Distribution of bound ferulate and candidate gene expression in Brachypodium distachyon. *Frontiers in Plant Science*, *4*. https://doi.org/10.3389/fpls.2013.00050.

Morrison, I. N. (1975). Ultrastructure of the cuticular membranes of the developing wheat grain. *Canadian Journal of Botany*, *53*, 2077–2087.

Morrison, W. H., Hartley, R. D., & Himmelsbach, D. S. (1992). Synthesis of substituted truxillic acids from p-coumaric and ferulic acid: Simulation of photodimerization in plant cell walls. *Journal of Agricultural and Food Chemistry*, *40*(5), 768–771. https://doi.org/10.1021/jf00017a012.

Mota, T. R., de Souza, W. R., Oliveira, D. M., Martins, P. K., Sampaio, B. L., Vinecky, F., et al. (2020). Suppression of a BAHD acyltransferase decreases p-coumaroyl on arabinoxylan and improves biomass digestibility in the model grass Setaria viridis. *Plant Journal*, 1–15. https://doi.org/10.1111/tpj.15046.

Neukom, H., & Markwalder, H. U. (1978). Oxidative gelation of wheat-flour pentosans - new way of cross-linking polymers. *Cereal Foods World*, *23*, 374–376.

Ng, A., Greenshields, R. N., & Waldron, K. W. (1997). Oxidative cross-linking of corn bran hemicellulose: Formation of ferulic acid dehydrodimers. *Carbohydrate Research*, *303*(4), 459–462. https://doi.org/10.1016/S0008-6215(97)00193-6.

Nyström, L., Lampi, A. M., Andersson, A. A., Kamal-Eldin, A., Gebruers, K., Courtin, C. M., et al. (2008). Phytochemicals and dietary fiber components in rye varieties in the HEALTHGRAIN Diversity Screen. *Journal of Agricultural and Food Chemistry*, *56*(21), 9758–9766. https://doi.org/10.1021/jf801065r.

Obel, N., Porchia, A. C., & Scheller, H. V. (2003). Intracellular feruloylation of arabinoxylan in wheat: Evidence for feruloyl-glucose as precursor. *Planta*, *216*(4), 620–629. https://doi.org/10.1007/s00425-002-0863-9.

Oparka, K. J., & Gates, P. (1981). Transport of assimilates in the developing caryopsis of rice (Oryza sativa L.). *Planta*, *152*, 388–396.

Patel, M., & Naik, S. (2004). Gamma-oryzanol from rice bran oil—A review. *Journal of Scientific & Industrial Research*, *63*, 569–578.

Pellny, T. K., Lovegrove, A., Freeman, J., Tosi, P., Love, C. G., Knox, J. P., et al. (2012). Cell walls of developing wheat starchy endosperm: Comparison of composition and RNA-Seq transcriptome. *Plant Physiology*, *158*(2), 612–627. https://doi.org/10.1104/pp.111.189191.

Perkins, M., Smit, R. A., & Samuels, L. (2019). The transport of monomers during lignification in plants: Anything goes but how? *Current Opinion in Biotechnology*, *56*, 69–74. https://doi.org/10.1016/j.copbio.2018.09.011.

Petrik, D. L., Karlen, S. D., Cass, C. L., Padmakshan, D., Lu, F., Liu, S., et al. (2014). p-Coumaroyl-CoA:monolignol transferase (PMT) acts specifically in the lignin biosynthetic pathway in Brachypodium distachyon. *The Plant Journal*, 77(5), 713–726. https://doi.org/10.1111/tpj.12420.

Peyron, S., Abecassis, J., Autran, J.-C., & Rouau, X. (2001). Enzymatic oxidative treatments of wheat bran layers: Effects on ferulic acid composition and mechanical properties. *Journal of Agricultural and Food Chemistry*, *49*(10), 4694–4699. https://doi.org/10.1021/jf0106805.

Peyron, S., Abecassis, J. L., Autran, J. C., & Rouau, X. (2002). Influence of UV exposure on phenolic acid content, mechanical properties of bran, and milling behavior of durum wheat (Triticum durum Desf.). *Cereal Chemistry*, *79*(5), 726–731.

Philippe, S., Robert, P., Barron, C., Saulnier, L., & Guillon, F. (2006). Deposition of cell wall polysaccharides in wheat endosperm during grain development: Fourier transform-infrared microspectroscopy study. *Journal of Agricultural and Food Chemistry*, *54*(6), 2303–2308. https://doi.org/10.1021/jf052922x.

Philippe, S., Tranquet, O., Utille, J.-P., Saulnier, L., & Guillon, F. (2007). Investigation of ferulate deposition in endosperm cell walls of mature and developing wheat grains by using a polyclonal antibody. *Planta*, *225*(5), 1287–1299. https://doi.org/10.1007/s00425-006-0422-x.

Piber, M., & Koehler, P. (2005). Identification of dehydro-ferulic acid-tyrosine in rye and wheat: Evidence for a covalent cross-link between arabinoxylans and proteins. *Journal of Agricultural and Food Chemistry*, *53*(13), 5276–5284. https://doi.org/10.1021/jf050395b.

Piston, F., Uauy, C., Fu, L. H., Langston, J., Labavitch, J., & Dubcovsky, J. (2010). Down-regulation of four putative arabinoxylan feruloyl transferase genes from family PF02458 reduces ester-linked ferulate content in rice cell walls. *Planta*, *231*(3), 677–691. https://doi.org/10.1007/s00425-009-1077-1.

Ralph, J., Bunzel, M., Marita, J. M., Hatfield, R. D., Lu, F., Kim, H., et al. (2004). Peroxidase-dependent cross-linking reactions of p-hydroxycinnamates in plant cell walls. *Phytochemistry Reviews*, *3*(1–2), 79–96. https://doi.org/10.1023/B:PHYT.0000047811.13837.fb.

Ralph, J., Grabber, J. H., & Hatfield, R. D. (1995). Lignin-ferulate crosslinks in grasses: Active incorporation of ferulate polysaccharide esters into ryegrass lignins. *Carbohydrate Research*, *275*, 167–178. https://doi.org/10.1016/0008-6215(95)00237-N.

Ralph, J., Hatfield, R. D., Quideau, S., Helm, R. F., Grabber, J. H., & Jung, H.-J. G. (1994). Pathway of p-coumaric acid incorporation into maize lignin as revealed by NMR. *American Chemical Society*, *116*(21), 9448–9456. https://doi.org/10.1021/ja00100a006.

Ralph, J., Lundquist, K., Brunow, G., Lu, F., Kim, H., Schatz, P., et al. (2004). Lignins: Natural polymers from oxidative coupling of 4-hydroxyphenyl-propanoids. *Phytochemistry Reviews*, *3*(1–2), 29–60. https://doi.org/10.1023/B:PHYT.0000047809.65444.a4.

Ralph, J., Quideau, S., Grabber, J. H., & Hatfield, R. D. (1994). Identification and synthesis of new ferulic acid dehydrodimers present in grass cell-walls. *Journal of the Chemical Society, Perkin Transactions*, *1*(23), 3485–3498.

Rautengarten, C., Ebert, B., Ouellet, M., Nafisi, M., Baidoo, E. E. K., Benke, P., et al. (2012). Arabidopsis deficient in cutin ferulate encodes a transferase required for feruloylation of hydroxy fatty acids in cutin polyester. *Plant Physiology*, *158*(2), 654–665. https://doi.org/10.1104/pp.111.187187.

Reem, N. T., Pogorelko, G., Lionetti, V., Chambers, L., Held, M. A., Bellincampi, D., et al. (2016). Decreased polysaccharide feruloylation compromises plant cell wall integrity and increases susceptibility to necrotrophic fungal pathogens. *Frontiers in Plant Science*, 7, 630. https://doi.org/10.3389/fpls.2016.00630.

Reynolds, A., Mann, J., Cummings, J., Winter, N., Mete, E., & Te Morenga, L. (2019). Carbohydrate quality and human health: A series of systematic reviews and meta-analyses. *The Lancet*, *393*(10170), 434–445. https://doi.org/10.1016/S0140-6736(18)31809-9.

Rosa, N. N., Aura, A.-M., Saulnier, L., Holopainen-Mantila, U., Poutanen, K., & Micard, V. (2013). Effects of disintegration on in vitro fermentation and conversion patterns of wheat aleurone in a metabolical colon model. *Journal of Agricultural and Food Chemistry*, *61*(24), 5805–5816. https://doi.org/10.1021/jf4001814.

Rosa, N. N., Barron, C., Gaiani, C., Dufour, C., & Micard, V. (2013). Ultra-fine grinding increases the antioxidant capacity of wheat bran. *Journal of Cereal Science*, *57*(1), 84–90. https://doi.org/10.1016/j.jcs.2012.10.002.

Rosa, N. N., Dufour, C., Lullien-Pellerin, V., & Micard, V. (2013). Exposure or release of ferulic acid from wheat aleurone: Impact on its antioxidant capacity. *Food Chemistry*, *141*(3), 2355–2362. https://doi.org/10.1016/j.foodchem.2013.04.132.

Rosa-Sibakov, N., Poutanen, K., & Micard, V. (2015). How does wheat grain, bran and aleurone structure impact their nutritional and technological properties? *Trends in Food Science & Technology*, *41*(2), 118–134. https://doi.org/10.1016/j.tifs.2014.10.003.

Rose, D. J., Patterson, J. A., & Hamaker, B. R. (2010). Structural differences among alkali-soluble arabinoxylans from maize (Zea mays), rice (Oryza sativa), and wheat (Triticum aestivum) brans influence human fecal fermentation profiles. *Journal of Agricultural and Food Chemistry*, *58*(1), 493–499. https://doi.org/10.1021/jf9020416.

Rouau, X., Cheynier, V., Surget, A., Gloux, D., Barron, C., Meudec, E., et al. (2003). A dehydrotrimer of ferulic acid from maize bran. *Phytochemistry*, *63*(8), 899–903.

Salazar-López, N. J., González-Aguilar, G. A., Loarca-Piña, G., Cinco-Moroyoqui, F. J., Rouzaud-Sández, O., Domínguez-Avila, J. A., et al. (2017). Contribution and interactions of hydroxycinnamic acids found in bran and wholegrain Sorghum (Sorghum bicolor L. Moench): Effects on the antioxidant capacity and inhibition of human erythrocyte hemolysis. *Oxidative Medicine and Cellular Longevity*, *2017*, 8219023. https://doi.org/10.1155/2017/8219023.

Saulnier, L., Crépeau, M.-J., Lahaye, M., Thibault, J.-F., Garcia-Conesa, M. T., Kroon, P. A., et al. (1999). Isolation and structural determination of two 5,5′-diferuloyl oligosaccharides indicate that maize heteroxylans are covalently cross-linked by oxidatively coupled ferulates. *Carbohydrate Research*, *320*(1–2), 82–92. https://doi.org/10.1016/S0008-6215(99)00152-4.

Saulnier, L., Guillon, F., & Chateigner-Boutin, A.-L. (2012). Cell wall deposition and metabolism in wheat grain. *Journal of Cereal Science*, *56*(1), 91–108. https://doi.org/10.1016/j.jcs.2012.02.010.

Saulnier, L., Guillon, F., Sado, P.-E., Chateigner-Boutin, A.-L., & Rouau, X. (2013). Plant cell wall polysaccharides in storage organs: Xylans (food applications). In *Reference Module in Chemistry, Molecular Sciences and Chemical Engineering* (pp. 1–32). Elsevier. https://doi.org/10.1016/B978-0-12-409547-2.01493-1.

Saulnier, L., Marot, C., Elgorriaga, M., Bonnin, E., & Thibault, J.-F. F. (2001). Thermal and enzymatic treatments for the release of free ferulic acid from maize bran. *Carbohydrate Polymers*, *45*(3), 269–275. https://doi.org/10.1016/S0144-8617(00)00259-9.

Saulnier, L., Vigouroux, J., & Thibault, J.-F. (1995). Isolation and partial characterization of feruloylated oligosaccharides from maize bran. *Carbohydrate Research*, *272*(2), 241–253. https://doi.org/10.1016/0008-6215(95)00053-V.

Schefer, S., Oest, M., & Rohn, S. (2021). Interactions between phenolic acids, proteins, and carbohydrates—influence on dough and bread properties. *Food, 10*(11), 2798. https://doi.org/10.3390/foods10112798.

Schendel, R. R., Becker, A., Tyl, C. E., & Bunzel, M. (2015). Isolation and characterization of feruloylated arabinoxylan oligosaccharides from the perennial cereal grain intermediate wheat grass (Thinopyrum intermedium). *Carbohydrate Research, 407*, 16–25. https://doi.org/10.1016/j.carres.2015.01.006.

Schendel, R. R., Meyer, M. R., & Bunzel, M. (2016). Quantitative profiling of feruloylated arabinoxylan side-chains from graminaceous cell walls. *Frontiers in Plant Science, 6*(January), 1–11. https://doi.org/10.3389/fpls.2015.01249.

Schöneberg, T., Kibler, K., Sulyok, M., Musa, T., Bucheli, T. D., Mascher, F., et al. (2018). Can plant phenolic compounds reduce Fusarium growth and mycotoxin production in cereals? *Food Additives & Contaminants: Part A, 35*(12), 2455–2470. https://doi.org/10.1080/19440049.2018.1538570.

Shewry, P. R. (2009). Wheat. *Journal of Experimental Botany, 60*(6), 1537–1553. https://doi.org/10.1093/jxb/erp058.

Shewry, P. R., Hawkesford, M. J., Piironen, V., Lampi, A. M., Gebruers, K., Boros, D., et al. (2013). Natural variation in grain composition of wheat and related cereals. *Journal of Agriculture and Food Chemistry, 61*(35), 8295–8303. https://doi.org/10.1021/jf3054092.

Shewry, P. R., Piironen, V., Lampi, A.-M., Edelmann, M., Kariluoto, S., Nurmi, T., et al. (2010). The HEALTHGRAIN wheat diversity screen: Effects of genotype and environment on phytochemicals and dietary fiber components. *Journal of Agricultural and Food Chemistry, 58*(17), 9291–9298. https://doi.org/10.1021/jf100039b.

Shewry, P. R., Piironen, V., Lampi, A.-M., Nyström, L., Li, L., Rakszegi, M., et al. (2008). Phytochemical and fiber components in oat varieties in the HEALTHGRAIN diversity screen. *Journal of Agricultural and Food Chemistry, 56*(21), 9777–9784. https://doi.org/10.1021/jf801880d.

Sibout, R., Le Bris, P., Legée, F., Cézard, L., Renault, H., & Lapierre, C. (2016). Structural redesigning Arabidopsis lignins into alkali-soluble lignins through the expression of p-coumaroyl-CoA:Monolignol transferase PMT. *Plant Physiology, 170*(3), 1358–1366. https://doi.org/10.1104/pp.15.01877.

Smith, M. M., & Hartley, R. D. (1983). Occurrence and nature of ferulic acid substitution of cell-wall polysaccharides in graminaceous plants. *Carbohydrate Research, 118*(C), 65–80. https://doi.org/10.1016/0008-6215(83)88036-7.

Snelders, J., Dornez, E., Delcour, J. A., & Courtin, C. M. (2013). Ferulic acid content and appearance determine the antioxidant capacity of arabinoxylanoligosaccharides. *Journal of Agricultural and Food Chemistry, 61*(42), 10173–10182. https://doi.org/10.1021/jf403160x.

Tinikul, R., Chenprakhon, P., Maenpuen, S., & Chaiyen, P. (2018). Biotransformation of Plant-Derived Phenolic Acids. *Biotechnology Journal, 13*(6), 1700632. https://doi.org/10.1002/biot.201700632.

Tranquet, O., Saulnier, L., Utille, J.-P., Ralph, J., & Guillon, F. (2009). Monoclonal antibodies to p-coumarate. *Phytochemistry, 70*(11–12), 1366–1373. https://doi.org/10.1016/j.phytochem.2009.06.019.

Väisänen, E., Takahashi, J., Obudulu, O., Bygdell, J., Karhunen, P., Blokhina, O., et al. (2020). Hunting monolignol transporters: Membrane proteomics and biochemical transport assays with membrane vesicles of Norway spruce. *Journal of Experimental Botany, 71*(20), 6379–6395. https://doi.org/10.1093/jxb/eraa368.

Vermaas, J. V., Dixon, R. A., Chen, F., Mansfield, S. D., Boerjan, W., Ralph, J., et al. (2019). Passive membrane transport of lignin-related compounds. *Proceedings of the National Academy of Sciences of the United States of America, 116*(46), 23117–23123. https://pubmed.ncbi.nlm.nih.gov/31659054/.

Wan, X., Wu, S., Li, Z., An, X., & Tian, Y. (2020). Lipid metabolism: Critical roles in male fertility and other aspects of reproductive development in plants. *Molecular Plant, 13*(7), 955–983. https://doi.org/10.1016/j.molp.2020.05.009.

Waterstraat, M., & Bunzel, M. (2019). A stable isotope dilution approach to analyze ferulic acid oligomers in plant cell walls using liquid chromatography-tandem mass spectrometry. *Analytical and Bioanalytical Chemistry, 411*(20), 5047–5062. https://doi.org/10.1007/s00216-019-01924-w.

Wilkerson, C. G., Mansfield, S. D., Lu, F., Withers, S., Park, J.-Y. Y., Karlen, S. D., et al. (2014). Monolignol ferulate transferase introduces chemically labile linkages into the lignin backbone. *Science, 344*(6179), 90–93. https://doi.org/10.1126/science.1250161.

Withers, S., Lu, F., Kim, H., Zhu, Y., Ralph, J., & Wilkerson, C. G. (2012). Identification of grass-specific enzyme that acylates monolignols with p-coumarate. *Journal of Biological Chemistry, 287*(11), 8347–8355. https://doi.org/10.1074/jbc.M111.284497.

Wu, G., Johnson, S. K., Bornman, J. F., Bennett, S. J., Clarke, M. W., Singh, V., et al. (2016). Growth temperature and genotype both play important roles in sorghum grain phenolic composition. *Scientific Reports, 24*(6), 21835. https://doi.org/10.1038/srep21835.

Yeats, T. H., & Rose, J. K. (2013). The formation and function of plant cuticles. *Plant Physiology, 163*(1), 5–20. https://doi.org/10.1104/pp.113.222737.

Yi Chou, E., Schuetz, M., Hoffmann, N., Watanabe, Y., Sibout, R., & Samuels, A. L. (2018). Distribution, mobility, and anchoring of lignin-related oxidative enzymes in Arabidopsis secondary cell walls. *Journal of Experimental Botany, 69*(8), 1849–1859. https://doi.org/10.1093/jxb/ery067.

Yin, N. W., Wang, S. X., Jia, L. D., Zhu, M. C., Yang, J., Zhou, B. J., et al. (2019). Identification and characterization of major constituents in different-colored rapeseed petals by UPLC-HESI-MS/MS. *Journal of Agricultural and Food Chemistry, 67*(40), 11053–11065. https://doi.org/10.1021/acs.jafc.9b05046.

Yoshida-Shimokawa, T., Yoshida, S., Kakegawa, K., & Ischii, T. (2001). Enzymic feruloylation of arabinoxylan-trisaccharide by ferloyl-CoA: Arabinoxylan-trisaccharide O-hydroxycinnamoyl transferase from Oryza sativa. *Planta, 212*, 470–474.

Zee, S.-Y., & O'Brien, T. P. (1970). A special type of tracheary element associated with 'Xylem discontinuity' in the floral axis of wheat. *Australian Journal of Biological Sciences, 23*, 783–791.

Zhao, Z., & Moghadasian, M. H. (2008). Chemistry, natural sources, dietary intake and pharmacokinetic properties of ferulic acid: A review. *Food Chemistry, 109*(4), 691–702. https://doi.org/10.1016/j.foodchem.2008.02.039.

Zhao, Q., Nakashima, J., Chen, F., Yin, Y., Fu, C., Yun, J., et al. (2013). LACCASE is necessary and nonredundant with PEROXIDASE for lignin polymerization during vascular development in Arabidopsis. *The Plant Cell, 25*(10), 3976–3987. https://doi.org/10.1105/tpc.113.117770.

Zhu, F. (2018). Anthocyanins in cereals: Composition and health effects. *Food Research International, 109*, 232–249. https://doi.org/10.1016/j.foodres.2018.04.015.

CHAPTER SIX

In situ imaging of lignin and related compounds by Raman, Fourier-transform infrared (FTIR) and fluorescence microscopy

Fabienne Guillon[a,]*, Notburga Gierlinger[b], Marie-Françoise Devaux[a], and András Gorzsás[c]
[a]INRAE, UR BIA, Nantes, France
[b]Institute of Biophysics, Department of Nanobiotechnology, University of Natural Resources and Life Sciences Vienna (BOKU), Vienna, Austria
[c]Department of Chemistry, Umeå University, Umea, Sweden
*Corresponding author: e-mail address: fabienne.guillon@inrae.fr

Contents

Advances in Botanical Research, Volume 104
ISSN 0065-2296
https://doi.org/10.1016/bs.abr.2022.03.009

Abstract

Spectral imaging technologies simultaneously record spectral and spatial information about plant tissues in a noninvasive way. Differences in techniques result in different selection rules and spatial resolutions. This article introduces the basic principles of Raman, Fourier-transform Infrared (FTIR) and autofluorescence imaging and finally compares their strength and drawbacks. The methods result in spectral datasets as bases for image generation. Spectral preprocessing together with univariate and multivariate data analysis approaches are essential for informative lignin imaging and analysis and therefore also briefly illustrated and discussed. Examples of imaging lignin and other aromatic components in a broad range of plant tissues show the potential as well as the limitations of microspectroscopic imaging.

1. Introduction

Lignin is the second most abundant plant biopolymer after cellulose, being present in almost all land plants. It results from the oxidative radical polymerization of three major *p*-hydroxycinnamyl alcohols (*p*-coumaryl, coniferyl, and sinapyl alcohols) leading to the formation of guaiacyl (G), syringyl (S), and hydroxyphenyl (H) subunits. The amount and chemical composition vary between species. For example, lignin accounts for 15–35% of dry matter in woody stems and 5–20% in stems of herbaceous plants. Gymnosperms have a lignin consisting mainly of guaiacyl units (G), with some traces of *p*-hydroxyphenyl units (H) while dicotyledonous angiosperms have a lignin consisting of both G and syringyl (S) units with traces of H units. Gramineae possess all three H, S and G units, the last two representing between 85% and 96% of the total. Lignin content is variable within population plants of the same species as well. For example, the average lignin content ranges from 15.7% to 27.9% and the molar S/G ratio, from 1 to 3 within a population of natural *populus* variants (Studer, DeMartini, Davis, et al., 2011). Beside these main monolignols, a number of other molecules can participate in the formation of lignin (Del Rio et al., 2020; see also the chapter "Unconventional lignin monomers: Extension of the lignin paradigm" by del Rio et al. in this book; Ralph, Lapierre, & Boerjan, 2019). For instance, in grasses, *para*-coumaric (*p*C) and ferulic acids are naturally present in relatively high amount (Del Rio et al., 2020; Hatfield & Marita, 2010; Ralph et al., 2019). The majority of *p*Cs acylates the S-units while ferulic acid are found ester linked to arabinoxylans, the major hemicellulose in grass cell wall and ether linked to G unit in lignin (see chapter "Ferulic and coumaric acids in the cereal grain: Occurrence, biosynthesis, biological functions" by Saulnier & Boutin in this book).

Within one plant, the lignin amount and composition also vary according to tissue type, the developmental state of the cell and tissue and plant growth environment (Campbell & Sederoff, 1996; El Hage et al., 2018; He & Terashima, 1990, 1991; Liu, Luo, & Zheng, 2018; Terashima, 2013; Terashima & Fukushima, 1988). In hardwood, the fiber secondary wall contains predominantly S units while the secondary cell wall of vessel consists mostly of G units (Terashima, Fukushima, & Tsuchiya, 1986). In growing maize internode, rind vascular tissue lignified earlier and to a greater extent than pith tissues (Morrison, Jung, Buxton, & Hatfield, 1998). The first lignin appearing in parenchyma is of the G type preceding the incorporation of S units in the later stages while in vascular bundles, the incorporation of S units is visible at the earliest stage of lignification (Joseleau & Ruel, 1997). *para*-Coumaric acid accumulation was shown to parallel the incorporation of lignin (Hatfield & Chaptman, 2009; Hatfield, Marita, & Frost, 2008; He & Terashima, 1990; Zhang, Legland, El Hage, et al., 2019), while ferulate remains relatively constant as a proportion of cell wall in developing stem tissues (Hatfield et al., 2008).

At the scale of the cell walls, the middle lamella and primary cell wall are highly lignified relative to the secondary cell wall (Donaldson, 2019; Gierlinger & Schwanninger, 2006; Terashima, 2013; Terashima et al., 1986). In hardwood, difference in composition has been highlighted, with G units generally found in the secondary vessel walls and middle lamella of all cell types and S units in the secondary fiber walls (Donaldson, 2001, 2019).

The amount and distribution of lignin influence the biomass properties, strongly impacting e.g. resistance and end use (Akin, 2008; Barrière et al., 2004; Christensen & Rasmussen, 2019; Devaux et al., 2018; Ding et al., 2012; Jung, Samac, & Sarath, 2012; Karthaüser, Biziks, Mai, & Militz, 2021; Lohrasebi, Mabee, & Roy, 1999; Yuan et al., 2021). Lignin composition also impacts the end use, affecting e.g. pulping and processing of wood (Tsutsumi, Kondo, Sakai, & Imamura, 1995), or saccharification yields (via their solubility in alkaline medium under mild conditions, Lapierre, 1993; Lapierre, Jouin, & Monties, 1989; Yang & Wyman, 2008).

With lignin's profound effects on the properties and use of lignocellulosic biomass, a better understanding of lignification and lignin interactions with other cell wall compounds *in planta* is of utmost importance and requires appropriate analytical tools. Wet chemical techniques provide qualitative and quantitative information on the amount and structural features of lignin, break down the structure of the sample. Therefore, no chemical information is gained in context with the micro- and nanostructure of the sample. Microscopic approaches reveal the tissue and cell wall structure and by chemical imaging the microstructure and chemical composition is revealed at once and related.

To obtain information on lignin location, different histochemical methods including phloroglucinol staining (Wiesner reagent) revealing lignin hydroxycinnamaldehyde functions (Liljegren, 2010; Nakano & Meshitsuka, 1992; Vallet, Chabbert, Czaninski, & Monties, 1996), Mäule staining providing indications on the S/G ratio (Meshitsuka & Nakano, 1979), safranin O (Baldacci-Cresp et al., 2020; Bond, Donaldson, Hill, & Hitchcock, 2008) and safranin/alcian blue staining, the FASGA method have been used for a long time (Tolivia & Tolivia, 1987). These dyes react with functional groups which can be present in different molecules, and thus may interact with compounds other than lignins in cells. Specific probes such as polyclonal or monoclonal antibodies against lignin motifs or hydroxycinnamic acids have been developed and used to detect them in situ (Kiyoto et al., 2013; Philippe, Tranquet, Utille, Saulnier, & Guillon, 2007; Ruel, Faix, & Joseleau, 1994; Tranquet, Saulnier, Utille, Ralph, & Guillon, 2009). However, they are few in number and not largely disseminated. In vivo chemistry based on biorthogonal reactions are very promising techniques for investigating lignification dynamics (Simon, Lion, Biot, Gierlinger, & Hawkins, 2018). However, this technology relies on the capacity of the cell wall machinery to incorporate tag-bearing monolignol surrogates (chemical reporters) and is thus restricted to living cells or tissues.

Spectral imaging techniques could be an alternative or complementary to all the above lignin imaging techniques to map the distribution of lignin and give chemical insight. The property of lignin to absorb UV light have been used in UV transmission microspectroscopy to detect lignin in situ. It has provided the main first UV spectra and images of lignin at microscopic scales (Fergus, Procter, Scott, & Urvey, 1969; He & Terashima, 1991; Scott, Procter, Fergus, & Goring, 1969) and is still used even if considered as more time consuming or less accessible with regard to the microscopy in the far UV (Day et al., 2005). The autofluorescence properties of lignin under UV and blue light excitation has made the fluorescence imaging technique very common to visualize the distribution of lignin, especially in wood (Donaldson, 2013, 2020; Donaldson, Radotić, Kalauzi, Djikanović, & Jeremić, 2010; Donaldson & Williams, 2018). Beside UV imaging techniques,

Fourier transform infrared (FTIR) and Raman microspectroscopies have been successfully applied to lignocellulosic material to map the distribution and structural features the cell wall polymers in situ including lignin (e.g., Béc, Grabska, Bonn, Popp, & Huck, 2020; Gierlinger, 2018; Simon et al., 2018).

These last three methods have the advantage of requiring little tissue preparation and, more importantly, no external agents (labels, dyes, markers). They are nondestructive and can be used to study real time changes during degradation, mechanical load, etc.

All three methods rely on the interaction of the molecules with electromagnetic radiation. Molecules can be excited by radiation to higher energy states and emit radiation when returning to their initial level via fluorescence (Fig. 1) (García-Plazaola et al., 2015). Molecules can also be excited to virtual levels and instead of returning to their ground states, they may return to an excited vibrational state (Fig. 1). The resulting energy gap manifests as a frequency shift of the incident light (Stokes shift). Molecules already in their excited states may relax to their ground state, resulting in a higher frequency scattered light than the incoming radiation (anti-Stokes shift, Fig. 1). Both of these shifts can be probed by Raman spectroscopy (Smith & Dent, 2005). If the sample is excited by energy in the mid-infrared region of the spectrum, absorbing this energy will result in the transition from its ground vibrational state to the first excited state, forming the basis of FTIR spectroscopy (Fig. 1) (Siebert & Hildebrandt, 2008; Smith & Dent, 2005).

Thus, all three techniques follow transitions between ground and excited energy states, albeit via different routes. Due to the different underlying

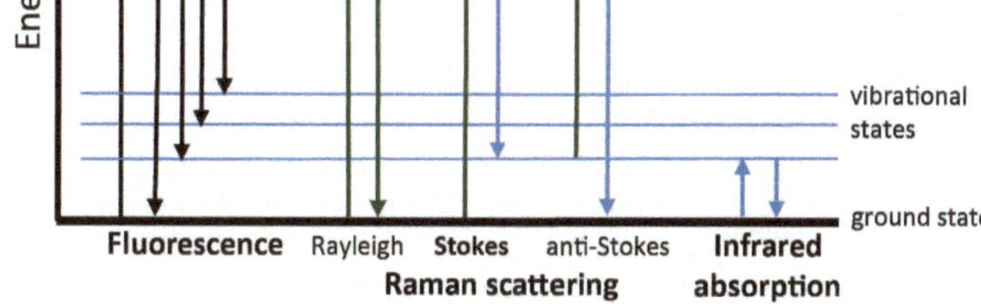

Fig. 1 Schematic overview of the physical basis of fluorescence, Raman and infrared imaging, showing the energy levels and transitions involved.

principles (absorption of infrared light vs inelastic (Raman) scattering vs fluorescence emission), different selection rules apply, and thus the techniques can be considered complementary and each method has its own advantages and disadvantages. For example, polar groups (like —OH, C=O) are more pronounced in IR spectra, whereas bands with high polarization potential, e.g. carbon backbones (C—C, C=C), are more pronounced in Raman spectra (Gierlinger, 2018; Siebert & Hildebrandt, 2008). While fluorescence can disturb Raman spectroscopic measurements, it allows further conclusions to be drawn regarding the chemical composition of plant tissues via fluorescence imaging. Autofluorescence of lignin and hydroxycinnamic acid allows them to be located unequivocally compared to other nonautofluorescent wall compounds. Its sensitivity and the possibility to probe from the whole organ to the cell wall scale makes it very attractive. FTIR imaging has high sensitivity and high information content. However, the presence of water can obscure the signal, the spatial resolution can be inherently limited by the diffraction of the light. In Raman microspectroscopy, the optimization of the measurement conditions can be more delicate (heat damage induced in the sample due to energy delivered, obscuration of the spectra signal by stimulated autofluorescence emission from some compounds), but it allows to reach a high spatial resolution and provide chemical, structural and even information on molecule's orientation.

2. Data analysis

Raman, FTIR or fluorescence imaging result in multi- or hyperspectral images that provide both spatial and spectral information to identify and localize chemical compounds and draw chemical. Multi- or hyperspectral images can be regarded as data cubes with two dimensions corresponding to the pixel spatial coordinates and one dimension corresponding to the spectral variables or channels. In hyperspectral images, an entire spectrum is obtained for each pixel, leading to more than 100 spectral variables. In multispectral images, the number of channels is small (<30), e.g. when using fluorescence microscopes or macroscopes. The number of pixels depends on the area covered and the pixel size or the step width.

2.1 Multi- and hyperspectral image analysis

The objective of multi- or hyperspectral image analysis can be qualitative (e.g., locating or colocating lignin in the sample, revealing lignin type and structure, etc.), or (semi)quantitative. Tissues, cells, or subcellular domains (e.g., cell wall layers) may be compared. In terms of data analysis, these objectives can be translated to identifying pixels showing lignin spectral bands, alone or in conjunction with bands assigned to other biochemical compounds and the drawing of their distribution maps; comparing pixel lignin spectra from different regions in the images or from different images; identifying regions in images showing different spectra; measuring the amount of lignin from pixel spectra, or counting the number of pixels corresponding to a given kind of lignin.

Though the spectral nature will largely drive the analysis, the complete multi- or hyperspectral image analysis workflow generally involves both spectral and image analysis techniques (Fig. 2) (e.g., De Juan et al., 2014; Gautam, Vanga, Ariese, & Umapathy, 2015; Grahn & Geladi, 2007).

On the spectral side, chemometrics-based multivariate spectral analyses are predominant. The image nature of multi- or hyperspectral images, i.e. the spatial relationships between the spectra, is generally not considered. Multivariate analyses are applied to reveal the spectral variations and/or to highlight the chemical constituents giving rise to the spectral signal. On the image analysis side, tools are used for the spatial representation of the multi- or hyperspectral images and of the spectral analysis results, for selection of pixels, or image segmentation. The visualization of data (both in the spectral and image dimensions) and results at each stage of the processing is also an important part of the analysis. Methods used in FTIR and Raman hyperspectral images are practically the same since the signals and the nature of the data are very comparable. Common workflows and software scripts are available for both kinds of data (Felten et al., 2015; Gierlinger, 2018). Multispectral fluorescence images are different in the sense that the data is mostly in the form of images and fluorescence spectra present only a few large bands compared to vibrational spectra. While most techniques have been developed to process one image at a time, multiset and multimodal hyperspectral image analyses have also been proposed to analyze series of images in a consistent way (Corcel, 2017; Courcoux, Devaux, & Bouchet, 2002; Ghaffari et al., 2019; Piqueras et al., 2015).

The analysis of multi- or hyperspectral images can be performed in a supervised or unsupervised manner. The spectral image is generally considered as a set of *n* pixels, temporarily ignoring the spatial relations between pixels. In supervised approaches, the localization of pixels is considered through the manual selection and labeling of sets of pixels (e.g., corresponding to particular tissues/cells) or of the so called *regions of interest* in the images. Selected spectra are then analyzed by the appropriate

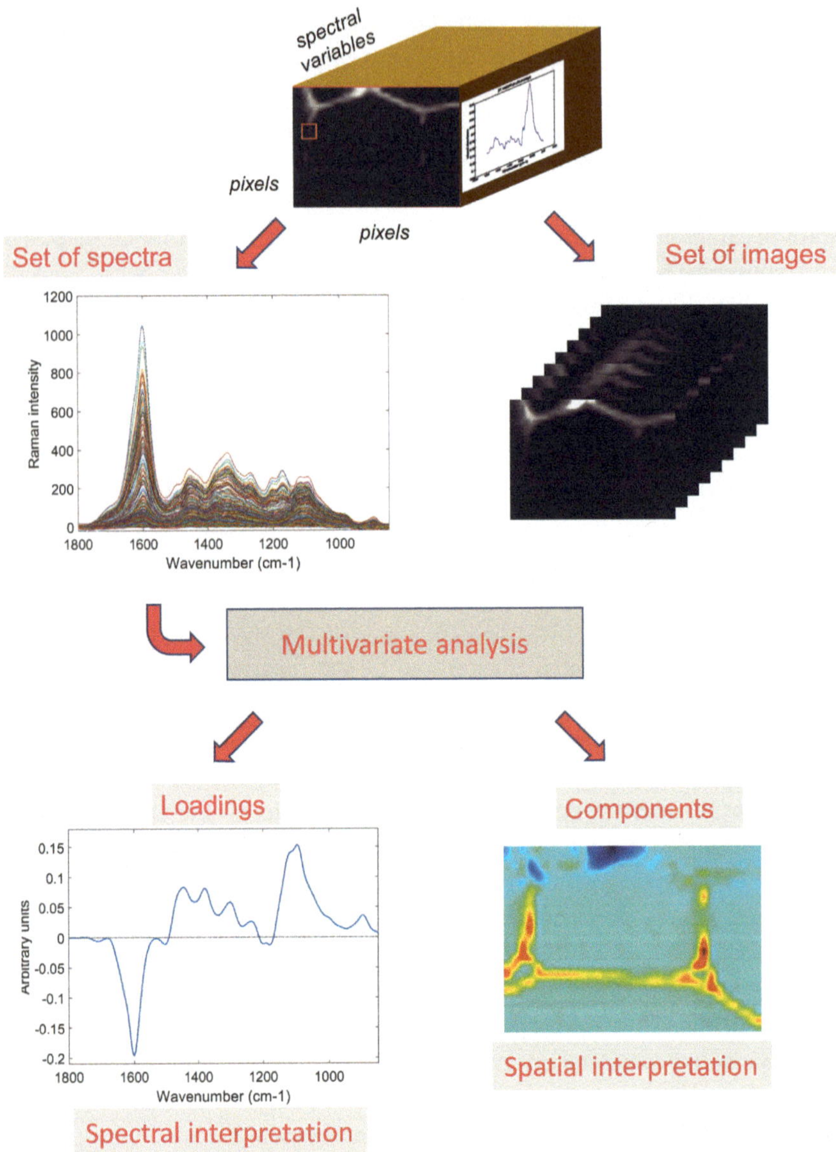

Fig. 2 Multi- and hyperspectral images: spectral and spatial dimensions, multivariate analysis.

multivariate tools. For example, discriminant approaches such as PLS-DA, or deep-learning strategies, are particularly suitable for finding classes based on the spectral properties of the pixels and representing them as regions in the images. Unsupervised approaches do not use manual assignments and are preferred when the objective is to reveal and study the variability in the images. In practice, the approaches are not mutually exclusive and often both are applied successively, generally starting with an unsupervised analysis.

2.2 Spectral image preprocessing

Preprocessing aims at removing as much irrelevant variation as possible from the dataset. Common signal defects are noise, baseline and background variations, uncontrolled intensity variations and cosmic rays in Raman and fluorescence spectra. Most preprocessing is applied independently on individual spectra and is thus very similar to those usually applied in the respective spectroscopies (see for example Afseth & Kohler, 2012; Felten et al., 2015; Gautam et al., 2015 for the main preprocessing steps used in Raman and FTIR spectroscopies; and Russ & Brent Neal, 2016, for multispectral fluorescence images). The use of pretreatment steps is optional and should be decided on an individual basis. For example, spectral normalization can correct for illumination heterogeneities (Corcel, Devaux, Guillon, & Barron, 2016), but pixels belonging to the background/empty areas should not be normalized as signals. One solution is to exclude pixels with intensities below a given threshold (spectral filtering). Alternatively, specific regions of interest can be defined manually, excluding e.g. empty image areas (spatial filtering).

2.3 Analyzing the spectra: Key multivariate analysis techniques

2.3.1 Principal Component Analysis

Principal Component Analysis is often applied as the first multivariate analysis step, as it requires no prior knowledge. All spectra (n pixels), and all spectral bands (p spectral variables) are simultaneously considered. The main variations within the entire dataset are highlighted by computing a few synthetic and uncorrelated variables, called the *Principal Components* (Jolliffe, 2002). The Principal Component *Scores* ($\mathbf{T}$) are computed for the n pixels of the original data matrix ($\mathbf{X}$, of dimension $n \times p$):

$$\mathbf{T} = \mathbf{XL}$$

where $\mathbf{L}$ is a matrix of weights with a dimension of $p \times p$, whose columns are the eigenvectors of the variance-covariance matrix of the spectral variables $\mathbf{V} = \mathbf{X}^t\mathbf{X}$. The normalized vectors of $\mathbf{L}$ are called *Loadings*. Due to the high correlation between the spectral bands, a significant reduction in dimensionality is usually achieved. Depending on the dataset, usually no more than 20 Principal Components are interpreted.

The Scores can be used to draw chemical maps, while the Loadings provide spectroscopic interpretation by showing the corresponding spectral bands (positive and negative peaks corresponding to correlated and anti-correlated bands). Principal Component Analysis is generally applied as the first tool to analyze multivariate data, as it very efficiently reveals the main contrasting regions in the image (the largest variation), together with the spectral bands responsible for this contrast.

2.3.2 Linear unmixing

Based on the Beer-Lambert law, and considering that any spectrum in a mixture can be described as a linear combination of the spectra of the pure compounds contributing to the mixture, linear unmixing methods have been proposed. The most popular ones are Vertex Component Analysis (VCA, Nascimento & Dias, 2005), Nonnegative Matrix Factorization (NNMF, Li, Adal, Wang, Emge, & Cichocki, 2007) and Multivariate Curve Resolution—Alternating Least Square (MCR-ALS) (Felten et al., 2015; Piqueras, Duponchel, Tauler, & de Juan, 2011).

Linear unmixing is based on the following bilinear model:

$$\mathbf{S} = \mathbf{C}\mathbf{U}^{t} + \mathbf{E}$$

where **S** is the original set of spectra, **U** is the unmixed set of "pure" component spectra, **C** is the corresponding concentration profiles of the "pure" components, and **E** is the residual not accounted for by the model (error, or lack of fit). It has to be noted that "pure" does not necessarily mean pure in a chemical sense, but "unique" in a sense that the component cannot be further resolved by the current analysis. The "pure" components are also called end-members. While Principal Component Analysis operates with a strict noncorrelation constraint, unmixing uses constraints related to the characteristics of the spectra and the mixtures (the most common being spectral and concentration nonnegativity, i.e., only positive spectral bands and only positive concentrations are allowed).

Before unmixing, the number of "pure" components has to be decided, e.g. by Principal Component Analysis, or from a priori knowledge (e.g., regarding the presence or absence of components, especially in certain regions of the image, De Juan, Maeder, Hancewicz, & Tauler, 2008). This step is followed by iterative algorithms estimating sequentially or simultaneously both the **U** and the **C** matrices. In practice, this step is often repeated several times by choosing different numbers of "pure" components and comparing the obtained results. The final solution is often accepted if it is meaningful both biologically and chemically, not only mathematically fitting.

2.3.3 (Semi)quantitative analysis

Quantitative analyses of spectra are preferentially done via calibration approaches. Only a few applications to hyperspectral image data are available in the field of plant sciences. For instance, Jamme, Robert, Saulnier, Dumas, and Guillon (2008) developed a Principal Component model using polysaccharide mixtures of arabinoxylans and β-glucans and projected synchrotron FTIR microscopy data of wheat grain aleurone cell walls to reveal their chemical heterogeneity. Li, Wei, Xu, Xu, and He (2018) developed a classical partial least square regression model from bamboo powder FTIR spectra to quantify the amount of cellulose, lignin and hemicellulose in hyperspectral images, after a proper calibration transfer procedure.

2.4 The image analysis aspect

Visualization is an essential step to recover the image nature of multi- or hyperspectral images and reveal the location of given spectra or spectral analysis results. This is generally the first step in the image analysis workflow: the result is a chemical map.

2.4.1 Visualization of multi- and hyperspectral images as chemical maps

Visualizing multi- and hyperspectral data is not a straightforward task (Gorzsás, 2017; Liao, Qian, Zhou, & Tang, 2016). Visualization is easiest to perform by selecting some spectral variables, i.e. individual channels that correspond to band selection or by integrating the signal over a spectral range, several channels or the entire spectra. This provides a quick overview of spatial heterogeneities that is easy to interpret. For example, Gierlinger and Schwanninger (2006) showed the chemical variations within a few cells of tension wood using Raman imaging. Gou, Park, Yu, Miller, and Liu (2008) published distributions of lignin, polysaccharides and oils in poplar stems using FTIR imaging. Computing band ratios is the natural extension and has been proposed to e.g. characterize plant fluorescent compounds (Mitsumoto, Yabusaki, & Aoyagi, 2009). Band ratios excel at visualizing relative chemical variations as they constitute a particular spectral normalization. Chemical maps based on band intensities or band ratio are particularly relevant when the bands are specific of a single component ("diagnostic bands"). When the spectral bands overlap or when several constituents must be analyzed together, multivariate image analysis methods are more suitable, as described in the previous section. The Principal Component Scores or the concentration profiles of "pure" components can be used to generate chemical maps that are based on the entire spectral information of all pixels in the image, not only selected bands. These maps are less susceptible to overlaps, shifts and nondiagnostic bands.

2.4.2 Image segmentation

Objective comparison between tissues or images may not be easy from visual observation of chemical maps alone. Image segmentation is an additional step that aims to define regions with similar and homogeneous spectral signature. Using these regions, it is then possible to identify certain features, or deduce/extract representative spectra for those regions. In most cases, image segmentation corresponds to the multivariate clustering of spectra followed by a representation of the results in the form of an image. Though raw multi- or hyperspectral data can be used, segmentation is often performed by using Principal Component Scores or contributions obtained by linear unmixing. Several techniques are available for this purpose, either supervised or unsupervised, depending on a priori knowledge (e.g., De Juan et al., 2014; Felten et al., 2015; Gautam et al., 2015).

Supervised analysis is particularly powerful to reveal differences between sample modalities or within sample regions that are a priori known or can be easily identified. In discriminant approaches, certain pixels are assigned to predefined classes and a model is built to separate these predefined classes. Several techniques have been designed to achieve the best possible separation, such as Factorial Discriminant Analysis or Partial Least Square—Discriminant Analysis (PLS-DA) (Brereton & Lloyd, 2014). A modified algorithm (orthogonal projection to latent structure—discriminant analysis, OPLS-DA) was used to contrast different xylem cell types in aspen wood (Gorzsás, Stenlund, Persson, Trygg, & Sundberg, 2011).

However, the most popular method in multi- and hyperspectral image segmentation is the unsupervised k-means clustering, which consist of iteratively assigning pixels to the class with the most similar center. New centers are computed until no changes are observed. The number of classes has to be a priori fixed and a measure of distance between pixels has to be chosen (Piqueras et al., 2011; Saporta, 2006). Using Raman images, Perera, Schmidt, Schuck, and Adams (2011) could cluster wood cell walls into layers based on lignin/cellulose variations by k-means clustering of Principal Component Scores. Hierarchical clustering aims to establish nested clusters, and it allows for easy comparison of several numbers of clusters. Using hierarchical clustering on Raman images, cell wall layers in poplar wood could be separated (Gierlinger, 2018). The drawback of hierarchical clustering is its high computation cost, which limits its application to small images. Corcel (2017) circumvented this by using k-means techniques to segment a series of large multispectral autofluorescence images of maize stems, together with a pyramid representation of the spectral image.

Evaluation of the segmentation quality, however, is less than straightforward. One method is to validate chemical relevance by comparing the average spectra of the classes and the homogeneities of spectra within each class (Felten et al., 2015).

3. Fluorescence imaging

3.1 Overview

Fluorescence refers to the light (luminescence) emitted by a molecule during the period they are excited by photons. During excitation, molecules absorb photons with the energy that allows the electrons to move from the ground to the excited state (with every molecule possessing a series of electronic energy levels with associated vibrational states, Fig. 1). The energy difference between two levels is in the range of photons in the UV (250–400nm) and visible region (400–750nm) of the electromagnetic spectrum. The excited-state electrons rapidly lose their energy and return to the lowest level of the first excited state. From there, electrons may return to one of the sublevels of the ground state, emitting photons with less energy (longer wavelength) than the absorbed photon (Stokes fluorescence). To observe fluorescence, the molecule must have delocalizable electrons (not strongly involved in bonding) that store the energy without ionizing.

The vast majority of fluorescent compounds in plants are aromatic, particularly phenolics (Donaldson, 2020; García-Plazaola et al., 2015; Talamond, Verdeil, & Conéjéro, 2015). A fluorophore is generally characterized by its excitation and emission spectra, quantum yield and lifetime. Most unsubstituted aromatic compounds exhibit intense fluorescence in the UV and visible region. The fluorescence quantum yield is defined as the number of emitted photons relative to the number of absorbed photons. Substances with high quantum yield display the brightest emissions. The lifetime corresponds to the average time that a molecule spends in the excited state before returning to the ground state. It is influenced by the fluorophore composition and environment (Chabbert et al., 2018; Donaldson & Radotić, 2013).

The main advantage of autofluorescence is that it does not require any staining or sample preparation (that could generate artifacts) and thus living organisms/cells and fresh tissue can be directly studied. There are fewer fluorescent compounds than absorbing ones and two compounds that absorb at the same wavelength are unlikely to emit at the same wavelength. However, fluorescence depends on environmental factors (such as pH, ionic state), which must be clearly defined when setting up an experiment. In addition, UV light may cause photochemical changes or destroy the fluorophore. Quenching may also occur, due to high concentrations of the fluorophore or by a deactivation process resulting from the interaction between a fluorophore and another compound present in the sample. Some fluorescent compounds are also temperature sensitive.

3.2 Sample preparation and image acquisition

Sample preparation step is crucial to obtain high quality sections that are compatible with the planned labeling and observation techniques, while preserving the structure and properties (in this case, autofluorescence) of the sample. Hand-cut fresh samples or cryosectioned cryofixed samples are the best options whenever possible. Samples are frequently stored and coated or wax or resin embedded prior to sectioning. For light microscopy, the most popular fixative is FAA (90:5:5 absolute ethanol:glacial acetic acid, formaldehyde). This fixative does not influence the fluorescence properties of wood, but some autofluorescent compounds such as chlorophyll or extractives may be removed (Donaldson et al., 2010). Fixed samples are washed and sectioned, often using either a sledge microtome or a vibrating blade microtome. An alternative to FFA fixation is to store the samples in 70°C ethanol, which also removes extractives and chlorophyll. If the sections are also stored in ethanol, they must be rehydrated before observation. The thickness of the sections varies according to the cutting instrument and the samples (10–20 μm for cryosections, 30–160 μm for sections made with a sledge microtome or a vibrating blade microtome). When different imaging modalities are to be used on the same section, the preparation protocol has to consider the constraints of each method. For example, paraffin embedding was chosen by Allouche et al. (2012) as it allowed the production of 10 μm thick sections from maize stem, a thickness well suited for the acquisition of fluorescence, FTIR and Raman hyperspectral images. Fixation with FAA and embedding in LR white acrylic resin was chosen by Donaldson and Knox (2012) to combine lignin autofluorescence and immunofluorescent labelling for cell wall characterization in normal and compression wood.

The mounting media is also important. Lignin autofluorescence is relatively weak in terms of quantum yield, and thus image acquisition within reasonable times and with good signal to noise ratios could be challenging. Glycerol/buffer pH 9 is recommended to image lignin especially under visible excitation (Donaldson, 2013), since lignin is sensitive to pH and has increased brightness at pH 9. Corcel et al. (2016) compared the fluorescence properties of cross sections and powders prepared from wheat grain mounted either in air or water, and showed that the autofluorescence intensity was lower in water, which could be ascribed to light absorption, and response varied according to tissues.

Filter equipped fluorescence devices such as wide field fluorescence macro/microscopes, confocal laser scanning (CLSM), two-photon excitation microscopes (2PEM) are the most common instruments for fluorescence imaging. The conventional sources of illumination are mercury, xenon arc-discharge, or metal halide lamps or lasers, covering the UV to near infrared spectral range. While lignin can be excited at wavelengths below 350 nm, it is not accessible with current laboratory equipment. However, the DISCO (Dichroism, Imaging, mass Spectrometry for Chemistry and biOlogy) beamline at the SOLEIL synchrotron radiation facility (Gif sur Yvette, France) has developed deep UV wide field microscopy, which provides access to this range for plant research (Jamme et al., 2013). Filter sets must be carefully selected based on the properties of the fluorophores in the specimen, as well as on the spectral filtering properties (Lichtman & Concello, 2006; Sanderson, Smith, Parker, & Bootman, 2014). The observation conditions of the same field of view can be multiplied to acquire a series of spectral images, also called multispectral images. With confocal or two photon microscopes equipped with spectral detectors, a multispectral image is directly recorded. Hyperspectral images (when a complete spectrum is acquired at every pixel) are recorded using microspectrofluorometers.

3.3 Fluorescence imaging of lignin and related compounds

3.3.1 The origin of the fluorescence of lignin and related compounds

Lignin can be excited with both UV and visible light and has a broad range of fluorescence emission: blue, green and red fluorescence (Djikanović, Kalauzi, Radotić, & Jeremić, 2007; Donaldson & Knox, 2012; Donaldson et al., 2010;

Donaldson, Singh, Yoshinaga, & Takabe, 1999; Radotić et al., 2006). The exact source of fluorescence in lignin is difficult to pinpoint because of its structural complexity. The substituted benzene rings are the main sources of fluorescence but substituents on the rings (including their position) and conjugations may produce variation in the fluorophore emission. Different molecular environments can also introduce small shifts of the maximum emission peaks.

The fluorescence of lignin model compounds has been studied using fluorescence spectroscopy combined with mathematical models for deconvolution of emission spectra plus fluorescence lifetime or FTIR and Raman spectroscopy. These studies formed the basis of current interpretations. Candidates proposed for the 240–320 excited UV emitting lignin fluorophores include anion coniferyl alcohol (Albinsson, Li, Lundquist, & Stomberg, 1999; Lundquist, Josefsson, & Nyquist, 1978), stilbene structures (Albinsson et al., 1999; Lundquist et al., 1978) and phenylcoumarone structure (Albinsson et al., 1999). The 340–400 nm-excited visible emitting lignin fluorophore remain more elusive: it is unclear whether lignin fluorescence is produced by a charge transfer mechanism (Barsberg, Elder, & Felby, 2003) or by distinct molecular species within the lignin polymer (Djikanović et al., 2007, 2012). Lignin fluorescence with visible light excitation potentially hints at multiple fluorescent structures and resonant energy transfer phenomena (Donaldson, 2020).

Since different lignin types may show differences in emission spectra, Billa, Pastou, Monries, Romero, and Koukios (2000) could discriminate wood of juvenile and adult eucalyptus tree by their fluorescence profile. Donaldson (2013) studied the fluorescence profile of lignin from pine (G-rich) and poplar (S-rich). He found that the two lignin types exhibited somewhat different emissions under blue excitation, with the fluorescence emission of poplar being greener than the yellow emission of pine. This shift in emission was ascribed to the presence of *p*-hydroxybenzoate groups in poplar lignin. In the compression wood of pine, increased levels of *p*-hydroxy type lignin produced a shift in maxima from violet to blue wavelength under UV excitation (Donaldson et al., 2010).

Ferulic acid and *p*-coumaric acid both emit blue fluorescence under UV excitation at approximately 350 nm (Harris & Hartley, 1976; Lang, Stober, & Lichtenthaler, 1991; Vidot et al., 2019). The fluorescence of ferulic acid and its ester derivatives are sensitive to pH. When excited by UV radiation, they emit a blue fluorescence with a maximum around 440–460 nm at acidic pH, which changes to green under basic pH (Harris & Hartley, 1976).

3.3.2 In situ lignin and related compounds analysis by fluorescence spectral imaging

Lignin is one of the most studied autofluorescent molecules in plants (alongside chlorophyll). Its autofluorescence has been used in light microscopy for its detection and mapping from the organ scale to the cell wall scale, especially in woody species.

It is possible to visualize the spatial distribution of lignin and its relative abundance by a single spectral image, as the spatial variation of fluorescence intensity is correlated to the relative abundance of lignin (Donaldson et al., 1999). Blue-excited emission should be preferred for this purpose, as it is more discriminant than the commonly used UV-excited emission, which is shared with other phenolics found in plants. Donaldson et al. (1999) used the lignin fluorescence intensity to estimate semiquantitative lignin distribution in the tracheid cell wall of mild compression wood of pine, while Dickson et al. (2017) investigated the stages of tracheid differentiation for comparatively fast and slow-growing pine trees. Concerning the impact of pretreatments on lignin distribution in lignocellulosic biomass, Chimenez, Gehlen, Marabezi, and Curvelo (2014) visualized the removal of lignin in sugarcane bagasse following chemical bleaching, and Herbaut et al. (2018) investigated changes in lignin distribution in poplar, miscanthus and wheat straw, following dilute acid, hot water, sodium chlorite and ionic liquid pretreatments. Lignin autofluorescence was also used to estimate wall porosity. Here, the quenching of lignin autofluorescence by carbohydrates bearing a nitrophenyl substituent was exploited.

De Micco and Aronne (2007) combined lignin UV autofluorescence with staining by safranine O to detect differences in lignification in the cell walls of three woody species. This combination enhanced the color contrast between the secondary walls, and the cell corners and middle lamellae of tracheary elements, which was interpreted as a difference in the monolignol composition of the lignin.

Ferulic acid and *p*-coumaric acid occurs in cell walls of two major groups of angiosperm commelinid monocotyledons, including grasses and the Caryophyllales (Harris & Trethewey, 2010). Their UV fluorescence properties have been used to detect them in plant organs or derived tissues (Buschmann, Langsdorf, & Lichtenthaler, 2000; Fulcher, O'Brien, & Lee, 1972; Harris & Hartley, 1976; Rudall & Caddick, 1994; Symons & Dexter, 1996).

In the majority of cases, the lignin or hydroxycinnamic acid distribution is examined against specific compounds. These compounds could be autofluorescent but most of them are not. Therefore, lignin autofluorescence are often combined with staining (Donaldson & Vaidya, 2017; Singh, Vaidya, Donaldson, & Singh, 2016) or fluorescent immunolabeling of cell wall polysaccharides (Donaldson & Knox, 2012). The fluorophores can then be chosen so that their spectral properties are different from those of lignin. In this case, two or three fluorescence images are acquired for the same field of view under well-defined preselected excitation-emission conditions.

The simple way of representing the results of this kind of multispectral imaging is overlaying the images acquired in the different excitation-emission conditions to produce for example an RGB image, where each channel corresponds to an excitation-emission condition. If only qualitative information of colocalization is required, the channels can be processed to have the same average brightness (to stop the brightest signal dominating the overlay image). This visualization approach has been used to identify lignin distribution relative to other natural fluorophores in pine needles (Donaldson & Williams, 2018), to detect changes in polysaccharides and lignin distribution between normal and compression wood (Donaldson & Knox, 2012; Donaldson, Kroese, Hill, & Franich, 2015; Donaldson et al., 2010; Mast et al., 2009) or to study the spatial distribution of enzymes or pathogens relative to lignin (Czymmek et al., 2007; Donaldson & Vaidya, 2017; Luterbacher, Moran-Mirabal, Burkholder, & Walker, 2015; Singh et al., 2016).

The same strategy can be used to track multiple autofluorescent compounds, when the fluorophores excited at the same wavelength emit at different emission wavelengths. For example, cell wall phenolic compounds and tryptophan (an amino acid present in proteins) can both be excited at 280 nm, but while the emission maxima fall in the 400–475 nm range for cell wall phenolics, it is at 360 nm for tryptophan (Jamme et al., 2013). By appropriate band filters, Chabbert et al. (2017) and Devaux et al. (2018) were able to follow both the location of enzymes and the degradation of the cell wall within the same section during enzymatic reactions. Beaugrand et al. (2022) took advantage of the different emission of proteins, hydroxycinnamic acids and lignin under UV excitation to study the evolution of their distribution in the tissues of the developing flax stem and including the application of 90 tilt bending.

While these types of studies are not yet common, multi- and hyperspectral imaging can be used to identify different fluorophores that are not fully discriminated or are overlapping, to reveal their location or colocalization.

When multispectral images are acquired using a spectral detector, a color image can be calculated as the sum of certain channels or all the channels. Donaldson et al. (2010) collected images in the violet (420–460 nm), blue (460–520 nm) and green (520-580 nm) range after UV excitation and then created overlay images for comparing normal and compression wood in pine. Color representation of the spectral images can be computed, and such composite color images were used by Auxenfrans, Terryn, and Paës (2017), Coletta, Rezende, da Conceição, Polikarpov, and Guimarães (2013) and Zoghlami, Refahi, Terryn, and Paes (2019) to visualize changes in lignin amount and distribution in lignocellulosic biomass after pretreatments. They showed that lignin emission extended in the green and yellow ranges with increasing severity of the acidic pretreatment. These shifts were attributed to changes in composition and arrangement of lignin in the cell wall. In case of NaOH treatment, no such extension to the visible range was observed (Coletta et al., 2013). This was explained by the partial removal of lignin, as a decrease in its concentration might reduce the probability of energy migration between aggregate states. A redistribution of lignin throughout the cell wall of sugarcane bagasse could also be observed with the inner region appearing blue-green and the outer one yellowish, whereas in the absence of pretreatment the wall appeared uniformly blue.

Multivariate image analyses can also be applied to reveal fluorophores and their locations. Principal Component Analysis was used to compare samples or different regions in samples (either by averaging spectra within regions or by selecting spectra in identified regions). Saadi, Lempereur, Sharonov, Autran, and Manfait (1998) could propose a 3D spatial distribution map of the main phenolics present in the outer layers of wheat grains by analyzing hyperspectral images of wheat grain acquired under UV excitation with a confocal microspectrofluorometer. Courcoux et al. (2002) adapted three-mode component analysis to compare multispectral images from wheat and barley grains acquired under four excitation wavelengths with a confocal microscope. The method made it possible to identify three main fluorescence behaviors, based on the presence of ferulic acid, lignin and cutin, and highlighted the strong similarities between the tissues of the two cereals. Principal Component Analysis was also applied by Corcel (2017) on multispectral images of maize stem cross-sections acquired with a fluorescence macroscope, obtaining color images by four filter blocks. Considering both the 4 excitation wavelengths and the RGB channels of the color images, the multispectral images were composed in total of 12 channels. Principal Component Analysis applied to a series of 40 images of cross and longitudinal sections from 10 maize stems could highlight lignified and nonlignified tissues based on their spectral behavior. Two lignin fluorescence phenotypes depending on individual stems were also identified.

Spectral unmixing was used by Donaldson et al. (2010) on confocal fluorescence images. Spectral deconvolution into a varying number of log-normal components showed that the fluorescence of lignin was complex and originated from several fluorophores. Nonnegative Matrix Factorization on multispectral images acquired from 500 to 700 nm using blue excitation at 488 nm was used to localize different lignin types in poplar wood (Donaldson, 2013). In the resulting unmixed image, the characteristic fluorescence localized at the inner secondary wall of fibers could be assigned to syringyl-hydroxybenzoate lignin. Ghaffari et al. (2019) used MCR-ALS to analyze multiexcitation hyperspectral images of wheat grain at different stages of development acquired by confocal autofluorescence microscopy. Four to six pure components were detected depending on the stages of development. Based on their fluorescence profiles, they were assigned to chlorophyll, flavonoid pigments, hydroxycinnamic acids, lignin and cuticle. In particular, lignin was detected in the pericarp in the late stages of development.

Segmentation of multi- and hyperspectral fluorescence images is a further step to visualize spatial heterogeneities. Discriminant analysis was applied by Baldwin et al. (1997) to differentiate outer tissue in wheat grain and milled flours, based on lignin and hydroxycinnamic fluorescence. Corcel (2017) tested both supervised and nonsupervised classifications on a series of maize stem section images. Supervised classification was performed by linear discriminant analysis after selecting individual pixels in images. An unsupervised clustering algorithm using multiresolution k-means was employed to develop the model from a 40-image series. Both models allowed the identification of the main tissue constituents of the stem: lignified and nonlignified parenchyma, sclerenchyma sheaths of bundles in the pith and in the rind, epidermis, and xylem fibers (Fig. 3).

Different imaging modalities of the same area in the sample can be coupled to fully characterize complex samples. Coletta et al. (2013) associated fluorescence lifetime imaging and autofluorescence imaging to fully characterize changes in lignin composition and environment as a function of pretreatment. Allouche et al. (2012) coupled spectral images acquired on the same section by three modalities (fluorescence, Raman and FTIR spectroscopies) and even analyzed the spectral data jointly by means of multiblock methods to characterize variation in cell wall composition in the vascular bundle of maize stem.

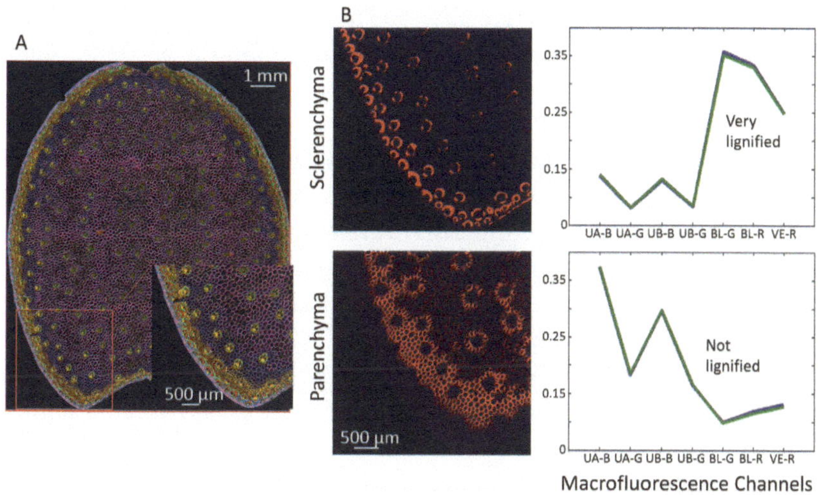

Fig. 3 Example of a segmentation of a multispectral fluorescence image of a maize stem cross-section. (A) Image of the whole section and zoom on a region. The multiscale k-means clustering model was used to segment the main stem tissues: sclerenchyma sheaths of vascular bundles, epidermis, fibers of xylem, lignified and nonlignified parenchyma, epidermis, xylem fibers, lignified and nonlignified parenchyma, etc. (B) Visualization of two groups of pixels and presentation of the associated pseudospectra corresponding to the parenchyma under the rind and around the bundles, and to the sclerenchyma sheath of the vascular bundles. *Adapted from Devaux, M.-F., Barron, C., Corcel, M., & Guillon, F. (2017). Autofluorescence variability in maize stems by multispectral image analysis of series of large images at the macroscopic scale. In:* 1st International plant spectroscopy conference, August 29–30, 2017, Umeå, Sweden.

4. Raman imaging

4.1 Overview

Raman imaging visualizes changes of molecular vibrations based on inelastic light scattering induced on the plant tissue by monochromatic laser light. Most of the light will either pass through or get reflected, and only a very small portion (ca. 10^{-6}) will be scattered elastically with unchanged frequency (Rayleigh scattering, Fig. 1). The even weaker inelastic scattering of photons was first proven experimentally by C.V. Raman and thus denoted as the Raman effect (Raman & Krishnan, 1928). With the invention of lasers in the 1970s and the availability of high-sensitivity charge-coupled devices (CCDs) shorter acquisition times paved the way for real-time monitoring and "hyperspectral" Raman imaging (Dieing & Hollricher, 2008). Raman excitation frequencies can range from the ultraviolet (e.g., 270 nm) to visible (e.g., 532, 633 nm) to near infrared (e.g., 785, 1064 nm) and influence Raman signal intensity (Smith & Dent, 2005), resonance enhancement effects (e.g., carotenoids at 514 and 532 nm (Gill, Kilponen, & Rimai, 1970)), as well as sample degradation and autofluorescence of plant tissues (Prats-Mateu, Bock, Schroffenegger, Toca-Herrera, & Gierlinger, 2018). Together with the numerical aperture of the microscope objective, the laser wavelength determines the laser spot size and the spatial resolution in Raman imaging experiments (Lee, 2012). The sampling area (laser spot size) of a single measurement in point mapping is often approximated with an airy disk, with a diameter of $d = 1.22\lambda/\mathrm{N.A.}$, where λ is the wavelength of the laser and N.A. is the numerical aperture of the lens. Therefore, experiments with high numerical aperture oil immersion objectives (e.g., N.A. = 1.4) in combination with visible laser wavelengths (e.g., 532 nm) result in a laser spot size of ca. 460 nm, and a spatial resolution according to the Rayleigh criterion of 230 nm ($0.61\lambda/\mathrm{N.A.}$) (Lee, 2012). Thus, choosing the right laser wavelength and experimental parameters is essential for high resolution Raman imaging of plant cells. Scanning a plant sample and acquiring thousands of Raman spectra across plant cells gives a chemical fingerprint at every pixel. The molecular vibrations are probed based on the frequency difference, the Raman shift, commonly expressed as wavenumbers in cm^{-1} units (Smith & Dent, 2005) (Fig. 1). Theoretical molecular calculations can help assigning individual peaks in the Raman spectrum of small molecules, e.g. lignin model monomers (Larsen & Barsberg, 2010, 2011). However, it becomes impossible for complex polymers like lignin, as the exact chemical structure is diverse and thus band assignments rely on the analysis of reference components and characteristic group frequencies. Often not every peak in the Raman spectra of plant spectra can be assigned, but the abundance and intensities of diagnostic peaks may be enough to differentiate classes of plant compounds, like aromatic components (Agarwal & Ralph, 1997; Gierlinger, Keplinger, & Harrington, 2012; Himmelsbach, Khahili, & Akin, 1999).

Special Raman imaging approaches, like Coherent anti-Stokes Raman Scattering (CARS) microscopy, Stimulated Raman Scattering (SRS) and Surface Enhanced Raman Spectroscopy (SERS) are gaining popularity in plant applications to overcome fluorescence problems and/or increase signal intensity. In SERS, Raman scattering is enhanced by adsorbing molecules on rough metal surfaces or nanostructures (Sharma, Frontiera, Henry, Ringe, & Van Duyne, 2012). CARS is a nonlinear process that requires a pump beam and a Stokes beam and the frequency difference leads to molecular vibrations when coinciding with the Raman resonance of the targeted molecules. SRS is an alternative to CARS that also relies on a pump and a Stokes laser and detects subtle changes in the intensities of the excitation fields as Stimulated Raman gain or Raman loss (Fu et al., 2012; Mansfield et al., 2013; Pohling et al., 2014; Rodriguez, Lockett, & Holtom, 2006; Zeng et al., 2010).

4.2 Sample preparation and Raman spectra acquisition

As the technique is based on scattering and not absorption, no optimal sample thickness of the plant tissue is predefined. Nevertheless, for good imaging results, plane surfaces are preferred, even though many modern instruments can automatically track uneven surfaces. Furthermore, the tissue orientation in relation to stem axis and/or plant surfaces is crucial, as the Raman signal can be orientation sensitive (due to polarized lasers and Raman polarizability changes along bond axes), informing about molecular order and orientation of e.g. cellulose (Gierlinger et al., 2010; Sasani, Bock, Felhofer, & Gierlinger, 2021). Sample preparation by micro-, ultra- or cryomicrotomy of native or embedded plant samples gives the best reproducible results as orientation can be controlled and flat surfaces are achieved (Gierlinger et al., 2012). While woody tissues can be cut in the native state, soft tissues, like the cambium, or nonwoody stems and leaves often need to be embedded in polymers or resins. As the embedding media can impregnate the plant tissue and their Raman signal superimpose with those from the tissue (Coste et al., 2021), water-soluble polyethylene glycol (PEG) is recommended (Gierlinger et al., 2012). However, the best alternative for microtoming of soft tissues is cryomicrotomy, which allows cutting native samples in the frozen state without any treatment. The microtome cut surface of the sample block as well as the resulting microsections can be used for Raman imaging experiments. Microsections can be washed with water or solvents, if cell contents and/or extractives don't need to be retained. To avoid and/or minimize sample fluorescence and sample degradation, the microsections are then embedded in water or heavy water without air bubbles (Prats-Mateu, Bock, & Gierlinger, 2020). Depending on lignin content and/or composition, it might be necessary to work with low laser powers, low integration (exposure or acquisition) times and/or to move toward near infrared excitation to avoid fluorescence backgrounds, chemical changes or even degradation of the plant tissue (Gierlinger et al., 2012; Prats-Mateu et al., 2018). NIR lasers at 1064 nm can eliminate fluorescence problems of lignocellulosic materials (Agarwal, 2014; Meyer, Lupoi, & Smith, 2011), but they have inherently lower Raman intensity and lower spatial resolution (closer to that of (synchrotron) FTIR setups). Thus, they are rarely used in Raman imaging studies of plant cells. Excitation at 785 nm provides a compromise between reasonable Raman signal and spatial resolution, particularly when 532 nm excitation is not possible due to fluorescence problems and sample degradation. Furthermore, the complementary use of different laser wavelength can be beneficial to discriminate different components and help spectra interpretation (Bock & Gierlinger, 2019).

Due to susceptibility of plant tissue to laser degradation, it is of utmost importance to check sample stability by acquiring spectra very fast in a time series (0.01 s time resolution). If a sample cannot sustain laser power for the length of a mapping experiment, then laser power, point distance in mapping and/or exposure (acquisition/integration) time need to be adjusted. Although shorter exposure (lower integration times) means noisier spectra, it is a compromise worth making, as average spectra with good signal to noise ratios can be extracted from chemically similar regions (e.g., cell corners) in the end, or spectral quality improved by multivariate noise filtering. Controlled laser bleaching experiments might help to understand the aromatic nature or the components or allow to remove them partly (e.g., chlorophyll bleaching) and to get access to underlying components (e.g., carbohydrates) (Prats-Mateu et al., 2018).

4.3 Raman imaging of changes in lignin amount and composition in wood

Many of the Raman imaging studies focus on trees (wood) as the most important renewable bioresource. These are usually more difficult to study than nonwoody stems, due to the higher amount of lignin and other aromatic components resulting in fluorescence problems. Especially impregnation with extractives (e.g., flavonoids) can induce a high fluorescence background, obscuring the Raman signal. Therefore, woody species with high extractive content, e.g. oak wood (Felhofer et al., 2021), robinia or yew, are more difficult to measure due to high fluorescence backgrounds and susceptibility to laser degradation. Wood species with low extractive content, e.g. poplar wood (Gierlinger & Schwanninger, 2006; Ma, Zhang, Zhou, & Xu, 2014; Ma, Zhou, Zhang, & Xu, 2013; Ozparpucu et al., 2018, 2017; Perera, Schmidt, Chiang, Schuck, & Adams, 2012; Segmehl et al., 2019; Zeng et al., 2015) or spruce wood (Agarwal, 2006; Agarwal, Mcsweeny, & Ralph, 2011; Gierlinger et al., 2010; Hänninen,

Kontturi, & Vuorinen, 2011) are easier to measure. Even in its pure form, lignin shows considerable fluorescence background and only a few clear Raman bands, as shown by the example spectra of G-dehydrogenation polymer (G-DHP) lignin and milled wood lignin from birch (Fig. 4). These two examples represent G-type and GS-lignin, and show similar Raman spectra.

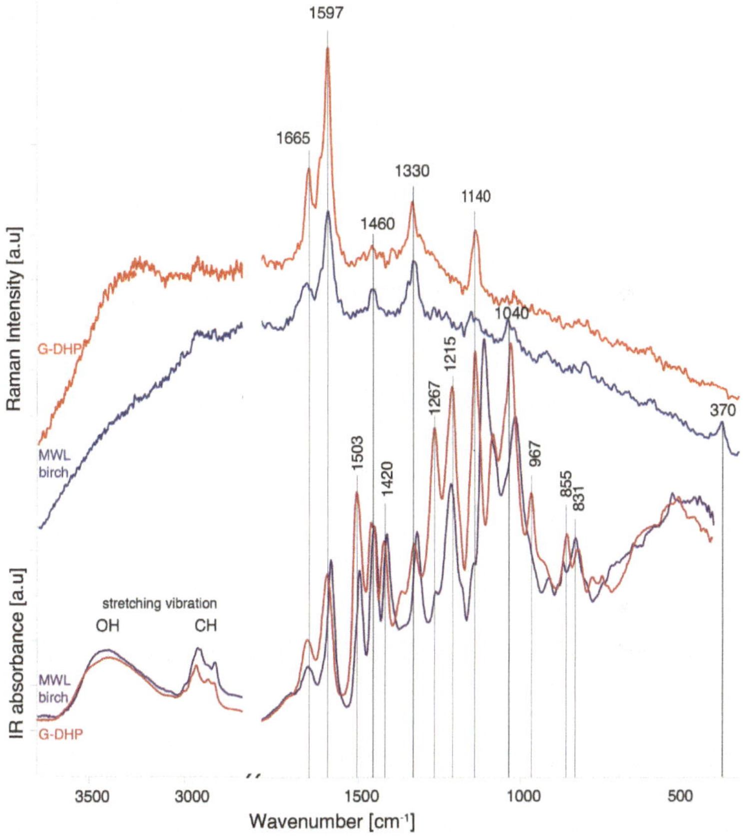

Fig. 4 Raman (top) and FTIR (bottom) spectra of milled wood lignin from birch (MWL birch) and G-dehydrogenation polymer (G-DHP). While the noisier Raman spectra show the typical high background and only a few lignin bands, the FTIR spectra show more distinct lignin bands across the whole fingerprint region. The 1597 cm^{-1} aromatic stretching vibration is the strongest marker band for lignin in the Raman spectra. In the FTIR spectra, the 1503 cm^{-1} band is the most diagnostic one. The Raman spectra have been acquired with 532 nm laser excitations and the FTIR spectra in Attenuated Total Reflection (ATR) mode, which can cause relative intensity distortions (due to penetration depth differences across the spectral range) and slight shifts in band positions compared to transmission or diffuse reflectance measurements. The marked bands labeled by their position are explained in Table 1.

The most prominent band for both lignins occurs around 1600 cm^{-1} and is assigned to the aromatic ring stretch of G- and S-rings (Bock et al., 2020) (Table 1). This band is stronger in G-DHP than in milled wood lignin of birch (Fig. 4) and is usually accompanied by a band around 1665 cm^{-1}, which is related to the C=O stretch of conjugated carbonyls (i.e., coniferyl aldehyde, carbonyls in α-position, carbonyls of quinone methide) and the C=C stretching of coniferyl alcohol. Additionally, a shoulder around 1620 cm^{-1} assigned to C=O of coniferyl aldehyde can be found (Bock et al., 2020). Medium intensity bands are observed at 1460, 1339, 1140 and 1040 cm^{-1} in both presented lignin model components (Fig. 4, Table 1). Based on a spectral deconvolution method and regression analysis, the bands between 1220 and 1530 cm^{-1} were used to determine S/G ratio quantitatively (Lupoi et al., 2014; Lupoi, Singh, Parthasarathi, Simmons, & Henry, 2015; Lupoi & Smith, 2012; Sun et al., 2012). The weak band at 370 cm^{-1} in birch lignin (Fig. 4), was found to be associated with the S unit and was also used to estimate the S content of hardwoods (Agarwal et al., 2011; Agarwal, Ralph, Padmakshan, Liu, & Foster, 2019).

Depending on lignin content and composition, Raman spectra of wood can be more or less dominated by lignin bands and fluorescence background. The strongest Raman band of lignin around 1600 cm^{-1} was used in the first Raman imaging studies to visualize lignin distribution at the cell wall layer level (Agarwal, 2006; Gierlinger & Schwanninger, 2006). A recent comprehensive analysis of lignin substructures revealed that coniferyl aldehyde and -alcohol are the most enhanced Raman scattering components (due to the enhancement of the Raman signal by extended electron clouds and charge transfers) and thus are responsible for most of the Raman signal in wood spectra (Bock & Gierlinger, 2019; Bock et al., 2020). That lignin and aromatic components are responsible for the fluorescence background becomes clear when comparing the background intensity (representing fluorescence, Fig. 5A) and the 1600 cm^{-1} band intensity (representing lignin, Fig. 5B). Both fluorescence intensity and 1600 cm^{-1} band intensity show highest values in the cell corners, followed by medium intensity in the vessels and lowest intensity in the fiber cell walls (Fig. 5A and B). This pattern holds true for almost all wood species, except for leatherwood, where an atypical lignification pattern was revealed by Raman imaging (Mottiar et al., 2020). Lignin impregnates the cell wall against water, a waterproofing that can be visualized by imaging water distribution via the OH-stretching (Fig. 5C). This shows the highest water content in the lumen, followed by fiber cell walls, then vessels and finally cell corners (Fig. 5C). This approach also

Table 1 Selected key bands below 2000 cm^{-1} in the Raman and FTIR spectra of lignin (see Fig. 4 for examples).

Position (cm^{-1})	Origin	Comments
1650–1700	C=O	Position depends on the electronegativity of the substituents on the aromatic ring. Relative intensity can vary depending on the lignin type
1590–1610	Aromatic C=C	Very strong in Raman spectra, but not only from lignin, other aromatic components can contribute. In FTIR spectra, its relative intensity tends to be higher for S type lignin than for G type lignin (Faix, 1991), but not in Raman spectra
1500–1515	Aromatic C=C	One of the most diagnostic bands for lignin (being fully diagnostic if no other aromatic compounds are present) in FTIR spectra, where it is strong and well resolved. Absent in Raman Higher relative intensity in G type lignin than in S type lignin in FTIR spectra (Faix, 1991) The ratio between the 1597 and 1505 bands can inform about S/G ratio in the sample, or more generally about lignin cross-linking (the higher this ratio, the lower the cross-linking is) in FTIR spectra
1450–1470	C—H	Can have contributions from =C—H, stronger in FTIR spectra than in Raman
1420–1430	C—H and aromatic C=C	Generally weaker than the band around 1460 cm^{-1} in FTIR spectra and can be absent in Raman spectra
1360–1370	Aliphatic C—H	Weak in lignins, as —OCH_3 groups do not contribute. Thus, a band in this position is often considered more representative (or even diagnostic) of cellulose
1320–1340	S-ring, condensed G-ring	Present in S- and G-lignin and can be masked by nonaromatic compounds in the matrix

Continued

Table 1 Selected key bands below 2000 cm^{-1} in the Raman and FTIR spectra of lignin (see Fig. 4 for examples).—cont'd

Position (cm^{-1})	Origin	Comments
1250–1270	Composite, including G-ring and C—O	Usually stronger in G than in S lignin, but can be masked by nonaromatic compounds in the matrix. Can be prominent in FTIR spectra, missing in Raman spectra
1210–1250	Composite, including C—O	More intensive in condensed G than in etherified G. Can be hard to quantify due to overlaps from other compounds in plant samples. Missing in Raman spectra
1140–1170	C—O—C	Usually heavily overlapping with bands from polysaccharidic compounds in plant samples and can appear as a shoulder only for lignin. Position can shift significantly
1120–1140	Composite, including aromatic C—H in lignins, sugar ring vibrations in carbohydrates (including polysaccharides)	Usually strong in FTIR spectra, but heavily overlaps with carbohydrate-related bands in plant samples
1030–1040	Aromatic C—H in lignins, sugar ring vibrations in carbohydrates (including polysaccharides)	In pure lignin samples this band tends to be stronger for G than for S. But it is usually completely masked by bands from polysaccharidic compounds in plant samples, which dominate this area of FTIR spectra. Much weaker in Raman spectra
960–990	Out of plane deformation of aromatic C—H	Could be just visible in FTIR imaging works, as FPA MCT detectors can have cut-offs around this limit. Very weak in Raman spectra, if visible at all
830–860	C—H out of plane vibrations of the aromatic rings	Several bands, positions and relative intensities in FTIR spectra can be informative of the subunit types (S, G or H), but often cut-off in imaging work, due to the range limits of FPA MCT detectors. Weak or even absent in Raman spectra
360–370	C—O—CH_3 bending? (Bock et al., 2020)	Outside of the mid-IR range, only visible in Raman spectra. Can be used to estimate S content if not influenced by cellulose band at 380 cm^{-1}

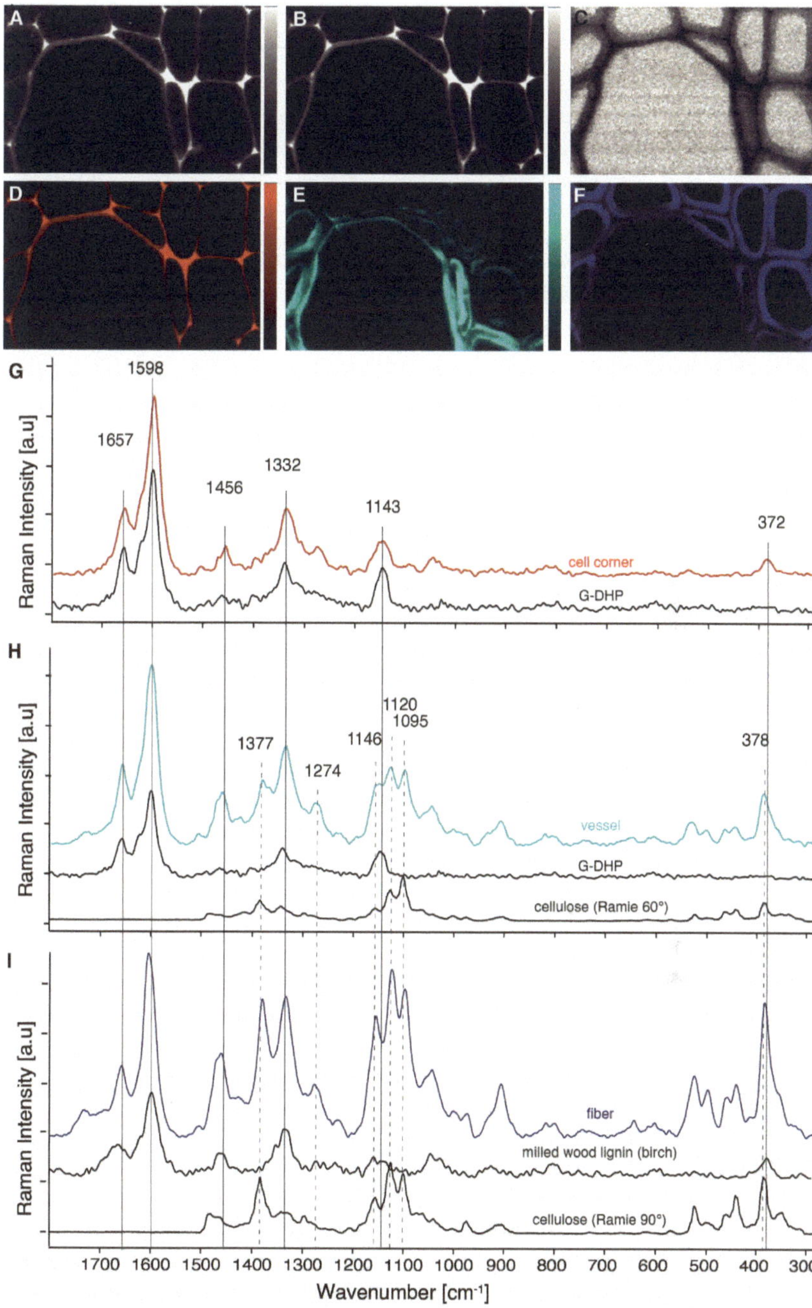

Fig. 5 Raman images of poplar wood based on univariate and multivariate image analyses, and the derived average spectra. (A) The intensity in a region without bands (1880–1940 cm^{-1}), corresponding to fluorescence background. (B) The intensity of the lignin band (integral between 1552 and 1630 cm^{-1}), highlighting the more lignified cell corners. (C) The intensity of the OH-stretching band (integral between 3050 and 3550 cm^{-1}) visualizing water within the sample. (D–F) True component analysis separating cell corners, vessel and fiber walls, based on (G–I) spectral differences in lignin content as well as composition (for experimental details, see Mottiar, Gierlinger, Jeremic, Master, & Mansfield, 2020).

showed higher volumes of water in the cell walls of transgenic compared to wild-type aspen, indicating an increase in the hydrophilicity of the cell wall in the mutants with different and reduced lignin composition (Horvath et al., 2012).

Besides analyzing lignin distribution, the functional groups of lignin in wood have also been investigated. The amounts of coniferyl alcohol and aldehyde groups were compared to the total amount of lignin in the Raman images of pine (*Pinus sylvestris*) and spruce (*Picea abies*) wood (Hänninen et al., 2011). While no significant differences could be seen in lignin and cellulose distribution between these samples, a clear distinction was observed in the distribution of coniferyl alcohols and coniferyl aldehyde, based on the ratios of the 1623 and 1660 cm^{-1} bands toward the 1600 cm^{-1} band. It should be noted that all aromatic components and lignin substructures contribute to the 1600 cm^{-1} band in different proportions (Bock & Gierlinger, 2019; Bock et al., 2020) and thus it is very difficult to correlate its absolute intensity to lignin amounts if the lignin composition changes. In other words, compositional changes of aromatic components in plants render the 1600 cm^{-1} band intensity unsuitable for absolute lignin quantification on its own. A change in coniferyl aldehyde content can be tracked clearly by the change of the 1620 cm^{-1} band (Ozparpucu et al., 2018, 2017), but the 1600 cm^{-1} band will also be influenced by this change and it is therefore no longer a suitable measure for the overall lignin content. This is also true when other molecules with extended delocalized π-electrons (cinnamic acids, flavonoids, stilbenes) induce bands with very high Raman intensities (Schmid & Brosa, 1971; Schmid & Topsom, 1981). Stilbenes, for example, are so strong Raman scatterers (due to the conjugated aromatic structures) that they can be visualized in situ in pine heartwood, despite their low amounts (Belt, Keplinger, Hanninen, & Rautkari, 2017; Felhofer, Prats-Mateu, Bock, & Gierlinger, 2018).

The assignment of Raman bands to chemical components is often hindered by the fact that a vibrational band relates not to a molecule, but to a vibrational mode of functional groups (e.g., aromatic stretching, CH-bending, see Table 1), and the same functional groups can be present in different molecules in the sample. While different aromatic substructures make up the 1600 cm^{-1} band, carbohydrates can also overlap with aromatics, for instance in the case of CH-bending or C—O stretching vibrations. This limits the use of single bands or wavenumber regions to image changes in amounts or in composition (e.g., S/G ratio) and to interpret spectral changes in terms of amounts and composition. Due to the multicomponent, multimolecular nature of plant tissues, the analysis and interpretation of Raman spectra and Raman images must be comprehensive and backed up by multivariate tools.

As shown in Section 2, many different multivariate approaches can be used for image generation and spectral analysis. Here we show an example of poplar wood (Fig. 5), using true component analysis; i.e. a function calculating the hyperspectral dataset as a linear combination of the most different spectra with a basis analysis algorithm (Dieing & Ibach, 2018). The number of components is stepwise increased as long as different and meaningful spectra are retrieved. This method differentiated the cell corner (Fig. 5D) from vessel (Fig. 5E) and fiber cell walls (Fig. 3F). The corresponding average spectra were clearly different (Fig. 5G–I) and were modeled with reference Raman spectra using Orthogonal Matching Pursuit (Pati, Rezaiifar, & Krishnaprasad, 1993). This algorithm compares the spectra of the distinguished cell corner and cell walls with spectra of known compounds from a reference library, in this case including 326 entities, from aromatics to lipids to carbohydrates. The matching shows that the cell corner and vessel cell wall represent mainly G-type lignin (Fig. 5G–H), whereas the fiber cell wall has more S-units (with milled wood lignin of birch chosen as the most similar reference, Fig. 5I). The contribution of celluloses is also included in the cell wall spectra, pointing to different orientations of the cellulose fibrils in the vessel and fiber walls (Fig. 5H and I). The overlap of cellulose and lignin bands, especially in the region relevant for S/G ratio determination (1220–1530 and 370–380 cm^{-1}) clearly shows why most of these bands are unsuitable for quantitative analyses in case of compositional changes.

Due to its nondestructive nature and high spatial resolution, Raman imaging can follow in-situ wood modification on the cellular level. This can provide important insights into genetic modifications of wood, as well as in chemical and enzymatic treatments (e.g., Horvath et al., 2012; Keplinger et al., 2015; Merk, Chanana, Gierlinger, Hirt, & Burgert, 2014; Ozparpucu et al., 2018, 2017; Segmehl et al., 2019; Segmehl, Studer, Keplinger, & Burgert, 2018).

4.4 Imaging cinnamic acids and lignin in nonwoody plant tissues

Wang, Keplinger, Gierlinger, and Burgert (2014) used Raman imaging when comparing cell wall chemistry of bamboo with wood and explaining mechanical performance. Bamboo showed significantly higher lignin content in cell corners compared to cell walls. Based on extracted average spectra, significant differences in lignin structure were revealed. In the spectrum of bamboo cell walls, the strong aryl ring stretching at 1600 cm^{-1} was a doublet, with a second peak at 1631 cm^{-1}, accompanied by a weak peak at 1683 cm^{-1}. This doublet peak, as well as the band at 1173 cm^{-1}, clearly indicated the presence of hydroxycinnamic acids (ferulic and *p*-coumaric acids) (Wang et al., 2014).

Different lignin composition was also visualized in Arabidopsis stems, in fibers and vessels and a change in cinnamic acids in a trichome of the stem (Mateu, Hauser, Heredia, & Gierlinger, 2016). To study the genetic mechanisms controlling lignin quantity and composition in different lignified cell types and tissues, Blaschek, Nuoendagula, Bacsik, Kajita, and Pesquet (2020) validated in situ Raman microspectroscopic characterizations of total lignin content, S/G ratio, and coniferyl alcohol content, directly in plant biopsies. Multiple lignin biosynthesis loss-of-function mutants revealed that the genetic regulation of lignin is tissue specific (Blaschek et al., 2020). In wild-type and lignin-downregulated alfalfa lines, Coherent Anti-Stokes Raman Scattering (CARS) microscopy was used to image the general trends of lignin content, based on the 1600-cm^{-1} Raman mode. Greater reduction of lignin content was observed in the cell corners and suggested to account for the enhanced susceptibility to chemical and enzymatic hydrolysis (Zeng et al., 2010). *Miscanthus* × *giganteus*, a potential energy crop and a model lignocellulosic material, was imaged as a function of NaOH treatment. Lignin colocalized with cellulose was completely removed upon treatment (Chu, Masyuko, Sweedler, & Bohn, 2010). Rice straw was also investigated after treatment to optimize biomass conversion efficiency and promote extensive utilization as biofuel (Li et al., 2020).

Raman imaging on the big leaves of New-Zealand flax (*Phormium tenax*) also revealed the high heterogeneity in lignification in nonwoody stem tissues. Only slight lignification was observed in the parenchyma and collenchyma in the top of the leaf and a high variability (7–44%) in the sclerenchyma. The sclerenchymatic fibers surrounding the vascular tissue had the highest lignin levels and potentially acting as a barrier and protection of the vascular tissue (Richter, Mussig, & Gierlinger, 2011). In spruce needles, Raman imaging was used to follow the transition of aromatic components (lignin, cinnamic acids, flavonoids) from the cuticle to the epidermal layer to get a better understanding of plant surfaces (Bock, Felhofer, Mayer, & Gierlinger, 2021; Sasani et al., 2021).

In horticultural products, lignification triggered by biotic or abiotic stresses plays an important role, as it hardens fruits and vegetables and eventually influences their consumer appeal. The deposition of lignin in loquat flesh was visualized in this context from tissue to single-cell level, by combining the advantages of stimulated Raman scattering (SRS) and spontaneous Raman microscopy (Zhu, Wu, & Chen, 2018; Zhu, Yang, Ji, Wu, & Chen, 2019). The effect of low temperature on lignification was also investigated, using the herb *Soldanella pusilla*, grown at 0 °C (Korner et al., 2019).

5. Fourier-transform infrared (FTIR) spectroscopic imaging

5.1 Overview

Electromagnetic radiation in the mid-infrared region (corresponding to 4000–400 cm^{-1} when expressed as wavenumbers) is absorbed by molecular vibrations altering dipole moments, and causing the transition from the ground vibrational state to the first excited state (Fig. 1).[a] Since these molecular vibrations are inherently tied to functional groups and structures, specificity depends on unique (diagnostic) features of compounds, much like in the case of Raman spectroscopy. Lignin is unique among the building blocks of xylem cell walls, as it is the only aromatic polymer in appreciable amounts. Thus, aromatic C=C vibrations can be considered diagnostic of lignin in this case, though they may not be in a different chemical matrix (when other aromatic compounds are also present in sufficient quantities). In terms of recording speed, FTIR imaging can be superior to its Raman counterparts, as focal plane array (FPA) detectors are capable of recording thousands of high-quality, full-range spectra in a matter of seconds over an area of thousands of square micrometers. However, lateral resolution tends to be considerably lower than in the case of Raman imaging, generally falling in the 4–20 μm range in the 4000–1000 cm^{-1} spectral region[b] on standard, nonsynchrotron-based systems. This is due to diffraction limits (Lasch & Naumann, 2006), which also results in higher wavenumber bands having higher spatial resolution within the same image. This is a very important, albeit often overlooked aspect, which generally means that FTIR imaging can provide cell specific information, but not cell wall layer specific data (Gorzsás et al., 2011). Synchrotron sources are capable of improving lateral

[a] Transitions in the rotational energy levels are ignored here, as they are primarily relevant for gases.

[b] This is the most common range used for the imaging of lignocellulosic materials. There are important bands below 1000 cm^{-1} as well (see Table 1), but high-performance focal plane array (FPA) detectors have cut-offs around that limit.

resolution (e.g., Dučić, Thieme, & Polle, 2012; Dumas, Carr, & Williams, 2000; Vongsvivut et al., 2019; Wetzel, Eilert, Pietrzak, Miller, & Sweat, 1998; Yu, 2004), but not to the level of Raman imaging (e.g., Gierlinger, 2018; Gierlinger & Schwanninger, 2007).[c] This limitation needs to be kept in mind when designing experiments and determining the scientific questions that can be answered by FTIR imaging, also with respect to lignin content and composition. For example, changes in the chemical composition of the lignin-rich middle lamella can easily be masked ("diluted") in FTIR imaging by the spectral contributions of thick cell wall layers that will dominate the spectrum of the whole cell. Nevertheless, comparisons between cells (and cell types) are possible even when subcellular comparisons are blurred by limitations in lateral resolution.

5.2 Sample preparation

FTIR microspectroscopy is an optical method, thus the optophysical properties of the sample are critically important for the results. Most measurements are performed in transmission mode as it gives the best visual representation. Attenuated total reflectance (ATR) could potentially improve spatial resolution (due to higher numerical apertures but requires very tight contact between the sample and the internal reflection element, which heavily distorts plant samples). Transmission mode measurements require thin (ideally below 10 μm), flat sections, on infrared transparent carriers (often CaF_2 or BaF_2), in order for the infrared light to pass through. Additionally, FTIR imaging is very sensitive to water which means that samples require drying (unless water content itself is the subject of the investigation). For woody tissues, this normally does not pose a major problem, as they already contain very little water and any further drying has limited (and often only localized) effect on anatomy and optical properties. For wet tissues, however, the drying step during sample preparation can be particularly challenging, as it frequently distorts (cracks, folds, curls, etc.) the samples. This is especially true for thin samples, needed for transmission mode measurements. Thus, sectioning, especially when followed with a drying step can compromise sample integrity, making sample preparation more challenging

[c] Augmented techniques, such as FTIR nanospectroscopy (or nano-FTIR for short), when vibrational spectra are collected at the tip of an atomic force microscope (Knoll & Keilmann, 1999), can further improve spatial resolution (~20 nm). However, these are not common setups (yet), and they change the way spectra are recorded and analyzed. Thus, although very elegant plant applications are emerging (e.g., Zancajo et al., 2020), they will not be discussed further here.

than for Raman imaging. However, the speed, ease of use, sensitivity, versatility, and the information content of FTIR imaging often makes up for these inconveniences.

5.3 Applications toward lignin and related compounds

FTIR microspectroscopy of plants in general has been extensively reviewed before (e.g., Dokken, Davis, & Marinkovic, 2005; Gierlinger, 2018; Kumar, Lahlali, Liu, & Karunakaran, 2016; Türker-Kaya & Huck, 2017; Wetzel, 2009; Zhao, Man, Wen, Guo, & Lin, 2019), thus we discuss FTIR imaging only in the context of lignin.

5.3.1 Bands related to lignin in FTIR spectra

To be able to image lignin, the bands related to lignin in FTIR spectra need to be known. There are very thorough fundamental works of excellent quality in this area, most notably by Oskar Faix (e.g., Faix, 1986, 1988, 1991; Faix, Andersons, & Zakis, 1998; Faix, Argyropoulos, Robert, & Neirinck, 1994). Starting from monomers and working up to polymeric models of lignin, Faix was able to assign key bands of FTIR spectra and determine their specificity with respect to the subunit types (H, G and S) (Faix, 1986, 1988, 1991). A summary of the most important bands is given in Table 1 and example spectra are shown in Fig. 4. Please note that these are of pure lignin samples, and thus several of the listed bands can overlap with bands of other cell wall components in native plant samples (see, e.g., Gorzsás & Sundberg, 2014). Thus, not all the listed bands are diagnostic for lignin. The most diagnostic band is originating from aromatic C=C vibrations, located around 1505 cm^{-1} (Table 1, Fig. 4). The intensity of this band shows a quantitative linear correlation to Klason lignin content (at least when lignin content is in the range of 5–30%, Faix, 1988). When only lignin is present in the sample, using this band for normalization can lead to excellent quantitative correlations between spectral features and subunit types (Faix, 1986). Generally, the easiest S/G ratio determination for in situ lignin analyses in plant samples using FTIR spectroscopy and imaging is based on the ratio of the 1595/1505 cm^{-1} aromatic bands (Fig. 4). While more bands show varying intensities with the subunit type (see Table 1), they are often masked by bands from other components in complex matrices (such as native plant cell walls), making it hard to reliably determine their intensities.

5.3.2 FTIR imaging of lignin and related compounds in different plant tissues

When it comes to analyzing lignin in situ, microspectroscopy and imaging must be clarified. Microspectroscopy covers all kinds of spectroscopic measurement through a microscope, even if only distinct points in the sample are recorded (i.e., the microscope is only used for precise sampling, making sure that the correct position is measured, but no image data is collected). Imaging is a subcategory of microspectroscopy, referring to measurements when a continuous area is covered, resulting in a multidimensional hyperspectral data cube (where there are at least two spatial dimensions, in addition to the spectral dimension(s)). Thus, imaging always brings spatial context to the analyses.

Of the potential sample types containing lignin, xylem is particularly suitable for FTIR imaging, as it is inherently dry, relatively easy to section, and the resulting sections remain stable over time. One of the major advantages of FTIR imaging is that all compounds are monitored simultaneously, nondestructively and without any external agent (marker, dye or label). Since all the major classes of biopolymers of woody cell walls (cellulose, lignin, pectins, and hemicelluloses) are infrared active, the recorded spectra will contain information about all of them, even if lignin is the primary target of the analysis. However, it also means that data evaluation can be complicated by the presence of overlapping bands from compounds that are not the focus of the analysis. Consequently, while lignin has unique, diagnostic features (especially the aforementioned aromatic C=C band at $1505\,cm^{-1}$) that could be sufficient to localize and even (semi)quantify its (relative) amounts, multivariate imaging approaches were shown to provide better results, both qualitatively (Gorzsás et al., 2011) and quantitatively (Li et al., 2018). A large number of FTIR imaging studies of xylem focus on the chemical compositional differences as a function of anatomical/developmental features (e.g., Gorzsás et al., 2011; Gou et al., 2008; Song, Yin, Salmén, Xiao, & Jiang, 2014), often in combinations with genetic or environmental factors. Comparing different cell types has the distinct advantage of being able to rely on the same image, assuming a large enough area can be recorded that covers the different cell types with enough spectra. This eliminates the need for comparisons between images, which is very complicated and is often done by extracting individual spectra representative of classes to be compared, and not using the image data as such. (For example, extracting spectra corresponding to the same cell types in different images of transgenic and wild type trees, and comparing the spectra, not the images.

In other words, microspectroscopy, not imaging.) Reports using FTIR imaging to elucidate the effects of e.g. different kinds of physical and chemical treatments (Himmelsbach, Khalili, & Akin, 2002) as well environmental factors on wood composition are available, with implications in very diverse areas, such as construction (Guo et al., 2018) or art preservation (Capobianco et al., 2017) and even ecology, forestry and agriculture (Raab & Vogel, 2004). With the impact of climate change on forests (particularly boreal forests), applications of FTIR imaging in this area are likely to gain momentum in the coming years.

We show a case study of FTIR imaging in xylem (Fig. 6).

A 64 × 64 FPA detector was used to collect FTIR spectra from a 10 μm-thick poplar wood cross-section, providing 4096 spectra in a minute, covering an area of ca. 170 × 170 μm (highlighted in Fig. 6A). The total intensity map (Fig. 6B) can be used to set thresholds to remove empty areas (lumen) or bad pixels (abnormally low or high intensity responses). Such thresholding helps to minimize noise and remove some of the unwanted variation, thereby "cleaning up" PRINCIPAL COMPONENT ANALYSIS (as principal components will not be used to describe such variation). While this is the recommended course of action (Gorzsás et al., 2011), we deliberately avoid thresholding here for demonstration purposes. As discussed in Section 2, thresholding is needed in order to avoid upscaling noise when normalizing spectral intensities. (Normalization is a necessary step even in FTIR imaging to account for varying sample thickness, pixel coverage and optical distortions.) For example, mapping the intensity of the diagnostic lignin band at 1505 cm^{-1} (Fig. 6C) after total area normalization shows blurred boundaries. This is largely due to the upscaling of low intensity signals, though there are several other contributing factors. One is the quality of the section: cracks, fissures and unevenness result in stray light that "contaminate" pixels with spectral contributions from other pixels (i.e., optical effects). Another contribution could be biophysical (e.g., sectioning can "smear" the contents of cells over empty areas), even though this is not expected for this particular sample (dry xylem). Finally, the resolution of FTIR imaging is very dependent on the wavelength of the light. Thus, a band at 1500 cm^{-1} has considerably worse spatial resolution than a band at 1600 cm^{-1}. Indeed, mapping the 1595 cm^{-1} aromatic C=C band results in a cleaner image (Fig. 6D), even though this is not only the result of the resolution differences. The 1595 cm^{-1} band is often better resolved than the 1505 cm^{-1}, less influenced by stray light and since it is less pronounced in vessels, it is less likely to "contaminate" neighboring empty pixels of lumen. Another factor to consider is the Z-scaling of the intensity

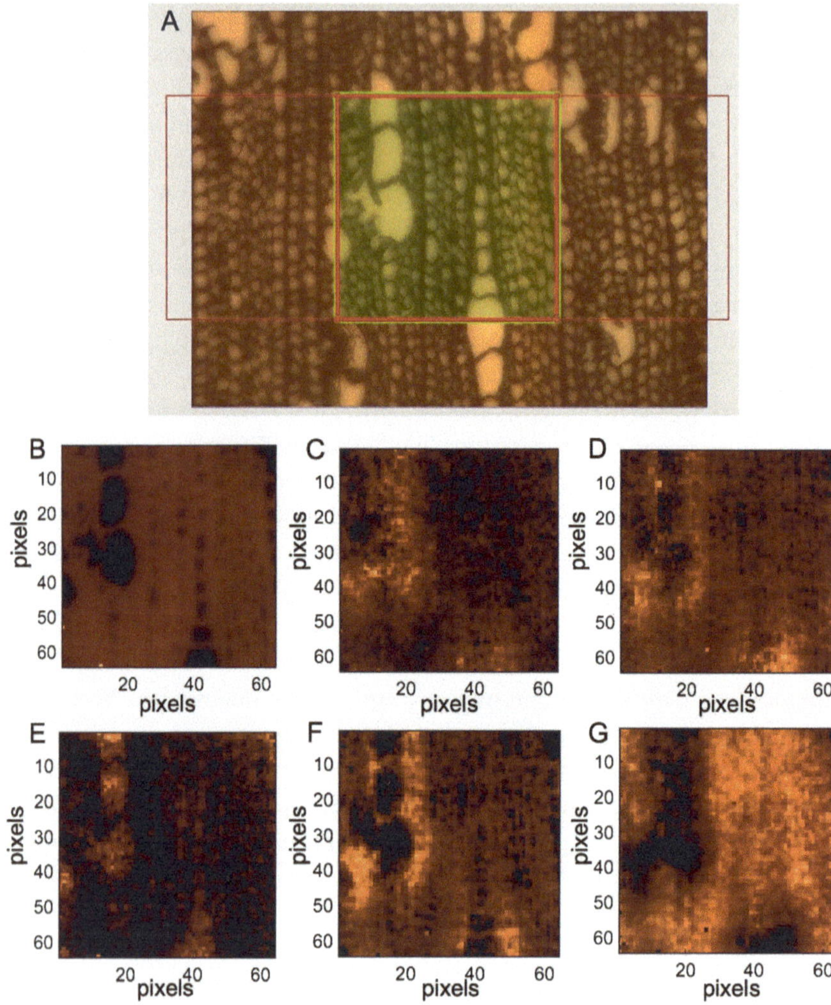

Fig. 6 (A) White light image of a 10 μm-thick cross-section of a wildtype poplar tree. The area of FPA FTIR imaging is highlighted as the shaded square in the center. Vessels, known to be rich in lignin (especially G-type) are visible, together with fibers (containing higher proportions of S-type lignin) and rays. (B) Total intensity map, showing the sum of integrals under all bands of the spectra. (C) Map of the 1505 cm^{-1} band intensity (aromatic C=C), after total area normalization. (D) Map of the 1595 cm^{-1} band intensity (aromatic C=C), after total area normalization. (E–G) Scores maps of principal components 1–3, respectively. In all maps, lighter colors represent higher intensity.

maps. Here, equal scaling is used, but if large differences manifest in certain areas, it renders other areas "flat," and contrast is lost there. While multivariate analysis can handle a wide range of spectral changes, thresholding is still important to avoid upscaling artifacts and to "waste" principal components on describing noise/empty areas. For instance, when mapping the first three principal components (Scores maps) in same dataset, it becomes clear that the largest variation (the first principal component, PC1) describes the lumen vs cell wall differences (Fig. 6E). Thus, thresholding would remove a significant unwanted variation from the dataset and would allow the Principal Component Analysis to "zoom-in" on the variation between cell walls. The second largest variation (the second principal component, PC2) considers differences between vessels and fibers in general, highlighted vessel areas (Fig. 6F). The corresponding Loadings (not shown) indicate that PC2 is positively correlated to lignin (meaning that higher proportions of lignin are found in vessels) and negatively correlated to carbohydrates (i.e., lower proportions of polysaccharides are found in vessels). In other words, the second largest variation in the dataset (after lumen-cell wall differences) is the lignin/polysaccharide ratio, which is the key difference between vessel and fiber spectra. In particular, the 1505 cm^{-1} lignin band is very prominent in the Loadings of PC2, indicating increased G-lignin proportions in vessels (Table 1). This can also be indirectly deduced by the increased intensity of the 1505 cm^{-1} band in vessels (Fig. 6C), which is not as pronounced for the 1595 cm^{-1} band (Fig. 6D). That is: vessels contain more lignin in general than fibers, but especially more G-type. The map of the third principal component (PC3) (Fig. 6G) is essentially a blurry inverse of the map of PC2 (Fig. 6F). The Loadings of this component are dominated by cellulose-related bands, explaining why they Scores map is the inverse and also why it is blurrier: carbohydrate ring vibrations are predominantly found below 1100 cm^{-1} in the spectrum, a region that has the worst spatial resolution of them all. Using thresholding and more supervised approaches, a single component map could replace the Scores maps of PCs 1–3 in this example, while retaining (and even refining) all chemical information found in their Loadings (see Gorzsás et al., 2011).

After xylem, bark is another tissue that is often rich in lignin and related compounds in woody species. While it generally constitutes a significantly lower proportion of available biomass than wood, it has great potential, as it is often an underused resource within the forestry industry (e.g., Lima et al., 2013). Raman imaging of tree bark is problematic as high concentration of tannins, lignin, suberin and other aromatics and colored materials cause high levels of fluorescence and absorption. FTIR spectroscopy, on the other hand is well suited for this task. Unfortunately, to date most studies are done without imaging (e.g., Lima et al., 2013). Other regions and cell types in stems, e.g. phloem, parenchyma, etc. are also seldom studied by FTIR imaging, but that is likely due to difficulties in sectioning and drying with maintained integrity and acceptable optophysical properties. Since these regions contain low levels of lignin, they are not discussed further here.

Roots, being soft tissues, pose similar challenges, in terms of sectioning and subsequent drying during sample preparation. In addition, small features can be hard to follow due to the diffraction limited lateral resolution of common instrumental setups. Thus, Dokken and Davis analyzed maize and sunflower roots at different distances from the root tip by means of synchrotron-FTIR imaging (followed by Principal Component Analysis) and found that the major spectral factor differentiating the root epidermis of these two species was H-type lignin, based on the band at $1635\,cm^{-1}$(Dokken & Davis, 2007). It is worth pointing out that this band appears in the spectra only as a shoulder on the main amide I band at ca. $1650\,cm^{-1}$, as root epidermis is also rich in proteins. Nevertheless, Principal Component Analysis Loadings clearly identified this band in the plethora of features originating from lignin, lipids, polysaccharides and proteins. Recently, the roots of *P. tremuloides* were imaged without a synchrotron source (Victor, Delpratt, Cseke, Miller, & Cseke, 2017). However, this study focused on the rhizosphere around the roots in a simulated environment, where the distribution of nitrate and sucrose could be tracked in combination with fungal colonization, up to a distance of several millimeters from the root.

In terms of FTIR imaging of H-lignin, a special case of leaf analysis gains increasing attention: grasses. These represent an important class of bioresources, both for potential new biomaterials production (e.g., biofuels, bioplastics) and as feedstock. In both aspects, their lignin content is considered one of the vital negative quality markers, as it lowers yields/increase costs for biofuels (Sills & Gossett, 2012)[d] and lowers digestibility for animals (Aiken, Himmelsbach, Carl, Hanna, & Barton, 1993). Accordingly, lignin (along with other major compounds) has been traced in grasses, in situ, qualitatively and (semi)quantitatively with these application areas in mind, using FTIR imaging (Aiken et al., 1993; Lupoi, Smith-Moritz, et al., 2015; Wetzel et al., 1998). However, the majority of FTIR spectroscopic work in these fields focuses on bulk analysis and does not involve imaging. This is generally true for FTIR spectroscopic quality assessments of leaves in plants where they are the primary products of use, e.g. tobacco, salads, etc. Readers are referred to Chen, Sun, and Zhou (2013) for an example of what multivariate FTIR imaging can bring to the chemical compositional

[d] Note, however, that lignin may not have an overall detrimental effect on thermochemical product yield for bioethanol production according to Lynd, Cushman, Nichols, and Wyman (1991).

analysis of leaves outside the field of lignin-research: selective detection of e.g. resin-like compounds, proteins (assigning higher relative abundance to secretory cavities and mesophylls), cellulose and even calcium oxalate in veins.

FTIR imaging has also been used for studying tissues involved in plant reproduction, from pollens to seeds, grains and fruits, representing the full reproductive cycle.

Pollens range from a few to about a hundred micrometer in size. This range that suits the spatial resolution of common FTIR imaging setups, as it allows spectra to be collected from individual pollen grains. However, resolving different components/compartments (e.g., the pollen coat or the cargo) is often only possible for larger pollens. Unfortunately, those pollens are usually too thick for transmission mode imaging, which necessitates either sectioning, or ATR based imaging, the latter resulting in data relating to surfaces only. Consequently, the majority of FTIR spectroscopic work on pollens is done as bulk, not as imaging, even when a microscopy accessory is used. The main compounds of interest here are sporopollenins, which contain aromatics and produce a band at around 1515 cm^{-1}. However, they are complex chemicals with often poor signals in FTIR spectra (especially when coupled to stray light and other optical problems due to pollen morphology). Despite these difficulties, chemical components of pollens, including sporopollenins, have been mapped by FTIR imaging, taking advantage of clever preprocessing and multivariate analyses (Zimmermann, Bağcıoğlu, Sandt, & Kohler, 2015).

Seeds vary in size even more than pollens do, covering over 10 orders of magnitudes in weight from orchids to coconuts (Harper, Lovell, & Moore, 1970), often containing three parts: the embryo, the seed coat and the endosperm. Thus, similarly to pollens, discerning different parts of seeds via FTIR imaging is restricted to suitably sized specimens, even though it can be pushed to physical limits by using synchrotron sources (Bonwell, Fisher, Fritz, & Wetzel, 2008; Theodoridou, Vail, & Yu, 2014; Withana-Gamage et al., 2013; Yu et al., 2003). Often the endosperm is studied in detail, since its high protein content makes it ideally suited for FTIR spectroscopy, allowing even tertiary structures to be derived (e.g., Yu, 2004, 2011; Yu et al., 2003). While these tertiary structures can inform about the quality of seeds and grains, and are therefore important for quality determination in e.g. agricultural applications,[e] they are not particularly relevant in terms of lignin and related compounds. Additionally, lipid and carbohydrate (including but not limited to starch and other polysaccharides) distribution can also be tracked within the same

[e] Applications of FTIR imaging to this area were pioneered by Yu and Wetzel and coworkers (Bonwell et al., 2008; Koch & Wetzel, 2007; Yu, 2004; Yu et al., 2003).

chemical images, providing fast means for (co)localization of vital chemical constituents (see Wetzel, 2009; Yu, 2011 and references therein for examples). Of particular importance in this context is that pigment layers do not disturb the measurements, and thus even colored seeds/pigmented tissue areas can easily be studied. Himmelsbach et al. could assign specific bands to tannin type aromatics and lignin type aromatics in cotton seed coats, for example, which are notoriously difficult by Raman imaging (Himmelsbach, Akin, Kim, & Hardin, 2003). In addition to images based on simple band intensities, recent works started to feature multivariate approaches even in this field, assigning unique chemical profiles to individual regions (Chen, Sun, & Zhou, 2016).

Finally, a particular case of lignin imaging is fungal attacks on plant tissues. These might consume lignin (e.g., white rot) or leave lignin (e.g., brown rot), but in either case the proportion of lignin is altered. Potentially, even lignin structure is affected, as the entire chemical matrix of the cell wall gets disrupted, thus interactions between components also change. Monitoring different kinds of fungal attacks is of vital importance to understand the process and provide means of interventions for agriculture and sylviculture as well as for archaeology, art and construction (preservation and protection). The power of FTIR imaging in this field has been demonstrated by Naumann, Navarro-González, Peddireddi, Kües, and Polle (2005), visualizing fungal attack on wood based on FPA FTIR spectra (*Schizophyllum commune* and *Trametes versicolor* in the lumen of Beech wood cells). For softwood, the degradation of spruce by brown-rot fungi (*Gloeophyllum trabeum* and *Poria placenta*) were mapped by images generated from FTIR microspectroscopy data, using Principal Component Analysis and PLS-DA based multivariate analyses (Fackler et al., 2010). A year later, the same authors expanded the study to the simultaneous (by *T. versicolor*) and selective (by *Ceriporiopsis subvermispora*) white rot decay of spruce (Fackler et al., 2011). However, that not only destructive actions of fungi can be monitored. Symbiotic relationships (especially with plant roots) are important too, and FTIR imaging was used to follow nutrient distribution in the rhizosphere around poplar root cells in the presence of *Laccaria bicolor* (Victor et al., 2017).

6. Summary

The three different imaging methods discussed in this chapter offer complementary tools for imaging lignin and related compounds in situ. They differ in sensitivity toward different molecules, have different spatial resolutions and may require different sample preparation. Thus, they each come with their own set of advantages and limitations. In Table 2, we

Table 2 Comparison of techniques.

Feature	Raman imaging	FTIR imaging	Autofluorescence imaging
Physical basis	Molecular vibrations via inelastic scattering	Molecular vibrations via absorption	Autofluorescence via visible light
Common spatial resolution	Nanometer range 230–1000 nm	Micrometer range (diffraction limited to ca. 4–20 μm in transmission mode on standard desktop systems)	Nanometer range Conventional confocal scanning microscope when optimally used Lateral resolution ≥180 nm Axial resolution ≥500 nm
Confocal measurements/ surface sensitivity	Yes/yes	No/possible via attenuated total reflectance (ATR)	Yes/yes
Common speed (spectra/time)	0.04–1 s/spectrum	Thousands per second simultaneously with full spectral range (via focal plane array detectors)	Wide field image acquisition: Multispectral mode less than 1 min for 10 channels for one field of view Hyperspectral mode for one wavelength: around 2 min Microspectroscopic mode: 1 s per spectrum
Anisotropy detection (orientation sensitiveness)	Yes	No (albeit theoretically possible)	No
Compounds easily detected in plants	Lignin, cellulose, pectin, carotenoids, stilbenes, minerals	All major cell wall components and extractives	Commonly aromatic compounds: phenolics (anthocyanins, flavonoids, tannins, cinnamic acids), alkaloids (betalains), terpenoids and porphyrins
Problematic compounds in plants	Fluorescent compounds may disturb, overenhancement of conjugated molecules (e.g., carotenoids) can suppress others	Water	Not all the compounds are autofluorescent

Continued

Table 2 Comparison of techniques.—cont'd

Feature	Raman imaging	FTIR imaging	Autofluorescence imaging
Sample preparation	Usually plane surfaces are necessary (microtomy), measurements in water preferred, glass slides for sample preparation	Critical, can be laborious. Drying and thin sections are required, infrared transparent or reflective carriers are needed	Fresh samples or cryosections Avoid fixatives and embedding resins
Need for markers, labels, dyes	No	No	No
Simultaneous detection of compounds	Yes	Yes	Yes

provide a comparison chart to aid in selecting the most suitable for a given scientific project. However, they are best used in combination, when they can compensate for each other's shortcomings and promote each other's strong sides. Ideally, their results are also analyzed simultaneously, or at least in combination, to maximize information gains about lignin content and composition in situ. A recent example (albeit not specifically focusing on lignin) combines the information from FTIR and Raman spectroscopies (as well as surface enhanced Raman scattering and matrix assisted laser desorption/ionization mass spectrometry) to characterize the pollens of Poa alpine grass grown under different conditions (Diehn et al., 2020). Similar approaches would benefit lignin imaging in situ, as the different techniques highlight different aspects of lignin chemistry and structure with different sensitivities, providing valuable complementary information.

Author contributions

All authors contributed to writing and revising the entire manuscript, with specific responsibilities for the certain sections as follows: Section 4—N.G., Section 5—A.G., Section 3—F.G., Section 2—M.-F.D.

References

Afseth, N. K., & Kohler, A. (2012). Extended multiplicative signal correction in vibrational spectroscopy, a tutorial. *Chemometrics and Intelligent Laboratory Systems*, *117*, 92–99.

Agarwal, U. P. (2006). Raman imaging to investigate ultrastructure and composition of plant cell walls: Distribution of lignin and cellulose in black spruce wood (*Picea mariana*). *Planta*, *224*, 1141–1153.

Agarwal, U. P. (2014). 1064 nm FT-Raman spectroscopy for investigations of plant cell walls and other biomass materials. *Frontiers in Plant Science*, *5*, 490.

Agarwal, U. P., Mcsweeny, J. D., & Ralph, S. A. (2011). FT-Raman investigation of milled-wood lignins: Softwood, hardwood, and chemically modified black spruce lignins. *Journal of Wood Chemistry and Technology*, *31*(4), 324–344.

Agarwal, U. P., & Ralph, S. A. (1997). FT-Raman spectroscopy of wood: Identifying contributions of lignin and carbohydrate polymers in the spectrum of black spruce (*Picea mariana*). *Applied Spectroscopy*, *51*(11), 1648–1655.

Agarwal, U. P., Ralph, S. A., Padmakshan, D., Liu, S., & Foster, C. E. (2019). Estimation of syringyl units in wood lignins by FT-Raman spectroscopy. *Journal of Agricultural and Food Chemistry*, *67*, 4367–4374.

Aiken, D. E., Himmelsbach, D. S., Carl, R. T., Hanna, W. W., & Barton, F. B. (1993). Mid-infrared microspectroscopy to assess lignin in plant tissues related to digestibility. *Agronomy Journal*, *85*, 171–175.

Akin, D. E. (2008). Plant cell wall aromatics: Influence on degradation of biomass. *Biofuels, Bioproducts & Biorefining*, *2*, 288–303.

Albinsson, B., Li, S., Lundquist, K., & Stomberg, R. (1999). The origin of lignin fluorescence. *Journal of Molecular Structure*, *508*, 19–27.

Allouche, F., Hanafi, M., Jamme, F., Robert, P., Barron, C., Guillon, F., et al. (2012). Coupling hyperspectral image data having different spatial resolutions using multiple co-inertia analysis. *Chemometrics and Intelligent Laboratory Systems*, *117*, 200–212.

Auxenfrans, T., Terryn, C., & Paës, G. (2017). Seeing biomass recalcitrance. *Scientific Reports*, 7, 8838.

Baldacci-Cresp, F., Spriet, C., Twyffels, L., Blervacq, A.-S., Neutelings, G., Baucher, M., et al. (2020). A rapid and quantitative safranine-based fluorescent microscopy method to evaluate cell wall lignification. *Plant Journal*, *102*, 1074–1089.

Baldwin, P. M., Bertrand, D., Novales, B., Bouchet, B., Collobert, G., & Gallant, D. J. (1997). Chemometric labeling of cereal tissues in multichannel fluorescence microscopy images using discriminant analysis. *Analytical Chemistry*, *69*, 4339–4348.

Barrière, Y., Emile, J. C., Traineau, R., Surault, F., Briand, M., & Gallais, A. (2004). Genetic variation for organic matter and cell wall digestibility in silage maize. Lessons from 34-year long experiment with sheep in digestibility crates. *Maydica*, *49*, 115–126.

Barsberg, S., Elder, T., & Felby, C. (2003). Lignin-quinone interactions: Implication for optical properties of lignin. *Chemistry of Materials*, *15*, 649–655.

Beaugrand, J., Goudenhooft, C., Alvarado, C., Devaux, M. F., Rivard, C., Durand, S., et al. (2022). Evolution of flax cell wall composition during development after gravitropism by synchrotron fluorescence imaging. *Industrial Crops and Products*, *175*, 114256.

Béc, K., Grabska, J., Bonn, G. K., Popp, M., & Huck, X. W. (2020). Principle and applications of vibrational spectroscopic imaging in plant science: A review. *Frontiers in Plant Science*, *11*, 1226.

Belt, T., Keplinger, T., Hanninen, T., & Rautkari, L. (2017). Cellular level distributions of Scots pine heartwood and knot heartwood extractives revealed by Raman spectroscopy imaging. *Industrial Crops and Products*, *108*, 327–335.

Billa, E., Pastou, A., Monries, B., Romero, J., & Koukios, E. G. (2000). Multivariate chemometric analysis of the fluorescence spectra of eucalyptus wood. *Industrial Crops and Products*, *11*, 187–196.

Blaschek, L., Nuoendagula, N., Bacsik, Z., Kajita, S., & Pesquet, E. (2020). Determining the genetic regulation and coordination of lignification in stem tissues of Arabidopsis using semiquantitative Raman microspectroscopy. *ACS Sustainable Chemistry and Engineering*, *8*, 4900–4909.

Bock, P., Felhofer, M., Mayer, K., & Gierlinger, N. (2021). A guide to elucidate the hidden multicomponent layered structure of plant cuticles by Raman imaging. *Frontiers in Plant Science*, *12*, 793330.

Bock, P., & Gierlinger, N. (2019). Infrared and Raman spectra of lignin substructures: Coniferyl alcohol, abietin, and coniferyl aldehyde. *Journal of Raman Spectroscopy*, *50*(6), 778–792.

Bock, P., Nousiainen, P., Elder, T., Blaukopf, M., Amer, H., Zirbs, R., et al. (2020). Infrared and Raman spectra of lignin substructures: Dibenzodioxocin. *Journal of Raman Spectroscopy*, *51*(3), 422–431.

Bond, J., Donaldson, L., Hill, S., & Hitchcock, K. (2008). Safranine fluorescent staining of wood cell walls. *Biotechnic and Histochemistry*, *83*, 161–171.

Bonwell, E. S., Fisher, T., Fritz, A., & Wetzel, D. (2008). Determination of endosperm protein secondary structure in hard wheat breeding lines using synchrotron infrared microspectroscopy. *Vibrational Spectroscopy*, *48*, 76–81.

Brereton, R. G., & Lloyd, G. R. (2014). Partial least squares discriminant analysis: Taking the magic away. *Journal of Chemometrics*, *28*(4), 213–225.

Buschmann, C., Langsdorf, G., & Lichtenthaler, H. K. (2000). Imaging of blue, green and red fluorescence emission of plants: On overview. *Photosynthetica*, *38*, 483–491.

Campbell, M. M., & Sederoff, R. (1996). Variation in lignin content and composition. *Plant Physiology*, *110*, 3–13.

Capobianco, G., Bracciale, M. P., Sali, D., Sbardella, F., Belloni, P., Bonifazi, G., et al. (2017). Chemometrics approach to FT-IR hyperspectral imaging analysis of degradation products in artwork cross-section. *Microchemical Journal*, *132*, 69–76.

Chabbert, B., Habrant, A., Herbaut, M., Foulon, L., Aguié-Béghin, V., Garajova, S., et al. (2017). Action of lytic polysaccharide monooxygenase on plant tissue is governed by cellular type. *Scientific Reports*, *7*, 17792.

Chabbert, B., Terryn, C., Herbaut, M., Vaidya, A., Habrant, A., Paes, G., et al. (2018). Fluorescence technique can reveal cell wall organization and predict saccharification in pretreated wood biomass. *Industrial Crops and Products*, *123*, 84–92.

Chen, J. B., Sun, S. Q., & Zhou, Q. (2013). Direct observation of bulk and surface chemical morphologies of Ginkgo biloba leaves by Fourier transform mid- and near-infrared microspectroscopic imaging. *Analytical and Bioanalytical Chemistry*, *405*, 9385–9400.

Chen, J., Sun, S., & Zhou, Q. (2016). Chemical morphology of Areca nut characterized directly by Fourier transform near-infrared and mid-infrared microspectroscopic imaging in reflection modes. *Food Chemistry*, *42*, 469–475.

Chimenez, T. A., Gehlen, M. H., Marabezi, K., & Curvelo, A. A. S. (2014). Characterization of sugarcane bagasse by autofluorescence microscopy. *Cellulose*, *21*, 653–664.

Christensen, C. S. L., & Rasmussen, S. K. (2019). Low lignin mutants and reduction of lignin content in grasses for increased utilization of lignocellulose. *Agronomy*, *9*, 256.

Chu, L. Q., Masyuko, R., Sweedler, J. V., & Bohn, P. W. (2010). Base-induced delignification of Miscanthus x giganteus studied by three-dimensional confocal Raman imaging. *Bioresource Technology*, *101*(13), 4919–4925.

Coletta, V. C., Rezende, C. A., da Conceição, F. R., Polikarpov, I., & Guimarães, F. E. G. (2013). Mapping the lignin distribution in pretreated sugarcane bagasse by confocal and fluorescence lifetime imaging microscopy. *Biotechnology for Biofuels*, *6*, 43.

Corcel, J. (2017). Imagerie multispectrale en macrofluorescence en vue de la prédiction de l'origine tissulaire de particules de tiges de maïs. In *Dissertation/master's thesis* Université de Nantes.

Corcel, M., Devaux, M.-F., Guillon, F., & Barron, C. (2016). Comparison of UV and visible autofluorescence of wheat grain tissues in macroscopic images of cross-sections and particles. *Computers and Electronics in Agriculture*, *127*, 281–288.

Coste, R., Soliman, M., Bercu, N. B., Potiron, S., Lasri, K., Aguié-Béghin, V., et al. (2021). Unveiling the impact of embedding resins on the physicochemical traits of wood cell walls with subcellular functional probing. *Composites Science and Technology*, *201*, 108485.

Courcoux, P., Devaux, M.-F., & Bouchet, B. (2002). Simultaneous decomposition of multivariate images using three-way data analysis—Application to the comparison of cereal grains by confocal laser scanning microscopy. *Chemometrics and Intelligent Laboratory Systems*, *62*, 103–113.

Czymmek, K. J., Fogg, M., Powell, D. H., Sweigard, J., Park, S.-Y., & Kang, S. (2007). In vivo time-lapse documentation using confocal and multi-photon microscopy reveals the mechanisms of invasion into the Arabidopsis root vascular system by Fusarium oxysporum. *Fungal Genetics and Biology*, *44*, 1011–1023.

Day, A., Ruel, K., Neutelings, G., Cronier, D., David, H., Hawkins, S., et al. (2005). Lignification in the flax stem: Evidence for an unusual lignin in bast fibers. *Planta*, *222*, 234–245.

De Juan, A., Maeder, M., Hancewicz, T., & Tauler, R. (2008). Use of local rank-based spatial information for resolution of spectroscopic images. *Journal of Chemometrics*, *22*, 291–298.

De Juan, A., Piqueras, S., Maeder, M., Hancewicz, T., Duponchel, L., & Tauler, R. (2014). Chemometric tools for image analysis. In R. Salzer, & H. W. Siesler (Eds.), *Infrared and Raman spectroscopic imaging* (2nd ed., pp. 57–110). Weinheim, DE: Wiley-VCH Verlag GmbH & Co. KaA.

Del Rio, J. C., Rencoret, J., Gutierrez, A., Elder, T., Kim, H., & Ralph, J. (2020). Lignin monomers from beyond canonical monolignol biosynthetic pathway: Another brick in the wall. *ACS Sustainable Chemistry and Engineering*, *8*, 4997–5012.

De Micco, V., & Aronne, G. (2007). Combined histochemistry and autofluorescence for identifying lignin distribution in cell walls. *Biotechnic & Histochemistry*, *82*, 209–216.

Devaux, M.-F., Jamme, F., Andre, W., Bouchet, B., Alvarado, C., Durand, S., et al. (2018). Synchrotron time-lapse imaging of lignocellulosic biomass hydrolysis: Tracking enzyme localization by protein autofluorescence and biochemical modification of cell walls by microfluidic infrared microspectroscopy. *Frontiers in Plant Science*, *9*, 200.

Dickson, A., Nanayakkara, B., Sellier, D., Meason, D., Donaldson, L. A., & Brownlie, R. (2017). Fluorescence imaging of cambial zones to study wood formation in *Pinus radiata* D. Don. *Trees*, *31*, 479–490.

Diehn, S., Zimmermann, B., Tafintseva, V., Seifert, S., Bagcioglu, M., Ohlson, M., et al. (2020). Combining chemical information from grass pollen in multimodal characterization. *Frontiers in Plant Science*, *10*, 1788.

Dieing, T., & Hollricher, O. (2008). High-resolution, high-speed confocal Raman imaging. *Vibrational Spectroscopy*, *48*, 22–27.

Dieing, T., & Ibach, W. (2018). Software requirements and data analysis in confocal Raman microscopy. In T. Dieing, O. Hollrichter, & J. Toporski (Eds.), *Confocal Raman microscopy* (2nd ed., pp. 89–120). Heidelberg: Springer.

Ding, S. Y., Liu, Y. S., Zeng, Y., Himmel, M. E., Baker, J. O., & Bayer, E. A. (2012). How does plant cell wall nanoscale architecture correlate with enzymatic digestibility? *Science*, *338*(6110).

Djikanović, D., Kalauzi, A., Radotić, K., & Jeremić, M. (2007). Deconvolution of lignin fluorescent spectra: A contribution to the comparative structural studies of lignins. *Russian Journal of Physical Chemistry A*, *81*, 1425–1428.

Djikanović, D., Simonović, J., Savić, A., Ristić, I., Bajuk-Bogdanović, D., Kalauzi, A., et al. (2012). Structural differences between lignin model polymers synthesized from various monomers. *Journal of Polymers and the Environment*, *20*, 607–617.

Dokken, K., & Davis, M. (2007). Infrared imaging of sunflower and maize root anatomy. *Journal of Agricultural and Food Chemistry*, *55*, 10517–10530.

Dokken, K. M., Davis, L. C., & Marinkovic, N. S. (2005). Use of infrared microspectroscopy in plant growth and development. *Applied Spectroscopy Reviews*, *40*, 301–326.

Donaldson, L. A. (2001). Lignification and lignin topochemistry—An ultrastructural view. *Phytochemistry*, *57*, 859–873.

Donaldson, L. A. (2013). Softwood and hardwood lignin fluorescence spectra of wood cell walls in different mounting media. *IAWA Journal*, *34*, 3–19.

Donaldson, L. A. (2019). Wood cell wall ultrastructure. The key to understanding wood properties and behaviour. *IAWA Journal*, *40*, 645–672.

Donaldson, L. A. (2020). Autofluorescence in plants. *Molecules*, *25*, 2393.

Donaldson, L. A., & Knox, J. P. (2012). Localization of cell wall polysaccharides in normal and compression wood of radiata pine: Relationships with lignification and microfibril orientation. *Plant Physiology*, *158*, 642–653.

Donaldson, L. A., Kroese, H. W., Hill, S. J., & Franich, R. A. (2015). Detection of wood cell wall porosity using small carbohydrate molecules and confocal fluorescence microscopy. *Journal of Microscopy*, *259*, 228–236.

Donaldson, L. A., & Radotić, K. (2013). Fluorescence lifetime imaging of lignin autofluorescence in normal and compression wood. *Journal of Microscopy*, *251*, 178–187.

Donaldson, L. A., Radotić, K., Kalauzi, A., Djikanović, D., & Jeremić, M. (2010). Quantification of compression wood severity in tracheids of Pinus radiata D. Don using confocal fluorescence imaging and spectral deconvolution. *Journal of Structural Biology*, *169*, 106–115.

Donaldson, L. A., Singh, A. P., Yoshinaga, A., & Takabe, K. (1999). Lignin distribution in mild compression wood of Pinus radiata D. Don. *Canadian Journal of Botany*, 77, 41–50.

Donaldson, L. A., & Vaidya, A. (2017). Visualizing recalcitrance by colocalization of cellulase, lignin and cellulose in pretreated pine biomass using fluorescence microscopy. *Scientific Reports*, 7, 44386.

Donaldson, L., & Williams, N. (2018). Imaging and spectroscopy of natural fluorophores in pine needles. *Plants*, 7, 10.

Dučić, T., Thieme, J., & Polle, A. (2012). Phosphorus compartmentalization on the cellular level of Douglas fir root as affected by Mn toxicity: A synchrotron-based FTIR approach. *Journal of Spectroscopy*, *27*, 265–272.

Dumas, P., Carr, G. L., & Williams, G. P. (2000). Enhancing the lateral resolution in infrared microspectrometry by using synchrotron radiation: Applications and perspectives. *Analysis*, *28*, 68–74.

El Hage, F., Legland, D., Borrega, N., Jacquemot, M.-P., Griveau, Y., Coursol, S., et al. (2018). Tissue lignification, cell wall p-coumaroylation and degradability of maize stems depend on water status. *Journal of Agricultural and Food Chemistry*, *66*, 4800–4808.

Fackler, K., Stevanic, J. S., Ters, T., Hinterstoisser, B., Schwanninger, M., & Salmén, L. (2010). Localisation and characterisation of incipient brown-rot decay within spruce wood cell walls using FT-IR imaging microscopy. *Enzyme and Microbial Technology*, *47*, 257–267.

Fackler, K., Stevanic, J. S., Ters, T., Hinterstoisser, B., Schwanninger, M., & Salmén, L. (2011). FT-IR imaging microscopy to localise and characterise simultaneous and selective white-rot decay within spruce wood cells. *Holzforschung*, *65*, 411–420.

Faix, O. (1986). Investigation of lignin polymer models (DHP's) by FTIR spectroscopy. *Holzforschung*, *40*, 273–280.

Faix, O. (1988). Practical uses of FTIR spectroscopy in wood science and technology. *Mikrochimica Acta*, *1*, 21–25.

Faix, O. (1991). Classification of lignins from different botanical origins by FT-IR spectroscopy. *Holzforschung*, *45*(Suppl), 21–27.

Faix, O., Andersons, B., & Zakis, G. (1998). Determination of carbonyl groups of six round robin lignin by modified oximation and FTIR spectroscopy. *Holzforschung*, *52*, 268–274.

Faix, O., Argyropoulos, D. S., Robert, D., & Neirinck, V. (1994). Determination of hydroxyl groups in lignins evaluation of ^{1}H-, ^{13}C-, ^{31}P-NMR, FTIR and wet chemical methods. *Holzforschung*, *48*, 387–394.

Felhofer, M., Bock, P., Xiao, N. N., Preimesberger, C., Lindemann, M., Hansmann, C., et al. (2021). Oak wood drying: Precipitation of crystalline ellagic acid leads to discoloration. *Holzforschung*, *75*, 712–720.

Felhofer, M., Prats-Mateu, B., Bock, P., & Gierlinger, N. (2018). Antifungal stilbene impregnation: Transport and distribution on the micron-level. *Tree Physiology*, *38*, 1526–1537.

Felten, J., Hall, H., Jaumot, J., Tauler, R., de Juan, A., & Gorzsás, A. (2015). Vibrational spectroscopic image analysis of biological material using multivariate curve resolution–alternating least squares (MCR-ALS). *Nature Protocols*, *10*, 217–240.

Fergus, B. J., Procter, A. R., Scott, J. A., & Urvey, N. D. (1969). The distribution of lignin in spruce wood as determined by ultraviolet microscopy. *Wood Science and Technology*, *3*, 117–138.

Fu, D., Lu, F. K., Zhang, X., Freudiger, C., Pernik, D. R., Holtom, G., et al. (2012). Quantitative chemical imaging with multiplex stimulated Raman scattering microscopy. *Journal of the American Chemical Society*, *134*, 3623–3626.

Fulcher, R., O'Brien, T., & Lee, J. (1972). Studies on the aleurone layer I. Conventional and fluorescence microscopy of the cell wall with emphasis on phenol-carbohydrate complexes in wheat. *Australian Journal of Biological Sciences*, *25*, 23–34.

García-Plazaola, J. I., Fernández-Marín, B., Duke, S. O., Hernández, A., López-Arbeloa, F., & Becerril, J. M. (2015). Autofluorescence: Biological functions and technical applications. *Plant Science*, *236*, 136–145.

Gautam, R., Vanga, S., Ariese, F., & Umapathy, S. (2015). Review of multidimensional data processing approaches for Raman and infrared spectroscopy. *EPJ Techniques and Instrumentation*, *2*, 8.

Ghaffari, M., Chateigner-Boutin, A.-L., Guillon, F., Devaux, M. F., Abdollahi, H., & Duponchel, L. (2019). Multi-excitation hyperspectral autofluorescence imaging for the exploration of biological samples. *Analytica Chimica Acta*, *1062*, 47–59.

Gierlinger, N. (2018). New insights into plant cell walls by vibrational microspectroscopy. *Applied Spectroscopy Reviews*, *53*, 517–551.

Gierlinger, N., Keplinger, T., & Harrington, M. (2012). Imaging of plant cell walls by confocal Raman microscopy. *Nature Protocols*, *7*, 1694–1708.

Gierlinger, N., Luss, S., Konig, C., Konnerth, J., Eder, M., & Fratzl, P. (2010). Cellulose microfibril orientation of *Picea abies* and its variability at the micron-level determined by Raman imaging. *Journal of Experimental Botany*, *61*, 587–595.

Gierlinger, N., & Schwanninger, M. (2006). Chemical imaging of poplar wood cell walls by confocal Raman microscopy. *Plant Physiology*, *140*, 1246–1254.

Gierlinger, N., & Schwanninger, M. (2007). The potential of Raman microscopy and Raman imaging in plant research—Review. *Spectroscopy*, *21*, 69–89.

Gill, D., Kilponen, R. G., & Rimai, L. (1970). Resonance Raman scattering of laser radiation by vibrational modes of carotenoid pigment molecules in intact plant tissues. *Nature*, *227*, 743–744.

Gorzsás, A. (2017). Chemical imaging of xylem by Raman microspectroscopy. *Methods in Molecular Biology*, *1544*, 133–178.

Gorzsás, A., Stenlund, H., Persson, P., Trygg, J., & Sundberg, B. (2011). Cell-specific chemotyping and multivariate imaging by combined FT-IR microspectroscopy and orthogonal projections to latent structures (OPLS) analysis reveals the chemical landscape of secondary xylem. *Plant Journal*, *66*, 903–914.

Gorzsás, A., & Sundberg, B. (2014). Chemical fingerprinting of Arabidopsis using Fourier transform infrared (FT-IR) spectroscopic approaches. *Methods in Molecular Biology*, *1062*, 317–352.

Gou, J. Y., Park, S., Yu, X. H., Miller, L. M., & Liu, C. J. (2008). Compositional characterization and imaging of "wall-bound" acylesters of Populus trichocarpa reveal differential accumulation of acyl molecules in normal and reactive woods. *Planta*, *229*, 15–24.

Grahn, H. F., & Geladi, P. (2007). *Techniques and applications of hyperspectral image analysis*. Chichester, GB: John Wiley & Sons Ltd.

Guo, J., Zhou, H., Stevanic, J. S., Dong, M., Yu, M., Salmén, L., et al. (2018). Effects of ageing on the cell wall and its hygroscopicity of wood in ancient timber construction. *Wood Science and Technology*, *52*, 131–147.

Hänninen, T., Kontturi, E., & Vuorinen, T. (2011). Distribution of lignin and its coniferyl alcohol and coniferyl aldehyde groups in *Picea abies* and *Pinus sylvestris* as observed by Raman imaging. *Phytochemistry, 72*, 1889–1895.

Harper, J. L., Lovell, P. H., & Moore, K. G. (1970). The shapes and sizes of seeds. *Annual Review of Ecology and Systematics, 1*, 327–356.

Harris, H., & Hartley, R. D. (1976). Detection of bound ferulic acid in cell walls of the Gramineae by ultraviolet fluorescence microscopy. *Nature, 259*, 508.

Harris, P. J., & Trethewey, J. A. K. (2010). The distribution of ester-linked ferulic acid in the cell walls of angiosperm. *Phytochemistry Reviews, 9*, 19–33.

Hatfield, R. D., & Chaptman, A. K. (2009). Comparing corn types for differences in cell wall characteristics and p-coumaroylation of lignin. *Journal of Agricultural and Food Chemistry, 57*, 4243–4249.

Hatfield, R. D., & Marita, J. M. (2010). Enzymatic process involved in the incorporation of hydroxycinnamates into grass walls. *Phytochemistry Reviews, 9*, 5–45.

Hatfield, R. D., Marita, J. M., & Frost, K. (2008). Characterization of p-coumarate accumulation, p-coumaroyl transferase and cell wall changes during the development of corn stems. *Journal of the Science of Food and Agriculture, 88*, 2529–2537.

He, L., & Terashima, N. (1990). Formation and structure of lignin in monocotyledons. III Heterogeneity of sugarcane (*Saccharum officinarum L.*) lignin with respect to the composition of structural units in different morphological region. *Journal of Wood Chemistry and Technology, 10*, 435–459.

He, L., & Terashima, N. (1991). Formation and structure of lignin in monocotyledons. IV Deposition process and structural diversity of the lignin in the cell wall of sugarcane and rice plant studied by ultraviolet microscopy spectroscopy. *Holzforschung, 45*, 191–198.

Herbaut, M., Zoghlami, A., Falourd, X., Foucat, L., Chabbert, B., & Paes, G. (2018). Multimodal analysis of pretreated biomass species highlights generic markers of lignocellulose recalcitrance. *Biotechnology for Biofuels, 11*, 52.

Himmelsbach, D. S., Akin, D. E., Kim, J., & Hardin, R. (2003). Chemical structural investigation of the cotton fiber base and associated seed coat: Fourier-transform infrared mapping and histochemistry. *Textile Research Journal, 73*, 281–288.

Himmelsbach, D. S., Khahili, S., & Akin, D. E. (1999). Near-infrared–Fourier-transform–Raman microspectroscopic imaging of flax stems. *Vibrational Spectroscopy, 19*, 361–367.

Himmelsbach, D. S., Khalili, S., & Akin, D. E. (2002). The use of FT-IR microspectroscopic mapping to study the effects of enzymatic retting of flax (*Linum usitatissimum L*) stems. *Journal of the Science of Food and Agriculture, 82*, 685–696.

Horvath, L., Peszlen, I., Gierlinger, N., Peralta, P., Kelley, S., & Csoka, L. (2012). Distribution of wood polymers within the cell wall of transgenic aspen imaged by Raman microscopy. *Holzforschung, 66*, 717–725.

Jamme, F., Kascakova, S., Villette, S., Allouche, F., Pallu, S., Rouam, V., et al. (2013). Deep UV autofluorescence microscopy for cell biology and tissue histology. *Biology of the Cell, 105*, 277–288.

Jamme, F., Robert, P., Saulnier, L., Dumas, P., & Guillon, F. (2008). Aleurone cell walls of wheat grain: High spatial resolution investigation using synchrotron infrared microspectroscopy. *Applied Spectroscopy, 62*, 895–900.

Jolliffe, I. T. (2002). *Principal component analysis* (2nd ed.). New York: Springer-Verlag.

Joseleau, J.-P., & Ruel, K. (1997). Study of lignification by noninvasive technique in growing maize internodes. *Plant Physiology, 114*, 1123–1133.

Jung, H. J. G., Samac, D. A., & Sarath, G. (2012). Modifying crops to increase cell wall digestibility. *Plant Science, 185–186*, 65–77.

Karthaüser, J., Biziks, V., Mai, C., & Militz, H. (2021). Lignin and lignin derived compounds for wood applications—A review. *Molecules, 26*, 2533.

Keplinger, T., Cabane, E., Chanana, M., Hass, P., Merk, V., Gierlinger, N., et al. (2015). A versatile strategy for grafting polymers to wood cell walls. *Acta Biomaterialia, 11*, 256–263.
Kiyoto, S., Yoshinaga, A., Tanaka, N., Wada, M., KamitakaharaA, H., & Takabe, K. (2013). Immunolocalization of 8-5' and 8-8' linked structures of lignin in cell walls of *Chamaecyparis obtusa* using monoclonal antibodies. *Planta, 237*, 705–715.
Knoll, B., & Keilmann, F. (1999). Near-field probing of vibrational absorption for chemical microscopy. *Nature, 399*, 134–137.
Koch, H., & Wetzel, D. L. (2007). Imaging local chemical microstructure of germinated wheat with synchrotron infrared microspectroscopy. *Spectroscopy, 22*, 24–32.
Korner, C., Riedl, S., Keplinger, T., Richter, A., Wiesenbauer, J., Schweingruber, F., et al. (2019). Life at 0 degrees C: The biology of the alpine snowbed plant *Soldanella pusilla. Alpine Botany, 129*, 63–80.
Kumar, S., Lahlali, R., Liu, X., & Karunakaran, C. (2016). Infrared spectroscopy combined with imaging: A new developing analytical tool in health and plant science. *Applied Spectroscopy Reviews, 51*, 466–483.
Lang, M., Stober, F., & Lichtenthaler, H. K. (1991). Fluorescence emission spectra of plant leaves and plant constituents. *Radiation and Environmental Biophysics, 30*, 333–347.
Lapierre, C. (1993). Applications of new methods for the investigation of lignin structure. In H. G. Jung, D. R. Buxton, R. D. Hatfield, & J. Ralph (Eds.), *Forage cell wall structure and digestibility* (pp. 133–163). Madison: American Society of Agronomy Inc.
Lapierre, C., Jouin, D., & Monties, B. (1989). On the molecular origin of the alkali solubility of Gramineae lignins. *Phytochemistry, 28*, 1401–1403.
Larsen, K. L., & Barsberg, S. (2010). Theoretical and Raman spectroscopic studies of phenolic lignin model monomers. *Journal of Physical Chemistry B, 114*(23), 8009–8021.
Larsen, K. L., & Barsberg, S. (2011). Environmental effects on the lignin model monomer, vanillyl alcohol, studied by Raman spectroscopy. *Journal of Physical Chemistry B, 115*(39), 11470–11480.
Lasch, P., & Naumann, D. (2006). Spatial resolution in infrared microspectroscopic imaging of tissues. *Biochimica et Biophysica Acta. Biomembranes, 1758*, 814–829.
Lee, E. (2012). Imaging modes. In A. Zoubir (Ed.), *Raman imaging*. Berlin, Heidelberg: Springer Verlag.
Li, H., Adal, T., Wang, W., Emge, D., & Cichocki, A. (2007). Non-negative matrix factorization with orthogonality constraints and its application to Raman spectroscopy. *Journal of VLSI Signal Processing, 48*, 83–97.
Li, X. L., Sha, J. J., Xia, Y. H., Sheng, K. C., Liu, Y. F., & He, Y. (2020). Quantitative visualization of subcellular lignocellulose revealing the mechanism of alkali pretreatment to promote methane production of rice straw. *Biotechnology for Biofuels, 13*, 8.
Li, X., Wei, Y., Xu, J., Xu, N., & He, J. (2018). Quantitative visualization of lignocellulose components in transverse sections of moso bamboo based on FTIR macro- and micro-spectroscopy coupled with chemometrics. *Biotechnology for Biofuels, 11*, 263.
Liao, D., Qian, Y., Zhou, J., & Tang, Y. Y. (2016). A manifold alignment approach for hyperspectral image visualization with natural color. *IEEE Transactions on Geoscience and Remote Sensing, 54*(6), 3151–3162.
Lichtman, J. W., & Concello, J. A. (2006). Fluorescence microscopy. *Nature Methods, 2*, 910–919.
Liljegren, S. (2010). Phloroglucinol stain for lignin. *Cold Spring Harbor Protocols, 2010*. pdb. prot4954.
Lima, M. A., Lavorente, G. B., da Silva, H. K. P., Bragatto, J., Rezende, C. A., Bernardinelli, O. D., et al. (2013). Effects of pretreatment on morphology, chemical composition and enzymatic digestibility of eucalyptus bark: A potentially valuable source of fermentable sugars for biofuel production. *Biotechnology for Biofuels, 6*, 75.

Liu, Q., Luo, L., & Zheng, L. (2018). Lignins: Biosynthesis and biological functions in plants. *International Journal of Molecular Sciences*, *19*, 335.

Lohrasebi, H., Mabee, W. E., & Roy, D. N. (1999). Chemistry and pulping feasibility of compression wood in black spruce. *Journal of Wood Chemistry and Technology*, *19*, 13–25.

Lundquist, K., Josefsson, B., & Nyquist, G. (1978). Analysis of lignin products by fluorescence spectroscopy. *Holzforschung*, *32*, 27–32.

Lupoi, J. S., Singh, S., Davis, M., Lee, D. J., Shepherd, M., Simmons, B. A., et al. (2014). High-throughput prediction of eucalypt lignin syringyl/guaiacyl content using multivariate analysis: A comparison between mid-infrared, near-infrared, and Raman spectroscopies for model development. *Biotechnology for Biofuels*, 7, 93.

Lupoi, J. S., Singh, S., Parthasarathi, R., Simmons, B. A., & Henry, R. J. (2015). Recent innovations in analytical methods for the qualitative and quantitative assessment of lignin. *Renewable & Sustainable Energy Reviews*, *49*, 871–906.

Lupoi, J. S., & Smith, E. A. (2012). Characterization of woody and herbaceous biomasses lignin composition with 1064 nm dispersive multichannel Raman spectroscopy. *Applied Spectroscopy*, *66*, 903–910.

Lupoi, J. S., Smith-Moritz, A., Singh, S., McQualter, R., Scheller, H. V., Simmons, B. A., et al. (2015). Localization of polyhydroxybutyrate in sugarcane using Fourier-transform infrared microspectroscopy and multivariate imaging. *Biotechnology for Biofuels*, *8*, 98.

Luterbacher, J. S., Moran-Mirabal, J. M., Burkholder, E. W., & Walker, L. P. (2015). Modeling enzymatic hydrolysis of lignocellulosic substrates using fluorescent confocal microscopy II: Pretreated biomass. *Biotechnology and Bioengineering*, *112*, 32–42.

Lynd, L. R., Cushman, J. H., Nichols, R. J., & Wyman, C. E. (1991). Fuel ethanol from cellulosic biomass. *Science*, *251*, 1318–1323.

Ma, J., Zhang, X., Zhou, X., & Xu, F. (2014). Revealing the changes in topochemical characteristics of poplar cell wall during hydrothermal pretreatment. *Bioenergy Research*, 7, 1358–1368.

Ma, J. F., Zhou, X., Zhang, X., & Xu, F. (2013). Label-free in situ Raman analysis of opposite and tension wood in *Populus nigra*. *Bioresources*, *8*, 2222–2233.

Mansfield, J. C., Littlejohn, G. R., Seymour, M. P., Lind, R. J., Perfect, S., & Moger, J. (2013). Label-free chemically specific imaging in planta with stimulated Raman scattering microscopy. *Analytical Chemistry*, *85*, 5055–5063.

Mast, S. W., Donaldson, L., Torr, K., Philipps, L., Flint, H., West, M., et al. (2009). Biosynthesis of ß(1-4) galactan in *Pinus radiata* compression wood. *Plant Physiology*, *150*, 573–583.

Mateu, B. P., Hauser, M. T., Heredia, A., & Gierlinger, N. (2016). Waterproofing in Arabidopsis: Following phenolics and lipids in situ by confocal Raman microscopy. *Frontiers in Chemistry*, *4*, 10.

Merk, V., Chanana, M., Gierlinger, N., Hirt, A. M., & Burgert, I. (2014). Hybrid wood materials with magnetic anisotropy dictated by the hierarchical cell structure. *ACS Applied Materials & Interfaces*, *6*, 9760–9767.

Meshitsuka, G., & Nakano, J. (1979). Studies on the mechanism of lignin color reaction. XIII. Mäule color reaction. *Journal of the Japan Wood Research Society*, *25*, 588–594.

Meyer, M. W., Lupoi, J. S., & Smith, E. A. (2011). 1064 nm dispersive multichannel Raman spectroscopy for the analysis of plant lignin. *Analytica Chimica Acta*, *706*, 164–170.

Mitsumoto, K., Yabusaki, K., & Aoyagi, H. (2009). Classification of pollen species using autofluorescence image analysis. *Journal of Bioscience and Bioengineering*, *107*, 90–94.

Morrison, T. A., Jung, H. G., Buxton, D. R., & Hatfield, R. D. (1998). Cell-wall composition of maize internode of varying maturity. *Crop Science*, *38*, 445–460.

Mottiar, Y., Gierlinger, N., Jeremic, D., Master, E. R., & Mansfield, S. D. (2020). A typical lignification in eastern leatherwood (*Dirca palustris*). *New Phytologist*, *226*(3), 704–713.

Nakano, J., & Meshitsuka, G. (1992). The detection of lignin. In C. Dence, & S. Lin (Eds.), *Methods in lignin chemistry*. Berlin: Springer-Verlag.

Nascimento, J. M., & Dias, J. M. B. (2005). Vertex component analysis: A fast algorithm to unmix hyperspectral data. *IEEE Transactions on Geoscience and Remote Sensing, 43*, 898–910.

Naumann, A., Navarro-González, M., Peddireddi, S., Kües, U., & Polle, A. (2005). Fourier transform infrared microscopy and imaging: Detection of fungi in wood. *Fungal Genetics and Biology, 42*, 829–835.

Ozparpucu, M., Gierlinger, N., Burgert, I., Van Acker, R., Vanholme, R., Boerjan, W., et al. (2018). The effect of altered lignin composition on mechanical properties of CINNAMYL ALCOHOL DEHYDROGENASE (CAD) deficient poplars. *Planta, 247*, 887–897.

Ozparpucu, M., Ruggeberg, M., Gierlinger, N., Cesarino, I., Vanholme, R., Boerjan, W., et al. (2017). Unravelling the impact of lignin on cell wall mechanics: A comprehensive study on young poplar trees downregulated for CINNAMYL ALCOHOL DEHYDROGENASE (CAD). *Plant Journal, 91*, 480–490.

Pati, Y. C., Rezaiifar, R., & Krishnaprasad, P. S. (1993). Orthogonal matching pursuit: Recursive function approximation with applications to wavelet decomposition. In *Proceedings of 27th Asilomar conference on signals, systems and computers* (pp. 40–44). IEEE.

Perera, P. N., Schmidt, M., Chiang, V. L., Schuck, P. J., & Adams, P. D. (2012). Raman-spectroscopy-based non-invasive microanalysis of native lignin structure. *Analytical and Bioanalytical Chemistry, 402*, 983–987.

Perera, P. N., Schmidt, M., Schuck, P. J., & Adams, P. D. (2011). Blind image analysis for the compositional and structural characterization of plant cell walls. *Analytica Chimica Acta, 702*, 172–177.

Philippe, S., Tranquet, O., Utille, J.-P., Saulnier, L., & Guillon, F. (2007). Investigation of ferulate deposition in endosperm cell walls of mature and developing wheat grains by using a polyclonal antibody. *Planta, 225*(5), 1287–1299.

Piqueras, S., Duponchel, L., Tauler, R., & de Juan, A. (2011). Resolution and segmentation of hyperspectral biomedical images by multivariate curve resolution-alternating least squares. *Analytica Chimica Acta, 705*, 182–192.

Piqueras, S., Krafft, C., Beleites, C., Egodage, K., von Eggeling, F., Guntinas-Lichius, O., et al. (2015). Combining multiset resolution and segmentation for hyperspectral image analysis of biological tissues. *Analytica Chimica Acta, 881*, 24–36.

Pohling, C., Brackmann, C., Duarte, A., Buckup, T., Enejder, A., & Motzkus, M. (2014). Chemical imaging of lignocellulosic biomass by CARS microscopy. *Journal of Biophotonics, 7*, 126–134.

Prats-Mateu, B. P., Bock, P., & Gierlinger, N. (2020). Raman imaging of plant cell walls. In Z. A. Popper (Ed.), *The plant cell wall: Methods and protocols* (pp. 251–295). New York, NY: Springer New York.

Prats-Mateu, B., Bock, P., Schroffenegger, M., Toca-Herrera, J. L., & Gierlinger, N. (2018). Following laser induced changes of plant phenylpropanoids by Raman microscopy. *Scientific Reports, 8*, 11804.

Raab, T. K., & Vogel, J. P. (2004). Ecological and agricultural applications of synchrotron IR microscopy. *Infrared Physics & Technology, 45*, 393–402.

Radotić, K., Kalauzi, A., Djikanović, D., Jeremić, M., Leblanc, R. M., & Cerović, Z. G. (2006). Component analysis of the fluorescence spectra of a lignin model compound. *Journal of Photochemistry and Photobiology B: Biology, 83*, 1–10.

Ralph, J., Lapierre, C., & Boerjan, W. (2019). Lignin structure and its engineering. *Current Opinion in Biotechnology, 46*, 240–249.

Raman, C. V., & Krishnan, K. S. (1928). A new type of secondary radiation. *Nature, 121*, 501–502.

Richter, S., Mussig, J., & Gierlinger, N. (2011). Functional plant cell wall design revealed by the Raman imaging approach. *Planta*, *233*, 763–772.

Rodriguez, L. G., Lockett, S. J., & Holtom, G. R. (2006). Coherent anti-stokes Raman scattering microscopy: A biological review. *Cytometry. Part A*, *69A*(8), 779–791.

Rudall, P. J., & Caddick, L. R. (1994). Investigation of the presence of phenolics compounds in monocotyledonous cell walls, using UV fluorescence microscopy. *Annals of Botany*, *74*, 486–491.

Ruel, K., Faix, O., & Joseleau, J. P. (1994). New immunogold probes for studying the distribution of the different lignin types during plant cell wall biogenesis. *Journal of Trace and Microprobe Techniques*, *12*, 247–265.

Russ, J. C., & Brent Neal, F. (2016). *The image processing handbook* (7th ed.). Boca Raton, FL: CRC Press.

Saadi, A., Lempereur, I., Sharonov, S., Autran, J., & Manfait, M. (1998). Spatial distribution of phenolic materials in durum wheat grain as probed by confocal fluorescence spectral imaging. *Journal of Cereal Science*, *28*, 107–114.

Sanderson, M. J., Smith, I., Parker, I., & Bootman, M. (2014). Fluorescence microscopy. *Cold Spring Harbor Protocols*, *10*. pdb.top071795.

Saporta, G. (2006). *Probabilités, analyse des données et statistique*. Paris, FR: TECHNIP.

Sasani, N., Bock, P., Felhofer, M., & Gierlinger, N. (2021). Raman imaging reveals in-situ microchemistry of cuticle and epidermis of spruce needles. *Plant Methods*, *17*, 17.

Schmid, E. D., & Brosa, B. (1971). Raman intensity and conjugation. 1. Substituent dependence of Raman intensities of 1600 1/cm ring vibrations of monosubstituted benzene derivatives. *Berichte der Bunsen-Gesellschaft Fur Physikalische Chemie*, *75*, 1334–1343.

Schmid, E. D., & Topsom, R. D. (1981). Raman intensity and conjugation. 5. A quantitative relationship between Raman intensity and the length of conjugation and an analysis of the Raman intensities of some substituted benzenes and biphenyls. *Journal of American Chemical Society*, *103*, 1628–1632.

Scott, J. A., Procter, A. R., Fergus, B. J., & Goring, D. A. I. (1969). The application of ultraviolet microscopy to the distribution of lignin in wood description and validity of the technique. *Wood Science and Technology*, *3*, 73–92.

Segmehl, J. S., Keplinger, T., Krasnobaev, A., Berg, J. K., Willa, C., & Burgert, I. (2019). Facilitated delignification in CAD deficient transgenic poplar studied by confocal Raman spectroscopy imaging. *Spectrochimica Acta A*, *206*, 177–184.

Segmehl, J. S., Studer, V., Keplinger, T., & Burgert, I. (2018). Characterization of wood derived hierarchical cellulose scaffolds for multifunctional applications. *Materials*, *11*(4), 517.

Sharma, B., Frontiera, R. R., Henry, A. I., Ringe, E., & Van Duyne, R. P. (2012). SERS: Materials, applications, and the future. *Materials Today*, *15*, 16–25.

Siebert, F., & Hildebrandt, P. (2008). *Vibrational spectroscopy in life science*. Weinheim: Wiley-VCH Verlag GmbH & Co. KGaA.

Sills, D. L., & Gossett, J. M. (2012). Using FTIR to predict saccharification from enzymatic hydrolysis of alkali-pretreated biomasses. *Biotechnology and Bioengineering*, *109*, 353–362.

Simon, C., Lion, C., Biot, C., Gierlinger, N., & Hawkins, S. (2018). Lignification and advances in plant cell walls. *Annual Plant Reviews*, *1*, 1–32.

Singh, T., Vaidya, A. A., Donaldson, L. A., & Singh, A. (2016). Improvement in the enzymatic hydrolysis of biofuel substrate by a combined thermochemical and fungal pretreatment. *Wood Science and Technology*, *50*, 1003–1014.

Smith, E., & Dent, G. (2005). *Modern Raman spectroscopy—A practical approach*. John Wiley & Sons Ltd.

Song, K., Yin, Y., Salmén, L., Xiao, F., & Jiang, X. (2014). Changes in the properties of wood cell walls during the transformation from sapwood to heartwood. *Journal of Materials Science*, *49*, 1734–1742.

Studer, M. H., DeMartini, J. D., Davis, M. F., et al. (2011). Lignin content in natural *Populus* variants affects sugar release. *Proceedings of the National Academy of Sciences of the United States of America*, *108*, 6300–6305.

Sun, L., Varanasi, P., Yang, F., Loque, D., Simmons, B. A., & Singh, S. (2012). Rapid determination of syringyl: Guaiacyl ratios using FT-Raman spectroscopy. *Biotechnology and Bioengineering*, *109*, 647–656.

Symons, S. J., & Dexter, J. E. (1996). Aleurone and pericarp fluorescence as estimators of mill stream refinement for various Canadian wheat classes. *Journal of Cereal Science*, *23*, 73–83.

Talamond, P., Verdeil, J.-L., & Conéjéro, G. (2015). Secondary metabolite localization by autofluorescence in living plant cells. *Molecules*, *20*, 5024–5037.

Terashima, N. (2013). Diversification of lignin supramolecular structure during the evolution of plants. *Mokuzai Gakkaishi*, *59*, 65–68.

Terashima, N., & Fukushima, K. (1988). Heterogeneity in formation of lignin—XI: An autoradiographic study of the heterogeneous formation and structure of pine lignin. *Wood Science and Technology*, *22*, 259–270.

Terashima, N., Fukushima, K., & Tsuchiya, S. (1986). Heterogeneity in formation of lignin. VII. An autoradiographic study on the formation of guaiacyl and syringyl lignin in poplar. *Journal of Wood Chemistry and Technology*, *6*, 495–504.

Theodoridou, K., Vail, S., & Yu, P. (2014). Explore protein molecular structure in endosperm tissues in newly developed black and yellow type canola seeds by using synchrotron-based Fourier transform infrared microspectroscopy. *Spectrochimica Acta A*, *120*, 421–427.

Tolivia, D., & Tolivia, J. (1987). Fasga: A new polychromatic method for simultaneous and differential staining of plant tissues. *Journal of Microscopy (Oxford)*, *148*, 113–117.

Tranquet, O., Saulnier, L., Utille, J.-P., Ralph, J., & Guillon, F. (2009). Monoclonal antibodies to *p*-coumarate. *Phytochemistry*, *70*, 1366–1373.

Tsutsumi, Y., Kondo, R., Sakai, K., & Imamura, H. (1995). The difference of reactivity between syringyl lignin and guaiacyl lignin in alkaline systems. *Holzforschung*, *49*, 423–428.

Türker-Kaya, S., & Huck, C. W. (2017). A review of mid-infrared and near-infrared imaging: Principles, concepts and applications in plant tissue analysis. *Molecules*, *22*, 168–187.

Vallet, C., Chabbert, B., Czaninski, Y., & Monties, B. (1996). Histochemistry of lignin deposition during sclerenchyma differentiation in alfalfa stems. *Annals of Botany*, *78*, 625–632.

Victor, T., Delpratt, N., Cseke, S. B., Miller, L. M., & Cseke, L. J. (2017). Nutrient distribution in the rhizosphere using FTIR imaging. *Analytical Chemistry*, *89*(9), 4831–4837.

Vidot, K., Devaux, M.-F., Alvarado, C., Guyot, S., Jamme, F., Gaillard, C., et al. (2019). Phenolic distribution in apple epidermal and outer cortex tissue by multispectral deep-UV autofluorescence cryo-imaging. *Plant Science*, *283*, 51–59.

Vongsvivut, J., Pérez-Guaita, D., Wood, B. R., Heraud, P., Khambatta, K., Hartnell, D., et al. (2019). Synchrotron macro ATR-FTIR microspectroscopy for high-resolution chemical mapping of single cells. *Analyst*, *144*, 3226–3238.

Wang, X. Q., Keplinger, T., Gierlinger, N., & Burgert, I. (2014). Plant material features responsible for bamboo's excellent mechanical performance: A comparison of tensile properties of bamboo and spruce at the tissue, fibre and cell wall levels. *Annals of Botany*, *114*(8), 1627–1635.

Wetzel, D. L. (2009). FT-IR microspectroscopic imaging of plant material. In R. Salzer, & H. W. Siesler (Eds.), *Infrared and Raman spectroscopic imaging*. Weinheim, Germany: Wiley-VCH Verlag GmbH & Co. KGaA.

Wetzel, D. L., Eilert, A. J., Pietrzak, L. N., Miller, S. S., & Sweat, J. A. (1998). Ultraspatially-resolved synchrotron infrared microspectroscopy of plant tissue *in situ*. *Cellular and Molecular Biology*, *44*, 145–168.

Withana-Gamage, T., Hegedus, D. D., Qiu, X., Yu, P., May, T., Lydiate, D., et al. (2013). Characterization of Arabidopsis thaliana lines with altered seed storage protein profiles using synchrotron-powered FT-IR spectromicroscopy. *Journal of Agricultural and Food Chemistry*, *5*, 22–29.

Yang, B., & Wyman, C. E. (2008). Pretreatment: The key to unlocking low-cost cellulosic ethanol. *Biofuels, Bioproducts & Biorefining*, *2*, 26–40.

Yu, P. Q. (2004). Application of advanced synchrotron radiation-based Fourier transform infrared (SR-FTIR) microspectroscopy to animal nutrition and feed science: A novel approach. *British Journal of Nutrition*, *92*, 869–885.

Yu, P. Q. (2011). Microprobing the molecular spatial distribution and structural architecture of feed-type sorghum seed tissue (Sorghum Bicolor L.) using the synchrotron radiation infrared microspectroscopy technique. *Journal of Synchrotron Radiation*, *18*, 790–801.

Yu, P. Q., McKinnon, J. J., Christensen, C. R., Christensen, D. A., Marinkovic, N. S., & Miller, L. M. (2003). Chemical imaging of microstructures of plant tissues within cellular dimension using synchrotron infrared microspectroscopy. *Journal of Agricultural and Food Chemistry*, *51*, 6062–6067.

Yuan, Y., Jiang, H., Wu, W., Wu, S., Yoncang, J., & Xiao, H. (2021). Recent advances in understanding the effects of lignin structural characteristics on enzymatic hydrolysis. *Biotechnology for Biofuels*, *14*, 205.

Zancajo, V. M. R., Lindtner, T., Eisele, M., Huber, A. J., Elbaum, R., & Kneipp, J. (2020). FTIR nanospectroscopy shows molecular structures of plant biominerals and cell walls. *Analytical Chemistry*, *92*, 13694–13701.

Zeng, Y. N., Saar, B. G., Friedrich, M. G., Chen, F., Liu, Y. S., Dixon, R. A., et al. (2010). Imaging lignin-downregulated alfalfa using coherent anti-Stokes Raman scattering microscopy. *Bioenergy Research*, *3*(3), 272–277.

Zeng, Y. N., Zhao, S., Wei, H., Tucker, M. P., Himmel, M. E., Mosier, N. S., et al. (2015). In situ micro-spectroscopic investigation of lignin in poplar cell walls pretreated by maleic acid. *Biotechnology for Biofuels*, *8*, 126.

Zhang, Y., Legland, D., El Hage, F., et al. (2019). Changes in cell walls lignification, feruloylation and p-coumaroylation throughout maize internode development. *PLoS One*, *14*(7), e02119923.

Zhao, Y., Man, Y., Wen, J., Guo, Y., & Lin, J. (2019). Advances in imaging plant cell walls. *Trends in Plant Science*, *24*, 867–878.

Zhu, N., Wu, D., & Chen, K. S. (2018). Label-free visualization of fruit lignification: Raman molecular imaging of loquat lignified cells. *Plant Methods*, *14*, 58.

Zhu, N., Yang, Y. F., Ji, M. B., Wu, D., & Chen, K. S. (2019). Label-free visualization of lignin deposition in loquats using complementary stimulated and spontaneous Raman microscopy. *Horticulture Research*, *6*, 72.

Zimmermann, B., Bağcıoğlu, M., Sandt, C., & Kohler, A. (2015). Vibrational microspectroscopy enables chemical characterization of single pollen grains as well as comparative analysis of plant species based on pollen ultrastructure. *Planta*, *242*, 1237–1250.

Zoghlami, A., Refahi, Y., Terryn, C., & Paes, G. (2019). Multimodal characterization of acid-pretreated poplar reveals spectral and structural parameters strongly correlate with saccharification. *Bioresource Technology*, *293*, 122015.

CHAPTER SEVEN

Spatio-temporal regulation of lignification

Maxime Chantreau* and Hannele Tuominen
Umeå Plant Science Centre (UPSC), Department of Forest Genetics and Plant Physiology, Swedish University of Agricultural Sciences, Umeå, Sweden
*Corresponding author: e-mail address: maxime.chantreau@slu.se

Contents

Abstract

Lignin is a poly-aromatic polymer found in plant cell walls. This polymer, mainly composed of three phenylpropanoid units, confers exceptional properties to the cell wall such as hydrophobicity, mechanical strength, or resistance against stresses. Thereby the cell wall deposition of lignin represents often the main molecular event that defines the biological function(s) of the lignified cell or tissue. The time and localization of lignin deposition as well as the composition of the polymer can be considered as the three essential components of the polymer that will define the biological function of the lignified tissue. In this review, we will cover the localizations, types and functions of lignin

Advances in Botanical Research, Volume 104
ISSN 0065-2296
https://doi.org/10.1016/bs.abr.2022.03.006

found in various part of land plants. Then, for the various lignified tissues, the mechanisms controlling the developmental deposition of lignin such as transcriptional regulation, intercellular coordinated control of lignification or posttranslational modification of proteins will be discussed. Finally, a focus will be made on the environmental cues that influence the lignification during radial tree growth, as well as the plant responses that these signals trigger in the regulation of lignification.

1. Introduction

Lignin is a poly-aromatic polymer encrusted in plant cell walls (CW). The polymer is mainly constituted of three subunits, the *p*-hydroxyphenyl (H), guaiacyl (G) and syringyl (S) units, which display different degrees of methoxylation. These three main lignin building blocks originate from the polymerization of the three monolignols *p*-coumaryl, coniferyl and sinapyl alcohols synthesized through the phenylpropanoid pathway (Bonawitz & Chapple, 2010; Umezawa, 2010). Biosynthesis of monolignols involves multiple enzymatic steps performed by multigene families which originated from gene duplication followed by subfunctionalization or neofunctionalization (Renault, Werck-Reichhart, & Weng, 2019). This evolutionary process partly explains the variation of lignin composition between species, with the most striking divergence being the general absence of the F5H (Ferulate-5-Hydroxylase) enzyme in gymnosperms, resulting in their inability to synthesize sinapyl alcohol (Li et al., 2001; Weng & Chapple, 2010).

The monolignols are then translocated to the apoplastic space where they are incorporated in the lignin polymer via radical coupling (Wang, Chantreau, Sibout, & Hawkins, 2013). Besides the three conventional lignin monomers, numerous other phenolic compounds can be found inside the polymer, either as naturally occurring true monomers or resulting from genetic engineering. Those extra lignin monomers are mostly derived from the general phenylpropanoid pathway (for recent papers see Del Río et al. (2020) and Rencoret et al. (2021)). The outcome of such a wide array of lignin monomers highlights the plasticity of the polymer, suggesting that almost any phenolic compounds translocated in the apoplastic space can be incorporated inside the lignin polymer via radical coupling (Del Río et al., 2020). Nonetheless, except for the caffeic alcohol based C lignin in the seed coat, the non-conventional monomers typically add up to only a small percentage of the total lignin polymer (Vanholme, De Meester, Ralph, & Boerjan, 2019), which makes the relative proportion of the three main lignin subunits, rather than the monomer diversity, the most important criteria that defines the structure and the physio-chemical characteristic of the polymer (Boerjan, Ralph, & Baucher, 2003; Mottiar, Vanholme, Boerjan, Ralph, & Mansfield, 2016; Ralph, 2010; Ralph et al., 2004; Vanholme et al., 2012; Vanholme, Morreel, Ralph, & Boerjan, 2008). Furthermore, the three main monomers are not homogenously distributed. Important variations in their respective relative abundance are observed between clades and species but also between different lignified tissues of an organism and between different cells of a lignified tissue (Barros, Serk, Granlund, & Pesquet, 2015; Donaldson, 2001). Altogether, this heterogeneity of the polymer suggests a tight control of the relative abundance of the different subunits and their localization.

Lignin is deposited in numerous different cell types (following cellular differentiation or in response to specific environmental signals) and fulfills several different functions. This review will cover the localization and type of lignin in various parts of land plants and how lignin deposition is regulated in a spatial and temporal manner to cover the different functions of lignin.

2. Localization, function, and composition of the lignins in plants

2.1 Xylem

The xylem comprises three main cell types which are all lignified: the tracheary elements, the xylary fibers and the parenchyma cells (Fig. 1). Mature tracheary elements are dead cells that fulfill the function of water transport. They are commonly divided into two cell types; the imperforate tracheids and the perforate vessel elements (Pittermann, 2010). Gymnosperms only possess tracheids while angiosperms have both tracheids and vessel elements. The vessel elements are axially connected to each other to form a long route for the transport of the water, and they resist high negative pressure and cell collapse thanks to their thick and lignified secondary CW (SCW) (Hacke, Sperry, Pockman, Davis, & McCulloh, 2001; Schmid & Evert, 2008; Schuetz, Smith, & Ellis, 2013). With their large lumen and relatively short size, the vessel elements are only negligibly involved in providing the mechanical support of the plant. This function is mainly attributed to the xylary fibers that are long and mostly dead cells with a small lumen size and a thick and lignified SCW (Schmid & Evert, 2008).

The wood of angiosperms is thus mainly constituted of two types of dead, lignified cells performing the two functions of transport and support, in contrast to the wood of gymnosperm where tracheids (up to 90% in volume) perform both functions of transport and support. The last type of xylem cells are the parenchyma cells that are living cells which can develop

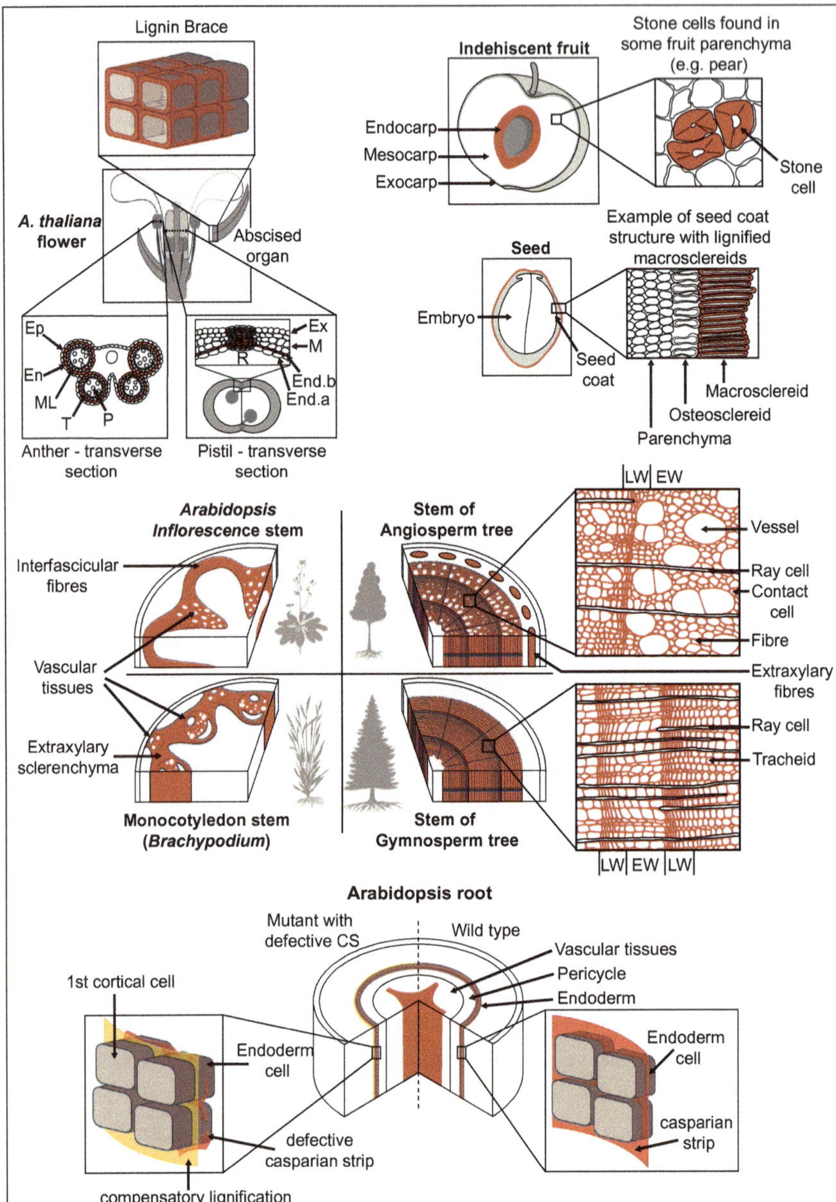

Fig. 1 Occurrence of lignification in plants. Developmental lignin and defense-lignin are represented in red and yellow, respectively. Ep: Epidermis; En: Endothecium; ML: Middle layer; T: Tapetum; P: Pollen grain; Ex: Exocarp; M: Mesocarp; End. a: Endocarp a; End.b: Endocarp b; R: Replum; CS: Casparian Strip; LW: Latewood; EW: Earlywood.

lignified SCW. These cells represent between 20% and 40% of the wood cells of eudicots. They are organized axially or radially (ray parenchyma cells) and serve different metabolic functions that are crucial for wood and tree functioning (Słupianek, Dolzblasz, & Sokołowska, 2021).

Heterogeneity of the lignin polymer, first hypothesized by Stone (1955), was then suggested by microscopy in the 1970s (Fergus & Goring, 1970a, 1970b). In 1986, Terashima, Fukushima, Tsuchiya, and Takabe (1986) administrated ^{3}H or ^{14}C-labeled precursors of monolignols to poplar and observed, by autoradiography, a heterogenous deposition of guaiacyl and syringyl lignin in differentiating xylem elements, which was later confirmed by bromination with TEM-EDXA (Transmission Electron Microscopy-Energy Dispersive X-ray Analysis) (Saka & Goring, 1988). This study showed that the S2 layer of the fiber SCW of white birch contained 12% guaiacyl residue and 88% syringyl residue, while the S2 layer of the vessel SCW consisted of 88% guaiacyl and only 12% syringyl-type lignin. Spectroscopy techniques have more recently demonstrated high potential for the CW characterization at a (sub)cellular level. For instance, the heterogeneity of the lignin polymer among the different xylem cell types with a clear predominance of G-lignin in vessel CW and S-lignin in the fibers was demonstrated in *Populus* wood by micro-Fourier-Transform Infrared Spectroscopy (FTIR) (Gorzsás, Stenlund, Persson, Trygg, & Sundberg, 2011). Fluorescence Lifetime Imaging Microscopy (FLIM) has also been developed as a read out for lignin chemical composition analysis. This new chemotyping method appears as particularly attractive for the in-situ lignin characterization (Escamez et al., 2021).

2.2 Extra-xylary sclerenchyma

Extra-xylary sclerenchyma tissues include a variety of cell types categorized as fibers and sclereids (Jarvis, 2012). The fibers are characterized by their elongated shape and their organization in bundles while the sclereids are much shorter with a thick SCW and can be found as single cells or in a small cluster at various locations in the plant body.

Extra-xylary fibers are elongated cells, often found in the stem. In dicotyledonous plants the extra-xylary fibers are associated in bundles within the stem cortex between the epidermis and the xylem core. Contrary to the fibers that are morphologically and functionally characterized by their high length, the sclereids are compact with a wide range of locations in the plant body. The brachysclereids, also called stone cells, are the smallest form of sclereids. They are formed from parenchyma cells and develop SCW via deposition of lignin and cellulose (Smith, 1935). They are found in the peduncle, where they bring strength and support (Horbens et al., 2015); seed coat; stem; root; endocarp; and in the flesh of some fruits (e.g., pear fruit) (Zhang et al., 2021). Nuclear Magnetic Resonance (NMR) and FTIR analyses showed that the pear stone cell lignin polymer, made up of G and S-units, with predominance of the G-units (Cai et al., 2010) is smaller and more branched than the typical wood poplar lignin polymer (Pu, Song, Xie, & Gu, 2002; Tefu, Luohua, & Qin, 2004), resulting in a comestible lignin. Astrosclereids originate from parenchyma cells that differentiate into radially branched, lignified structure. They are mainly found in the foliar parenchyma tissues of different plants and are believed to confer support to the structure. The last kind of sclereids, osteosclereids and macrosclereids are found in the thick and waterproof layer of the seed coat (also called testa) of certain plant species, such as *Trifolium pratense* L. or *Glycine* max L. (Fig. 1) (Algan & Bakar Büyükkartal, 2000; Qutob, Ma, Peterson, Bernards, & Gijzen, 2008). The macrosclereids (also called malpighian cells or palisade cells) are elongated cells at the outer layer of the seed coat which become thick-walled after deposition of lignin at the heart-shaped developmental embryo stage (Algan & Bakar Büyükkartal, 2000). A layer of I-shaped osteosclereids (Hourglass-cells) is sometimes located under the macrosclereids and support them. The osteosclereids have been described as lignified in *Baptisia leucophaea* (Pammel, 1899) but no lignin could be detected in other species (Zimmerman, 1936). In addition to providing mechanical strength, the seed coat lignin has been suggested to influence seed germination and water permeability in *Arabidopsis* and soybean (Capeleti et al., 2005; Liang, Davis, Gardner, Cai, & Wu, 2006). The majority of lignin in the seeds of *Brassicaceae* has been demonstrated to be present in the seed coat (Marles & Gruber, 2004), but little is known about the proportions of the different monolignols in the different cell layers of the seed coat (Francoz, Lepiniec, & North, 2018). An unusual catechyl lignin homopolymer called C-lignin, solely derived from caffeyl alcohol, was found as a specific feature of seed coat lignin in vanilla orchids and some members of the Cactaceae family (Barsberg, Lee, & Rasmussen, 2018; Chen, Tobimatsu, Havkin-Frenkel, Dixon, & Ralph, 2012; Tobimatsu et al., 2013).

2.3 Endodermis

The endodermis is a lignified plant tissue in the roots of vascular plants between the periderm and the cortex (Bonnett, 1968; Kroemer, 1903).

After specification, endodermal cells undergo differentiation characterized by local lignin deposition at the junction between adjacent cells, forming the Casparian strip (CS) (Fig. 1) (Alassimone, Naseer, & Geldner, 2010; Bonnett, 1968; Naseer et al., 2012). After the deposition of lignin, suberin lamellae are deposited at the inner surface of the lignified primary CW of the endodermis (Haas & Carothers, 1975; Kroemer, 1903; Robards, Jackson, Clarkson, & Sanderson, 1973). This second hydrophobic layer restricts the transcellular transport of water and nutrients in the endodermal cells already decorated with the CS (Doblas, Geldner, & Barberon, 2017). Due to the promiscuity of the two polymers of lignin and suberin, the chemical nature of the CS has been ambiguous for many years (Caspary, 1865; Delevoryas & Esau, 1960; Rosen, Meyer, Anderson, Böhning, & Bohning, 1960). An estimation of the chemical nature by thioacidolysis coupled with Gas Chromatography-Mass Spectrometry (GC-MS) on root tips of the *ahp6* (*Arabidopsis histidine phosphotransfer protein 6*) mutant, grown in conditions of delayed xylem differentiation, demonstrated the lignin nature of the endodermal apoplastic barrier (Naseer et al., 2012). This is in accordance with a more recent approach by Raman spectroscopy, which estimated that the CS lignin is highly similar to the xylem lignin composition, with a strong predominance of the G-units (Reyt et al., 2021).

2.4 Endothecium

A lignified tissue is also found in the male reproductive organ, the anther, that contains the pollen. The first stage of anther dehiscence in *Arabidopsis* is characterized by the SCW thickening in the endothecial cell layer (Hao et al., 2014; Wilson, Song, Taylor, & Yang, 2011). The presence of lignin has been demonstrated in the thickened SCW of the endothecium by phloroglucinol and acridine orange/ethidium bromide staining (Dawson et al., 1999; Yang et al., 2007). The composition of the lignin polymer in these very specialized cells is not known yet. It has been demonstrated by genetic analyses of indehiscent lignin-defective mutants (Dawson et al., 1999; Hao et al., 2014; Mitsuda, Seki, Shinozaki, & Ohme-Takagi, 2005; Sanders et al., 1999; Schilmiller et al., 2009; Steiner-Lange et al., 2003; Thévenin et al., 2011; Wang, Mao, Yang, & He, 2015; Weng, Mo, & Chapple, 2010; Yang et al., 2007) that the thickening and lignification of the endothecium layers provides the mechanical tension that leads to the stomium breakage, making these processes essential for anther opening and thereby for reproduction (Bonner & Dickinson, 1989; Keijzer, 1987; Mitsuda et al., 2005; Yang et al., 2007).

2.5 Endocarp

A lignified tissue, the endocarp, is commonly found adjacent to the seeds in several fruits. The endocarp varies in lignin composition and content, but also in function, depending on the species. In dehiscent fruits, such as the *Arabidopsis* silique, the lignification of the endocarp and the valve margin are important and necessary for fruit dehiscence and hence seed dispersal (Liljegren et al., 2000; Spence, Vercher, Gates, & Harris, 1996). In *Arabidopsis*, the endocarp has two cell layers; an inner epidermis with thin CW (endocarp a) and a subepidermal layer of small and longitudinally elongated cells with lignified CW at maturity (endocarp b) (Fig. 1) (Rajani & Sundaresan, 2001). Upon maturation of the silique, the lignified endocarp b and other adjoining tissues/cells of the silique, in coordination with the CW degrading enzyme released by the endocarp a, generate differential shrinkage which is translated into mechanical tension leading to the dehiscence of the silique (Ferrándiz, 2002; Liljegren et al., 2004; Ogawa, Kay, Wilson, & Swain, 2009; Rajani & Sundaresan, 2001). In some species like *Cardamine hirsuta*, the mechanical tension generated is so intense that it results in explosive seed dispersal (Hay et al., 2014; Rich, 1991). A comparative study between explosive and non-explosive species of *Brassicaceae* demonstrated that the explosive feature of seed dispersal is conferred by the asymmetrical deposition of lignin within the endocarp b cells, demonstrating subcellular regulation of lignin deposition as an evolutionary novel mechanism in seed dispersal (Hofhuis et al., 2016).

Numerous economically important species such as peach, cherry or coffee produce an indehiscent, fleshy fruit with a hardened endocarp that surrounds and protects the seed (Doster & Michailides, 1999). Recent characterization of lignin from endocarp tissues of several species demonstrated great variability in the content and composition of the lignin polymer. The nutshell of pistachio, walnut and almond was found to be rich in S-units with 77%, 64% and 59% of S-lignin, respectively, while hazelnut and macadamia endocarp was almost devoid of S-unit and had 95% and 93% of G-lignin, respectively (Landucci, Smith, Liu, Karlen, & Ralph, 2020). Lignin content is often higher in endocarp than in the woody tissues. For instance, the Brazil nutshell, macadamia and pecan display between 47% and 40% dry weight of lignin (Landucci et al., 2020) whereas the poplar woody tissues display between 14% and 25% dry weight of lignin (Porth et al., 2013; Ralph, Akiyama, Coleman, & Mansfield, 2012). The endocarp of certain palm fruit incorporates a significant amount of the hydroxystilbene piceatannol into the lignin. This non-conventional subunit induces a greater accumulation of condensed lignin, contributing to the endocarp hardening and might also contribute to seed protection through antiviral, antibacterial and antioxidant properties (Piotrowska, Kucinska, & Murias, 2012; Roupe, Remsberg, Yanez, & Davies, 2008).

2.6 Stress/defense lignin

Lignin is deposited in the CW following stresses (Cabane, Afif, & Hawkins, 2012; Hawkins & Boudet, 2003). For instance, increased lignification can be a way of maintaining osmotic balance and protecting CW integrity in the case of osmotic stress generated by drought or salt stress (Agarwal, Shukla, Gupta, & Jha, 2013; Ahsan, Nakamura, & Komatsu, 2012). In this case, the hydrophobic property and the rigidity of the lignin polymer is used by the plant to endure the difficult external conditions, but lignin can also be synthesized to counter unfavorable external conditions. In case of heavy metal pollution, the lignin polymer, through the presence of numerous functional groups in its structure, represents a chemical trap for heavy metal ions limiting their entry in the cells (Bhuiyan et al., 2007; Demirbas, 2004; Guo, Zhang, Shan, & quan., 2008; Kováčik & Bačkor, 2007). This has been observed in soybean roots under cadmium treatment (Bhuiyan et al., 2007), in the endodermal cells of the zinc hyperaccumulating plant *Thlaspi caervlescens* (Van De Mortel et al., 2008), and in the copper retention of *Oriza sativa* (Su et al., 2020). For several reasons, H-lignin has been considered characteristic of stress-lignin. H lignin is more dense because it is richer in condensed bonds, and less chemically reactive than the most common β-O-4 bond found in S and G-lignin (Aminzadeh, Zhang, & Henriksson, 2017). The synthesis of H-units requires fewer enzymatic steps, hence enabling a faster response for the lignin deposition. Finally, deposition of H-lignin has been observed following different abiotic stresses (Cabané et al., 2004; Pitre et al., 2007; Timell, 1986).

Recently, Reyt et al. (2021) characterized a form of "stress" lignin in the root of *Arabidopsis thaliana*, based on its high content of H-units. The observed polymer was not resulting from environmental stress, but from a developmental defect in the CS lignification whereby *Arabidopsis* mutants with a discontinuous, non-sealed CS displayed so-called compensatory lignification by depositing pure H-lignin, chemically distinct from the typical CS lignin, in the cell corners (cortex-facing side) of the endodermal cells (Fig. 1).

Plant pathogen infection also often activates the biosynthesis and deposition of lignin (Baayen, Ouellette, & Rioux, 1996; Bhuiyan, Selvaraj, Wei, & King, 2009; Lange, Lapierre, & Sandermann, 1995; Menden, Kohlhoff, & Moerschbacher, 2007; Nicholson & Hammerschmidt, 1992; Siegrist, Jeblick, & Kauss, 1994; Smit & Dubery, 1997; Vance, Kirk, & Sherwood, 1980). Lignin can act as a physical barrier to prevent pathogen spreading (Lee et al., 2019), as a CW protective agent against pathogen-derived compounds, such as toxins or CW-degrading enzymes, or directly through its antimicrobial activity (Alzagameem et al., 2019). The chemical nature of the lignin induced by pathogen infection can vary depending on the species and the infection. The wheat fusarium resistance was related to a higher S lignin content (Lionetti et al., 2015). A basal immune response, driven by MYB15, correlated with accumulation of G-lignin in *Arabidopsis* (Chezem, Memon, Li, Weng, & Clay, 2017), and, recently, a higher G/S ratio was considered more important than the total lignin content for the cotton Verticillium wilt resistance (Xiong et al., 2021).

The CW response following biotic and/or abiotic stress has been suggested to be initiated by the CW integrity (CWI) maintenance mechanism. Detailed mode of action of this system is still limited (Bacete & Hamann, 2020) but evidence suggests an initiation of a compensatory response following mechanical and/or osmotic stresses perceived by the distortion of the plasma membrane relative to the cell wall (Engelsdorf et al., 2018; Hamann, Bennett, Mansfield, & Somerville, 2009) or through the perception of CW-derived fragments (Bacete, Mélida, Miedes, & Molina, 2018). An integral part of CWI maintenance involves production and deposition of lignin in the primary cell wall (Cano-Delgado, Metzlaff, & Bevan, 2000; Denness et al., 2011; Ellis, Karafyllidis, Wasternack, & Turner, 2002; Ellis & Turner, 2001; Hamann et al., 2009; Tsang, Edmond, Harrington, & Nühse, 2011).

2.7 Abscission zone

The occurrence of a layer of lignified cells in the Abscission Zone (AZ) has been observed for some species (Merelo et al., 2017; Tabuchi, Ito, & Arai, 2000; Tsuchiya, Satoh, & Iwai, 2015). The "lignin brace," deposited in the AZ of flower petals in *A. thaliana* enables proper organ abscission, while protecting the plant against potential infection. Lee et al. (2018) demonstrated the expression of lignin-biosynthetic genes and accumulation of H_2O_2 in the flower petal abscission zone, which leads to the deposition of lignin in a honeycomb structure at the cell margins of 2–3 layers of cells in the abscission zone (Fig. 1). The composition of this lignin is still not known, but the lignin brace was demonstrated to play a role in organ separation by restricting the CW degrading enzymes at the location of cell-cell detachment and hence enabling proper organ abscission (Lee et al., 2018).

3. Developmental regulation of lignification

3.1 Transcriptional network controlling deposition of CW and lignin

Deposition of SCW and lignin in xylem cells is strictly controlled by a hierarchical transcription network constituted of three layers of transcription factors (TFs; Fig. 2).The first layer of TFs controlling the SCW and lignin deposition program consists of the secondary wall NAC TFs of the VNS (VND-, NST/SND- and SOMBRERO (SMB)-related proteins) family (Hussey, Mizrachi, Creux, & Myburg, 2013). The VND TFs are considered as master transcriptional regulators of xylem vessel differentiation (Kubo et al., 2005). They activate genes related to SCW biosynthesis and cell death (Nakano, Yamaguchi, Endo, Rejab, & Ohtani, 2015; Ohashi-Ito, Oda, & Fukuda, 2010; Ohtani & Demura, 2019; Yamaguchi et al., 2011; Zhong, Lee, & Ye, 2010). On the other hand, the members of the NST/SND subfamily are involved in the control of SCW formation in fibers. The *Arabidopsis* triple mutant *nst1 nst2 snd1*, with a pendant phenotype, is characterized by a complete depletion of SCW and lignin deposition in both xylary and interfascicular fiber cells (Zhong & Ye, 2015). Similarly, in hybrid aspen the inactivation of the 4 NST/SND subfamily members (*VNS09, VNS10, VNS11, VNS12*) prevents SCW deposition both in xylem and phloem fibers (Takata et al., 2019). Recently, *Arabidopsis* NAC TF SND2–5 were identified as SCW regulators with an intermediate hierarchical position between the first and the second layer TFs (Zhong, Lee, Haghighat, & Ye, 2021; Zhong, Lee, Zhou, McCarthy, & Ye, 2008).

The second layer TF correspond to MYB46 and MYB83 in *Arabidopsis* (Gui et al., 2019; Patzlaff et al., 2003; Yang et al., 2016), MYB4 in *Pinus taeda* (Morse, Whetten, Dubos, & Campbell, 2009) or MYB2, MYB3, MYB20 and MYB21 in *Populus trichocarpa* (McCarthy et al., 2010; Zhong, McCarthy, Haghighat, & Ye, 2013). The overexpression of the *Arabidopsis MYB46* or *MYB83* induces ectopic deposition of secondary wall cellulose, xylan and lignin in parenchyma cells, and their simultaneous inactivation results in the loss of secondary wall thickening (McCarthy, Zhong, & Ye, 2009; Zhong, Richardson, & Ye, 2007). These TFs bind to the Secondary Wall MYB-Responsive Element (SMRE) sequences and activate the expression of a set of downstream TFs corresponding to

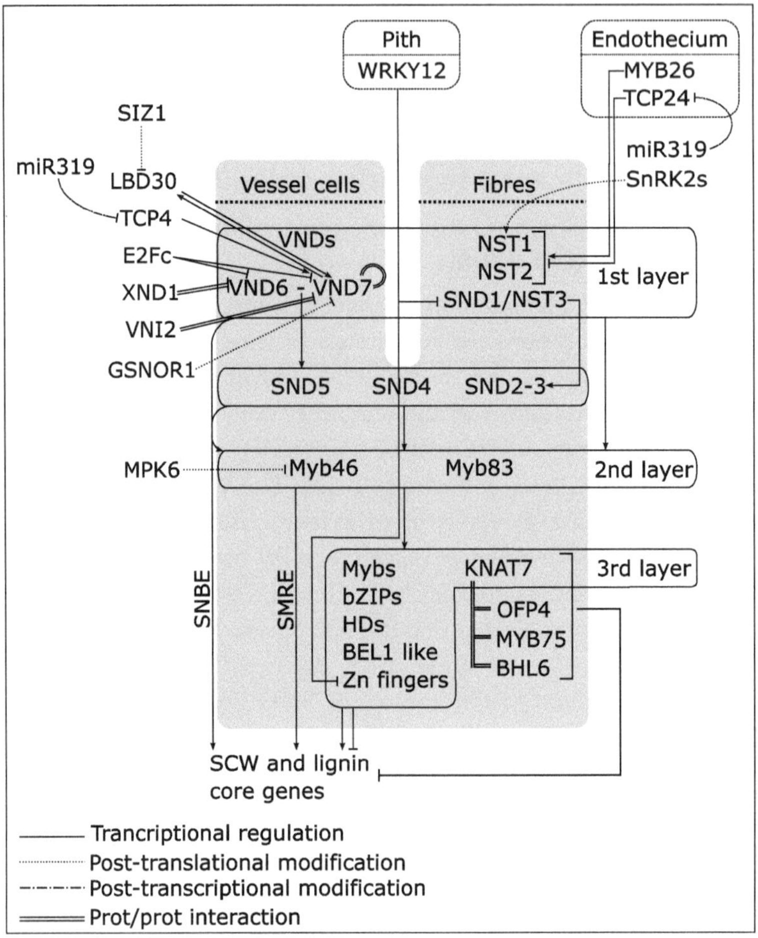

Fig. 2 Simplified view of the hierarchical transcription network controlling SCW and lignin deposition in *Arabidopsis thaliana.* Black arrows indicate activation or induction whereas blunt arrows indicate inactivation or repression.

the third layer master switches. They also activate the transcription of the core secondary wall biosynthetic genes and genes involved in CW modification and cell death (Zhong & Ye, 2012).

Numerous MYB TFs (e.g., in *Arabidopsis*: MYB3, MYB4, MYB7, MYB15, MYB26, MYB32, MYB41, MYB46, MYB58, MYB61, AtMYB63, MYB75, MYB85, and MYB103) regulate deposition of lignin (Kosma et al., 2014; Miyamoto, Tobimatsu, & Umezawa, 2020; Öhman et al., 2013; Zhao & Dixon, 2011; Zhou et al., 2017). However, only a few of them have been identified as specific regulators of lignin-biosynthetic genes (Geng et al., 2020; Perkins et al., 2020; Zhao et al., 2013). *MYB58* and *MYB63* are specifically expressed in the xylem vessel elements and fibers (Zhou, Lee, Zhong, & Ye, 2009). They bind to the AC elements in the promoters of monolignol-biosynthetic genes and induce their expression as well as the expression of the polymerizing enzyme *LACCASE4* (Zhou et al., 2009). Overexpression of *MYB58* and *MYB63* induces a dwarf phenotype and ectopic lignification (Perkins et al., 2020; Zhao et al., 2013). *MYB15* is a key driver of pathogen-induced lignification and is able to induce the expression of most of the lignin-biosynthetic genes (Kim et al., 2020). *MYB103* influences the expression of *F5H* and S-lignin biosynthesis even though it does not directly bind to *F5H* promoter (Öhman et al., 2013). MYB20, MYB42, MYB43, and MYB85 directly activate lignin and phenylalanine (Phe) biosynthetic genes, and they also induce the expression of *MYB4* which represses the expression of the *CHALCONE SYNTHASE* (CHS), the first enzyme of the flavonoid biosynthesis pathway (Jin, 2000).

This regulatory cascade has mainly been deciphered in the secondary xylem tissues. Nonetheless some surveys highlighted the occurrence of tissue-specific regulation. MYB26 is specifically expressed within the anther endothecium and its expression is essential to the formation and spatial arrangement of the SCW deposition in this tissue via induction of *NST1* and *NST2* (Fig. 2) (Dawson et al., 1999; Mitsuda et al., 2005; Steiner-Lange et al., 2003; Yang et al., 2017, 2007). MYB26 thus appears as a master switch activating the whole regulatory cascade of TFs required for the endothecium CW thickening and lignification. On the opposite, *TCP24*, whose expression in the endothecium occurs during early anther development, acts as a repressor of secondary wall thickening through a MYB26-independent inhibition of the first layer TF *NST1* and *NST2* (Fig.2) (Wang, Mao, et al., 2015).

In the pith of *Arabidopsis* and *Medicago truncatula* (devoid of SCW and lignin) the SCW and lignification program is under the control of the WRKY12 TF. This TF prevents the deposition of lignified SCW in these cells. It is expressed both in pith and cortical cells, but only the pith cells display a lignified SCW in the *wrky12* mutant, suggesting another mechanism of lignification in cortical cells (Fig. 2) (Wang et al., 2010).

A pear orthologue of MYB85, PbrMYB169, has been identified to positively regulate lignification of sclereid stone cells in pear fruit, which is in accordance with the lignin-specific activity of MYB85. PbrMYB169 activates the expression of several lignin-biosynthetic genes, and its overexpression in *Arabidopsis* stimulates deposition of lignin in xylem CW (Xue et al., 2019).

Nonetheless, the authors stipulate that the pear orthologues of MYB46 and MYB83, the second layer master switches, do not induce the expression of *PbrMYB169*. This suggest that despite a functional conservation, the expression of *PbrMYB169* would not be under the same transcriptional regulatory network as its orthologue *MYB85* (Xue et al., 2019). Other MYB TFs have also been reported to influence lignification in fruit. This is for instance the case of EjMYB1, EjMYB2 and EjMYB8 from *Eriobotrya japonica* (Wang et al., 2016; Xu et al., 2014), but their upstream regulation is still mostly undetermined. Recently, the MADS-box TF *AGL65* has been discovered to repress the lignin activator EjMYB8 (Ge et al., 2021).

Physical interaction between TFs is another mechanism influencing SCW and lignin gene expression. For instance, the Arabidopsis NIGHT LIGHT-INDUCIBLE AND CLOCK-REGULATED1 (LNK1) and LNK2 interact with MYB3 to facilitate the binding to the *C4H* promoter and suppression of lignin biosynthesis (Zhou et al., 2017). KNAT7, a regulator of SCW biosynthesis and lignification, interacts with numerous other TFs as OFP4, BLH6 and MYB75, forming a complex to inhibit SCW formation (Fig. 2) (Bhargava, Mansfield, Hall, Douglas, & Ellis, 2010; Li, Wang, Liu, Chen, & Douglas, 2011; Liu et al., 2014). The KNAT7/MYB75 complex has been found both in the xylem and the seed coat tissues, suggesting a shared regulatory complex of lignification between these two tissues (Bhargava et al., 2013).

The transcriptional regulation of lignification in non-xylem tissues suggests that the expression of the SCW and lignin genes is governed by the same cascade of TFs found in the xylem, but that the upstream regulators might differ between the different tissue types.

Finally, despite a large accumulation of data concerning mechanisms of CS lignification, the transcriptional regulation of monolignol biosynthesis in the endodermis is not known. Contrary to the other lignified tissues, CS lignification occurs without deposition of SCW. It is therefore plausible that the transcriptional network of CS lignification is rather simple. MYB36 is considered as the TF that initiates the differentiation of the endodermal cells by inducing the expression of the required machinery for the spatial coordination and deposition of the CS. Kamiya et al. (2015) demonstrated that MYB36 positively regulates 23 genes corresponding to CASPs, ESB-like genes, PER64 as well as some uncharacterized protein kinases or uncharacterized proteins but does not seem to activate any specific master regulators of SCW or lignin deposition. MYB36 is however required for deposition of CS lignin (Reyt et al., 2021). Interestingly, the *myb36* mutant displays ectopic lignification in the corners of the endodermal cells, which supposedly reflects presence of a compensatory mechanism of lignification to seal the cells in the case of discontinuous CS (Reyt et al., 2021).

3.2 Coordinated control and fine-tuning of lignification

Adequate timing and localization of developmental lignification is particularly important; lignin deficiency in the vasculature of the plant may lead to phenotypic alteration and weakening of plant fitness (Franke et al., 2002; Ha et al., 2016, 2019; Hoffmann et al., 2004; Jones, Ennos, & Turner, 2001; Leplé et al., 2007; Piquemal et al., 1998; Reddy et al., 2005; Schilmiller et al., 2009; Shadle et al., 2007; Van Acker et al., 2013; Voelker, Lachenbruch, Meinzer, & Strauss, 2011; Wang, Matthews, Williams, et al., 2018). An increase or premature lignification of the vasculature may also have a negative impact on plant physiology. This has been demonstrated nicely in the *Arabidopsis* plants modified in the *XND1* gene. Overexpression of *XND1* leads to extreme dwarfism with vessel cells lacking SCW thickening (Zhong, Kandasamy, & Ye, 2021) whereas a knock-out mutant displayed shorter vessel elements due to accelerated differentiation (Tang et al., 2018; Zhao, Avci, Grant, Haigler, & Beers, 2008), resulting in a reduction of plant size. *XND1* is specifically expressed in vessel cells, and co-localization experiments with VND6 suggested that XND1 physically interacts with VNDs to inhibit their function in xylem differentiation by sequestrating them in the cytoplasm (Fig. 2). This suggests that XND1 fine-tunes the timing of SCW and lignin deposition through the inhibition of the VNDs to ensure sufficient expansion of the vessel elements before their full differentiation (Zhong, Kandasamy, & Ye, 2021). This highlights the prime importance of a proper spatial and temporal regulation of the onset of lignification in a heterogeneous tissue like the xylem, composed of cells with different lifespan.

The vessel elements differentiate rapidly and undergo complete autolysis to guarantee efficient water conductivity (Bollhöner, Prestele, & Tuominen, 2012). In contrast, fiber, tracheid, and ray parenchyma cells live substantially longer. The lifetime of xylem fibers varies significantly depending on the cambium activity, external factors, and their position in the xylem. Finally, the ray parenchyma cells are often the longest-living cells of the xylem, with a lifespan up to 42 years (Tulik, Jura-Morawiec, Bieniasz, & Marciszewska, 2019).

Given the developmental differences among the different xylem cells, it is legitimate to consider that they might display distinct mechanisms of lignification (Smith et al., 2013). The main difference in the process of lignification among the xylem cells is their ability to independently perform their CW lignification (cell-autonomous lignification) or their dependency on the neighboring cells to fully accomplish or regulate this process (non-cell-autonomous lignification). The fact that the vessel elements die quickly to accomplish their function in water transport raised the question of the involvement of other cells, remaining alive longer, in vessel lignification by providing them with monolignols and Reactive Oxygen Species (ROS) necessary for lignification (Barros et al., 2015). The use of radio-labeled lignin monomers demonstrated that lignification can occur post mortem in the CW of xylem cells (Pickett-Heaps, 1968; Terashima et al., 1986; Terashima & Fukushima, 1988). The *post-mortem* lignification of vessel-like tracheary elements was demonstrated using xylogenic cell culture of *Zinnia elegans* (Hosokawa, Suzuki, Umezawa, & Sato, 2001; Pesquet et al., 2013). Lignin content increased in the cell cultures even after death of the lignified tracheary elements, and pharmacological inhibition of lignification also impacted the tracheary elements lignification after their cell death (Hosokawa et al., 2001; Pesquet et al., 2013). These results can be explained only if the tracheary elements lignification was assisted by the living, non-lignified cells of the culture, providing the monolignols and ROS necessary for lignification. This assumption is supported by the observation that non-lignifying, neighboring cells adjacent to lignified cell types frequently express genes of the lignin pathway (Baghdady et al., 2006; Bevan, Shufflebottom, Edwards, Jefferson, & Schuch, 1989; Blokhina et al., 2019; Chen et al., 2000; Hauffe et al., 1991; Hu et al., 1999; Lacombe, Van Doorsselaere, Boerjan, Boudet, & Grima-Pettenati, 2000; Lauvergeat et al., 2002) and produce monolignols and glycoconjugates (Blokhina et al., 2019). Finally, the work from Smith et al. (2013, 2017) provided *in planta* evidence for the occurrence of non-cell autonomous lignification of the tracheary elements and cell autonomous lignification of the interfascicular fibers (Fig. 3). The inactivation of *CCR1* under the control of the *CESA7* promoter enabled the suppression of lignin biosynthesis specifically in cells producing SCW. This resulted in plants with a pendant phenotype due to a drastic reduction of lignification but still, the tracheary elements and xylary fibers appeared normal and displayed a wild type level of phloroglucinol staining (Smith et al., 2013). The subsequent inhibition of CCR1 using the promoter of AtPRX47, specific of the xylary parenchyma cells, demonstrated that these cells were at least partially responsible of

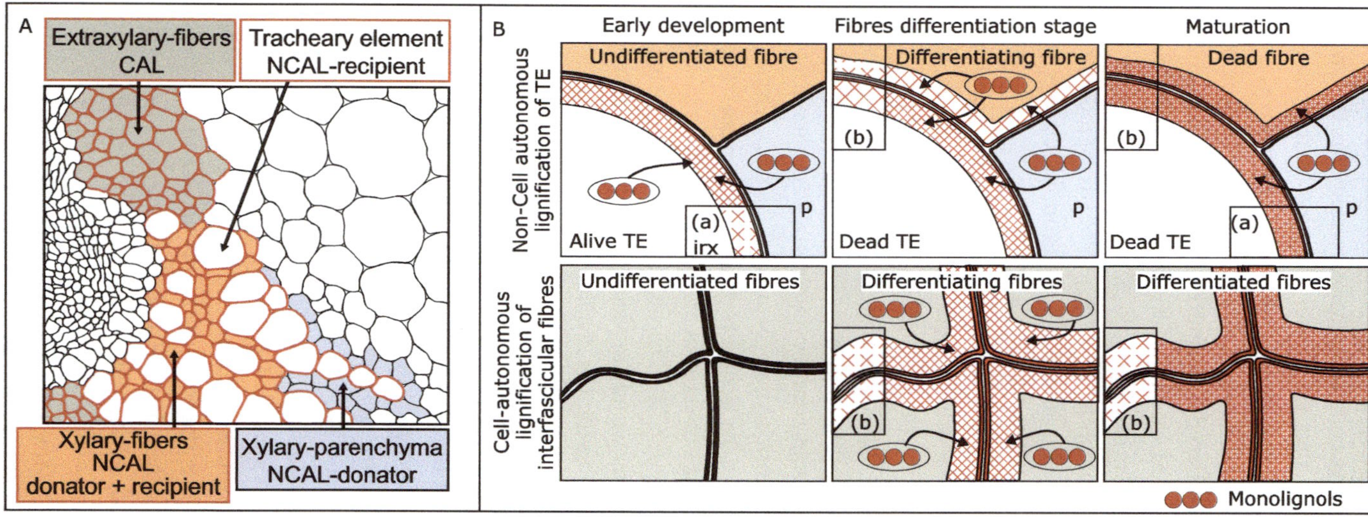

Fig. 3 Lignification in the inflorescence of *Arabidopsis thaliana.* (A) Cell population contributing to lignification. (B) Schematic representation of monolignol movement during non-cell autonomous lignification of tracheary elements and cell-autonomous lignification of interfascicular fibers during early development, fibers differentiation and maturation stages. (a) and (b): cell wall phenotype in *pAtPRX47::miRNA CCR1* and *pAtPRX64::miRNA CCR* plants, respectively. The density of the red crosses in the cell wall corresponds to the lignification degree. A loosened mesh represented by individually, non-connected crosses correspond to a low lignification profile and a tight mesh correspond to a mature profile of lignified cell wall. The panel B represents the biological process that has been deciphered thanks to the work of Smith et al. (2017). The small panels (a) and (b) represent the cell wall phenotypes of the mutants generated in Smith et al. (2017). CAL: Cell-autonomous lignification; NCAL: Non-cell autonomous lignification; p: parenchyma cell; TE: tracheary element.

the proper lignification of the tracheary elements (Smith et al., 2017). Indeed, this mutant displayed collapsed tracheary elements that were only slightly lignified suggesting that the vessel elements participate in their own lignification while they are alive, but they need assistance of other cells after their death. And finally, the use of the promoter AtPRX64 to specifically supress *CCR1* in the fibers demonstrated that xylary fibers also use the pool of monolignols provided by the xylary parenchyma cells, because at maturity, the vascular bundles displayed the same lignification level as the wild type, whereas the interfascicular fibers displayed strong reduction in lignin phloroglucinol staining and hence a cell-autonomous mechanism of lignification (Smith et al., 2017).

Lignification of vessel elements relies on both a cell autonomous and non-cell autonomous process during their lifetime (Smith et al., 2017, 2013), yet in *Arabidopsis* the tracheary elements die early (Ohashi-Ito et al., 2010), suggesting that their lignification is mainly performed by xylary-parenchyma and xylary fibers (Smith et al., 2017, 2013). Nonetheless, vessel CW still displays a specific lignin composition, mainly composed of G-units, different than fiber CW enriched in S-units (Chapple, Vogt, Ellis, & Somerville, 1992). This raises the question of the mechanism of regulation of lignification that enables the vessel elements to conserve their G-type lignin in an environment rich in S-type CW. This question is even more legitimate if we consider the case of a proper woody model plant like *Populus*. Recent analysis of poplar *ccr2* mutant expressing the *Arabidopsis* CCR1 under the control of the vessel and parenchyma ray cell specific promoter SNBE demonstrated that monolignols are able to migrate over multiple cell layers (De Meester et al., 2021). Furthermore, it was observed that the CW of vessel-neighboring cells, in the transition zone, display a CW composition more similar to vessel CW to that of fiber CW in fiber rich areas (Gorzsás et al., 2011). This suggests the occurrence of a regulatory mechanism of lignification to control the G-type composition of vessel CW.

A potential candidate for this regulation of lignification pattern of the dead tracheary element was first identified using the xylogenic cell culture of *Z. elegans*. In this system, expression of a *PIRIN (PRN)* gene was suppressed when the lignification of the tracheary elements was chemically repressed (Pesquet et al., 2013). A further characterization of the *PRN* genes demonstrated the expression of *PRN2* in vessel neighboring cells both in *Arabidopsis* and in *Populus* as well as the ability of the Arabidopsis PRN2 to repress the deposition of S-type lignin subunits in a non-cell-autonomous fashion (Zhang et al., 2020). This gene, specifically expressed in the vessel-neighboring cells, is not directly involved in the supply of monolignols, but it regulates (mainly by repression of *F5H*) their biosynthesis to maintain a vessel specific lignification profile (Zhang, Sztojka, et al., 2020).

Breakthrough works conducted by Smith et al. (2013, 2017) and the identification of *PRN2* suggest that the lignification of the xylem and the regulation of its lignin composition is a collective effort of the xylem cells, fine-tuned along the plant lifespan by the long-living cells of the xylem. Further characterization of the non-cell autonomous lignification mechanisms (transport/diffusion of monolignols; lifespan of the xylem cell types; availability and activity of the polymerizing enzymes) as well as the identification of new regulators of this process will certainly shed light on the spatial regulation of lignin monomer patterning in the plant vasculature of angiosperm and gymnosperm as well as its fine tuning along the plant lifetime.

As demonstrated with the interfascicular fibers of *Arabidopsis*, some cells contribute autonomously to their own lignification (Smith et al., 2017). This is also the case in the root endodermal cells, as shown by the lack of CS lignification in transgenic *Arabidopsis* plants where lignin biosynthesis was inhibited specifically in the endodermal cells by overexpression of *MYB4*, *MYB3* and *MYB32* (Andersen et al., 2021). However, the deposition of the CS lignin is regulated in a non-cell autonomous manner. The precise spatial deposition of lignin at the site of CS is mainly controlled by the CS membrane domain proteins (CASPs) (Roppolo et al., 2011) which predict the site of CS formation (Stöckle & Vermeer, 2020) at the CS membrane domain (CSD). CASPs organize into a scaffold structure, at the CSD, and gather the molecular actors required for the lignin deposition, including peroxidases64 and NADPH oxidase RBOHF (Lee, Rubio, Alassimone, & Geldner, 2013; Pfister et al., 2014; Rojas-Murcia et al., 2020) as well as the dirigent protein ESB1 (Hosmani et al., 2013). Other proteins, such as the *MYB36*-regulated Uclacyanin proteins (UCC1 and UCC2) (Reyt et al., 2020) and some laccase proteins (Rojas-Murcia et al., 2020; Zhuang, Zuo, Tao, Cai, & Li, 2020) are found in CSD but their function in CS formation is still unclear (Reyt et al., 2020; Stöckle & Vermeer, 2020; Zhuang et al., 2020).

Recent works have identified a surveillance system sensing the integrity of the hydrophobic barrier and participating in its complete lignification (Fig. 4) (Alassimone et al., 2016; Andersen et al., 2021; Doblas et al., 2017; Fujita et al., 2020; Nakayama et al., 2017; Reyt et al., 2021). This system is composed of the peptides CASPARIAN STRIP INTEGRITY FACTORS 1 and 2 (CIF1 and CIF2) (Doblas, Smakowska-Luzan, et al., 2017; Nakayama et al., 2017), the leucine-rich repeat receptor-like kinase

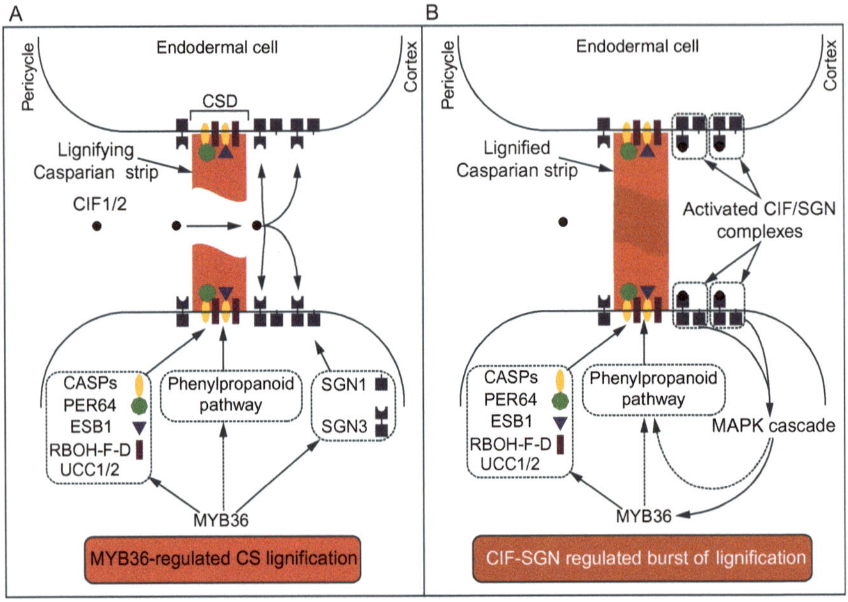

Fig. 4 Mechanism of CS lignification. (A) MYB36 initiates the differentiation of endodermal cells. It regulates expression of the machinery necessary for initiation of CS lignification. CASP proteins spatially localize PER64, ASB1, RBOH-F/D and UCC1/2 at the CSD for CS lignification. At this stage CIF1/2 proteins, expressed in the stele, can still diffuse through the CS and encounter the receptor-like kinase SGN3. (B) The direct interaction of CIF1/2 with SGN3 activates the protein SGN1 which initiates a phosphorylation cascade inducing ROS production as well as induction of gene expression, including MYB36. The complete sealing of the CS, obtained after the metabolic burst generated by activation of the CIF/SGN pathway, blocks the CIF peptides diffusion, thus inactivating the SCHENGEN pathway. Dot lines correspond to potential pathways.

SCHENGEN3 (SGN3) (Pfister et al., 2014) and the protein kinase SGN1 (Alassimone et al., 2016). The spatial organization of these three actors is particularly important for the CS integrity. The CIFs proteins are produced in the stele and migrate via apoplastic pathway to the endodermis (Fujita, 2021). SGN1 is a plasma membrane-anchored protein and SGN3 is a single membrane spanning receptor like kinase. Both are localized on the endodermal plasma membrane but while SGN3 is flanked to both sides of the CSD, SGN1 is only localized on the cortex side of the CSD. SGN3 is activated by the CIFs which induces the phosphorylation of SGN1 (Alassimone et al., 2016). The activation of this CIF-SGN3-SGN1 pathway induces the activation of RBOHF and D as well as the expression of downstream genes, leading to ROS production necessary for lignin polymerization (Alassimone et al., 2016; Fujita et al., 2020; Lee et al., 2013). The sealing of the CS, completed by the lignification boost brought by the SCHENGEN pathway, blocks the diffusion of the CIF proteins, which in turn inactivates the pathway (Andersen et al., 2021). CS lignification matches developmentally with the differentiation and lignification of the protoxylem (Reyt et al., 2021), suggesting that spatial and temporal control of lignification is coordinated in these two different tissues.

3.3 Post-translational modifications that regulate lignification

Post-Translational Modifications (PTM) are biochemical reactions performed on a protein, altering its properties. The range of reactions is vast (e.g., phosphorylation, cleavage, ubiquitination) and can influence several parameters, such as stability, location or binding affinity (Walsh, 2006). The involvement of the PTMs in the control of lignification has been demonstrated relatively recently and some of these modifications represent the last molecular events controlling the spatio-temporal deposition of lignin (Sulis & Wang, 2020).

Some lignin biosynthetic enzymes (PAL; AldOMT2) (Allwood, Davies, Gerrish, Ellis, & Bolwell, 1999; Wang et al., 2015) as well as some putative lignin-related peroxidases (PXP1–6 (Christensen, Bauw, Welinder, Van Montagu, & Boerjan, 1998); ZePRX33; 34; 44; 44; 70 (Gabaldón, López-Serrano, Pedreño, & Barceló, 2005; Gabaldón et al., 2006)) are subjected to PTMs (see Sulis & Wang, 2020). While the impact of these modifications onto the plant lignification is relatively low and/or still need to be clarified, some other PTMs have been demonstrated as of first importance in the regulation of lignin deposition. The second layer master switch regulator AtMYB46, as well as the loblolly pine (*Pinus teada*) ortholog PtMYB4 are functionally regulated by phosphorylation (Fig. 2) (Im et al., 2021; Morse et al., 2009). In both cases, the TF phosphorylation is performed by a mitogen-activated protein kinase 6 (MAPK6), resulting in a decrease in MYB protein activity *in planta*. However, the mechanism of the negative regulation by phosphorylation differs between *Arabidopsis* and *Pinus taeda*. In *Arabidopsis*, the phosphorylation is suggested to induce the degradation of AtMYB46 through the ubiquitin mediated proteasome pathway (Im et al., 2021) whereas in loblolly pine, the phosphorylation of PtMYB4 decreases its transcriptional activation activity without altering its DNA-binding ability (Morse et al., 2009). In loblolly pine, the phosphorylation of PtMYB4 has been suggested as temporally controlling the deposition and lignification of SCW. Indeed, kinase activities have been shown to depend on the developmental stage, dividing cells having higher kinase activity than the differentiating xylem elements. This suggests that in the early stage of xylogenesis, active MAPK6 participates in the inhibition of the PtMYB4 to restrict its activity whereas, later in the development, a decrease in MAPK activity releases PtMYB4 activity and induces SCW and lignin deposition (Morse et al., 2009; Sulis & Wang, 2020).

The phosphorylation of TF can also switch on the CW and lignin deposition program by inhibition of a repressor. This is the case for the lignin biosynthesis-associated transcription factor 1 (LTF1) in *Populus* which shares high sequence similarity with *Arabidopsis* MYB TFs from the group 4 (Gui et al., 2019). LTF1 is a transcriptional repressor, predominantly expressed in the xylem, that directly binds to the promoter of key lignin-biosynthetic genes. Gui et al. (2019) showed that phosphorylation negatively impacted the stability of LTF1 making it more susceptible to proteasome degradation.

Sumoylation of LBD30 (LOB Domain-containing protein 30) as well as S-nitrosylation of VND7 are other PTM that regulate the xylem cell differentiation program (Fig. 2), but do not directly regulate lignification. On the contrary, ubiquitination of PAL (Yu, Kim, Yun, Suh, & Lee, 2019) and CCR (Borah & Khurana, 2018) by Kelch domain containing F-box (KFB) has been reported to influence lignification through regulation of the protein turnover rate (Sulis & Wang, 2020). The effect of these PTM on the spatio-temporal control of lignification rather than on the overall lignification machinery still needs to be investigated.

Recently, the N-myristoylation of ADR (ANTHER DEHISCENCE REPRESSOR) has also been demonstrated as necessary for the control of ROS accumulation involved in the lignification of the *Arabidopsis* endothecium (Dai, Hsu, & Yang, 2019). *ADR* is specifically expressed in the anthers during the early anther development. Plant overexpressing ADR (*35S::ADR*) suffers from male sterility despite viable pollen grains. A close examination of the anthers demonstrated the absence of lignin in the CW thickening of the endothecium cells. This suggests that the overexpression of ADR inhibited the endothecium lignification, which is responsible for the male sterile phenotype. An estimation of ROS accumulation revealed a significant reduction of H_2O_2 accumulation, and the external application of H_2O_2 restored the dehiscence of the anthers. These observations suggest that ADR restricts the accumulation of ROS in the early anther development phase to prevent early anther dehiscence. Sequence analysis of the protein also revealed the occurrence of a N-myristoylation site. As most of the ROS species are concentrated into peroxisome membrane-bound organelles, the authors suggested that the N-myristoylation of ADR is necessary to its biological function corresponding to the restriction of lignification. This was validated by the overexpression of a mutated version of ADR without the N-myristoylation site (*35SS::ADR-Gly*). Indeed, contrary to the *35S::ADR*, the *35S::ADR-Gly* lines are not sterile and display normal anther development, which is explained by the inability of the overexpressed ADR protein to inhibit the ROS accumulation into the peroxisome because of the lack of N-myristoylation. In this very interesting example, the lignification is regulated at different levels. The place and time of ADR expression are very specific, suggesting a tight regulation by other molecular actors. Then, the PTM of the ADR protein regulates its cellular localization, which has been demonstrated to be a determinant in the regulation of the biological process. Finally, through the observed decreased expression of *NST1* and *NST2* in *35S::ADR*, the authors suggested a supplementary regulation of gene expression by the accumulation of H_2O_2, a process called retrograde signaling (Jung et al., 2013; Kimura, Yoshizumi, Manabe, Yamamoto, & Matsui, 2001; Yabuta, Maruta, Yoshimura, Ishikawa, & Shigeoka, 2004). All these differential layers of control serve the same purpose, the correct deposition of lignin in the endothecium at the correct time. The occurrence of such a highly complex and costly regulatory system of lignification is explained by the prime importance of a precise (in time and space) deposition of lignin that defines the reproductive success of the plant. Similarly to the endothecium and the endodermis, localization of ROS production as well as H_2O_2-generating enzymes has been considered as a mechanism spatially and temporally controlling xylem lignification throughout stem growth (Hoffmann, Benske, Betz, Schuetz, & Lacey Samuels, 2020; Lee et al., 2013; Schuetz et al., 2014; Yi Chou et al., 2018). Thereby, the occurrence of a similar system of control of ROS subcellular localization, correlating with the precise CW localization of lignin polymerizing enzymes (Hoffmann et al., 2020; Yi Chou et al., 2018) can be considered in the long living cells of the xylem tissues.

4. Environmental regulation of lignin deposition

The radial stem growth of temperate trees displays a typical tree ring structure that reflects the seasonal influence onto the cambial activity (Rathgeber, Cuny, & Fonti, 2016). This influence of seasonality is characterized in the structure of the wood by the presence of annual growth rings composed of two morphologically different types of wood, the early and the latewood. The earlywood is formed during the growing season when the climatic conditions are optimal, and characterized by a low density wood with large lumen cells and thin CW compare to the latewood (Arend & Fromm, 2007). The latewood, formed in late summer or autumn when the cambial activity declines, is characterized by cells with smaller lumen, smaller radial diameters and thicker CW than earlywood cells (Zimmerman & Brown, 1971). Initial bulk analysis of lignin composition reported a higher lignin content in earlywood than in latewood for several gymnosperms (Bergander, 2001; Bertaud & Holmbom, 2004; Gindl, 2001; Khattak & Bot, 1986), which has been explained by the high prevalence of the highly lignified middle lamella and primary CW in the earlywood compare to the latewood, mainly composed of the thick S2 layer of the SCW. Also, UV microspectroscopy analysis of the S2 layer of the SCW revealed a higher relative lignin content in earlywood cells than those of latewood cells, both in conifer and angiosperm species (Fergus, Procter, Scott, & Goring, 1969; Fredriksson, Pedersen, & Thygesen, 2018; Takabe, Miyauchi, Tsunoda, & Fukazawa, 1992). Furthermore, Takabe et al. (1992) observed a higher amount of G-moieties in latewood fibers compared to the earlywood fiber CW of Japanese beech.

In Norway spruce, the cambial reactivation is followed by expression of genes encoding CW and lignin biosynthesis as well as genes related to Program Cell Death (PCD; Jokipii-Lukkari et al., 2018). Interestingly, the transcripts related to monolignol biosynthesis accumulated beyond the period of cambial activity (Emiliani, Traversi, Anichini, Giachi, & Giovannelli, 2011; Jokipii-Lukkari et al., 2018). After a rapid decrease in gene expression in autumn, Jokipii-Lukkari et al. (2018) observed an expression peak of a cluster of highly co-expressed monolignol biosynthetic genes in midwinter. This winter boost of monolignol biosynthetic genes was however not correlated with an increase of lignin deposition, which might be explained by the absence of laccase and peroxidase expression during cambial dormancy, suggesting that lignification occurs during the active growth phase and latewood formation while monolignol biosynthesis can also occur during cambial dormancy (Jokipii-Lukkari et al., 2018).

4.1 The light

Little is known about the effect of light on lignification, Kimura et al. (2003) reported an increased expression of lignin-biosynthetic genes in *Arabidopsis* plants exposed to high light intensity. The diurnal expression of the genes coding for lignin-biosynthetic enzymes is modulated by three inter-related stimuli; the circadian clock, the availability in hexose carbohydrate and the light (Rogers et al., 2005). In buckwheat, shade tolerant plants displayed more lignin that the sensitive ones (Wang et al., 2015), seedlings of *Ebenus cretica* grown in the light displayed substantially more lignin than seedlings grown in the dark (Syros, Yupsanis, & Economou, 2005) and an increase of lignification in maize has been reported in the mesocotyl elongation zone following a blue light treatment (Schopfer, Lapierre, & Nolte, 2001).

The first molecular actors regulating the light-induced CW deposition and lignification were identified in *Arabidopsis* (Zhang et al., 2018). The CRYPTOCHROME1 (CRY1) is a blue light receptor and the *cry1* mutant shows thinner fiber CW than the wild type while the overexpressor displays thicker fiber CW. These phenotypes occur when the plants are grown in white and blue lights but not in red light, suggesting a specific impact of the blue light. Analysis of gene expression revealed induction of *NST1* expression by blue light in wild-type plants but not in the *cry1* mutant, suggesting that the blue light signal perceived by CRY1 activates SCW thickening through the expression of *NST1*. A yeast one-hybrid screening approach finally identified MYC2 and MYC4, two bHLH TFs, as potential direct activators of *NST1* and responsible for the blue light mediated SCW thickening and lignification in *Arabidopsis* inflorescence stem (Zhang et al., 2018).

Recently, an interesting light-controlled mechanism was identified in the regulation of homeostasis between anthocyanin and lignin biosynthesis in apple fruit peel (Hu et al., 2021). The mechanism relies on the suppression of *CCR* by miR7125 in a light-dependent manner whereby *miR7125* is activated by light-induced MdMYB1 and suppressed by dark-induced MdMYB16. Therefore, light conditions increase *miR7125* levels, and through suppression of *CCR,* decrease lignification in favor of anthocyanin accumulation, while darkness vice versa suppresses *miR7125* and hence stimulates lignification. The involvement of a miRNA was earlier demonstrated also in the regulation of a laccase gene (Wang et al., 2014; Zhao et al., 2015) and *TCP4* (*TEOSINTE BRANCHED1/CYCLOIDEA/PCF),* a TF promoting lignin biosynthesis in vascular tissues of *Arabidopsis* (Fig. 2) (Sun et al., 2017). Light-dependent induction of miRNAs might be part of a more general mechanism to regulate CW deposition and lignification during the seasonal variation of photo-period and light intensity.

4.2 The temperature

The temperature has a direct influence on the seasonal tree growth because the warmer temperature in the spring triggers bud break (Sarvas, 1972, 1974) but also because a rapid decrease in temperature induces latewood formation in artificially reactivated cambium of conifer trees (Begum et al., 2012). In Norway spruce, the rate of CW deposition has also been strongly and positively correlated with the temperature but only during latewood formation (Cuny & Rathgeber, 2016).

Recently, Crivellaro and Büntgen (2020) postulated that lignification, which is the last cellular event in xylogenesis, is a temperature-limited process. Numerous reports of trees growing in extreme cold condition display discontinuous pattern of lignification with concentric rings in the stem characterized by a low lignin content (Büntgen et al., 2015; Doležal, Dvorský, Börner, Wild, & Schweingruber, 2018; Piermattei, Crivellaro, Carrer, & Urbinati, 2015; Schenker, Lenz, Körner, & Hoch, 2014). Crivellaro and Büntgen (2020) suggested, using the description of seasonal developmental cycle of the snowbell, *Soldanella pusilla* made by Körner et al. (2019), that lignification was limited by extremely low temperature while the meristematic activity of cell division and thereby the biosynthesis and deposition of primary CW was still effective, even if relatively diminished. This concept of rate-limited lignification by low temperature is in accordance with previous observation of a lower lignification in the last layer of the latewood cells that lignify last when the temperature is low (Gindl, Grabner, & Wimmer, 2000). Despite a clear impact of temperature fluctuation onto the CW deposition and lignification, our understanding of the lignin-related molecular events triggered by temperature is still relatively scare.

Most of the studies related to the effect of temperature onto the regulation of lignification have been done on loquat fruit (*E. japonica*), a chilling-sensitive fruit which undergoes chilling-induced lignification of stone cells during storage at low temperature (Xu et al., 2019). Recently, several molecular actors of cold-induced lignification have been identified by using this model plant. The TFs EjMYB1 and EjMYB8 have been identified as activators of lignin biosynthesis in low temperature while EjMYB2 has been identified as a transcriptional repressor (Wang et al., 2016; Xu et al., 2014). EjMYB8 can interact with EjERF39 and synergistically activate *Ej4CL1* expression in low temperature (Zhang et al., 2020) while the interaction of EjMYB1 and EjMYB2 with EjAP2–1, an AP2/ERF TF, *trans*-represses the expression of *Ej4CL1* (Zeng et al., 2015). The TF EjNAC3, whose expression is induced at 0 °C can physically interact with a *EjCAD*-like gene, which could explain the resulting lignification of loquat fruit in low temperature (Ge et al., 2017). EjHAT1, an HD-ZIP TF, has been shown to repress the expression of *EjCAD5* during low-temperature conditioning, a treatment for reducing chilling injury symptoms such as fruit lignification (Xue et al., 2019). The TF EjAP2-1 can also form a ternary complex with the TFs EjbHLH1 and EjMYB2 to repress the expression of the Ej4CL1 gene induced by low temperature (Xu, Li, et al., 2019). These important studies on the impact of low temperature onto the regulation of loquat fruit could help to identify new regulatory processes of CW deposition and lignification in tree species.

4.3 Gravitropic stimulation

In response to gravitational stimulation, trees develop reaction wood (Alméras & Clair, 2016; Groover, 2016; Timell, 1986; Yamamoto, Yoshida, & Okuyama, 2002). In conifers, the reaction wood, called compression wood (CoW) is located on the underside of the leaning stems (Timell, 1986) whereas in angiosperm trees the reaction wood is called tension wood (TW) and located on the upper side of the leaning stems (Côté, Day, & Timell, 1969; Scurfield, 1973). Numerous studies have focused their attention on the physiological and molecular characterization of reaction wood (Donaldson & Radotic, 2013; Donaldson & Singh, 2016; Du & Yamamoto, 2007; Felten & Sundberg, 2013; Mellerowicz & Gorshkova, 2012; Timell, 1986), and recent studies have brought particularly interesting information concerning the regulation of lignification in this gravitropically induced wood (Hiraide et al., 2021; Liu et al., 2021).

The TW is mainly characterized by a reduced lignin content and an increased cellulose quantity (Aoyama, Matsumura, Tsutsumi, & Nishida, 2001; Côté et al., 1969; Gerttula et al., 2015; Wu, Joshi, & Chiang, 2000), and transcriptomic analyses demonstrated a decrease expression of lignin biosynthesis genes in TW compare to normal and opposite wood (wood facing the TW) (Andersson-Gunnerås et al., 2006; Chen, Chen, & Zhang, 2015). Recently, Liu, Liu, et al. (2021) identified two TFs in *Populus trichocarpa* (PtrHSFB3–1 and PtrMYB092) involved in the transcriptional reprogramming of lignin biosynthesis during TW formation. They are among the most transcriptionally repressed genes during tension wood formation. Transactivation analysis highlighted that PtrHSFB3–1 and PtrMYB092 directly transactivate 8 and 11 monolignol pathway genes, respectively. These two TFs act independently of the previously established transcriptional regulatory network associated with wood formation in *P. trichocarpa* (Chen et al., 2019) and might thus be part of a transcriptional network specifically involved in TW formation (Liu, Liu, et al., 2021).

Contrary to the TW, the CoW is characterized by an increased lignin content and an increased proportion of H-units (Brennan, McLean, Altaner, Ralph, & Harris, 2012; Donaldson, 2001; Timell, 1986; Zhang et al., 2017). Hiraide et al. (2021), demonstrated the contribution of the oxidative enzymes in the control of lignin composition in CoW development. The authors identified a differential localization pattern for two laccase isoforms with different substrate specificity during CoW formation in *Chamaecyparis obtusa*. CoLac1, a CoW specific laccase, displaying a substrate specificity for precursors of H-units, was localized in H-rich cell wall region of CoW whereas CoLac3, a laccase involved in G-lignin polymerization in normal wood and CoW was specifically localized in G-rich region of CoW (Hiraide et al., 2021).

These studies on regulation of lignification in reaction wood demonstrated two different mechanisms of spatial regulation of lignification; one based on localized transcriptional control of lignin biosynthetic enzymes, and one based on differential localization pattern of lignin polymerizing enzyme with different substrate specificities.

4.4 Hormone: Abscisic acid (ABA)

Our knowledge about the hormonal control of lignification during tree growth remains elusive. Recently, Liu, Yu, Rao, Li, and Dixon (2021), in a breakthrough paper, revealed the mechanism of an ABA hormonal regulation of CW and lignin deposition in *Arabidopsis* that could be relevant to latewood formation. ABA is typically defined as a stress hormone, and is considered as the central regulator of abiotic stresses in plants (Sah, Reddy, & Li, 2016). Accumulation of ABA triggers physiological responses that tend to increase the plant stress adaptation (Finkelstein, 2013; Ng, Melcher, Teh, & Xu, 2014). Among these physiological response, ABA has been reported to induce the expression of lignin biosynthesis genes and to increase the level of lignification (Jensen et al., 2010; Xu et al., 2020).

The plant detection of ABA releases the constitutive inhibition of Snf1 (Sucrose nonfermenting1)-Related Kinase (SnRK) protein kinases which regulate diverse physiological processes through phosphorylation. A yeast two-hybrid screen identified NST1 as a direct interactor of SnRK2 proteins and further characterization validated that SnRK2s are able to phosphorylate NST1. *Trans*-activation assays using a modified version of NST1 without phosphorylation site demonstrated the significance of this ABA-dependent post-translational activation of NST1 as a transcriptional regulator (Liu, Yu, et al., 2021). This study demonstrated that ABA can regulate the SCW formation at a transcript level, which is in accordance with previous results (Jensen et al., 2010; Xu et al., 2020), but also that the ABA-signaling pathway driven by the SnRK proteins activates the SCW program through post-translational modification of the master regulator NST1. This mechanistic demonstration along with the observed increased of ABA level during latewood formation (Jenkins & Shepherd, 1974; Wodzicki & Wodzicki, 1980) point out ABA accumulation as an important key driver of SCW deposition and lignification in the early- to latewood transition. A further characterization of the environmental stimuli driving the accumulation of ABA, as well as the ABA interaction with other biochemical signals (e.g., auxin) might be of valuable interest to improve the prediction of radial tree growth.

5. Conclusion

The vast functional spectra of the lignin polymer as well as the diversity of lignified tissues gave rise to a multitude of regulatory processes of lignification to cover the different functions of lignin.

Deposition of lignin in the heterogeneous tissues of xylem composed of different cell types, and of different developmental stages requires a coordinated control of lignification at the tissue scale, with dead tracheary element cells that are dependent on their neighboring cells for monolignol supply as well as for the fine tuning of their own lignification profile. On the contrary the lignification of tissues composed of a single cell type, as the interfascicular zone of *Arabidopsis* or the endodermis, has so far been only characterized as a cell autonomous process.

Recently, analyses of specialized lignified tissues have demonstrated their ability to uncover new molecular actors as well as new regulatory mechanisms of lignification. Study of stone cells lignification has identified numerous lignin-related TF and new interconnection between TF families (Ge et al., 2021). The investigations of CS lignification have been particularly informative about the mechanisms controlling precise spatial deposition of lignification at a micrometer-scale (Fujita et al., 2020; Lee et al., 2013). And, analysis of the regulation of the endothecium lignification unraveled a ROS-inhibitory based mechanism that precisely regulates the timing of lignin deposition (Dai et al., 2019).

These newly discovered actors and regulatory mechanisms of lignification, in conjunction with recent studies aimed to characterize the physico-mechanical properties of the lignin polymer depending on its molecular composition (Blaschek, Nuoendagula, Bacsik, Kajita, & Pesquet, 2020; Ménard et al., 2021), are together of prime importance for a better appreciation and understanding of the impact of lignification in plants.

Acknowledgments

The authors gratefully acknowledge Drs L. Samuels, R. Sibout, N. Hoffmann and S. Escamez for their helpful comments on the earlier version of the manuscript. The authors gratefully acknowledge Dr. Daria Chrobok (https://www.dariasciart.com/) for her contribution to the Figs. 1 and 3. This work is supported by the Umeå Plant Science Centre (UPSC)-Department of Forest Genetics and Plant Physiology and the Swedish research council Formas.

References

Agarwal, P. K., Shukla, P. S., Gupta, K., & Jha, B. (2013). Bioengineering for salinity tolerance in plants: State of the art. *Molecular Biotechnology*, *54*(1), 102–123.

Ahsan, N., Nakamura, T., & Komatsu, S. (2012). Differential responses of microsomal proteins and metabolites in two contrasting cadmium (cd)-accumulating soybean cultivars under cd stress. *Amino Acids*, *42*(1), 317–327.

Alassimone, J., Fujita, S., Doblas, V. G., Van Dop, M., Barberon, M., Kalmbach, L., et al. (2016). Polarly localized kinase SGN1 is required for Casparian strip integrity and positioning. *Nature Plants*, *2*(8), 1–10.

Alassimone, J., Naseer, S., & Geldner, N. (2010). A developmental framework for endodermal differentiation and polarity. *Proceedings of the National Academy of Sciences of the United States of America*, *107*(11), 5214–5219.

Algan, G., & Bakar Büyükkartal, H. N. (2000). Ultrastructure of seed coat development in the natural tetraploid *Trifolium pratense* L. *Journal of Agronomy and Crop Science*, *184*(3), 205–213.

Allwood, E. G., Davies, D. R., Gerrish, C., Ellis, B. E., & Bolwell, G. P. (1999). Phosphorylation of phenylalanine ammonia-lyase: Evidence for a novel protein kinase and identification of the phosphorylated residue. *FEBS Letters*, *457*(1), 47–52.

Alméras, T., & Clair, B. (2016). Critical review on the mechanisms of maturation stress generation in trees. *Journal of the Royal Society Interface*, *13*, 20160550, (Royal Society of London).

Alzagameem, A., Klein, S. E., Bergs, M., Do, X. T., Korte, I., Dohlen, S., et al. (2019). Antimicrobial activity of lignin and lignin-derived cellulose and chitosan composites against selected pathogenic and spoilage microorganisms. *Polymers*, *11*(4), 670–687.

Aminzadeh, S., Zhang, L., & Henriksson, G. (2017). A possible explanation for the structural inhomogeneity of lignin in LCC networks. *Wood Science and Technology*, *51*(6), 1365–1376.

Andersen, T. G., Molina, D., Kilian, J., Franke, R. B., Ragni, L., & Geldner, N. (2021). Tissue-autonomous phenylpropanoid production is essential for establishment of root barriers. *Current Biology*, *31*(5), 965–977.e5.

Andersson-Gunnerås, S., Mellerowicz, E. J., Love, J., Segerman, B., Ohmiya, Y., Coutinho, P. M., et al. (2006). Biosynthesis of cellulose-enriched tension wood in Populus: Global analysis of transcripts and metabolites identifies biochemical and developmental regulators in secondary wall biosynthesis. *Plant Journal, 45*(2), 144–165.

Aoyama, W., Matsumura, A., Tsutsumi, Y., & Nishida, T. (2001). Lignification and peroxidase in tension wood of Eucalyptus viminalis seedlings. *Journal of Wood Science, 47*(6), 419–424.

Arend, M., & Fromm, J. (2007). Seasonal change in the drought response of wood cell development in poplar. *Tree Physiology, 27*(7), 985–992.

Baayen, R. P., Ouellette, G. B., & Rioux, D. (1996). Compartmentalization of decay in carnations resistant to *fusarium oxysporum* f. sp. *dianthi. Phytopathology, 86*(10), 1018–1031.

Bacete, L., & Hamann, T. (2020). The role of mechanoperception in plant cell wall integrity maintenance. *Plants, 9*(5), E574.

Bacete, L., Mélida, H., Miedes, E., & Molina, A. (2018). Plant cell wall-mediated immunity: Cell wall changes trigger disease resistance responses. *Plant Journal, 93*(4), 614–636.

Baghdady, A., Blervacq, A. S., Jouanin, L., Grima-Pettenati, J., Sivadon, P., & Hawkins, S. (2006). *Eucalyptus gunnii CCR* and *CAD2* promoters are active in lignifying cells during primary and secondary xylem formation in *Arabidopsis thaliana. Elsevier, 44*(11 – 12), 674–683.

Barros, J., Serk, H., Granlund, I., & Pesquet, E. (2015). The cell biology of lignification in higher plants. *Annals of Botany, 115*(7), 1053–1074.

Barsberg, S. T., Lee, Y. I., & Rasmussen, H. N. (2018). Development of C-lignin with G/S-lignin and lipids in orchid seed coats - an unexpected diversity exposed by ATR-FT-IR spectroscopy. *Seed Science Research, 28*(1), 41–51.

Begum, S., Nakaba, S., Yamagishi, Y., Yamane, K., Islam, M. A., Oribe, Y., et al. (2012). A rapid decrease in temperature induces latewood formation in artificially reactivated cambium of conifer stems. *Annals of Botany, 110*(4), 875–885.

Bergander, A. (2001). *Local variability in chemical and physical properties of spruce wood fibers. (ProQuest Dissertations and Theses).*

Bertaud, F., & Holmbom, B. (2004). Chemical composition of earlywood and latewood in Norway spruce heartwood, sapwood and transition zone wood. *Wood Science and Technology, 38*(4), 245–256.

Bevan, M., Shufflebottom, D., Edwards, K., Jefferson, R., & Schuch, W. (1989). Tissue- and cell-specific activity of a phenylalanine ammonia-lyase promoter in transgenic plants. *EMBO Journal, 8*(7), 1899–1906.

Bhargava, A., Ahad, A., Wang, S., Mansfield, S. D., Haughn, G. W., Douglas, C. J., et al. (2013). The interacting MYB75 and KNAT7 transcription factors modulate secondary cell wall deposition both in stems and seed coat in Arabidopsis. *Planta, 237*(5), 1199–1211.

Bhargava, A., Mansfield, S. D., Hall, H. C., Douglas, C. J., & Ellis, B. E. (2010). MYB75 functions in regulation of secondary cell wall formation in the arabidopsis inflorescence stem. *Plant Physiology, 154*(3), 1428–1438.

Bhuiyan, N. H., Liu, W., Liu, G., Selvaraj, G., Wei, Y., & King, J. (2007). Transcriptional regulation of genes involved in the pathways of biosynthesis and supply of methyl units in response to powdery mildew attack and abiotic stresses in wheat. *Plant Molecular Biology, 64*(3), 305–318.

Bhuiyan, N. H., Selvaraj, G., Wei, Y., & King, J. (2009). Gene expression profiling and silencing reveal that monolignol biosynthesis plays a critical role in penetration defence in wheat against powdery mildew invasion. *Journal of Experimental Botany, 60*(2), 509–521.

Blaschek, L., Nuoendagula, Bacsik, Z., Kajita, S., & Pesquet, E. (2020). Determining the genetic regulation and coordination of lignification in stem tissues of Arabidopsis using semiquantitative Raman microspectroscopy. *ACS Sustainable Chemistry & Engineering, 8*(12), 4900–4909.

Blokhina, O., Laitinen, T., Hatakeyama, Y., Delhomme, N., Paasela, T., Zhao, L., et al. (2019). Ray parenchymal cells contribute to lignification of tracheids in developing xylem of Norway spruce. *Plant Physiology, 181*(4), 1552–1572.

Boerjan, W., Ralph, J., & Baucher, M. (2003). Lignin biosynthesis. *Annual Review of Plant Biology, 54*(1), 519–546.

Bollhöner, B., Prestele, J., & Tuominen, H. (2012). Xylem cell death: Emerging understanding of regulation and function. *Journal of Experimental Botany, 63*(3), 1081–1094.

Bonawitz, N. D., & Chapple, C. (2010). The genetics of lignin biosynthesis: Connecting genotype to phenotype. *Annual Review of Genetics, 44*, 337–363.

Bonner, L. J., & Dickinson, H. G. (1989). Anther dehiscence in *Lycopersicon esculentum* mill. I. Structural aspects. *New Phytologist, 113*(1), 97–115.

Bonnett, H. T. (1968). The root endodermis: Fine structure and function. *The Journal of Cell Biology, 37*(1), 199–205. (https://rupress.org/jcb/article-pdf/37/1/199/445615/199.pdf).

Borah, P., & Khurana, J. P. (2018). The OsFBK1 E3 ligase subunit affects anther and root secondary cell wall thickenings by mediating turnover of a cinnamoyl-CoA reductase. *Plant Physiology, 176*(3), 2148–2165.

Brennan, M., McLean, J. P., Altaner, C. M., Ralph, J., & Harris, P. J. (2012). Cellulose microfibril angles and cell-wall polymers in different wood types of Pinus radiata. *Cellulose, 19*(4), 1385–1404.

Büntgen, U., Hellmann, L., Tegel, W., Normand, S., Myers-Smith, I., Kirdyanov, A. V., et al. (2015). Temperature-induced recruitment pulses of Arctic dwarf shrub communities. *Journal of Ecology, 103*(2), 489–501.

Cabane, M., Afif, D., & Hawkins, S. (2012). Lignins and abiotic stresses. *Advances in Botanical Research, 61*, 219–262.

Cabané, M., Pireaux, J. C., Léger, E., Weber, E., Dizengremel, P., Pollet, B., et al. (2004). Condensed Lignins are synthesized in poplar leaves exposed to ozone. *Plant Physiology, 134*(2), 586–594.

Cai, Y., Li, G., Nie, J., Lin, Y., Nie, F., Zhang, J., et al. (2010). Study of the structure and biosynthetic pathway of lignin in stone cells of pear. *Scientia Horticulturae, 125*(3), 374–379.

Cano-Delgado, A. I., Metzlaff, K., & Bevan, M. W. (2000). The eli1 mutation reveals a link between cell expansion and secondary cell wall formation in Arabidopsis thaliana. *Development, 127*(15), 3395–3405.

Capeleti, I., Bonini, E. A., Ferrarese, M. D. L. L., Teixeira, A. C. N., Krzyzanowski, F. C., & Ferrarese-Filho, O. (2005). Lignin content and peroxidase activity in soybean seed coat susceptible and resistant to mechanical damage. *Acta Physiologiae Plantarum, 27*(1), 103–108.

Caspary, R. (1865). Bemerkungen über die Schutzscheide und die Bildung des Stammes und der Wurzel. *Jahrbücher Für Wissenschaftliche Botanik, 4*, 101–124.

Chapple, C. C. S., Vogt, T., Ellis, B. E., & Somerville, C. R. (1992). An arabidopsis mutant defective in the general phenylpropanoid pathway. *Plant Cell, 4*(11), 1413–1424.

Chen, J., Chen, B., & Zhang, D. (2015). Transcript profiling of Populus tomentosa genes in normal, tension, and opposite wood by RNA-seq. *BMC Genomics, 16*(1), 1–16.

Chen, C., Meyermans, H., Burggraeve, B., De Rycke, R. M., Inoue, K., De Vleesschauwer, V., et al. (2000). Cell-specific and conditional expression of caffeoyl-coenzyme A-3-O-methyltransferase in poplar. *Plant Physiology, 123*(July), 853–867.

Chen, F., Tobimatsu, Y., Havkin-Frenkel, D., Dixon, R. A., & Ralph, J. (2012). A polymer of caffeyl alcohol in plant seeds. *Proceedings of the National Academy of Sciences of the United States of America, 109*(5), 1772–1777.

Chen, H., Wang, J. P., Liu, H., Li, H., Lin, Y. C. J., Shi, R., et al. (2019). Hierarchical transcription factor and chromatin binding network for wood formation in populus trichocarpa. *Plant Cell, 31*(3), 602–626.

Chezem, W. R., Memon, A., Li, F. S., Weng, J. K., & Clay, N. K. (2017). SG2-type R2R3-MYB transcription factor MYB15 controls defense-induced lignification and basal immunity in *Arabidopsis*. *Plant Cell*, *29*(8), 1907–1926.
Christensen, J. H., Bauw, G., Welinder, K. G., Van Montagu, M., & Boerjan, W. (1998). Purification and characterization of peroxidases correlated with lignification in poplar xylem. *Plant Physiology*, *118*(1), 125–135.
Côté, W. A., Day, A. C., & Timell, T. E. (1969). A contribution to the ultrastructure of tension wood fibers. *Wood Science and Technology*, *3*(4), 257–271.
Crivellaro, A., & Büntgen, U. (2020). New evidence of thermally constrained plant Cell Wall lignification. *Trends in Plant Science*, *25*(4), 322–324. Elsevier Ltd.
Cuny, H. E., & Rathgeber, C. B. K. (2016). Xylogenesis: Coniferous trees of temperate forests are listening to the climate tale during the growing season but only remember the last words! *Plant Physiology*, *171*(1), 306–317.
Dai, S. Y., Hsu, W. H., & Yang, C. H. (2019). The gene *ANTHER DEHISCENCE REPRESSOR (ADR)* controls male fertility by suppressing the ROS accumulation and anther Cell Wall thickening in *Arabidopsis*. *Scientific Reports*, *9*(1), 1–14.
Dawson, J., Sözen, E., Vizir, I., Van Waeyenberge, S., Wilson, Z. A., & Mulligan, B. J. (1999). Characterization and genetic mapping of a mutation (*ms35*) which prevents anther dehiscence in *Arabidopsis thaliana* by affecting secondary wall thickening in the endothecium. *New Phytologist*, *144*(2), 213–222.
De Meester, B., Vanholme, R., de Vries, L., Wouters, M., Van Doorsselaere, J., & Boerjan, W. (2021). Vessel- and ray-specific monolignol biosynthesis as an approach to engineer fiber-hypolignification and enhanced saccharification in poplar. *Plant Journal*, *108*(3), 752–765.
Del Río, J. C., Rencoret, J., Gutiérrez, A., Elder, T., Kim, H., & Ralph, J. (2020). Lignin monomers from beyond the canonical monolignol biosynthetic pathway: Another brick in the wall. *ACS Sustainable Chemistry & Engineering*, *8*(13), 4997–5012. American Chemical Society.
Delevoryas, T., & Esau, K. (1960). Anatomy of seed plants. *AIBS Bulletin*, *10*(3), 40.
Demirbas, A. (2004). Adsorption of lead and cadmium ions in aqueous solutions onto modified lignin from alkali glycerol delignication. *Journal of Hazardous Materials*, *109*(1–3), 221–226.
Denness, L., McKenna, J. F., Segonzac, C., Wormit, A., Madhou, P., Bennett, M., et al. (2011). Cell wall damage-induced lignin biosynthesis is regulated by a reactive oxygen species- and jasmonic acid-dependent process in arabidopsis. *Plant Physiology*, *156*(3), 1364–1374.
Doblas, V. G., Geldner, N., & Barberon, M. (2017). The endodermis, a tightly controlled barrier for nutrients. *Current Opinion in Plant Biology*, *39*, 136–143.
Doblas, V. G., Smakowska-Luzan, E., Fujita, S., Alassimone, J., Barberon, M., Madalinski, M., et al. (2017). Root diffusion barrier control by a vasculature-derived peptide binding to the SGN3 receptor. *Science*, *355*(6322), 280–284.
Doležal, J., Dvorský, M., Börner, A., Wild, J., & Schweingruber, F. H. (2018). Anatomy, age and ecology of High Mountain plants in Ladakh, the Western Himalaya. In *Anatomy, age and ecology of High Mountain plants in Ladakh, the Western Himalaya*.
Donaldson, L. A. (2001). Lignification and lignin topochemistry - an ultrastructural view. *Phytochemistry*, *57*(6), 859–873.
Donaldson, L. A., & Radotic, K. (2013). Fluorescence lifetime imaging of lignin autofluorescence in normal and compression wood. *Journal of Microscopy*, *251*(2), 178–187.
Donaldson, L. A., & Singh, A. P. (2016). Reaction wood. In *Secondary xylem biology: Origins, functions, and applications* (pp. 93–110).
Doster, M. A., & Michailides, T. J. (1999). Relationship between shell discoloration of pistachio nuts and incidence of fungal decay and insect infestation. *Plant Disease*, *83*(3), 259–264.

Du, S., & Yamamoto, F. (2007). An overview of the biology of reaction wood formation. *Journal of Integrative Plant Biology*, *49*(2), 131–143.
Ellis, C., Karafyllidis, I., Wasternack, C., & Turner, J. G. (2002). The Arabidopsis mutant cev1 links cell wall signaling to jasmonate and ethylene responses. *Plant Cell*, *14*(7), 1557–1566.
Ellis, C., & Turner, J. G. (2001). The Arabidopsis mutant cev1 has constitutively active jasmonate and ethylene signal pathways and enhanced resistance to pathogens. *Plant Cell*, *13*(5), 1025–1033.
Emiliani, G., Traversi, M. L., Anichini, M., Giachi, G., & Giovannelli, A. (2011). Transcript accumulation dynamics of phenylpropanoid pathway genes in the maturing xylem and phloem of *Picea abies* during latewood formation. *Journal of Integrative Plant Biology*, *53*(10), 783–799.
Engelsdorf, T., Gigli-Bisceglia, N., Veerabagu, M., McKenna, J. F., Vaahtera, L., Augstein, F., et al. (2018). The plant cell wall integrity maintenance and immune signaling systems cooperate to control stress responses in Arabidopsis thaliana. *Science Signaling*, *11*(536), 3070.
Escamez, S., Terryn, C., Gandla, M. L., Yassin, Z., Scheepers, G., Näsholm, T., et al. (2021). Fluorescence lifetime imaging as an in situ and label-free readout for the chemical composition of lignin. *BioRxiv* (457748).
Felten, J., & Sundberg, B. (2013). Biology, chemistry and structure of tension wood. *Plant Cell Monographs*, *20*, 203–224.
Fergus, B. J., & Goring, D. A. I. (1970a). The distribution of lignin in birch wood as determined by ultraviolet microscopy. *Holzforschung*, *24*(4), 118–124.
Fergus, B. J., & Goring, D. A. I. (1970b). The location of Guaiacyl and Syringyl Lignins in birch xylem tissue. *Holzforschung*, *24*(4), 113–117.
Fergus, B. J., Procter, A. R., Scott, J. A. N., & Goring, D. A. I. (1969). The distribution of lignin in sprucewood as determined by ultraviolet microscopy. *Wood Science and Technology*, *3*(2), 117–138.
Ferrándiz, C. (2002). Regulation of fruit dehiscence in *Arabidopsis*. *Journal of Experimental Botany*, *53*(377), 2031–2038.
Finkelstein, R. (2013). Abscisic acid synthesis and response. *The Arabidopsis Book*, *11*, e0166.
Francoz, E., Lepiniec, L., & North, H. M. (2018). Seed coats as an alternative molecular factory: Thinking outside the box. *Plant Reproduction*, *31*(3), 327–342. Springer Verlag.
Franke, R., Humphreys, J. M., Hemm, M. R., Denault, J. W., Ruegger, M. O., Cusumano, J. C., et al. (2002). The Arabidopsis *REF8* gene encodes the 3-hydroxylase of phenylpropanoid metabolism. *Plant Journal*, *30*(1), 33–45.
Fredriksson, M., Pedersen, N. B., & Thygesen, L. G. (2018). The cell wall composition of Norway spruce earlywood and latewood revisited. *International Wood Products Journal*, *9*(2), 80–85.
Fujita, S. (2021). CASPARIAN STRIP INTEGRITY FACTOR (CIF) family peptides - regulator of plant extracellular barriers. *Peptides*, *143*.
Fujita, S., De Bellis, D., Edel, K. H., Köster, P., Andersen, T. G., Schmid-Siegert, E., et al. (2020). SCHENGEN receptor module drives localized ROS production and lignification in plant roots. *The EMBO Journal*, *39*, e103894.
Gabaldón, C., López-Serrano, M., Pedreño, M. A., & Barceló, A. R. (2005). Cloning and molecular characterization of the basic peroxidase isoenzyme from *Zinnia elegans*, an enzyme involved in lignin biosynthesis. *Plant Physiology*, *139*(3), 1138–1154.
Gabaldón, C., López-Serrano, M., Pomar, F., Merino, F., Cuello, J., Pedreño, M. A., et al. (2006). Characterization of the last step of lignin biosynthesis in *Zinnia elegans* suspension cell cultures. *FEBS Letters*, *580*(18), 4311–4316.
Ge, H., Shi, Y. N., Zhang, M. X., Li, X., Yin, X. R., & Chen, K. S. (2021). The MADS-box transcription factor *EjAGL65* controls loquat flesh lignification *via* direct transcriptional inhibition of *EjMYB8*. *Frontiers in Plant Science*, *12*, 652959.

Ge, H., Zhang, J., Zhang, Y. J., Li, X., Yin, X. R., Grierson, D., et al. (2017). *EjNAC3* transcriptionally regulates chilling-induced lignification of loquat fruit via physical interaction with an atypical *CAD*-like gene. *Journal of Experimental Botany, 68*(18), 5129–5136.
Geng, P., Zhang, S., Liu, J., Zhao, C., Wu, J., Cao, Y., et al. (2020). MYB20, MYB42, MYB43, and MYB85 regulate phenylalanine and lignin biosynthesis during secondary cell wall formation. *Plant Physiology, 182*(3), 1272–1283.
Gerttula, S., Zinkgraf, M., Muday, G. K., Lewis, D. R., Ibatullin, F. M., Brumer, H., et al. (2015). Transcriptional and hormonal regulation of gravitropism of woody stems in populus. *Plant Cell, 27*(10), 2800–2813.
Gindl, W. (2001). The effect of varying latewood proportion on the radial distribution of lignin content in a pine stem. *Holzforschung, 55*(5), 455–458.
Gindl, W., Grabner, M., & Wimmer, R. (2000). The influence of temperature on latewood lignin content in treeline Norway spruce compared with maximum density and ring width. *Trees - Structure and Function, 14*(7), 409–414.
Gorzsás, A., Stenlund, H., Persson, P., Trygg, J., & Sundberg, B. (2011). Cell-specific chemotyping and multivariate imaging by combined FT-IR microspectroscopy and orthogonal projections to latent structures (OPLS) analysis reveals the chemical landscape of secondary xylem. *Plant Journal, 66*(5), 903–914.
Groover, A. (2016). Gravitropisms and reaction woods of forest trees—Evolution, functions and mechanisms. *The New Phytologist, 211*(3), 790–802, Blackwell Publishing Ltd.
Gui, J., Luo, L., Zhong, Y., Sun, J., Umezawa, T., & Li, L. (2019). Phosphorylation of LTF1, an MYB transcription factor in Populus, acts as a sensory switch regulating lignin biosynthesis in wood cells. *Molecular Plant, 12*(10), 1325–1337.
Guo, X., Zhang, S., Shan, X., & quan. (2008). Adsorption of metal ions on lignin. *Journal of Hazardous Materials, 151*(1), 134–142.
Ha, C. M., Escamilla-Trevino, L., Yarce, J. C. S., Kim, H., Ralph, J., Chen, F., et al. (2016). An essential role of caffeoyl shikimate esterase in monolignol biosynthesis in *Medicago truncatula. The Plant Journal : For Cell and Molecular Biology, 86*(5), 363–375.
Ha, C. M., Fine, D., Bhatia, A., Rao, X., Martin, M. Z., Engle, N. L., et al. (2019). Ectopic defense gene expression is associated with growth defects in *Medicago truncatula* lignin pathway mutants. *Plant Physiology, 181*(1), 63–84.
Haas, D. L., & Carothers, Z. B. (1975). Some ultrastructural observations on endodermal cell development in *Zea mays* roots. *American Journal of Botany, 62*(4), 336–348. https://bsapubs.onlinelibrary.wiley.com/doi/abs/10.1002/j.1537-2197.1975.tb14056.x.
Hacke, U. G., Sperry, J. S., Pockman, W. T., Davis, S. D., & McCulloh, K. A. (2001). Trends in wood density and structure are linked to prevention of xylem implosion by negative pressure. *Oecologia, 126*(4), 457–461.
Hamann, T., Bennett, M., Mansfield, J., & Somerville, C. (2009). Identification of cell-wall stress as a hexose-dependent and osmosensitive regulator of plant responses. *Plant Journal, 57*(6), 1015–1026.
Hao, Z., Avci, U., Tan, L., Zhu, X., Glushka, J., Pattathil, S., et al. (2014). Loss of Arabidopsis *GAUT12/IRX8* causes anther indehiscence and leads to reduced G lignin associated with altered matrix polysaccharide deposition. *Frontiers in Plant Science, 5*(JUL).
Hauffe, K. D., Paszkowski, U., Schulze-Lefert, P., Hahlbrock, K., Dangl, J. L., & Douglas, C. J. (1991). A parsley 4CL-1 promoter fragment specifies complex expression patterns in transgenic tobacco. *Plant Cell, 3*(5), 435–443.
Hawkins, S., & Boudet, A. (2003). "Defence lignin" and hydroxycinnamyl alcohol dehydrogenase activities in wounded *Eucalyptus gunnii. Forest Pathology, 33*(2), 91–104.
Hay, A. S., Pieper, B., Cooke, E., Mandáková, T., Cartolano, M., Tattersall, A. D., et al. (2014). *Cardamine hirsuta*: A versatile genetic system for comparative studies. *Plant Journal, 78*(1), 1–15.

Hiraide, H., Tobimatsu, Y., Yoshinaga, A., Lam, P. Y., Kobayashi, M., Matsushita, Y., et al. (2021). Localised laccase activity modulates distribution of lignin polymers in gymnosperm compression wood. *New Phytologist, 230*(6), 2186–2199.

Hoffmann, N., Benske, A., Betz, H., Schuetz, M., & Lacey Samuels, A. (2020). Laccases and peroxidases co-localize in lignified secondary cell walls throughout stem development. *Plant Physiology, 184*(2), 806–822.

Hoffmann, L., Besseau, S., Geoffroy, P., Ritzenthaler, C., Meyer, D., Lapierre, C., et al. (2004). Silencing of hydroxycinnamoyl-coenzyme A shikimate/quinate hydroxycinnamoyltransferase affects phenylpropanoid biosynthesis. *Plant Cell, 16*(6), 1446–1465.

Hofhuis, H., Moulton, D., Lessinnes, T., Goriely, A., Smith, R., & Hay Correspondence, A. (2016). Morphomechanical innovation drives explosive seed dispersal. *Cell, 166*, 222–233.

Horbens, M., Branke, D., Gärtner, R., Voigt, A., Stenger, F., & Neinhuis, C. (2015). Multi-scale simulation of plant stem reinforcement by brachysclereids: A case study in apple fruit peduncles. *Journal of Structural Biology, 192*(1), 116–126.

Hosmani, P. S., Kamiya, T., Danku, J., Naseer, S., Geldner, N., Guerinot, M. L., et al. (2013). Dirigent domain-containing protein is part of the machinery required for formation of the lignin-based Casparian strip in the root. *Proceedings of the National Academy of Sciences of the United States of America, 110*(35), 14498–14503.

Hosokawa, M., Suzuki, S., Umezawa, T., & Sato, Y. (2001). Progress of lignification mediated by intercellular transportation of monolignols during tracheary element differentiation of isolated *Zinnia* mesophyll cells. *Plant and Cell Physiology, 42*(9), 959–968.

Hu, Y., Cheng, H., Zhang, Y., Zhang, J., Niu, S., Wang, X., et al. (2021). The MdMYB16/MdMYB1-miR7125-MdCCR module regulates the homeostasis between anthocyanin and lignin biosynthesis during light induction in apple. *New Phytologist, 231*(3), 1105–1122.

Hu, W. J., Harding, S. A., Lung, J., Popko, J. L., Ralph, J., Stokke, D. D., et al. (1999). Repression of lignin biosynthesis promotes cellulose accumulation and growth in transgenic trees. *Nature Biotechnology, 17*(8), 808–812.

Hussey, S. G., Mizrachi, E., Creux, N. M., & Myburg, A. A. (2013). Navigating the transcriptional roadmap regulating plant secondary cell wall deposition. *Frontiers in Plant Science, 4*(AUG).

Im, J. H., Ko, J. H., Kim, W. C., Crain, B., Keathley, D., & Han, K. H. (2021). Mitogen-activated protein kinase 6 negatively regulates secondary wall biosynthesis by modulating MYB46 protein stability in *Arabidopsis thaliana*. *PLoS Genetics, 17*(4).

Jarvis, M. C. (2012). Sclerenchyma. In *eLS* John Wiley & Sons, Ltd.

Jenkins, P. A., & Shepherd, K. R. (1974). Seasonal changes in levels of indole-acetic acid and abscisic acid in stem tissues of Pinus radiata. *New Zealand Journal of Forestry Science, 4*(3), 511–519.

Jensen, M. K., Kjaersgaard, T., Nielsen, M. M., Galberg, P., Petersen, K., O'shea, C., et al. (2010). The *Arabidopsis thaliana* NAC transcription factor family: Structure-function relationships and determinants of ANAC019 stress signalling. *Biochemical Journal, 426*(2), 183–196.

Jin, H. (2000). Transcriptional repression by AtMYB4 controls production of UV-protecting sunscreens in *Arabidopsis*. *The EMBO Journal, 19*(22), 6150–6161.

Jokipii-Lukkari, S., Delhomme, N., Schiffthaler, B., Mannapperuma, C., Prestele, J., Nilsson, O., et al. (2018). Transcriptional roadmap to seasonal variation in wood formation of Norway spruce. *Plant Physiology, 176*(4), 2851–2870.

Jones, L., Ennos, A. R., & Turner, S. R. (2001). Cloning and characterization of irregular xylem4 (irx4): A severely lignin-deficient mutant of *Arabidopsis*. *Plant Journal, 26*(2), 205–216.

Jung, K. W., Kim, Y. Y., Yoo, K. S., Ok, S. H., Cui, M. H., Jeong, B. C., et al. (2013). A cystathionine-β-synthase domain-containing protein, CBSX2, regulates endothecial secondary cell wall thickening in anther development. *Plant and Cell Physiology*, *54*(2), 195–208.

Kamiya, T., Borghi, M., Wang, P., Danku, J. M. C., Kalmbach, L., Hosmani, P. S., et al. (2015). The MYB36 transcription factor orchestrates Casparian strip formation. *Proceedings of the National Academy of Sciences of the United States of America*, *112*(33), 10533–10538.

Keijzer, C. J. (1987). The processes of anther dehiscence and pollen dispersal: ii. The formation and the transfer mechanism of pollenkitt, cell-wall development of the loculus tissues and a function of orbicules in pollen dispersal. *New Phytologist*, *105*(3), 499–507.

Khattak, T., & Bot, A. M. (1986). Estimation of lignin, holocellulose and alphacellulose content of earlywood and latewood among innerwood and and outerwood of blue pine, (*Pinus wallichiana*, A.B.Jacks.). *Pakistan Journal of Botany*, *18*(2), 235–241.

Kim, S. H., Lam, P. Y., Lee, M. H., Jeon, H. S., Tobimatsu, Y., & Park, O. K. (2020). The *Arabidopsis* R2R3 MYB transcription factor MYB15 is a key regulator of lignin biosynthesis in effector-triggered immunity. *Frontiers in Plant Science*, *11*, 583153.

Kimura, M., Yamamoto, Y. Y., Seki, M., Sakurai, T., Sato, M., Abe, T., et al. (2003). Identification of *Arabidopsis* genes regulated by high light-stress using cDNA microarray. *Photochemistry and Photobiology*, *77*(2), 226–233.

Kimura, M., Yoshizumi, T., Manabe, K., Yamamoto, Y. Y., & Matsui, M. (2001). Arabidopsis transcriptional regulation by light stress via hydrogen peroxide-dependent and -independent pathways. *Genes to Cells*, *6*(7), 607–617.

Körner, C., Riedl, S., Keplinger, T., Richter, A., Wiesenbauer, J., Schweingruber, F., et al. (2019). Life at 0 °C: The biology of the alpine snowbed plant *Soldanella pusilla*. *Alpine Botany*, *129*(2), 63–80.

Kosma, D. K., Murmu, J., Razeq, F. M., Santos, P., Bourgault, R., Molina, I., et al. (2014). AtMYB41 activates ectopic suberin synthesis and assembly in multiple plant species and cell types. *Plant Journal*, *80*(2), 216–229.

Kováčik, J., & Bačkor, M. (2007). Phenylalanine ammonia-lyase and phenolic compounds in chamomile tolerance to cadmium and copper excess. *Water, Air, and Soil Pollution*, *185*(1–4), 185–193.

Kroemer, K. (1903). Wurzelhaut, Hypodermis und Endodermis der Angiospermenwurzel. *Bibliotheca Botanica*, *59*, 1–148. (http://kbd.kew.org/kbd/detailedresult.do?id=2978).

Kubo, M., Udagawa, M., Nishikubo, N., Horiguchi, G., Yamaguchi, M., Ito, J., et al. (2005). Transcription switches for protoxylem and metaxylem vessel formation. *Genes and Development*, *19*(16), 1855–1860.

Lacombe, E., Van Doorsselaere, J., Boerjan, W., Boudet, A. M., & Grima-Pettenati, J. (2000). Characterization of cis-elements required for vascular expression of the cinnamoyl CoA reductase gene and for protein-DNA complex formation. *Plant Journal*, *23*(5), 663–676.

Landucci, L., Smith, R. A., Liu, S., Karlen, S. D., & Ralph, J. (2020). Eudicot nutshells: Cell-wall composition and biofuel feedstock potential. *Energy and Fuels*, *34*(12), 16274–16283.

Lange, B. M., Lapierre, C., & Sandermann, H. (1995). Elicitor-induced spruce stress lignin: Structural similarity to early developmental lignins. *Plant Physiology*, *108*(3), 1277–1287.

Lauvergeat, V., Rech, P., Jauneau, A., Guez, C., Coutos-Thevenot, P., & Grima-Pettenati, J. (2002). The vascular expression pattern directed by the Eucalyptus gunnii cinnamyl alcohol dehydrogenase *EgCAD2* promoter is conserved among woody and herbaceous plant species. *Plant Molecular Biology*, *50*(3), 497–509. https://link.springer.com/article/10.1023/A:1019817913604.

Lee, M., Jeon, H. S., Kim, S. H., Chung, J. H., Roppolo, D., Lee, H., et al. (2019). Lignin-based barrier restricts pathogens to the infection site and confers resistance in plants. *The EMBO Journal, 38*, e101948.

Lee, Y., Rubio, M. C., Alassimone, J., & Geldner, N. (2013). A mechanism for localized lignin deposition in the endodermis. *Cell, 153*(2), 402–412.

Lee, Y., Yoon, T. H., Lee, J., Jeon, S. Y., Lee, J. H., Lee, M. K., et al. (2018). A lignin molecular brace controls precision processing of cell walls critical for surface integrity in *Arabidopsis. Cell, 173*(6), 1468–1480 (e9).

Leplé, J. C., Dauwe, R., Morreel, K., Storme, V., Lapierre, C., Pollet, B., et al. (2007). Downregulation of cinnamoyl-coenzyme a reductase in poplar: Multiple-level phenotyping reveals effects on cell wall polymer metabolism and structure. *Plant Cell, 19*(11), 3669–3691.

Li, L., Cheng, X. F., Leshkevich, J., Umezawa, T., Harding, S. A., & Chiang, V. L. (2001). The last step of syringyl monolignol biosynthesis in angiosperms is regulated by a novel gene encoding sinapyl alcohol dehydrogenase. *Plant Cell, 13*(7), 1567–1585.

Li, E., Wang, S., Liu, Y., Chen, J. G., & Douglas, C. J. (2011). OVATE FAMILY PROTEIN4 (OFP4) interaction with KNAT7 regulates secondary cell wall formation in Arabidopsis thaliana. *Plant Journal, 67*(2), 328–341.

Liang, M., Davis, E., Gardner, D., Cai, X., & Wu, Y. (2006). Involvement of AtLAC15 in lignin synthesis in seeds and in root elongation of *Arabidopsis. Planta, 224*(5), 1185–1196.

Liljegren, S. J., Ditta, G. S., Eshed, Y., Savidge, B., Bowmant, J. L., & Yanofsky, M. F. (2000). *SHATTERPROOF* MADS-box genes control dispersal in *Arabidopsis. Nature, 404*(6779), 766–770.

Liljegren, S. J., Roeder, A. H. K., Kempin, S. A., Gremski, K., Østergaard, L., Guimil, S., et al. (2004). Control of fruit patterning in *Arabidopsis* by INDEHISCENT. *Cell, 116*(6), 843–853.

Lionetti, V., Giancaspro, A., Fabri, E., Giove, S. L., Reem, N., Zabotina, O. A., et al. (2015). Cell wall traits as potential resources to improve resistance of durum wheat against *fusarium gramineanum. BMC Plant Biology, 15*(1).

Liu, B., Liu, J., Yu, J., Wang, Z., Sun, Y., Li, S., et al. (2021). Transcriptional reprogramming of xylem cell wall biosynthesis in tension wood. *Plant Physiology, 186*(1), 250–269.

Liu, Y., You, S., Taylor-Teeples, M., Li, W. L., Schuetz, M., Brady, S. M., et al. (2014). BEL1-LIKE HOMEODOMAIN6 and KNOTTED ARABIDOPSIS THALIANA7 interact and regulate secondary cell wall formation via repression of REVOLUTA. *Plant Cell, 26*(12), 4843–4861.

Liu, C., Yu, H., Rao, X., Li, L., & Dixon, R. A. (2021). Abscisic acid regulates secondary cell-wall formation and lignin deposition in *Arabidopsis thaliana* through phosphorylation of NST1. *Proceedings of the National Academy of Sciences of the United States of America, 118*(5), e2010911118.

Marles, M. A. S., & Gruber, M. Y. (2004). Histochemical characterisation of unextractable seed coat pigments and quantification of extractable lignin in the Brassicaceae. *Journal of the Science of Food and Agriculture, 84*(3), 251–262.

McCarthy, R. L., Zhong, R., Fowler, S., Lyskowski, D., Piyasena, H., Carleton, K., et al. (2010). The poplar MYB transcription factors, PtrMYB3 and PtrMYB20, are involved in the regulation of secondary wall biosynthesis. *Plant and Cell Physiology, 51*(6), 1084–1090.

McCarthy, R. L., Zhong, R., & Ye, Z. H. (2009). MYB83 is a direct target of SND1 and acts redundantly with MYB46 in the regulation of secondary cell wall biosynthesis in *Arabidopsis. Plant and Cell Physiology, 50*(11), 1950–1964.

Mellerowicz, E. J., & Gorshkova, T. A. (2012). Tensional stress generation in gelatinous fibres: A review and possible mechanism based on cell-wall structure and composition. *Journal of Experimental Botany, 63*(2), 551–565.

Ménard, D., Blaschek, L., Kriechbaum, K., Lee, C. C., Zhu, C., Nuoendagula, B., et al. (2021). Specific and dynamic lignification at the cell-type level controls plant physiology and adaptability. *BioRxiv*, 1–11.

Menden, B., Kohlhoff, M., & Moerschbacher, B. M. (2007). Wheat cells accumulate a syringyl-rich lignin during the hypersensitive resistance response. *Phytochemistry*, *68*(4), 513–520.

Merelo, P., Agustí, J., Arbona, V., Costa, M. L., Estornell, L. H., Gómez-Cadenas, A., et al. (2017). Cell wall remodeling in abscission zone cells during ethylene-promoted fruit abscission in citrus. *Frontiers in Plant Science*, *8*(FEBRUARY).

Mitsuda, N., Seki, M., Shinozaki, K., & Ohme-Takagi, M. (2005). The NAC transcription factors NST1 and NST2 of *Arabidopsis* regulate secondary wall thickenings and are required for anther dehiscence. *Plant Cell*, *17*(11), 2993–3006.

Miyamoto, T., Tobimatsu, Y., & Umezawa, T. (2020). MYB-mediated regulation of lignin biosynthesis in grasses. *Current Plant Biology*, *24*, 100174.

Morse, A. M., Whetten, R. W., Dubos, C., & Campbell, M. M. (2009). Post-translational modification of an R2R3-MYB transcription factor by a MAP kinase during xylem development. *New Phytologist*, *183*(4), 1001–1013.

Mottiar, Y., Vanholme, R., Boerjan, W., Ralph, J., & Mansfield, S. D. (2016). Designer lignins: Harnessing the plasticity of lignification. *Current Opinion in Biotechnology*, *37*, 190–200.

Nakano, Y., Yamaguchi, M., Endo, H., Rejab, N. A., & Ohtani, M. (2015). NAC-MYB-based transcriptional regulation of secondary cell wall biosynthesis in land plants. *Frontiers in Plant Science*, *6*(MAY). Frontiers Research Foundation.

Nakayama, T., Shinohara, H., Tanaka, M., Baba, K., Ogawa-Ohnishi, M., & Matsubayashi, Y. (2017). A peptide hormone required for Casparian strip diffusion barrier formation in *Arabidopsis* roots. *Science*, *355*(6322), 284–286.

Naseer, S., Lee, Y., Lapierre, C., Franke, R., Nawrath, C., & Geldner, N. (2012). Casparian strip diffusion barrier in *Arabidopsis* is made of a lignin polymer without suberin. *Proceedings of the National Academy of Sciences of the United States of America*, *109*(25), 10101–10106.

Ng, L. M., Melcher, K., Teh, B. T., & Xu, H. E. (2014). Abscisic acid perception and signaling: Structural mechanisms and applications. *Acta Pharmacologica Sinica*, *35*(5), 567–584.

Nicholson, R. L., & Hammerschmidt, R. (1992). Phenolic compounds and their role in disease resistance. *Annual Review of Phytopathology*, *30*, 369–389.

Ogawa, M., Kay, P., Wilson, S., & Swain, S. M. (2009). *Arabidopsis* dehiscence zone Polygalacturonase1 (ADPG1), ADPG2, and Quartet2 are polygalacturonases required for cell separation during reproductive development in *Arabidopsis*. *Plant Cell*, *21*(1), 216–233.

Ohashi-Ito, K., Oda, Y., & Fukuda, H. (2010). *Arabidopsis* VASCULAR-RELATED NAC-DOMAIN6 directly regulates the genes that govern programmed cell death and secondary wall formation during xylem differentiation. *Plant Cell*, *22*(10), 3461–3473.

Öhman, D., Demedts, B., Kumar, M., Gerber, L., Gorzsas, A., Goeminne, G., et al. (2013). MYB103 is required for FERULATE-5-HYDROXYLASE expression and syringyl lignin biosynthesis in Arabidopsis stems. *Plant Journal*, *73*(1), 63–76.

Ohtani, M., & Demura, T. (2019). The quest for transcriptional hubs of lignin biosynthesis: Beyond the NAC-MYB-gene regulatory network model. *Current Opinion in Biotechnology*, *56*, 82–87.

Pammel, L. H. (1899). *Anatomical characters of the seeds of leguminosae, chiefly genera of gray's manual*. Shaw School of Botany (173 pp).

Patzlaff, A., McInnis, S., Courtenay, A., Surman, C., Newman, L. J., Smith, C., et al. (2003). Characterisation of a pine MYB that regulates lignification. *Plant Journal*, *36*(6), 743–754.

Perkins, M. L., Schuetz, M., Unda, F., Smith, R. A., Sibout, R., Hoffmann, N. J., et al. (2020). Dwarfism of high-monolignol *Arabidopsis* plants is rescued by ectopic LACCASE overexpression. *Plant Direct*, *4*(9), 1–16.

Pesquet, E., Zhang, B., Gorzsas, A., Puhakainen, T., Serk, H., Escamez, S., et al. (2013). Non-cell-autonomous postmortem lignification of tracheary elements in *Zinnia elegans*. *The Plant Cell*, *25*(4), 1314–1328.

Pfister, A., Barberon, M., Alassimone, J., Kalmbach, L., Lee, Y., Vermeer, J. E. M., et al. (2014). A receptor-like kinase mutant with absent endodermal diffusion barrier displays selective nutrient homeostasis defects. *eLife*, *3*, e03115.

Pickett-Heaps, J. D. (1968). Xylem wall deposition - radioautographic investigations using lignin precursors. *Protoplasma*, *65*(1–2), 181–205.

Piermattei, A., Crivellaro, A., Carrer, M., & Urbinati, C. (2015). The "blue ring": Anatomy and formation hypothesis of a new tree-ring anomaly in conifers. *Trees - Structure and Function*, *29*(2), 613–620.

Piotrowska, H., Kucinska, M., & Murias, M. (2012). Biological activity of piceatannol: Leaving the shadow of resveratrol. *Mutation Research, Reviews in Mutation Research*, *750*(1), 60–82.

Piquemal, J., Lapierre, C., Myton, K., O'Connell, A., Schuch, W., Grima-Pettenati, J., et al. (1998). Down-regulation of cinnamoyl-CoA reductase induces significant changes of lignin profiles in transgenic tobacco plants. *Plant Journal*, *13*(1), 71–83.

Pitre, F. E., Pollet, B., Lafarguette, F., Cooke, J. E. K., Mackay, J. J., & Lapierre, C. (2007). Effects of increased nitrogen supply on the lignification of poplar wood. *Journal of Agricultural and Food Chemistry*, *55*(25), 10306–10314.

Pittermann, J. (2010). The evolution of water transport in plants: An integrated approach. *Geobiology*, *8*(2), 112–139.

Porth, I., Klápště, J., Skyba, O., Lai, B. S. K., Geraldes, A., Muchero, W., et al. (2013). Populus trichocarpa cell wall chemistry and ultrastructure trait variation, genetic control and genetic correlations. *New Phytologist*, *197*(3), 777–790.

Pu, J., Song, J., Xie, Y., & Gu, R. (2002). Characteristics of lignin structure of triploid clones of *Populus tomentosa* Carr. *Beijing Linye Daxue Xuebao/Journal of Beijing Forestry University*, *24*(5–6), 211–215.

Qutob, D., Ma, F., Peterson, C. A., Bernards, M. A., & Gijzen, M. (2008). Structural and permeability properties of the soybean seed coat. *Botany*, *86*(3), 219–227.

Rajani, S., & Sundaresan, V. (2001). The *Arabidopsis* myc/bHLH gene alcatraz enables cell separation in fruit dehiscence. *Current Biology*, *11*(24), 1914–1922.

Ralph, J. (2010). Hydroxycinnamates in lignification. *Phytochemistry Reviews*, *9*(1), 65–83.

Ralph, J., Akiyama, T., Coleman, H. D., & Mansfield, S. D. (2012). Effects on lignin structure of coumarate 3-hydroxylase downregulation in poplar. *Bioenergy Research*, *5*(4), 1009–1019.

Ralph, J., Lundquist, K., Brunow, G., Lu, F., Kim, H., Schatz, P. F., et al. (2004). Lignins: Natural polymers from oxidative coupling of 4-hydroxyphenyl- propanoids. *Phytochemistry Reviews*, *3*(1–2), 29–60.

Rathgeber, C. B. K., Cuny, H. E., & Fonti, P. (2016). Biological basis of tree-ring formation: A crash course. *Frontiers in Plant Science*, 7(MAY2016). Frontiers Research Foundation.

Reddy, M. S. S., Chen, F., Shadle, G., Jackson, L., Aljoe, H., & Dixon, R. A. (2005). Targeted down-regulation of cytochrome P450 enzymes for forage quality improvement in alfalfa (*Medicago sativa* L.). *Proceedings of the National Academy of Sciences of the United States of America*, *102*(46), 16573–16578.

Renault, H., Werck-Reichhart, D., & Weng, J. K. (2019). Harnessing lignin evolution for biotechnological applications. In *Vol. 56. Current opinion in biotechnology* (pp. 105–111). Elsevier Ltd.

Rencoret, J., Rosado, M. J., Kim, H., Timokhin, V. I., Gutiérrez, A., Bausch, F., et al. (2021). Flavonoids naringenin chalcone, naringenin, dihydrotricin, and tricin are lignin monomers in papyrus. *Plant Physiology, 188*(1), 208–219.

Reyt, G., Chao, Z., Flis, P., Salas-González, I., Castrillo, G., Chao, D. Y., et al. (2020). Uclacyanin proteins are required for lignified nanodomain formation within Casparian strips. *Current Biology, 30*(20), 4103–4111.e6.

Reyt, G., Ramakrishna, P., Salas-González, I., Fujita, S., Love, A., Tiemessen, D., et al. (2021). Two chemically distinct root lignin barriers control solute and water balance. *Nature Communications, 12*(1).

Rich, T. C. G. (1991). Crucifers of Great Britain and Ireland. In *B.S.B.I. handbook no.6*, Botanical Society of the British Isles.

Robards, A. W., Jackson, S. M., Clarkson, D. T., & Sanderson, J. (1973). The structure of barley roots in relation to the transport of ions into the stele. *Protoplasma: An International Journal of Cell Biology*, 77(2), 291–311.

Rogers, L. A., Dubos, C., Cullis, I. F., Surman, C., Poole, M., Willment, J., et al. (2005). Light, the circadian clock, and sugar perception in the control of lignin biosynthesis. *Journal of Experimental Botany, 56*(416), 1651–1663.

Rojas-Murcia, N., Hématy, K., Lee, Y., Emonet, A., Ursache, R., Fujita, S., et al. (2020). High-order mutants reveal an essential requirement for peroxidases but not laccases in Casparian strip lignification. *Proceedings of the National Academy of Sciences of the United States of America, 117*(46), 29166–29177.

Roppolo, D., De Rybel, B., Tendon, V. D., Pfister, A., Alassimone, J., Vermeer, J. E. M., et al. (2011). A novel protein family mediates Casparian strip formation in the endodermis. *Nature, 473*(7347), 381–384.

Rosen, W. G., Meyer, B. S., Anderson, D. B., Böhning, R. H., & Bohning, R. H. (1960). Introduction to plant physiology. *AIBS Bulletin, 10*(3), 39.

Roupe, K., Remsberg, C., Yanez, J., & Davies, N. (2008). Pharmacometrics of stilbenes: Seguing towards the clinic. *Current Clinical Pharmacology, 1*(1), 81–101.

Sah, S. K., Reddy, K. R., & Li, J. (2016). Abscisic acid and abiotic stress tolerance in crop plants. *Frontiers in Plant Science*, 7(MAY2016). Frontiers Media S.A.

Saka, S., & Goring, D. A. I. (1988). The distribution of lignin in white birch wood as determined by bromination with TEM-EDXA. *Holzforschung, 42*(3), 149–153.

Sanders, P. M., Bui, A. Q., Weterings, K., McIntire, K. N., Hsu, Y. C., Lee, P. Y., et al. (1999). Anther developmental defects in *Arabidopsis thaliana* male-sterile mutants. *Sexual Plant Reproduction, 11*(6), 297–322.

Sarvas, R. (1972). Investigations on the annual cycle of development of forest trees. Active period. *Metsantutkimuslaitoksen Julkaisuja, 76*(3), 1–110.

Sarvas, R. (1974). Investigations on the annual cycle of development of Forest trees: II. Autumn dormancy and winter dormancy. *Communicationes Instituti Forestalis Fenniae, 84*(1), 1–101.

Schenker, G., Lenz, A., Körner, C., & Hoch, G. (2014). Physiological minimum temperatures for root growth in seven common European broad-leaved tree species. *Tree Physiology, 34*(3), 302–313.

Schilmiller, A. L., Stout, J., Weng, J. K., Humphreys, J., Ruegger, M. O., & Chapple, C. (2009). Mutations in the *cinnamate 4-hydroxylase* gene impact metabolism, growth and development in *Arabidopsis*. *Plant Journal, 60*(5), 771–782.

Schmid, R., & Evert, R. F. (2008). Esau's plant anatomy: Meristems, cells, and tissues of the plant body: Their structure, function, and development. *Taxon, 57*(2), 678.

Schopfer, P., Lapierre, C., & Nolte, T. (2001). Light-controlled growth of the maize seedling mesocotyl: Mechanical cell-wall changes in the elongation zone and related changes in lignification. *Physiologia Plantarum, 111*(1), 83–92.

Schuetz, M., Benske, A., Smith, R. A., Watanabe, Y., Tobimatsu, Y., Ralph, J., et al. (2014). Laccases direct lignification in the discrete secondary cell wall domains of protoxylem. *Plant Physiology*, *166*(2), 798–807.

Schuetz, M., Smith, R., & Ellis, B. (2013). Xylem tissue specification, patterning, and differentiation mechanisms. *Journal of Experimental Botany*, *64*(1), 11–31.

Scurfield, G. (1973). Reaction wood: Its structure and function. *Science*, *179*(4074), 647–655.

Shadle, G., Chen, F., Srinivasa Reddy, M. S., Jackson, L., Nakashima, J., & Dixon, R. A. (2007). Down-regulation of hydroxycinnamoyl CoA: Shikimate hydroxycinnamoyl transferase in transgenic alfalfa affects lignification, development and forage quality. *Phytochemistry*, *68*(11), 1521–1529.

Siegrist, J., Jeblick, W., & Kauss, H. (1994). Defense responses in infected and elicited cucumber (*Cucumis sativus* L.) hypocotyl segments exhibiting acquired resistance. *Plant Physiology*, *105*(4), 1365–1374.

Słupianek, A., Dolzblasz, A., & Sokołowska, K. (2021). Xylem parenchyma—Role and relevance in wood functioning in trees. *Plants*, *10*(6), 1247.

Smit, F., & Dubery, I. A. (1997). Cell wall reinforcement in cotton hypocotyls in response to a *verticillium dahliae* elicitor. *Phytochemistry*, *44*(5), 811–815.

Smith, W. W. (1935). The course of stone cell formation in pear fruits. *Plant Physiology*, *10*(4), 587–611. (https://www.ncbi.nlm.nih.gov/pmc/articles/PMC439172/).

Smith, R. A., Schuetz, M., Karlen, S. D., Bird, D., Tokunaga, N., Sato, Y., et al. (2017). Defining the diverse cell populations contributing to lignification in *Arabidopsis* stems. *Plant Physiology*, *174*(2), 1028–1036.

Smith, R. A., Schuetz, M., Roach, M., Mansfield, S. D., Ellis, B., & Samuels, L. (2013). Neighboring parenchyma cells contribute to *Arabidopsis* xylem lignification, while lignification of interfascicular fibers is cell autonomous. *Plant Cell*, *25*(10), 3988–3999.

Spence, J., Vercher, Y., Gates, P., & Harris, N. (1996). "Pod shatter" in *Arabidopsis thaliana*, *Brassica napus* and *B. juncea*. *Journal of Microscopy*, *181*(2), 195–203.

Steiner-Lange, S., Unte, U. S., Eckstein, L., Yang, C., Wilson, Z. A., Schmelzer, E., et al. (2003). Disruption of *Arabidopsis thaliana MYB26* results in male sterility due to non-dehiscent anthers. *Plant Journal*, *34*(4), 519–528.

Stöckle, D., & Vermeer, J. E. (2020). Plant biology: Journey to the Center of the Casparian Strip. *Current Biology*, *30*(20), R1273–R1275.

Stone, J. E. (1955). A Study of the Lignin Removed During a Neutral Sulphite Cook of Aspen. Undefined.

Su, N., Ling, F., Xing, A., Zhao, H., Zhu, Y., Wang, Y., et al. (2020). Lignin synthesis mediated by CCoAOMT enzymes is required for the tolerance against excess cu in *Oryza sativa*. *Environmental and Experimental Botany*, *175*, 104059.

Sulis, D. B., & Wang, J. P. (2020). Regulation of lignin biosynthesis by post-translational protein modifications. *Frontiers in Plant Science*, *11*, 914.

Sun, X., Wang, C., Xiang, N., Li, X., Yang, S., Du, J. C., et al. (2017). Activation of secondary cell wall biosynthesis by miR319-targeted TCP4 transcription factor. *Plant Biotechnology Journal*, *15*(10), 1284–1294.

Syros, T. D., Yupsanis, T. A., & Economou, A. S. (2005). Expression of peroxidases during seedling growth in *Ebenus Cretica* L. as affected by light and temperature treatments. *Plant Growth Regulation*, *46*(2), 143–151.

Tabuchi, T., Ito, S., & Arai, N. (2000). Structural changes in the abscission zones of pedicels at different ripening stages of tomato. *Journal of the Japanese Society for Horticultural Science*, *69*(6), 705–707.

Takabe, K., Miyauchi, S., Tsunoda, R., & Fukazawa, K. (1992). Distribution of Guaiacyl and Syringyl Lignins in Japanese beech (*Fagus Crenata*): Variation within an annual ring. *IAWA Journal*, *13*(1), 105–112.

Takata, N., Awano, T., Nakata, M. T., Sano, Y., Sakamoto, S., Mitsuda, N., et al. (2019). Populus NST/SND orthologs are key regulators of secondary cell wall formation in wood fibers, phloem fibers and xylem ray parenchyma cells. *Tree Physiology, 39*(4), 514–525.

Tang, N., Shahzad, Z., Lonjon, F., Loudet, O., Vailleau, F., & Maurel, C. (2018). Natural variation at XND1 impacts root hydraulics and trade-off for stress responses in *Arabidopsis. Nature Communications, 9*(1).

Tefu, Q., Luohua, H., & Qin, Z. (2004). The chemical groups and bonds characterization of juvenile wood and mature wood lignins of Chinese fir. *Scientia Silvae Sinica, 40*(2), 137–141. (https://europepmc.org/article/cba/400533).

Terashima, N., & Fukushima, K. (1988). Heterogeneity in formation of lignin-XI: An autoradiographic study of the heterogeneous formation and structure of pine lignin. *Wood Science and Technology, 22*(3), 259–270.

Terashima, N., Fukushima, K., Tsuchiya, S., & Takabe, K. (1986). Heterogeneity in formation of lignin. VII. An autoradiographic study on the formation of guaiacyl and syringyl lignin in poplar. *Journal of Wood Chemistry and Technology, 6*(4), 495–504.

Thévenin, J., Pollet, B., Letarnec, B., Saulnier, L., Gissot, L., Maia-Grondard, A., et al. (2011). The simultaneous repression of CCR and CAD, two enzymes of the lignin biosynthetic pathway, results in sterility and dwarfism in *Arabidopsis thaliana. Molecular Plant, 4*(1), 70–82.

Timell, T. E. (1986). Compression wood in gymnosperms. In *Compression wood in gymnosperms*.

Tobimatsu, Y., Chen, F., Nakashima, J., Escamilla-Treviño, L. L., Jackson, L., Dixon, R. A., et al. (2013). Coexistence but independent biosynthesis of catechyl and guaiacyl/syringyl lignin polymers in seed coats. *Plant Cell, 25*(7), 2587–2600.

Tsang, D. L., Edmond, C., Harrington, J. L., & Nühse, T. S. (2011). Cell wall integrity controls root elongation via a general 1-aminocyclopropane-1-carboxylic acid-dependent, ethylene-independent pathway. *Plant Physiology, 156*(2), 596–604.

Tsuchiya, M., Satoh, S., & Iwai, H. (2015). Distribution of XTH, expansin, and secondary-wall-related CesA in floral and fruit abscission zones during fruit development in tomato (Solanum lycopersicum). *Frontiers in Plant Science, 6*(MAY), 1–9.

Tulik, M., Jura-Morawiec, J., Bieniasz, A., & Marciszewska, K. (2019). How long do wood parenchyma cells live in the stem of a scots pine (*Pinus sylvestris* L.)? Studies on cell nuclei status along the radial and longitudinal stem axes. *Forests, 10*(11).

Umezawa, T. (2010). The cinnamate/monolignol pathway. *Phytochemistry Reviews, 9*(1), 1–17.

Van Acker, R., Vanholme, R., Storme, V., Mortimer, J. C., Dupree, P., & Boerjan, W. (2013). Lignin biosynthesis perturbations affect secondary cell wall composition and saccharification yield in *Arabidopsis thaliana. Biotechnology for Biofuels, 6*(1).

Van De Mortel, J. E., Schat, H., Moerland, P. D., Van Themaat, E. V. L., Van Der Ent, S., Blankestijn, H., et al. (2008). Expression differences for genes involved in lignin, glutathione and sulphate metabolism in response to cadmium in *Arabidopsis thaliana* and the related Zn/cd-hyperaccumulator *Thlaspi caerulescens. Plant, Cell and Environment, 31*(3), 301–324.

Vance, C. P., Kirk, T. K., & Sherwood, R. T. (1980). Lignification as a mechanism of disease resistance. *Annual Review of Phytopathology, 18*(1), 259–288.

Vanholme, R., De Meester, B., Ralph, J., & Boerjan, W. (2019). Lignin biosynthesis and its integration into metabolism. *Current Opinion in Biotechnology, 56*, 230–239. Elsevier Ltd.

Vanholme, R., Morreel, K., Darrah, C., Oyarce, P., Grabber, J. H., Ralph, J., et al. (2012). Metabolic engineering of novel lignin in biomass crops. *New Phytologist, 196*(4), 978–1000. John Wiley & Sons Ltd.

Vanholme, R., Morreel, K., Ralph, J., & Boerjan, W. (2008). Lignin engineering. *Current Opinion in Plant Biology, 11*(3), 278–285.

Voelker, S. L., Lachenbruch, B., Meinzer, F. C., & Strauss, S. H. (2011). Reduced wood stiffness and strength, and altered stem form, in young antisense 4CL transgenic poplars with reduced lignin contents. *New Phytologist, 189*(4), 1096–1109.

Walsh, C. (2006). *Posttranslational modification of proteins : Expanding nature's inventory* (p. 490). Roberts and Co. Publishers.

Wang, H., Avci, U., Nakashima, J., Hahn, M. G., Chen, F., & Dixon, R. A. (2010). Mutation of WRKY transcription factors initiates pith secondary wall formation and increases stem biomass in dicotyledonous plants. *Proceedings of the National Academy of Sciences of the United States of America, 107*(51), 22338–22343.

Wang, Y., Chantreau, M., Sibout, R., & Hawkins, S. (2013). Plant cell wall lignification and monolignol metabolism. *Frontiers in Plant Science, 4*, 220.

Wang, J. P., Chuang, L., Loziuk, P. L., Chen, H., Lin, Y. C., Shi, R., et al. (2015). Phosphorylation is an on/off switch for 5-hydroxyconiferaldehyde O-methyltransferase activity in poplar monolignol biosynthesis. *Proceedings of the National Academy of Sciences of the United States of America, 112*(27), 8481–8486.

Wang, C., Hu, D., Liu, X., She, H., Ruan, R., Yang, H., et al. (2015). Effects of uniconazole on the lignin metabolism and lodging resistance of culm in common buckwheat (*Fagopyrum esculentum* M.). *Field Crops Research, 180*, 46–53.

Wang, H., Mao, Y., Yang, J., & He, Y. (2015). TCP24 modulates secondary cell wall thickening and anther endothecium development. *Frontiers in Plant Science, 6*(JUNE), 1–10.

Wang, J. P., Matthews, M. L., Williams, C. M., ... Chiang, V. L. (2018). Improving wood properties for wood utilization through multi-omics integration in lignin biosynthesis. *Nature Communications, 9*(1), 1–16.

Wang, W. Q., Zhang, J., Ge, H., Li, S. J., Li, X., Yin, X. R., et al. (2016). EjMYB8 transcriptionally regulates flesh lignification in loquat fruit. *PLoS One, 11*(4).

Wang, C. Y., Zhang, S., Yu, Y., Luo, Y. C., Liu, Q., Ju, C., et al. (2014). MiR397b regulates both lignin content and seed number in *Arabidopsis* via modulating a laccase involved in lignin biosynthesis. *Plant Biotechnology Journal, 12*(8), 1132–1142.

Weng, J. K., & Chapple, C. (2010). The origin and evolution of lignin biosynthesis. *New Phytologist, 187*(2), 273–285.

Weng, J. K., Mo, H., & Chapple, C. (2010). Over-expression of F5H in COMT-deficient *Arabidopsis* leads to enrichment of an unusual lignin and disruption of pollen wall formation. *Plant Journal, 64*(6), 898–911.

Wilson, Z. A., Song, J., Taylor, B., & Yang, C. (2011). The final split: The regulation of anther dehiscence. *Journal of Experimental Botany, 62*(5), 1633–1649.

Wodzicki, T. J., & Wodzicki, A. B. (1980). Seasonal abscisic acid accumulation in stem cambial region of Pinus silvestris, and its contribution to the hypothesis of a late-wood control system in conifers. *Physiologia Plantarum, 48*(3), 443–447.

Wu, L., Joshi, C. P., & Chiang, V. L. (2000). A xylem-specific cellulose synthase gene from aspen (Populus tremuloides) is responsive to mechanical stress. *Plant Journal, 22*(6), 495–502.

Xiong, X. P., Sun, S. C., Zhu, Q. H., Zhang, X. Y., Li, Y. J., Liu, F., et al. (2021). The cotton lignin biosynthetic gene *Gh4CL30* regulates lignification and phenolic content and contributes to verticillium wilt resistance. *Molecular Plant-Microbe Interactions, 34*(3), 240–254.

Xu, M., Li, S., & jia, Liu, X. fen, Yin, X. ren, Grierson, D., & Chen, K. song. (2019). Ternary complex EjbHLH1-EjMYB2-EjAP2-1 retards low temperature-induced flesh lignification in loquat fruit. *Plant Physiology and Biochemistry, 139*, 731–737.

Xu, W., Tang, W., Wang, C., Ge, L., Sun, J., Qi, X., et al. (2020). SiMYB56 confers drought stress tolerance in transgenic Rice by regulating lignin biosynthesis and ABA signaling pathway. *Frontiers in Plant Science, 11*.

Xu, Q., Yin, X. R., Zeng, J. K., Ge, H., Song, M., Xu, C. J., et al. (2014). Activator-and repressor-type MYB transcription factors are involved in chilling injury induced flesh lignification in loquat via their interactions with the phenylpropanoid pathway. *Journal of Experimental Botany*, *65*(15), 4349–4359.

Xu, M., Zhang, M. X., Shi, Y. N., Liu, X. F., Li, X., Grierson, D., et al. (2019). *EjHAT1* participates in heat alleviation of loquat fruit lignification by suppressing the promoter activity of key lignin monomer synthesis gene *EjCAD5*. *Journal of Agricultural and Food Chemistry*, *67*(18), 5204–5211.

Xue, C., Yao, J. L., Xue, Y. S., Su, G. Q., Wang, L., Lin, L. K., et al. (2019). PbrMYB169 positively regulates lignification of stone cells in pear fruit. *Journal of Experimental Botany*, *70*(6), 1801–1814.

Yabuta, Y., Maruta, T., Yoshimura, K., Ishikawa, T., & Shigeoka, S. (2004). Two distinct redox signaling pathways for cytosolic APX induction under photooxidative stress. *Plant and Cell Physiology*, *45*(11), 1586–1594.

Yamaguchi, M., Mitsuda, N., Ohtani, M., Ohme-Takagi, M., Kato, K., & Demura, T. (2011). VASCULAR-RELATED NAC-DOMAIN 7 directly regulates the expression of a broad range of genes for xylem vessel formation. *Plant Journal*, *66*(4), 579–590.

Yamamoto, H., Yoshida, M., & Okuyama, T. (2002). Growth stress controls negative gravitropism in woody plant stems. *Planta*, *216*(2), 280–292.

Yang, C., Song, J., Ferguson, A. C., Klisch, D., Simpson, K., Mo, R., et al. (2017). Transcription factor MYB26 is key to spatial specificity in anther secondary thickening formation. *Plant Physiology*, *175*(1), 333–350.

Yang, C., Xu, Z., Song, J., Conner, K., Barrena, G. V., & Wilson, Z. A. (2007). *Arabidopsis MYB26/MALE STERILE35* regulates secondary thickening in the endothecium and is essential for anther dehiscence. *Plant Cell*, *19*(2), 534–548.

Yang, L., Zhao, X., Yang, F., Fan, D., Jiang, Y., & Luo, K. (2016). PtrWRKY19, a novel WRKY transcription factor, contributes to the regulation of pith secondary wall formation in *Populus trichocarpa*. *Scientific Reports*, *6*.

Yi Chou, E., Schuetz, M., Hoffmann, N., Watanabe, Y., Sibout, R., & Samuels, A. L. (2018). Distribution, mobility, and anchoring of lignin-related oxidative enzymes in Arabidopsis secondary cell walls. *Journal of Experimental Botany*, *69*(8), 1849–1859.

Yu, S.-I., Kim, H., Yun, D. J., Suh, M. C., & Lee, B.-H. (2019). Post-translational and transcriptional regulation of phenylpropanoid biosynthesis pathway by Kelch repeat F-box protein SAGL1. *Plant Molecular Biology*, *99*(1–2), 135–148.

Zeng, J.-K., Li, X., Xu, Q., Chen, J.-Y., Yin, X. R., Ferguson, I. B., et al. (2015). *EjAP2-1*, an *AP2/ERF* gene, is a novel regulator of fruit lignification induced by chilling injury, via interaction with *EjMYB* transcription factors. *Plant Biotechnology Journal*, *13*(9), 1325–1334.

Zhang, M., Lapierre, C., Nouxman, N. L., Nieuwoudt, M. K., Smith, B. G., Chavan, R. R., et al. (2017). Location and characterization of lignin in tracheid cell walls of radiata pine (Pinus radiata D. Don) compression woods. *Plant Physiology and Biochemistry*, *118*, 187–198.

Zhang, J., Li, J., Xue, C., Wang, R., Zhang, M., Qi, K., et al. (2021). The variation of Stone cell content in 236 germplasms of sand pear (*Pyrus pyrifolia*) and identification of related candidate genes. *Horticultural Plant Journal*, 7(2), 108–116.

Zhang, B., Sztojka, B., Escamez, S., Vanholme, R., Hedenström, M., Wang, Y., et al. (2020). PIRIN2 suppresses S-type lignin accumulation in a noncell-autonomous manner in *Arabidopsis* xylem elements. *New Phytologist*, *225*(5), 1923–1935.

Zhang, Q., Xie, Z., Zhang, R., Xu, P., Liu, H., Yang, H.-Q. H., et al. (2018). Blue light regulates secondary cell wall thickening via myc2/myc4 activation of the nst1-directed transcriptional network in *Arabidopsis*. *Plant Cell*, *30*(10), 2512–2528.

Zhang, J., Yin, X. R., Li, H., Xu, M., Zhang, M. X., Li, S. J., et al. (2020). ETHYLENE RESPONSE FACTOR39-MYB8 complex regulates low-temperature-induced lignification of loquat fruit. *Journal of Experimental Botany*, *71*(10), 3172–3184.

Zhao, C., Avci, U., Grant, E. H., Haigler, C. H., & Beers, E. P. (2008). *XND1*, a member of the NAC domain family in *Arabidopsis thaliana*, negatively regulates lignocellulose synthesis and programmed cell death in xylem. *Plant Journal, 53*(3), 425–436.
Zhao, Q., & Dixon, R. A. (2011). Transcriptional networks for lignin biosynthesis: More complex than we thought? *Trends in Plant Science, 16*(4), 227–233.
Zhao, Y., Lin, S., Qiu, Z., Cao, D., Wen, J., Deng, X., et al. (2015). MicroRNA857 is involved in the regulation of secondary growth of vascular tissues in *Arabidopsis. Plant Physiology, 169*(4), 2539–2552.
Zhao, Q., Nakashima, J., Chen, F., Yin, Y., Fu, C., Yun, J., et al. (2013). *LACCASE* is necessary and nonredundant with *PEROXIDASE* for lignin polymerization during vascular development in *Arabidopsis. Plant Cell, 25*(10), 3976–3987.
Zhao, Q., Tobimatsu, Y., Zhou, R., Pattathil, S., Gallego-Giraldo, L., Fu, C., et al. (2013). Loss of function of *cinnamyl alcohol dehydrogenase 1* leads to unconventional lignin and a temperaturesensitive growth defect in *Medicago truncatula. Proceedings of the National Academy of Sciences of the United States of America, 110*(33), 13660–13665.
Zhong, R., Kandasamy, M. K., & Ye, Z. H. (2021). XND1 regulates secondary wall deposition in xylem vessels through the inhibition of VND functions. *Plant and Cell Physiology, 62*(1), 53–65.
Zhong, R., Lee, C., Haghighat, M., & Ye, Z. H. (2021). Xylem vessel-specific SND5 and its homologs regulate secondary wall biosynthesis through activating secondary wall NAC binding elements. *New Phytologist, 231*(4), 1496–1509.
Zhong, R., Lee, C., & Ye, Z. H. (2010). Global analysis of direct targets of secondary wall NAC master switches in *Arabidopsis. Molecular Plant, 3*(6), 1087–1103.
Zhong, R., Lee, C., Zhou, J., McCarthy, R. L., & Ye, Z. H. (2008). A battery of transcription factors involved in the regulation of secondary cell wall biosynthesis in *Arabidopsis. Plant Cell, 20*(10), 2763–2782.
Zhong, R., McCarthy, R. L., Haghighat, M., & Ye, Z. H. (2013). The poplar MYB master switches bind to the SMRE site and activate the secondary wall biosynthetic program during wood formation. *PLoS One, 8*(7).
Zhong, R., Richardson, E. A., & Ye, Z. H. (2007). The MYB46 transcription factor is a direct target of SND1 and regulates secondary wall biosynthesis in Arabidopsis. *Plant Cell, 19*(9), 2776–2792.
Zhong, R., & Ye, Z. H. (2012). MYB46 and MYB83 bind to the SMRE sites and directly activate a suite of transcription factors and secondary wall biosynthetic genes. *Plant and Cell Physiology, 53*(2), 368–380.
Zhong, R., & Ye, Z. H. (2015). The arabidopsis NAC transcription factor NST2 functions together with SND1 and NST1 to regulate secondary wall biosynthesis in fibers of inflorescence stems. *Plant Signaling & Behavior, 10*(2).
Zhou, J., Lee, C., Zhong, R., & Ye, Z. H. (2009). MYB58 and MYB63 are transcriptional activators of the lignin biosynthetic pathway during secondary cell wall formation in Arabidopsis. *Plant Cell, 21*(1), 248–266.
Zhou, M., Zhang, K., Sun, Z., Yan, M., Chen, C., Zhang, X., et al. (2017). LNK1 and LNK2 corepressors interact with the MYB3 transcription factor in phenylpropanoid biosynthesis. *Plant Physiology, 174*(3), 1348–1358.
Zhuang, Y., Zuo, D., Tao, Y., Cai, H., & Li, L. (2020). Laccase3-based extracellular domain provides possible positional information for directing Casparian strip formation in *Arabidopsis. Proceedings of the National Academy of Sciences of the United States of America, 117*(27), 15400–15402.
Zimmerman, K. (1936). Zur physiologischen Anatomie der Leguminosentesta. *Landwirtschaftlichen Versuchs, 127*, 1–56.
Zimmerman, M., & Brown, C. (1971). *Trees: Structure and function.* Berlin Heidelberg: Springer.

CHAPTER EIGHT

Transcriptional regulation of secondary cell wall formation and lignification

Steven G. Hussey*
Department of Biochemistry, Genetics and Microbiology, Forestry and Agricultural Biotechnology Institute (FABI), University of Pretoria, Pretoria, South Africa
*Corresponding author: e-mail address: steven.hussey@fabi.up.ac.za

Contents

Abstract

Secondary xylem is the chief form of lignocellulosic biomass on the planet and consists mainly of fibers, vessels and tracheids with lignified secondary cell walls (SCWs). These cells develop from xylem mother cells produced from the meristematic vascular cambium, undergoing cell expansion, SCW deposition, programmed cell death and lignification within a narrow developmental zone. Understanding the transcriptional control of this complex process has been an activate area of research for over two decades,

Advances in Botanical Research, Volume 104
ISSN 0065-2296
https://doi.org/10.1016/bs.abr.2022.03.007

where significant contributions in the model annual *Arabidopsis thaliana* (Arabidopsis) and the model tree genus *Populus* have provided detailed insights into a semi-hierarchical transcriptional regulatory network (TRN) underlying SCW deposition and lignification. In this chapter, I summarize the classical three-tiered SCW TRN model and elaborate on its expanded structure, discussing key transcription factors underlying lignin formation in Arabidopsis. I present a comprehensive new model of the SCW TRN of poplar, with a commentary on the conservation and divergence of the SCW TRN across plant lineages and the possible existence of lineage-specific regulators that act as "accessory" components to the core TRN. Finally, I discuss key post-translational modifications that affect the activity of the transcription factors comprising the SCW TRN and recent insights into cooperation and antagonism between transcription factors through the formation of protein complexes. We are learning that the SCW TRN has undergone substantial re-wiring in well-studied models and that plants have adopted a myriad of combinatorial and post-translational mechanisms to modulate SCW transcriptional control according to particular biological contexts.

1. Introduction

Lignocellulose is the world's most abundant form of biomass, of which a major component comprises lignified secondary xylem (wood). Comprehensive recent estimates assert that approx. 450 gigatons (≈80%) of the organically bound carbon present in Earth's biomass resides in plants, of which nearly 70% is sequestered as woody biomass (Bar-On, Phillips, & Milo, 2018). The vast majority of the dry weight of this renewable material consists of lignified secondary cell walls (SCWs). While SCWs are deposited in a number of different plant cell types, including sclereids and phloem fibers, the bulk of SCW deposition takes place in the fibers and vessels comprising angiosperm xylem, or the tracheids making up most of that of gymnosperms (Zhong, Cui, & Ye, 2018). During secondary xylem development, these cells are born from the vascular cambium, a lateral meristem that gives rise to xylem mother cells on its centripetal side which in turn undergo cell expansion, SCW deposition, programmed cell death (PCD) and heartwood formation (Dejardin et al., 2010).

Lignification of xylem begins after the bulk of the polysaccharide component of the SCW, namely cellulose and hemicellulose, has been deposited. However, it continues post-mortem in a non-cell-autonomous fashion by so-called "good neighbor" cells such as xylem parenchyma, which exhibit prolonged lifespans in mature xylem and are able to supply monolignols to adjacent fibers and tracheary elements (Meents, Watanabe, & Samuels, 2018; Pesquet et al., 2013; Smith et al., 2013,

2017). Monolignol polymerization *in muro* is achieved by laccases such as LAC4 and LAC17 as well as a number of peroxidases (Barros, Serk, Granlund, & Pesquet, 2015; Berthet et al., 2011; Schuetz et al., 2014; Tobimatsu & Schuetz, 2019; Zhao et al., 2013). Like the biosynthesis of cellulose and major hemicelluloses such as glucuronoarabinoxylan, the biosynthesis of monolignols via the phenylpropanoid pathway is fairly well understood. Phenylalanine produced from the plastidic shikimate pathway enters the cytosolic phenylpropanoid pathway and is subjected to several ring hydroxylations and *O*-methylations. Following the deamination of phenylalanine to cinnamic acid by phenylalanine ammonia-lyase (PAL), it is hydroxylated via cinnamate 4-hydroxylase (C4H), thioesterified to CoA (4-coumarate:CoA ligase; 4CL), and diverted mainly to one of two pathways. The lignin-specific branch (1) results in the formation of *p*-hydroxyphenyl (H), guaiacyl (G) or synapyl (S) monolignols through the action of *p*-hydroxycinnamoyl-CoA: quinate/shikimate *p*-hydroxycinnamoyl transferase (HCT), caffeoyl-CoA *O*-methyltransferase (CCoAOMT) and cinnamoyl-CoA reductase (CCR), with cinnamyl alcohol dehydrogenase (CAD) responsible for the hydroxylation of each monolignol's aldehyde precursor into its alcohol derivative (Vanholme, Demedts, Morreel, Ralph, & Boerjan, 2010). Ferulate 5-hydroxylase (F5H), also known as coniferaldehyde 5-hydroxylase (Cald5H), is crucial for the formation of synapyl monolignol through 5-hydroxylation of coniferaldehyde or coniferyl alcohol and its methylation by caffeate *O*-methyltransferase (COMT). Alternatively, commitment to the flavonoid pathway (2) is achieved via chalcone synthase (CHS), which converts 4-coumaroyl-CoA and malonyl-CoA to naringenin chalcone (Deng & Lu, 2017; Fraser & Chapple, 2011). It should be appreciated however that the phenylpropanoid pathway also gives rise to a plethora of other aromatic metabolites such as lignans, coumarins and various hydroxycinnamic acids (Vogt, 2010). Coumarins play an important role in iron acquisition and their biosynthesis and transcriptional regulation is less well understood compared to other phenylpropanoid pathway derivatives such as lignin and flavonoids (reviewed by Robe, Izquierdo, Vignols, Rouached, & Dubos, 2021).

Plants have evolved sophisticated strategies to achieve precise spatio-temporal control of the lignin pathway, a process that is tightly linked to plant growth and defense (Xie et al., 2018). The transcriptional regulation of SCW deposition and lignification is a highly active area of research. While this process has been extensively studied in the model annual *Arabidopsis thaliana* (Arabidopsis) (Hussey, Mizrachi, Creux, & Myburg, 2013;

Nakano, Yamaguchi, Endo, Rejab, & Ohtani, 2015; Zhu & Li, 2021), significant progress has also been made in various hardwood and softwood tree species as well as grasses such as rice, switchgrass and *Brachypodium* (Coomey, Sibout, & Hazen, 2020; McCahill & Hazen, 2019; Miyamoto, Tobimatsu, & Umezawa, 2020; Zhang, Xie, Tuskan, Muchero, & Chen, 2018). Despite these advances, the similarities and differences of the transcriptional regulatory networks (TRNs) controlling SCW deposition in distinct cell types is far from resolved. We lack a thorough understanding of how these networks have changed during the course of plant evolution, and we are only beginning to understand the combinatorial regulation of SCW biosynthetic genes by the transcription factors (TFs) comprising this network.

In this chapter, an updated and expanded model of the classical three-tiered TRN underlying SCW deposition in Arabidopsis is presented. Next, evidence for both conservation and divergence of the SCW TRN structure and TF membership is discussed, illustrating detailed insights gained from recent studies of poplar SCW regulation. Finally, our current understanding of post-translational and combinatorial regulation of SCW biosynthesis is summarized, illustrating the crucial roles that post-translational modifications play on TF function and the flexibility of SCW and lignin-related TF complexes. Throughout the chapter, loci corresponding to species other than Arabidopsis will be denoted with an appropriate prefix (e.g., PtrWND1A—*Populus trichocarpa* WND1A).

2. A semi-hierarchical transcriptional regulatory network regulates secondary cell wall formation

Nearly two decades of research have revealed the existence of a multi-tiered TRN underlying SCW development. According to generalized models (e.g., Ohtani & Demura, 2019), an upper tier of NAC domain master regulators activates a middle tier of R2R3-type MYB TFs, followed by a panel of TFs from diverse families which in turn regulate the structural genes associated with SCW formation and PCD (Fig. 1). In this chapter, I will refer to this as the NAC-MYB-X model, where X represents the many TF families represented in the lowest tier. Broadly speaking, TFs can be positioned into network tiers by virtue of their direct gene targets, as well as the TFs which regulate them from higher tiers. While the naming of the tiers is arbitrary, it is useful to designate to the first tier those TFs which predominantly have "ground level" SCW structural genes as direct targets,

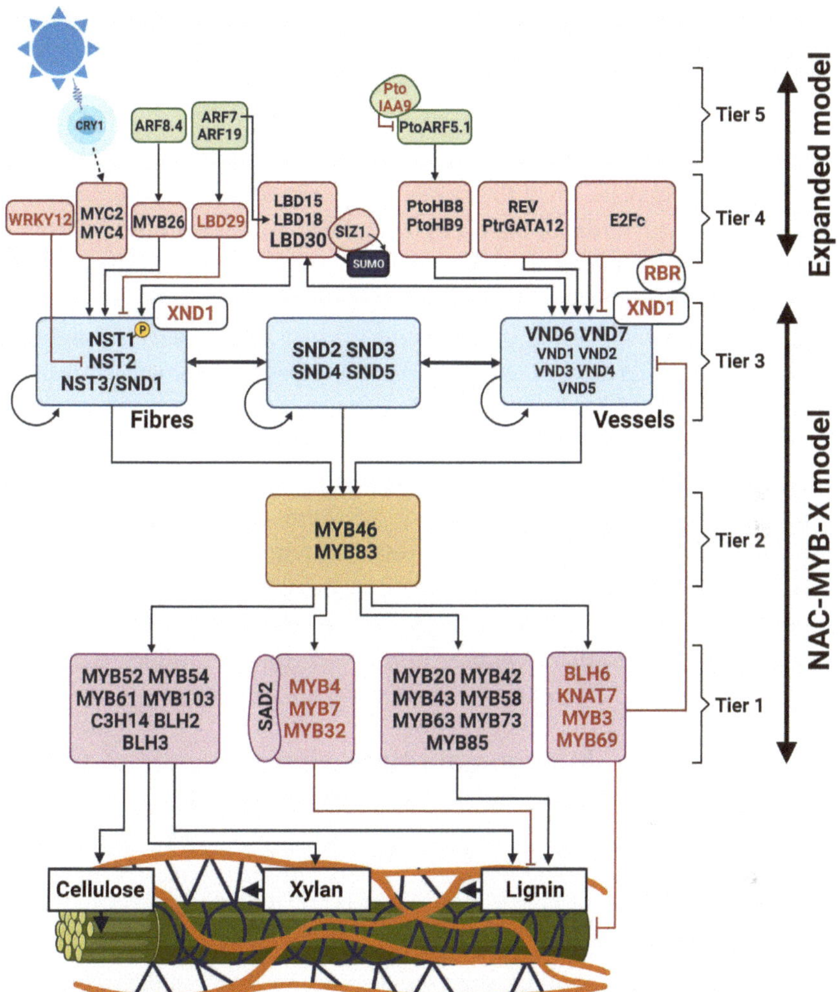

Fig. 1 Generalized model of secondary cell wall transcriptional regulation in Arabidopsis. Five tiers of transcription factors are shown to regulate the structural genes responsible for the biosynthesis of secondary cell wall cellulose, hemicelluloses and lignin (bottom). In the classical NAC-MYB-X model, the first three tiers comprise a diverse group of transcription factors from different families ("X", represented as Tier 1) which are regulated by R2R3-MYB hub regulators MYB46 and MYB83 (Tier 2), which are in turn regulated by master regulators of fiber, vessel and anther endothecium secondary cell wall formation from the NAC domain family (Tier 3). In this model, extensive self- and cross-regulation is thought to occur within Tier 3. An extended model incorporating evidence of transcription factors that regulate the NAC domain master regulators (Tier 4) and auxin response factors (ARFs) (Tier 5) that regulate Tier 4 are also shown. Arrows indicate positive interactions; blunted edges represent repressive interactions. Post-translational modifications such as phosphorylation of NST1 and sumoylation for LBD30 (Section 4.1) are also indicated. Created with BioRender.com.

followed by their regulators as the second tier, and those regulating them, in turn, as the third tier, much like the storeys of a building. This also allows for additional tiers to be appended to the model as further regulators of the top-tier master regulators are discovered.

While the NAC-MYB-X model is useful for understanding, it has several important caveats. First, the tier structure is neither strictly hierarchical nor is it necessarily ordinal, since TFs from tiers 2 and 3 may bind directly to the promoters of TFs as well as structural genes from subordinate tiers. Similarly, TFs from lower tiers may participate in extensive feedback regulation by directly activating or repressing TFs in higher tiers. Second, the positioning of different TFs into a rigid hierarchy is not biologically realistic given the plastic and inherently stochastic nature of biological networks, where TFs may activate different targets in specific contexts and where regulatory decisions are made by the cooperative behavior of the network as a whole (Balazsi, van Oudenaarden, & Collins, 2011; Levy, Landry, & Michnick, 2010). Finally, opinions vary on where each TFs should be positioned, especially since we still have a poor knowledge of direct gene targets for many of them. Nonetheless, the overall NAC-MYB-X model is well-established in the literature. In this section, we further explore the central properties and characteristics of the generalized model.

2.1 The NAC and MYB domain transcription factors of tiers 2 and 3: Master regulators of SCW deposition

Early transcriptomic studies of brassinolide-induced *in vitro* Arabidopsis vessel element transdifferentation revealed the upregulation of seven *VASCULAR-RELATED NAC DOMAIN* (*VND*) transcripts, *VND1* through *VND7* (Kubo et al., 2005). Of these, *VND6* and *VND7* were necessary and sufficient to induce meta- and protoxylem vessel differentiation, respectively, and are regarded as master regulators of tracheary element differentiation due to their ability to activate the full vessel developmental pathway when misexpressed or induced ectopically, including the formation of patterned, lignified SCW deposition and PCD (Ohashi-Ito, Oda, & Fukuda, 2010; Yamaguchi et al., 2010; Yamaguchi, Kubo, Fukuda, & Demura, 2008). Subsequently, the remaining VND paralogs have all been shown to similarly support vessel-specific differentiation (Endo et al., 2015; Tan et al., 2018; Zhou, Zhong, & Ye, 2014). Similarly, three NAC SECONDARY WALL THICKENING PROMOTING FACTOR (NST) proteins NST1 through NST3 have been shown to act redundantly as master regulators of xylary and interfascicular fiber development as well as silique valve SCW formation, and to some extent anther endothecium wall deposition (Ko, Yang, Park, Lerouxel, & Han, 2007; Mitsuda et al., 2007; Mitsuda & Ohme-Takagi, 2008; Mitsuda, Seki, Shinozaki, & Ohme-Takagi, 2005; Zhong, Richardson, & Ye, 2007b; Zhong & Ye, 2015). Some authors have also used the term "Secondary wall-associated NAC Domain" (SND), of which SND1 is synonymous with NST3 (Zhong, Demura, & Ye, 2006). More generally, Zhong, Lee, and Ye (2010a) refer to "secondary wall NACs" (SWNs) as NAC-domain SCW master regulators. Dexamethasone-induced activation of SND1 protein in transgenic suspension culture cells is sufficient for their transdifferentiation into smoothed-walled lignified cells reminiscent of fibers, with less pronounced activation of PCD genes compared to that following VND6 induction (Ohashi-Ito et al., 2010; Yamaguchi et al., 2008; Zhong et al., 2010a). Together, VND1 to VND7 (in vessels) and NST1, NST2 and SND1 (in fibers) are the third-tier master regulators of xylem SCW deposition (Fig. 1).

Other SND proteins are now thought to act jointly with the SWNs. Among these are SND2, SND3 and their poplar orthologs of which affect xylem production and SCW deposition when overexpressed or dominantly repressed (Grant, Fujino, Beers, & Brunner, 2010; Hussey et al., 2011; Zhong, Lee, Zhou, McCarthy, & Ye, 2008). Fusion of these proteins to the VP16 activation domain brings about ectopic SCW cellulose, xylan and lignin deposition and upregulates a suite of SCW biosynthetic genes in Arabidopsis (Zhong, Lee, Haghighat, & Ye, 2021). Another two Arabidopsis paralogs SND4 (ANAC075) and SND5 (ANAC099), together with their poplar orthologs, were recently found to regulate the full complement of structural and TF genes such as the SWN master regulators as well as themselves (Zhong, Lee, et al., 2021), confirming an earlier prediction that ANAC075 is a SCW-related TF (Shen, Yin, Chen, Xu, & Dixon, 2009). This auto-activating panel of SND2 through SND5 appears to act downstream of the SWNs and yet upstream of the second major tier, mutually activating and reinforcing SWN master regulators. Thus, they are tentatively included in the same tier in Fig. 1.

The second tier in the NAC-MYB-X model is occupied by the functionally redundant paralogs MYB46 and MYB83 (Fig. 1). These are directly activated by the SWNs and are considered mid-level master regulators of SCW formation in both fibers and vessels, activating cellulose, hemicellulose and lignin biosynthesis but not PCD (Kim, Kim, Ko, Kang, & Han, 2014; Ko, Kim, & Han, 2009; Zhong, Richardson, & Ye, 2007a). Despite their subordinate position in the model, the combined loss of function in the *myb46 myb83* double mutant leads to severe stunting and almost complete loss of SCW formation in vascular cells (McCarthy, Zhong, & Ye, 2009), compared to the mild drooping inflorescence phenotype due to loss of fiber SCW deposition in the *nst1 nst2 snd1* triple mutant (Zhong & Ye, 2015). This is in agreement with general observations that regulators essential for survival often occupy lower levels in the gene regulatory networks of other eukaryotes and prokaryotes (Yu & Gerstein, 2006). Thus, MYB46 and MYB83 act as the central hub regulators of SCW biosynthesis in the TRN (Ko, Jeon, Kim, Kim, & Han, 2014).

2.2 Tiers 1, 4 and 5 feature diverse families of TFs that are involved in SCW deposition and lignification

The lowest level, tier 1, comprises a diverse suite of TFs from different families such as R2R3-MYB (e.g., MYB3, 4, 7, 20, 32, 42, 43, 52, 54, 58, 61, 63, 69, 73, 85 and 103) (Cassan-Wang et al., 2013; Geng et al., 2020; Ko et al., 2009; Öhman et al., 2012; Taylor-Teeples et al., 2015; Zhong et al., 2008), C3H-type zinc finger C3H14 (Chai et al., 2015; Kim et al., 2014; Ko et al., 2009), and three-amino-acid-loop-extension (TALE) class homeoproteins such as BLH2 and BLH3 (Zhong & Ye, 2012), as well as KNAT7 and BLH6 which are discussed in Section 4.2 (Fig. 1). Tier 1 TFs are not generally considered to possess SCW master regulator properties (i.e., they are not sufficient to induce the SCW deposition program when expressed ectopically), but are rather co-opted by tier 2 and 3 master regulators to directly regulate the various structural genes involved in SCW biosynthesis and xylem development. Interestingly, tier 1 TFs frequently participate in positive and negative feedback loops with higher-tier TFs, presumably as a mechanism to fine-tune SCW deposition (reviewed in Hussey et al., 2013). Of particular note, R2R3-MYB subgroup 4 TFs MYB4, MYB7 and MYB32 repress SCW deposition through a series of negative feedback loops, including cross- and self-repressing interactions (Ko et al., 2009) (Fig. 1). These TFs contain a conserved ethylene responsive element binding factor-associated amphiphilic repression (EAR) domain which, when mutated, abolishes interaction with SAD2 which is required for their nuclear localization (Zhou et al., 2015).

Beyond the NAC-MYB-X model, evidence is emerging to support the existence of a fourth and even a fifth tier that regulate the SCW NAC master regulators of Tier 3 (Fig. 1; extended model). WRKY12 (Tier 4) and some of its orthologs inhibit SCW formation in the pith, in part by the direct repression of *NST2* (Wang et al., 2010; Yang et al., 2016; Yu et al., 2013). MYC-related basic helix-loop-helix TFs MYC2 and MYC4 also seemingly occupy Tier 4, both directly regulating the *NST1* promoter to activate the fiber SCW developmental program under blue light stimulation, facilitated by the photo-excitation of CRYPTOCHROME1 (CRY1) (Zhang et al., 2018). A member of the GATA family, PtrGATA12 from *P. trichocarpa*, was shown to directly regulate homologs of VND6/7 and MYB46 and activate all major components of SCW deposition in fibers and vessels, suggesting it acts upstream of the master regulators (Ren et al., 2021). From an enhanced yeast one-hybrid screen, the E2F TF E2Fc was found to interact with a large number of SCW-related promoters, directly activating or repressing the *VND7* promoter in a dose-dependent fashion as a potential tier 4 regulator (Taylor-Teeples et al., 2015). Other examples of Tier 4 members include R2R3-MYB, ASYMMETRIC LEAVES2-LIKE/LATERAL ORGAN BOUNDARIES DOMAIN (ASL/LBD) and HD-ZIPIII family TFs that are thought to mediate the transduction of auxin signaling to the SCW master regulators via a higher tier of auxin response factors (reviewed in Qu et al., 2021). MYB26 (Tier 4), for example, activates *NST1* and *NST2* during anther endothecium secondary wall thickening and lignification, while NST1 in turn activates *MYB26* through a positive feedback loop (not shown; Mitsuda et al., 2005; Yang et al., 2007). *MYB26* itself is activated by a splice variant of AUXIN RESPONSE FACTOR8.4 (ARF8.4) as an apparent Tier 5 regulator, which links auxin signaling to endothecium lignification through MYB26 (Ghelli et al., 2018). ASL11/LBD15, ASL19/LBD30 and ASL20/LBD18 promote vessel-specific differentiation through the activation of *VND7* (Endo et al., 2015; Ohashi-Ito, Iwamoto, & Fukuda, 2018; Soyano, Thitamadee, Machida, & Chua, 2008), while LBD30 was also shown to activate *NST1/SND1*-mediated fiber SCW deposition (Liu, Yu, & Li, 2019). While most evidence supports LBD family placement in Tier 4, VND6/VND7 form a positive feedback loop with LBD18/LBD30 (Soyano et al., 2008). LBD29, which is preferentially expressed in roots, leaves and young stems, also seems to occupy Tier 4. Its overexpression has a strong negative effect on fiber SCW deposition, and it represses *NST1, NST2* and *NST3/SND1* in an auxin-dependent fashion downstream of ARF7 and ARF19 (Lee, Du, Zhuo, Qi, & Wang, 2019). ARF7 and ARF19 positively regulate lateral root development via *LBD29*, while LBD29 may control cellular reprogramming, being important for callus formation and lateral root development (Dai et al., 2020; Feng, Sun, Wang, Liu, & Zhu, 2012; Liu, Wang, Feng, & Zhu, 2010; Okushima et al., 2005; Okushima, Fukaki, Onoda, Theologis, & Tasaka, 2007; Xu, Cao, Xu, Zhang, & Hu, 2018). Interestingly, ARF7 also appears to activate *LBD18* (Soyano et al., 2008). Finally, in *P. tomentosa*, a homolog of ARF5 promotes auxin-responsive secondary xylem differentiation by directly activating HD-ZIPIII TFs *PtoHB7* and *PtoHB8* (Tier 4), which in turn activate VND-clade homologs *PtoWND6A* and *PtoWND6B* (Tier 3) during vascular differentiation (Xu et al., 2019).

Some TFs have been notoriously difficult to place in the network. The NAC domain protein XND1 has an inhibitory effect on SCW deposition and PCD (Zhang, Luo, Zhong, He, & Li, 2020; Zhao, Avci, Grant, Haigler, & Beers, 2008), where overexpression of its poplar ortholog causes numerous inhibitory effects on xylem differentiation (Grant et al., 2010). Although it is a direct target of SND1 and VND7 (Zhong et al., 2010a), we still lack comprehensive knowledge of its targets. Furthermore, new data suggest that the inhibitory effect of XND1 may lie in heterodimerization with SWNs rather than intrinsic repression activity or interference with the ability of the heterodimer to bind to DNA, with VND6-XND1 complexes being shown to localize to the cytoplasm rather than the nucleus (Zhong, Kandasamy, & Ye, 2021). Co-expression of NST1 and XND1 does not seem to alter their nuclear localization, however (Zhang et al., 2020). The C-terminal domain of XND1 has convergently evolved linear motifs shared with a number of papillomaviruses and adenoviruses, which interact with the conserved cell cycle and differentiation regulator RETINOBLASTOMA-RELATED (RBR) (Zhao et al., 2017). RBR can form DREAM complexes involving MYB3R4 and E2Fc TFs, inhibiting the ability of E2Fs to activate the cell cycle and hence cell proliferation (Desvoyes & Gutierrez, 2020). Since XND1-RBR interaction is essential for its ability to attenuate tracheary element differentiation, and E2Fc regulates VND7 as discussed above, it may be the case that XND1 forms a feedback loop that regulates E2Fc through RBR, retroactively linking tiers 3 and 4 (Fig. 1).

2.3 Transcription factors regulating the phenylpropanoid pathway

Fine transcriptomic dissection of hybrid aspen developing xylem shows that monolignol biosynthetic genes are induced in phase with SCW master regulators as well as SCW cellulose and hemicellulose biosynthetic genes, while homologs of laccase enzymes known to play a role in lignin polymerization show a tight expression peak at the height of SCW deposition

(Sundell et al., 2017). Single-cell analysis of protoplasts isolated from developing secondary xylem, on the other hand, reveal that monolignol biosynthesis is more pronounced in ray parenchyma relative to fibers and vessels (Chen et al., 2021), demonstrating that the transcriptional regulation of monolignol biosynthesis can be uncoupled from cellulose and hemicellulose deposition and even PCD. Some R2R3-MYB-domain TFs have evidently evolved specialized roles in regulating the general and lignin-specific phenylpropanoid pathways, accompanied by many pleiotropic roles in biotic and abiotic stress responses. It is well established that AC-like elements are overrepresented in the promoters of genes encoding phenylpropanoid, lignin and flavonoid pathway enzymes which are bound by R2R3-MYB TFs (e.g., Hatton et al., 1995; Lois, Dietrich, Hahlbrock, & Schulz, 1989; Raes, Rohde, Christensen, Peer, & Boerjan, 2003; Sablowski et al., 1994; Zhong & Ye, 2012). These MYB TFs play an important role in regulating carbon flux into and downstream of the phenylpropanoid pathway, with some members activating or repressing both lignification and flavonoid branches simultaneously, while the majority regulate 4-coumaryl-CoA flux to either lignin or flavonoid production by playing contrasting roles in activating expression of the enzymes involved in one branch while repressing those in the other (Zhang, Yang, Li, Chiang, & Fu, 2021).

Among the more prominent of the activators of the lignin pathway are MYB20, MYB42, MYB43, MYB58, MYB61, MYB63 and MYB85 (Geng et al., 2020; Newman, Perazza, Juda, & Campbell, 2004; Romano et al., 2012; Zhong et al., 2008;). MYB58 functions redundantly with MYB63 to activate lignin deposition in Arabidopsis xylem (Zhou et al., 2009), while their rice ortholog is capable of transactivating both lignin (*OsCAldOMT1*) and cellulose (*OsCesA7*) gene promoters (Noda et al., 2015). Overexpression and transactivation studies showed that MYB85 activated the lignin-specific pathway but not cellulose or xylan deposition during SCW deposition, while the gene encoding MYB85 is activated by MYB46/83 and all of SND2 through SND5 (McCarthy et al., 2009; Nakano et al., 2010; Zhong et al., 2008; Zhong, Lee, et al., 2021). More recently, Geng et al. (2020) demonstrated that MYB20, MYB42, MYB43 and MYB85 directly activate not only lignin but also the general phenylpropanoid pathway through phenylalanine biosynthesis enzymes AROGENATE DEHYDRATASE 4 (ADT4) through ADT6. This MYB quartet preferentially promotes lignification by simultaneously repressing the flavonoid pathway through the direct activation of transcriptional repressors MYB4, MYB7 and MYB32, which in turn repress flavonoid-specific *CHS* (Geng et al., 2020; Wang et al., 2020).

The ectopic lignification resulting from overexpression of *MYB61* suggests a lignin-preferential role for this TF (Newman et al., 2004). However, reverse genetics analysis demonstrates a complex pleiotropic role in carbon allocation, with its loss- and gain-of-function affecting germination rate and leaf senescence, vessel number and SCW thickness, as well as root growth and the ratio of primary to lateral roots (Romano et al., 2012). A close *P. tomentosa* homolog PtoMYB74 promoted SCW deposition and lignification when overexpressed in Arabidopsis and *Populus*, where it activated cellulose- and lignin-specific SCW pathways but not xylan biosynthesis, as well as homologs of *SND2* and *MYB103* (Li, Ma, Yu, Fu, & Luo, 2018). This finding was corroborated by the functional analysis of a related homolog, PtrMYB121 (Liu et al., 2020). Thus, MYB61 and its homologs appear to regulate xylem development, cellulose and lignin deposition, and several other developmental processes.

Among the MYB TFs that negatively regulate lignin biosynthesis are MYB3, MYB4, MYB7, MYB32, and potentially MYB69. The latter has an unclear biological role, since yeast assays suggest that MYB69 has transcriptional activation activity and activates the *C3H1* and *4CL1* promoter, while it represses both early (*PAL1*, *HCT*) and late (*CCoAOMT1*, *F5H*) steps of the lignin pathway (Öhman et al., 2012; Zhong et al., 2008). Given the repressive effect of its poplar homologs (Section 3.2), it is most likely a repressor of SCW formation. MYB3 interacts with the *C4H*, *HCT* and *4CL1* promoters (Taylor-Teeples et al., 2015) and acts together with NIGHT LIGHT-INDUCIBLE AND CLOCK-REGULATED1 (LNK1) and LNK2 to co-repress the *C4H* promoter (Zhou et al., 2017). MYB4 represses the phenylpropanoid pathway through *C4H*, its downregulation allowing for synapate ester accumulation in response to UV-B exposure (Jin et al., 2000). *MYB7*, which is repressed by MYB4, also regulates UV sunscreens by repressing the flavonoid pathway and early phenylpropanoid pathway (Fornale et al., 2014), while *MYB32* is highly expressed in the anther tapetum and may affect pollen development by repressing *COMT* and altering phenylpropanoid pathway flux (Preston, Wheeler, Heazlewood, Li, & Parish, 2004). Although definitive evidence is lacking that MYB7 and MYB32 are important in regulating SCW formation, they along with MYB4 are activated by Tier 2 master regulators and can repress the *SND1* promoter *in vivo* (Wang, Zhao, Chen, Wang, & Dixon, 2011; Xiao, Zhang, Guo, Li, & Lu, 2021), suggesting a potentially complex role in SCW lignification.

3. Much of the same, yet one of a kind: Conservation and divergence of secondary cell wall transcriptional control

The identification and functional characterization of homologs of Arabidopsis SWN and MYB SCW master regulators in distantly related dicots such as *Populus* (McCarthy et al., 2010; Ohtani et al., 2011; Takata et al., 2019; Zhong, McCarthy, Lee, & Ye, 2011), monocots such as *Brachypodium, Musa, Oryza, Panicum* and *Zea* (Negi, Tak, & Ganapathi, 2017; Rao et al., 2019; Valdivia et al., 2013; Yoshida et al., 2013; Zhong et al., 2011, 2015) and gymnosperms such as *Pinus* (Akiyoshi et al., 2020; Pascual et al., 2018; Patzlaff et al., 2003, 2003; Yao et al., 2021) suggests that these transcriptional switches are largely conserved in vascular plants (Nakano et al., 2015). A wealth of functional studies from diverse woody angiosperm and gymnosperm lineages demonstrate largely conserved biological roles played by putative orthologs of the core SCW TRN (reviewed by Zhang, Xie, Tuskan, et al., 2018). However, recent data also suggest that a number of lineage-specific TFs with SCW-related functions as well as substantial network re-wiring exists across different plant lineages and species. In this section, I briefly discuss the evidence of a conserved "core" network, as well as the apparent existence of "accessory" network components that have evolved in some groups.

3.1 Evolution and conservation of the SCW-associated transcription factors in land plants

R2R3-MYB TFs are evolutionarily ancient proteins found across distantly related eukaryotes (Du et al., 2015). Bowman et al. (2017) showed that SCW-related MYB domain TFs evolved following extensive gene duplication in vascular plants from single-copy progenitors in moss (*Physcomitrella patens*). A broader taxonomic analysis by Jiang and Rao (2020) revealed that five single-copy moss ancestral genes gave rise to the clades containing all known SCW-related R2R3-MYB TFs. The common ancestor of the monophyletic group MYB20/40/42/43/85/99 (clade VIII-A-2) first appeared in bryophytes, as did the ancestor of clade VIII-A-1 which include MYB26, MYB46, MYB67 and MYB83. However, the earliest gene duplication events giving rise to MYB46/83 (subclade VIII-A-1-[s2]), MYB3/4/7/32 (subclade VIII-*E*-[s1]) and MYB52/54/69 (subfamily V) can be traced to their gymnosperm ancestors, while the earliest emergence of MYB58/63 (clade VIII-D-[s2]) can be traced to Monilophytes (ferns)

(Jiang & Rao, 2020). This suggests that SCW-related R2R3-MYB TFs diversified in vascular plants from single-copy ancestors in moss, where paralogs acquired SCW-related as well as non-SCW-related regulatory functions in derived lineages.

The first appearance of the NAC protein family is evident in extant charophycean algae such as *Coleochaete orbicularis* and *Spirogyra pratensis*, after which it expanded and diversified significantly in land plants (Bowman et al., 2017). The VNS members, encompassing VND, NST and SMB/BRN members of clade Ic (Zhu, Nevo, Sun, & Peng, 2012), have representatives in *C. orbicularis* and are conserved in bryophytes (*Physcomitrella patens*, *Marchantia polymorpha*), spikemoss (*Selaginella moellendorffii*), coniferous gymnosperms (*Picea abies*) and angiosperms (Ohtani, Akiyoshi, Takenaka, Sano, & Demura, 2017). The multiple VND homologs in *Physcomitrella patens* are important for the correct development of hydroid cells and wall thickening in gametophyte stereid cells (Xu et al., 2014). Furthermore, ectopic overexpression of PpVNS7 led to activation of SCW deposition, lignification and PCD developmental pathways in both moss and Arabidopsis, presumably assisted by the upregulation of the progenitors of MYB26, MYB46/83, MYB85 and MYB103 (Jiang & Rao, 2020; Xu et al., 2014). Similarly, peat moss (*Sphagnum palustre*) VNS homologs induced ectopic SCW deposition when overexpressed in the dicot *Nicotiana benthamiana* (Terada et al., 2021). Together, these data support the early evolution of VNS-mediated development of water-conducting cells in all land plants, followed by their diversification in vascular plants.

3.2 Divergence in secondary cell wall transcriptional regulation

Despite the apparent conservation of the NAC-MYB-X model in vascular plants, there are exceptions to the rule. Given the various and distinct whole-genome duplication histories of different angiosperm groups (Van de Peer, Mizrachi, & Marchal, 2017), it is reasonable to expect that gene duplication and loss events might have affected SCW TRN topology. In the case of the Salicaceae, which underwent a relatively recent salicoid whole genome duplication (WGD) event (Tuskan et al., 2006), it is unsurprising that *Populus trichocarpa* possesses a number of functionally redundant pairs of each of the VND, NST and SMB/BRN groups of the Ic clade (Hu et al., 2010; Zhong, Lee, & Ye, 2010b; Zhu et al., 2012). However, it also appears that this genus has evolved a number of unique modulations to the SCW network structure. I focus on three broad examples.

The first example highlights functional divergence following gene duplication. An expanded number of SCW-associated cellulose synthase

(*CesA*) genes in poplar (Kumar et al., 2009) has led to divergence in expression patterns between the paralogous promoters of *Pt×tCesA7-A* and *Pt×tCesA7-B,* as well as those of *Pt×tCesA8-A* and *Pt×tCesA8-B,* in hybrid *Populus tremula* × *P. tremuloides* (Takata & Taniguchi, 2015). Interestingly, promoter transactivation screening revealed that the duplicated paralogs are often regulated by different combinations MYB and NAC-domain TFs, demonstrating SCW transcriptional network re-wiring following the salicoid WGD event (Takata & Taniguchi, 2015). Similarly, functional divergence has also occurred between *Populus tomentosa* homeologs PtoMYB158 and PtoMYB189 since the salicoid WGD (Jiang et al., 2020). The *P. trichocarpa* ortholog of PtoMYB189, PtrMYB189, is an MYB69-related transcriptional repressor of SCW cellulose, xylan and lignin biosynthesis that has evolved a novel repression motif shared by PtrMYB158 but distinct from that of the related repressors MYB4/7/32 (Jiao, Zhao, Lu, Guo, & Luo, 2019). Yet, MYB69 does not share two aspartic acid residues in the PtrMYB189 repression motifs that is essential for repression (Jiao et al., 2019), suggesting novel repression mechanisms in the poplar homologs. Although overexpression of *PtoMYB158* and *PtoMYB189* both confer a similar phenotype of stunted growth and reduced SCW deposition, they have divergent expression patterns and do not repress the same SCW-related genes, while PtoMYB158 alone was shown to directly repress *PtoSND1-B2* (Jiang et al., 2020).

A second example relates to the evolution of lineage-specific post-transcriptional regulatory mechanisms. Some *Populus trichocarpa* members of both the VND group of SWNs (*PtrVND6-C1*) and the NST group (*PtrSND1-A2*) can produce intron-retaining alternative splice variants that encode truncated proteins which heterodimerise with their full-length counterparts and exert a dominant negative effect on TF complex function. The intron-retaining truncated splice variants not only heterodimerise with proteins from the opposite group to inhibit their self-activation, but also with paralogs within their own group, while PtrSND1-A2 and PtrVND6-C1 genes are themselves repressed by the heterodimerization of the alternatively spliced variants of each other in a reciprocal manner (Li et al., 2012; Lin et al., 2017) (Fig. 2). This was independently shown for *PtrSND1-A2* by Zhao, Sun, Xu, Zhang, and Li (2014) who also demonstrated that the intronic sequence encoding the splicing pattern is absent in Arabidopsis. Although the function of this cross-regulatory mechanism is unclear, it is postulated to assist with regulating the stoichiometric ratio between PtrVND6 versus PtrSND1 family members in *Populus* (Lin et al., 2017).

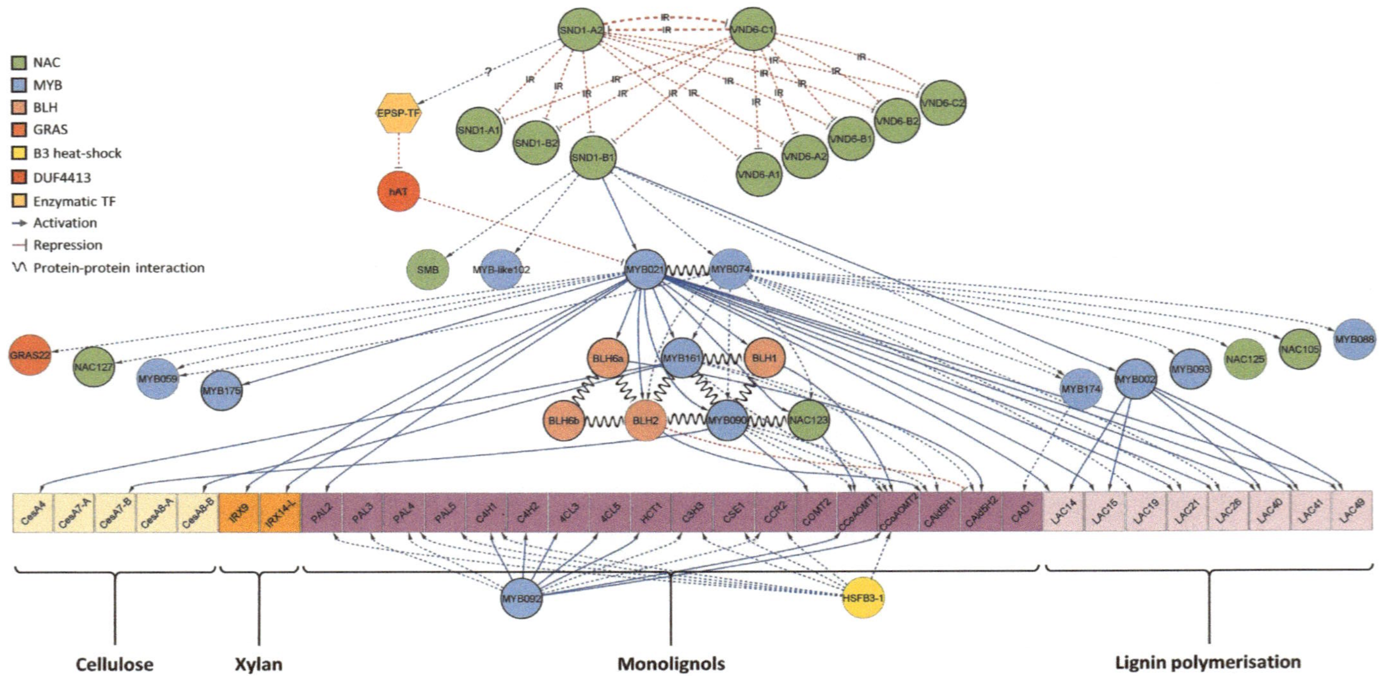

Fig. 2 Model of the transcriptional regulatory network underlying SCW deposition and lignification in *Populus*. Edges conserved with Arabidopsis are indicated by solid lines, while nodes with a solid border indicate those with known SCW-related orthologs in Arabidopsis. IR, intron-retaining alternative splicing variant. Evidence lines (direct targets only) were obtained from genome-wide top-down experiments by Lin et al. (2013; 2017), Chen et al. (2019) and Liu et al. (2021) for cellulose, xylan, monolignol biosynthesis and putative monolignol polymerization genes. Directed TF-target interactions reported in other studies (e.g., Gui et al., 2019; Jiao et al., 2019; McCarthy et al., 2010; Takata & Taniguchi, 2015; Tang et al., 2015; Wang et al., 2021; Yang et al., 2017; Zhong et al., 2010b; Zhong, McCarthy, Haghighat, & Ye, 2013; Zhong & Ye, 2010) have been omitted to avoid target selection bias. All interactions are assumed to be conserved across *Populus* spp.. Interactions supported exclusively by yeast one-hybrid evidence are not represented.

In the third example, it was evident from the work of Lin et al. (2013) and Chen et al. (2019) that the gene targets directly activated by PtrSND1-B1 in *Populus trichocarpa* secondary developing xylem-derived protoplasts are largely novel, having orthologs that are not known to be activated by SND1 in Arabidopsis (Zhong et al., 2010a). In fact, only *PtrMYB021*, a close homolog of *MYB46* from Arabidopsis, seemed to be common among the direct targets of the PtrSND1-B1-directed TRN (Chen et al., 2019). Similarly, the direct gene targets of PtrMYB021 differ substantially from those of MYB46 (Zhong & Ye, 2012), demonstrating that there has been significant network rewiring between the relevant species (Chen et al., 2019). A number of *in vivo* validations of the PtrSND1-B1 and PtrMYB021 direct targets, such as ChIP-qPCR and stable transgenic poplar experiments, strongly support this claim (Chen et al., 2019; Lin et al., 2013).

Figure 2 depicts a representation of the known direct targets of the poplar SCW TRN with the combined evidence lines from these and other studies. It also shows which nodes are conserved in the SCW TRN of Arabidopsis (permitting reasonable assumptions about orthology), along with TF-target interaction edges that are either shared with Arabidopsis or unique to *Populus* to the best of our knowledge. In particular, *MYB46/83* homologs *PtrMYB002* and *PtrMYB021* are highly conserved direct targets of PtrSND1-B1 (i.e., in common with SND1), where PtrMYB021 directly activates conserved network nodes *PtrNAC127* (a homolog of *SND4/5*), three close homologs of *MYB52/54* (*PtrMYB90*, *PtrMYB161* and *PtrMYB175*), *SND2/3* homolog *PtrNAC123*, as well as *PtrBLH6a* and *PtrBLH1* (homologs of *BLH6/7* and *BLH3/10* respectively). PtrMYB092 is closely related to MYB42 and MYB85. Non-conserved nodes include a MYB-like protein 102 with unclear homology, PtrMYB059 (a homolog of MYB107), PtrMYB174 (unclear homology), PtrMYB088 (homolog of MYB36), PtrNAC125 (a possible homolog of ANAC041), PtrSMB (homolog of AtSMB), a heat shock factor related to HSFB3, a new BEL1-like homeodomain protein related to BLH2, PtrGRAS22 (homolog of HAM3) and novel TFs PtrMYB074 and PtrEPSP-TF (discussed in Section 3.3).

3.3 Novel and "woody preferential" transcription factors involved in the transcriptional control of lignification

From Sections 3.1 and 3.2, we can broadly deduce that the NAC-MYB-X model of SCW regulation contains a relatively conserved core set of nodes that are widely shared among vascular plants as orthologous groups, although the topology of the SCW network may differ markedly between lineages. Is it possible, though, that transcription factors have evolved novel SCW-related regulatory functions in isolated lineages, or that SCW-related TFs have been preferentially retained in some lineages but lost in others? Given that woody perennials such as trees assimilate substantially more fixed carbon into xylogenesis than herbaceous annuals, and regulate the vascular cambium in response to seasonal and various abiotic influences over many years (Begum et al., 2017; Bhalerao & Fischer, 2017; Neale, Martinez-Garcia, De La Torre, Montanari, & Wei, 2017), the question emerges whether they have evolved additional TFs dedicated to boosting or fine-tuning the regulation of this process. Genome-wide studies of the large MYB and NAC TF families in woody plants provide some evidence for the existence of clades that are limited to, dominated by, or expanded in woody dicots (Camargo, Ployet, Cassan-Wang, Mounet, & Grima-Pettenati, 2019). These TFs may constitute "accessory" network components that are poorly conserved across vascular plants.

In their comparative genomics study of the R2R3-MYB TF family of *Eucalyptus grandis*, Soler et al. (2015) identified three subgroups (S5, S6, SAtMYB5) that were evidently expanded in woody representatives (*E. grandis*, *Populus trichocarpa*, *Vitis vinifera*) relative to non-woody dicots (Arabidopsis) and monocots (*O. sativa*), where the few Arabidopsis members were known to regulate proanthocyanidin and anthocyanin biosynthesis. Another five so-called Woody Plant Specific (WPS) subgroups were only present in representatives with true vascular cambiums (that is, gymnosperms and dicots, but not monocots, lycophytes or nonvascular plants). These subgroups were preferentially expanded in woody perennial dicots and gymnosperms but underrepresented, or completely absent (e.g., Brassicaceae), in nonwoody dicots. The preferential expression of *Eucalyptus* members of WPS-I, WPS-II and WPS-III in the vascular cambium and shoot tip suggest a specialized role in meristematic regulation of phenylpropanoid-derived compounds produced in differentiating xylem (Soler et al., 2015). Subsequently, EgrMYB88 (WPS-I) from *E. gunnii* was also found to increase the S:G lignin ratio and the concentrations of the flavonoid catechin and cinnamate-derived salicinoid phenolic glycosides when overexpressed in a woody background (poplar), but not a nonwoody background (Arabidopsis) (Soler et al., 2016). Similarly, in poplar, a focused phylogenetic revision of subgroup 13-related PtrMYB074 and its homologs across 69 species showed that it, too, groups preferentially with woody dicots and is only distantly related to MYB7, MYB17, MYB35 and

MYB50 from Arabidopsis (Chen et al., 2019). PtrMYB074, which was shown to interact with the MYB46 homolog PtrMYB021, activated a number of monolignol biosynthetic genes when overexpressed in developing xylem protoplasts, suggesting that this woody-preferential clade plays a unique role in regulating specialized aspects of xylogenesis (Chen et al., 2019) (Fig. 2).

Our previous work on the 189-member NAC domain family of *E. grandis* revealed conserved representatives of Arabidopsis SWNs in *Eucalyptus*, but evidence of substantial tandem duplication in four of the 22 identified subfamilies (Hussey, Saïdi, Hefer, & Grima-Pettenati, 2015). Interestingly, of these, subfamily IVc contained 21 *E. grandis* members versus only two in Arabidopsis (ANAC061 and ANAC090), most of them expressed preferentially in developing secondary xylem, phloem and/or roots. *EgrNAC141* and its *E. globulus* ortholog were preferentially expressed in stems and developing xylem and strongly upregulated during opposite wood formation, prompting speculation that it may be a novel regulator of xylogenesis-related processes in *Eucalyptus* (Hussey, Saïdi, et al., 2015). In an independent study, Sun et al. (2021) confirmed this hypothesis, showing that its overexpression in Arabidopsis increased interfascicular fiber production and upregulated most monolignol biosynthetic genes, with directed promoter transactivation assays demonstrating that EgrNAC141 directly activates the Arabidopsis *CCoAOMT1*, *CCR1*, *CAFFEOYL SHIKIMATE ESTERASE* (*CSE*), *COMT* and *CAD1* gene promoters. Since the expression patterns of Arabidopsis subfamily IVc representatives *ANAC061* and *ANAC090* from are leaf-preferential and the latter demonstrated to play a role in leaf senescence (with *ANAC061* upregulated in *anac090* mutants) (Kim et al., 2018), it appears that *EgrNAC141* and potentially other paralogs in this woody-expanded clade have evolved novel lignification-associated regulatory functions.

Finally, it appears that even enzymes can evolve unique transcriptional regulatory activities for lineage-specific SCW regulation. From a *Populus trichocarpa* association study, Xie et al. (2018) discovered a modified 5-enolpyruvylshikimate 3-phosphate synthase containing an N-terminal DNA-binding helix-turn-helix with dual enzymatic and transcriptional regulatory activity (henceforth called PtrEPSP-TF). Through a series of *in vitro* and *in planta* experiments, they demonstrated that PtrEPSP-TF is a transcriptional repressor that promotes lignin biosynthesis in poplar by directly repressing a *SLEEPER*-like transcriptional regulator hAT, which in turn directly represses *PtrMYB021* (Xie, Muchero, et al., 2018).

The helix-turn-helix motif of PtrEPSP-TF, which is surprisingly absent from its closest paralog, is largely restricted to dicot genomes (Xie, Muchero, et al., 2018). The enzymatic transcriptional regulator, which is thought to be directly activated by PtrSND1-A2 (also known as PtrWND1B), is capable of upregulating the lignin pathway in rice through homologs of *SLEEPER-like* homologs of *hAT*, even though rice lacks the helix-turn-helix motif (Xie et al., 2020). The same study identified five rice targets of EPSP-TF, namely *OsXND1*, *OsMYB48*, *OsMYB3R1*, *OsMYB3R3* and *OsNAC047*, which are directly repressed by EPSP-TF and thought to negatively regulate lignification in rice. Together, these interesting findings show that plants have evolved novel modifications to the core network regulating SCW and lignin formation, likely in response to dynamic genome evolution and yet to be understood adaptive pressures.

4. Post-translational and combinatorial regulation of lignification

4.1 Post-translational regulation of secondary cell wall associated transcription factors

Covalent and non-covalent post-translational modifications (PTMs) are well known to dynamically and reversibly affect the activity of key regulators of cellular differentiation during plant development (Aguilar-Hernández, Brito-Argáez, Galaz-Ávalos, & Loyola-Vargas, 2020). Enzymes and TFs involved in lignification and its regulation are subject to extensive post-translational regulation involving phosphorylation, glycosylation, S-nitrosylation, ubiquitination and sumoylation (McCahill & Hazen, 2019; Sulis & Wang, 2020).

Phosphorylation is required for the activation of SCW-associated cellulose synthases and certain lignin-associated biosynthetic enzymes such as PtrAldOMT2 (Speicher, Li, & Wallace, 2018; Wang et al., 2015). Recently, it was uncovered that the Arabidopsis NST1 SWN is phosphorylated at serine 316 through an abscissic acid (ABA) signaling cascade, crucially affecting its ability to activate SCW-related target genes (Liu, Yu, Rao, Li, & Dixon, 2021) (Fig. 1). Among others, ABA mediates physiological responses to abiotic stresses such as water deficit (Sah, Reddy, & Li, 2016). ABA binds to and activates Pyrabactin Resistance 1-like (PYL) proteins, which in turn interact with and inhibit type 2C protein phosphatases (PP2Cs) clade A, which are negative regulators of ABA signaling (reviewed by Ng, Melcher, Teh, & Xu, 2014). The inhibition of PP2C phosphatases leads to phosphorylation and activation of Snf1 (sucrose nonfermenting 1)-Related Kinase (SnRK2) proteins, which serve as positive regulators of ABA signaling by phosphorylating TFs and ion channels involved in the physiological response to ABA (Kulik, Wawer, Krzywinska, Bucholc, & Dobrowolska, 2011). A *snrk2.2/2.3/2.6* triple mutant exhibited a severe reduction in SCW deposition in inflorescence stem fibers and vessels, as did an ABA biogenesis mutant *aba2* (Liu, Yu, et al., 2021). The same study showed that SnRK2.2, SnRK2.3 and SnRK2.6 interact with NST1 *in vitro* and *in vivo*, and are capable of phosphorylating NST1 at Ser316 *in vitro*. Unlike wild-type NST1, the introduction of a phosphorylation null mutant $NST1_{(S316A)}$ failed to restore SCW deposition in the *snd1 nst1* double mutant, suggesting that the *in vivo* phosphorylation of this residue is crucial for its regulatory function. To prove this, Liu, Yu, et al. (2021) employed promoter transactivation assays in Arabidopsis mesophyll protoplasts to show that the mutant effector $NST1_{(S316A)}$ could not activate the promoters of *MYB46*, *MYB83* or its own promoter to the same extent as wild-type NST1. The Ser316 phosphorylation site is conserved in NST1 homologs across dicots but is noticeably absent in monocots as well as SND1 (Liu, Yu, et al., 2021). Together, these data suggest that ABA-mediated phosphorylation of NST1 is not only a requirement for normal SCW deposition in Arabidopsis xylem, but also that increased rates of NST1 phosphorylation during abiotic stress may lead to enhanced SCW deposition and lignification through the ABA signaling pathway in dicots.

The half-life of a TF, like all proteins, is largely determined by the difference in the rate of translation and proteolysis. Polyubiquitination of proteins by E3 ubiquitin ligases may tag them for 26S proteasome-mediated degradation, following the stepwise action of E1 ubiquitin-activating and E2 ubiquitin-conjugating enzymes (Glickman & Ciechanover, 2002). *Populus tomentosa* PtoMYB156 and *P. deltoides* PdMYB221, both co-orthologs of MYB4, are conserved repressors of SCW biogenesis (Tang et al., 2015; Yang et al., 2017). Zheng et al. (2019) demonstrated that the *P. tomentosa* E2 ubiquitin-conjugating enzyme 34 (PtoUBC34) interacts *in vivo* with both repressors, reducing their repression activities by translocating them to the endoplasmic reticulum. Whether or not they are proteolytically degraded following translocation is unresolved. Interestingly, it was shown that lignin associated transcription factor 1 (LTF1) from *Populus*, which is synonymous with PdMYB221, does indeed undergo ubiquitin-mediated degradation, but this is dependent on wound-responsive phosphorylation of Thr146 and Thr178 by the mitogen-activated protein kinase PdMPK6 (Gui et al., 2019). The molecular mechanism by which the phosphorylation-dependent degradation of PdMYB221 occurs remains to be understood.

SIZ1-mediated sumoylation, which involves the covalent attachment of ubiquitin-related Small Ubiquitin-like Modifier (SUMO), may inhibit or promote binding to an interacting protein, alter TF function through a conformation change, or act to recruit transcriptional inhibitory factors (Miura & Hasegawa, 2010). In the Arabidopsis SCW TRN, Liu et al. (2019) showed that *siz1* mutants are defective in inflorescence fiber, but not vessel, SCW deposition. They further demonstrated that SIZ1 interacts with LBD30, a protein generally associated with the activation of VND6 and VND7 during vessel differentiation (Section 2.2; Fig. 1). SIZ1-mediated sumoylation of LBD30 at lysine-226 is essential for the correct activation of NST1/SND1-directed fiber SCW deposition, with site-directed mutation of K226 resulting in reduced *NST1* and *SND1* promoter activation and a failure of $LBD30_{(K226R)}$-overexpressing plants to exhibit the severe dwarfing and ectopic lignification observed in wild-type LBD30-overexpressing plants (Liu et al., 2019).

For over a decade, nitric oxide (NO) has been implicated in the post-translational modification of SCW development (Correa-Aragunde, Lombardo, & Lamattina, 2008). More recently, it has been found to post-translationally regulate VND7. Dexamethasone-mediated post-translational induction of VND7 causes ectopic transdifferentiation of non-tracheary element cells into vessel-like elements (Yamaguchi et al., 2010). The abundance of VND7 protein affects not only the rate of vessel cell differentiation but also the qualities of the SCW and whether the PCD pathway will be activated (Hirai et al., 2019), suggesting that the fine regulation of VND7 protein activity is important for vessel differentiation. From a forward genetics screen, Kawabe, Ohtani, Kurata, Sakamoto, and Demura (2018) identified mutant Arabidopsis seedlings that were resistant to VND7-induced vessel transdifferentation. One so-called *suppressor of ectopic vessel cell differentiation induced by VND7* (*seiv*) mutant, *seiv1*, was found to have a loss-of-function point mutation in *S-NITROSOGLUTATHIONE REDUCTASE1* (*GSNOR1*), an enzyme that reduces an important NO donor molecule, *S*-nitrosoglutathione (GSNO) (Frungillo, Skelly, Loake, Spoel, & Salgado, 2014; Kawabe et al., 2018). The *siev1* and *gsnor1* mutant plants exhibited an accumulation of NO, displayed stunted growth, developed abnormal vessels and failed to undergo ectopic SCW deposition when VND7 was

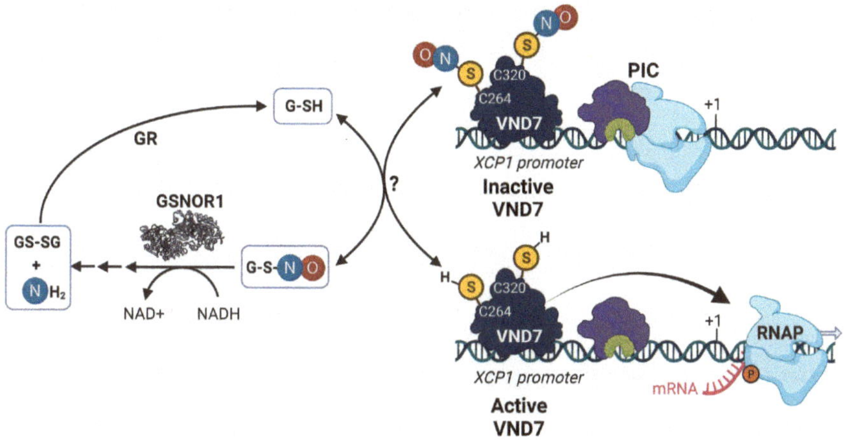

Fig. 3 Post-translational regulation of VND7 through *S*-nitrosylation. *S*-nitrosoglutathione (G-S-NO) serves as a major intracellular nitric oxide reservoir and is formed by the *S*-nitrosylation of reduced glutathione (G-SH). It can be degraded into an unstable intermediate by the reductase GSNOR1 and further converted into oxidized glutathione (GS-SG), which is then recycled into the reduced form by glutathione reductase (GR). G-S-NO can also serve as donor for the formation of high molecular weight *S*-nitrosothiols through the *S*-transnitrosation of cysteine sulfhydryl (-SH) groups on proteins (Corpas, Alche, & Barroso, 2013). Kawabe et al. (2018) propose that VND7 is *S*-transnitrosated through an unknown mechanism when G-S-NO abundance is elevated in the *gsnor1* mutant, leading to post-translational inactivation of VND7 that possibly prohibits its ability to activate the pre-initiation complex (PIC) at SCW-related promoters such as *XCP1*. Restoration of GSNOR1 activity reduces the G-S-NO metabolite pool and restores the reduced sulfhydryl groups of Cys264 and Cys320 of VND7, allowing for transcriptional activation. RNAP, RNA polymerase. Illustrated with BioRender.com.

post-translationally induced. The authors determined this was caused by reduced *S*-nitrosylation of Cys264 and Cys320 residues of VND7, which they could show resulted in the inhibition of PCD-related *XCP1* promoter transactivation (Kawabe et al., 2018). As illustrated in Fig. 3, the authors propose that the accumulation of GSNO in the *gsnor1* mutant leads to *S*-transnitrosation of Cys264 and Cys320 of VND7 by an unknown mechanism.

4.2 Combinatorial regulation of SCW transcriptional regulation

TFs do not act alone or independently in the transcriptional regulation of a gene; rather, they typically interact and cooperate with other TFs and regulatory proteins at gene promoters (Spitz & Furlong, 2012). The clustering of their cis-regulatory elements form cis-regulatory modules (CRMs) which improve the specificity of locus-specific transcriptional control, while flexibility of TF membership at CRMs dramatically expand the possible mechanisms by which TFs can regulate their target genes (Brkljacic & Grotewold, 2017). In this section, I discuss recent evidence of combinatorial regulation of SCW biogenesis through the formation of variable protein complexes. The reader should take note that while some complexes have been shown to exist *in vivo*, most others are hypothetical complexes derived from pairwise protein-protein interaction data that remain to be validated as such.

MYB-bHLH-WD40-repeat (MBW) complexes are well known to underlie the regulation of anthocyanins and proanthocyanidins which are derived from the flavonoid branch of the phenylpropanoid pathway, as well as root hair, seed coat mucilage and trichome development in eudicots (Gonzalez, Mendenhall, Huo, & Lloyd, 2009; Li et al., 2009; Serna & Martin, 2006; Zhang & Schrader, 2017). The WD40-repeat (WDR) protein TRANSPARENT TESTA GLABRA1 (TTG1) is generally associated with MBW complexes, with common basic helix-loop-helix (bHLH) proteins including TRANSPARENT TESTA 8 (TT8), GLABRA 3 (GL3), ENHANCER OF GLABRA 3 (EGL3) and MYC1, and R2R3-MYB proteins such as GL1, MYB5, MYB75/PAP1, MYB61 and TT2/MYB123 (Baudry et al., 2004; Broun, 2005; Gonzalez et al., 2009). Crucially, bHLH and WDR proteins associate with various R2R3-MYB species to regulate different developmental processes and distinct branches of the flavonoid pathway, and may undergo various modifications such as dimerization via their MYB or bHLH members (Xu, Dubos, & Lepiniec, 2015). In eudicots, some bHLH members of these complexes may engage in dynamic and competitive binding with either R2R3-MYB or WDR proteins, as is the case with GL3 which forms competing dimers with either GL1 or TTG1 to regulate different promoters during trichome differentiation, counteracted by alternative complex formation in the presence of the third complementary protein (Pesch et al., 2015). Other combinations are synergistic, such as the association of TT8 with TTG1 and R2R3-MYB proteins such as MYB75 to form stable activation complexes (Zhang, Chopra, Schrader, & Hulskamp, 2019) (Fig. 4C). MYB75 (also known as PAP1) is a positive regulator of the phenylpropanoid and anthocyanin pathway (Bhargava et al., 2013; Bhargava, Mansfield, Hall, Douglas, & Ellis, 2010; Borevitz, Xia, Blount, Dixon, & Lamb, 2000; Teng, Keurentjes, Bentsink, Koornneef, & Smeekens, 2005) that can form MBW complexes with TT8, GL3/EGL3 and TTG1 (Gonzalez, Zhao, Leavitt, & Lloyd, 2008;

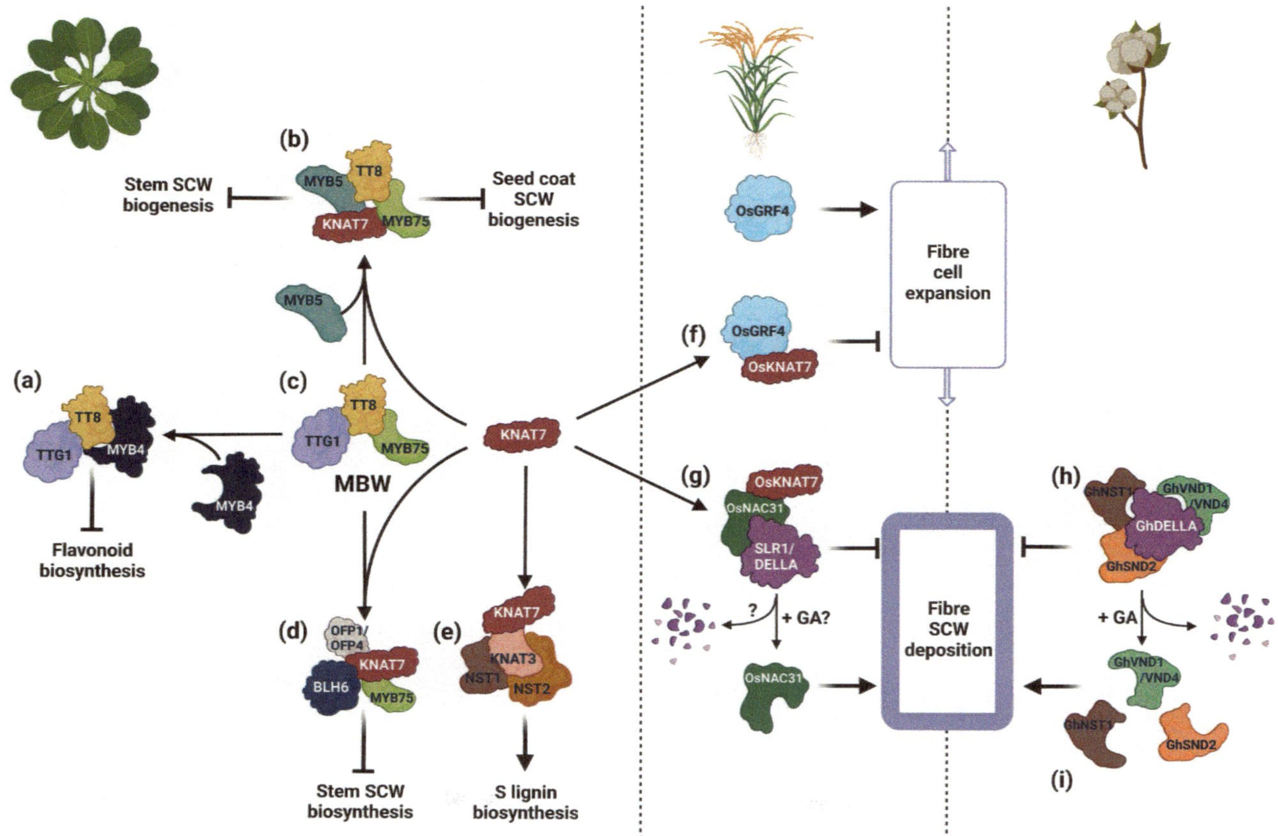

Fig. 4 Variable protein complexes are proposed to regulate secondary cell wall deposition, lignification, flavonoid biosynthesis and fiber development. Examples are discussed in *Arabidopsis thaliana* (left panel), *Oryza sativa* (middle panel) and *Gossypium hirsutum* (right panel) in Section 4.2. Illustrated with BioRender.com.

Ramsay & Glover, 2005; Zhang et al., 2019; Zhang, Gonzalez, Zhao, Payne, & Lloyd, 2003; Zimmermann, Heim, Weisshaar, & Uhrig, 2004). However, the association of the SCW transcriptional repressors MYB4, MYB7 or MYB32 with this complex inhibits MBW complex formation through their direct interaction with TT8 *in planta* and the subsequent repression of late flavonoid biosynthesis and pre-phenylpropanoid pathway genes such as *ADT6* (Dubos et al., 2008; Wang et al., 2020) (Fig. 4A).

KNOTTED-LIKE HOMEOBOX OF ARABIDOPSIS THALIANA7 (KNAT7) is a KNOTTED homeodomain box (KNOX) TF that often negatively regulates SCW deposition (Gong et al., 2014; Li et al., 2012), despite being directly regulated by the SWNs (Zhong et al., 2008) as well as MYB46/83 (Ko et al., 2009; McCarthy et al., 2009; Zhong & Ye, 2012). From over a decade of research, we now know that KNAT7 associates with several TFs including components of MBW complexes, resulting in different regulatory outcomes. KNAT7 interacts with MYB75 and MYB5 but not with TT8, while TT8 interacts with MYB5, possibly forming a complex that is thought to repress cell wall biosynthesis in the seed coat (Bhargava et al., 2010, 2013) (Fig. 4B). MYB75 loss-of-function analysis and overexpression analysis of MYB5 homologs in poplar (PtoMYB6), along with evidence of conserved interactions with poplar homologs of TT8 and KNAT7, further suggests that this complex also represses stem SCW biogenesis (Bhargava et al., 2010; Wang et al., 2019) (Fig. 4B). It remains to be proven whether the tetrameric complex exists *in vivo* and whether TTG1 also associates with it to form a stabilized MBW complex. KNAT7 may act independently of MYB5-TT8 by interacting with MYB75, OVATE FAMILY PROTEIN4 (OFP1) and OFP4 to repress fiber SCW formation (Bhargava et al., 2013; Li, Wang, Liu, Chen, & Douglas, 2011) while KNAT7 interaction with BLH6 appears to enhance its repression of SCW biosynthesis genes, in part by repressing the promoter of the HD-ZIP III TF *REVOLUTA/INTERFASCICULAR FIBERLESS1* (*REV/IFL1*) (Liu & Douglas, 2015) (Fig. 4D). It is still unknown whether these proteins constitute a separate complex to that proposed in Fig. 4B. Since BLH6, like KNAT7, is likely a Tier 1 regulator (Zhong & Ye, 2012) while REV/IFL1 is important for interfascicular fiber SCW deposition and is thought to act upstream of the NAC-MYB-X cascade (Endo et al., 2015; Lev-Yadun, Wyatt, & Flaishman, 2005; Talbert, Adler, Parks, & Comai, 1995; Zhong, Burk, & Ye, 2001; Zhong, Taylor, & Ye, 1997; Zhong & Ye, 1999) (Fig. 1), it is possible that a negative feedback loops exists between *REV* and the KNAT7-BLH6-containing protein complexes.

The assertion that KNAT7 is exclusively a repressor of SCW deposition is controversial. It is challenged by the *irregular xylem* (*irx*) phenotype observed in *atknat7* mutants due to vessel collapse despite thicker interfascicular fibers, that *KNAT7* overexpression has variable effects on xylary fiber and vessel SCW deposition and the observation that KNAT7 activates xylan biosynthetic genes (He et al., 2018; Li, Bhargava, et al., 2012; Liu & Douglas, 2015). This suggests that KNAT7 may regulate SCW deposition in a cell type-specific manner depending on the composition of the TF complex. For instance, Qin et al. (2020) reported that KNAT7 can heterodimerise with the broadly expressed KNAT3 TF, promoting the upregulation of the syringyl, but not the guaiacyl branch, of the lignin pathway through indirect *F5H* activation. KNAT3 may achieve this by interacting with SWNs NST1 and NST2, while KNAT7 itself is unable to interact with these SWNs (Qin et al., 2020) (Fig. 4E). Although it is not known whether KNAT3/NST1/NST2 and the KNAT7-associated components interact simultaneously, it is clear that KNAT7 may participate ad hoc in different complexes to promote or inhibit SCW deposition. In rice, it is instead the KNAT7 ortholog OsKNAT7 that interacts with SWNs, in this case orthologs of NST1 (OsNAC29) and VND7 (OsNAC31), where it counteracts the transcription activation of rice SCW-related TF genes *OsMYB61* and *OsMYB103* by OsNAC31 (Wang et al., 2019) (Fig. 4G). Remarkably, OsKNAT7 also interacts *in vivo* with GROWTH REGULATING FACTOR4 (OsGRF4), a cell growth regulator that promotes organ growth and grain size and activates expansin gene expression (Che et al., 2015; Duan et al., 2015), forming a complex that inhibits OsGRF4 transcriptional activation and reduces sclerenchyma fiber growth (Wang, Yang, et al., 2019) (Fig. 4F). Since *OsKNAT7*, *OsNAC31* and *OsGRF4* have largely overlapping expression in stem internodes and expanding panicles and are co-expressed in microdissected sclerenchyma fiber cells, it is proposed that OsKNAT7 can modulate fiber expansion as well as fiber SCW deposition by alternatively interacting with OsGRF4 or OsNAC31, respectively (Wang, Yang, et al., 2019) (Fig. 4, middle panel).

It is noteworthy that the activities of OsNAC29 and OsNAC31 can also be inhibited by a different mechanism than that involving KNAT7: in the absence of gibberellic acid (GA), SLENDER RICE1 (SLR1) can bind to the DNA binding domains of OsNAC29 and OsNAC31 and prevent their activation of second-tier SCW-related MYB TFs, whereas application of GA results in their release from SLR1 (Huang et al., 2015) (Fig. 4G). SLR1 is a DELLA transcriptional repressor of GA-dependent processes in plants

(Daviere & Achard, 2013). In cotton (*Gossypium hirsutum*), GhDELLA proteins such as GhGAI1D interact with and sequester cotton orthologs of SWNs VND1, VND4, NST1 and SND2, thus inhibiting stem xylem SCW formation and lignification (Fig. 4H), while application of GA results in the degradation of GhDELLAs and release of the SWNs to activate the SCW transcriptional program (Wang et al., 2021) (Fig. 4I). Presumably, a similar phenomenon occurs with SLR1 in rice (Fig. 4G). The finding that DELLAs also interact with KNAT1 as part of the GA signaling pathway regulating fiber differentiation in Arabidopsis (Felipo-Benavent et al., 2018) raises the interesting possibility that multiple KNOX and SWN TFs may associate with DELLA proteins in the post-translational control of SCW deposition.

5. Conclusion and future challenges

Since its humble beginnings less than 20 years ago when the master regulators of fiber and vessel SCW deposition were first identified, researchers have made substantial progress in understanding the biological functions of dozens of TFs comprising the SCW TRN and uncovering the hundreds of regulatory interactions that have been critical for inferring its multi-tiered structure. We see that while comparative genomics studies of network nodes suggested that the Arabidopsis SCW TRN is relatively well-conserved with woody angiosperms, monocots and even gymnosperms, the TRN has also evolved novel regulators and undergone substantial rewiring (Sections 3.2 and 3.3). This poses a challenge for translational research and suggests that we require substantial species-specific data on the physical structure of the SCW TRN of non-model target crops to be able to re-engineer SCW deposition.

This chapter focused on physical interactions comprising the SCW TRN and the biological functions of key network nodes. Although efforts have also been made to reconstruct gene regulatory network (GRN) models for wood formation and SCW deposition (e.g., Lin et al., 2013; Ployet et al., 2019; Quan et al., 2021; Taylor-Teeples et al., 2015; Tong, Chen, & Williams, 2021), we still lack robust and quantitative predictive models of how SCW-related TF dynamics affect target gene expression and impact xylem formation and wood property traits in the long run. It is useful to incorporate not just high-resolution transcriptomics for better co-expression network inference, but also to integrate orthogonal -omics data including both physical data (i.e., molecular interactions, many of which are purely incidental) and functional data (transcriptomes, metabolomes, epigenomes, genetic associations, phenotypes, etc.) into context-relevant quantitative GRN models (Hawe, Theis, & Heinig, 2019; Qian & Huang, 2020). High-throughput TF interaction mapping (O'Malley et al., 2016), transient protoplast-based TF perturbation assays (Bargmann et al., 2013; Brooks et al., 2019), co-expression data from large sets of diverse and high-resolution transcriptomes, and the integration of predictions from multiple algorithms has greatly improved our ability to quantitatively model the relationships between TF and target gene expression levels (Geng, Wang, Gong, Xu, & Ma, 2021; Haque, Ahmad, Clark, Williams, & Sozzani, 2019; Kulkarni, Vaneechoutte, Van de Velde, & Vandepoele, 2018; Marbach et al., 2012; Song et al., 2020). The development of effective protocols for developing xylem protoplast transfection and time series transcriptome analysis (Lin et al., 2014), coupled to improved ChIP protocols to identify direct *in vivo* TF-target interactions for this challenging tissue (Hussey, Mizrachi, Groover, Berger, & Myburg, 2015; Li et al., 2014; Liu, Missirian, Zinkgraf, Groover, & Filkov, 2014) has been particularly powerful in the comprehensive reconstruction of a hierarchical TRN of SCW formation in poplar in a relatively short period (Fig. 2). Furthermore, increased resolution of transcriptomic data from micrometer-scale cryosections (Jokipii-Lukkari et al., 2017; Sundell et al., 2017), FACS-sorted labeled nuclei (Shi et al., 2021), laser-capture microdissection (LCM) (Blokhina et al., 2016; Shi et al., 2017) and single protoplasts derived from developing secondary xylem (Chen et al., 2021; Li et al., 2021) provide valuable data for critical developmental phases and fine-scale spatial contexts. As a recent example, Yu et al. (2022) integrated RNA-seq analyses of LCM-dissected tension wood (TW) forming cells and that of TW from gene-edited trees with mutations in *PtrLBD22* and *PtrLBD39*, two poplar co-orthologs of *LBD4* which are induced during TW formation. Coupling this context-specific and fine-scale functional data with physical interaction data from a ChIP-seq analysis of PtrLBD39 allowed them to reconstruct a comprehensive TW-related TRN in which PtrLBD39 activates 26 reportedly novel TW-responsive TFs.

Finally, we are gradually gaining insight into the critical role of the post-translational and combinatorial regulation of SCW deposition. We have seen from protein-protein interaction data that a plethora of dynamic complexes may form, with variable effects on target gene activation depending on their composition and covalent modifications (Section 4). However, we largely lack empirical evidence of TF complex composition *in vivo* and the biological contexts in which they occur. Similarly, the functional relevance of the protein-protein interactions identified in the poplar SCW TRN (Fig. 2) is still largely unknown. Identification of the full spectrum of PTMs occurring in SCW-related TFs *in vivo* is an exciting area of future research, potentially revealing new avenues for the manipulation of lignocellulosic biomass.

References

Aguilar-Hernández, V., Brito-Argáez, L., Galaz-Ávalos, R. M., & Loyola-Vargas, V. M. (2020). Post-translational modifications drive plant cell differentiation. *Plant Cell, Tissue and Organ Culture (PCTOC), 143*(1), 1–12. https://doi.org/10.1007/s11240-020-01908-0.

Akiyoshi, N., Nakano, Y., Sano, R., Kunigita, Y., Ohtani, M., & Demura, T. (2020). Involvement of VNS NAC-domain transcription factors in tracheid formation in *Pinus taeda. Tree Physiology, 40*(6), 704–716. https://doi.org/10.1093/treephys/tpz106.

Balazsi, G., van Oudenaarden, A., & Collins, J. J. (2011). Cellular decision making and biological noise: From microbes to mammals. *Cell, 144*(6), 910–925. https://doi.org/10.1016/j.cell.2011.01.030.

Bar-On, Y. M., Phillips, R., & Milo, R. (2018). The biomass distribution on earth. *Proceedings of the National Academy of Sciences of the United States of America, 115*(25), 6506–6511. https://doi.org/10.1073/pnas.1711842115.

Bargmann, B. O., Marshall-Colon, A., Efroni, I., Ruffel, S., Birnbaum, K. D., Coruzzi, G. M., et al. (2013). TARGET: A transient transformation system for genome-wide transcription factor target discovery. *Molecular Plant, 6*(3), 978–980. https://doi.org/10.1093/mp/sst010.

Barros, J., Serk, H., Granlund, I., & Pesquet, E. (2015). The cell biology of lignification in higher plants. *Annals of Botany, 115*(7), 1053–1074. https://doi.org/10.1093/aob/mcv046.

Baudry, A., Heim, M. A., Dubreucq, B., Caboche, M., Weisshaar, B., & Lepiniec, L. (2004). TT2, TT8, and TTG1 synergistically specify the expression of *BANYULS* and proanthocyanidin biosynthesis in *Arabidopsis thaliana. The Plant Journal, 39*, 366–380.

Begum, S., Kudo, K., Rahman, M. H., Nakaba, S., Yamagishi, Y., Nabeshima, E., et al. (2017). Climate change and the regulation of wood formation in trees by temperature. *Trees, 32*(1), 3–15. https://doi.org/10.1007/s00468-017-1587-6.

Berthet, S., Demont-Caulet, N., Pollet, B., Bidzinski, P., Cezard, L., Le Bris, P., et al. (2011). Disruption of LACCASE4 and 17 results in tissue-specific alterations to lignification of *Arabidopsis thaliana* stems. *Plant Cell, 23*(3), 1124–1137. https://doi.org/10.1105/tpc.110.082792.

Bhalerao, R. P., & Fischer, U. (2017). Environmental and hormonal control of cambial stem cell dynamics. *Journal of Experimental Botany, 68*(1), 79–87. https://doi.org/10.1093/jxb/erw466.

Bhargava, A., Ahad, A., Wang, S., Mansfield, S. D., Haughn, G. W., Douglas, C. J., et al. (2013). The interacting MYB75 and KNAT7 transcription factors modulate secondary cell wall deposition both in stems and seed coat in *Arabidopsis. Planta, 237*, 1199–1211.

Bhargava, A., Mansfield, S. D., Hall, H. C., Douglas, C. J., & Ellis, B. E. (2010). MYB75 functions in regulation of secondary cell wall formation in the *Arabidopsis* inflorescence stem. *Plant Physiology, 154*(3), 1428–1438.

Blokhina, O., Valerio, C., Sokolowska, K., Zhao, L., Karkonen, A., Niittyla, T., et al. (2016). Laser capture microdissection protocol for xylem tissues of Woody plants. *Frontiers in Plant Science*, 7, 1965. https://doi.org/10.3389/fpls.2016.01965.

Borevitz, J. O., Xia, Y., Blount, J., Dixon, R. A., & Lamb, C. (2000). Activation tagging identifies a conserved MYB regulator of Phenylpropanoid biosynthesis. *The Plant Cell, 12*, 2383–2394.

Bowman, J. L., Kohchi, T., Yamato, K. T., Jenkins, J., Shu, S., Ishizaki, K., et al. (2017). Insights into land plant evolution garnered from the *Marchantia polymorpha* genome. *Cell, 171*(2), 287–304 e215. https://doi.org/10.1016/j.cell.2017.09.030.

Brkljacic, J., & Grotewold, E. (2017). Combinatorial control of plant gene expression. *Biochim Biophys Acta Gene Regul Mech, 1860*(1), 31–40. https://doi.org/10.1016/j.bbagrm.2016.07.005.

Brooks, M. D., Cirrone, J., Pasquino, A. V., Alvarez, J. M., Swift, J., Mittal, S., et al. (2019). Network walking charts transcriptional dynamics of nitrogen signaling by integrating validated and predicted genome-wide interactions. *Nature Communications, 10*(1), 1569. https://doi.org/10.1038/s41467-019-09522-1.

Broun, P. (2005). Transcriptional control of flavonoid biosynthesis: A complex network of conserved regulators involved in multiple aspects of differentiation in *Arabidopsis*. *Current Opinion in Plant Biology, 8*(3), 272–279. https://doi.org/10.1016/j.pbi.2005.03.006.

Camargo, E. L. O., Ployet, R., Cassan-Wang, H., Mounet, F., & Grima-Pettenati, J. (2019). Chapter seven—Digging in wood: New insights in the regulation of wood formation in tree species. In F. M. Cánovas (Ed.), *Vol. 89. Advances in botanical research* (pp. 201–233). Academic Press.

Cassan-Wang, H., Goué, N., Saidi, M. N., Legay, S., Sivadon, P., Goffner, D., et al. (2013). Identification of novel transcription factors regulating secondary cell wall formation in *Arabidopsis*. *Frontiers in Plant Science, 4*, 189.

Chai, G., Kong, Y., Zhu, M., Yu, L., Qi, G., Tang, X., et al. (2015). *Arabidopsis* C3H14 and C3H15 have overlapping roles in the regulation of secondary wall thickening and anther development. *Journal of Experimental Botany, 66*(9), 2595–2609. https://doi.org/10.1093/jxb/erv060.

Che, R., Tong, H., Shi, B., Liu, Y., Fang, S., Liu, D., et al. (2015). Control of grain size and rice yield by GL2-mediated brassinosteroid responses. *Nat Plants, 2*, 15195. https://doi.org/10.1038/nplants.2015.195.

Chen, H., Wang, J. P., Liu, H., Li, H., Lin, Y. J., Shi, R., et al. (2019). Hierarchical transcription factor and chromatin binding network for wood formation in black cottonwood (*Populus trichocarpa*). *Plant Cell, 31*(3), 602–626. https://doi.org/10.1105/tpc.18.00620.

Chen, Y., Tong, S., Jiang, Y., Ai, F., Feng, Y., Zhang, J., et al. (2021). Transcriptional landscape of highly lignified poplar stems at single-cell resolution. *Genome Biology, 22*(1), 319. https://doi.org/10.1186/s13059-021-02537-2.

Coomey, J. H., Sibout, R., & Hazen, S. P. (2020). Grass secondary cell walls, *Brachypodium distachyon* as a model for discovery. *The New Phytologist, 227*(6), 1649–1667. https://doi.org/10.1111/nph.16603.

Corpas, F. J., Alche, J. D., & Barroso, J. B. (2013). Current overview of S-nitrosoglutathione (GSNO) in higher plants. *Frontiers in Plant Science, 4*, 126. https://doi.org/10.3389/fpls.2013.00126.

Correa-Aragunde, N., Lombardo, C., & Lamattina, L. (2008). Nitric oxide: An active nitrogen molecule that modulates cellulose synthesis in tomato roots. *The New Phytologist, 179*(2), 386–396. https://doi.org/10.1111/j.1469-8137.2008.02466.x.

Dai, X., Liu, N., Wang, L., Li, J., Zheng, X., Xiang, F., et al. (2020). MYB94 and MYB96 additively inhibit callus formation via directly repressing *LBD29* expression in *Arabidopsis thaliana*. *Plant Science, 293*, 110323. https://doi.org/10.1016/j.plantsci.2019.110323.

Daviere, J. M., & Achard, P. (2013). Gibberellin signaling in plants. *Development, 140*(6), 1147–1151. https://doi.org/10.1242/dev.087650.

Dejardin, A., Laurans, F., Arnaud, D., Breton, C., Pilate, G., & Leple, J. C. (2010). Wood formation in angiosperms. *Comptes Rendus Biologies*, *333*(4), 325–334. https://doi.org/10.1016/j.crvi.2010.01.010.

Deng, Y., & Lu, S. (2017). Biosynthesis and regulation of Phenylpropanoids in plants. *Critical Reviews in Plant Sciences*, *36*(4), 257–290. https://doi.org/10.1080/07352689.2017.1402852.

Desvoyes, B., & Gutierrez, C. (2020). Roles of plant retinoblastoma protein: Cell cycle and beyond. *The EMBO Journal*, *39*(19), e105802. https://doi.org/10.15252/embj.2020105802.

Du, H., Liang, Z., Zhao, S., Nan, M. G., Tran, L. S., Lu, K., et al. (2015). The evolutionary history of R2R3-MYB proteins across 50 eukaryotes: New insights into subfamily classification and expansion. *Scientific Reports*, *5*, 11037. https://doi.org/10.1038/srep11037.

Duan, P., Ni, S., Wang, J., Zhang, B., Xu, R., Wang, Y., et al. (2015). Regulation of *OsGRF4* by OsmiR396 controls grain size and yield in rice. *Nat Plants*, *2*, 15203. https://doi.org/10.1038/nplants.2015.203.

Dubos, C., Le Gourrierec, J., Baudry, A., Huep, G., Lanet, E., Debeaujon, I., et al. (2008). MYBL2 is a new regulator of flavonoid biosynthesis in *Arabidopsis thaliana*. *The Plant Journal*, *55*(6), 940–953. https://doi.org/10.1111/j.1365-313X.2008.03564.x.

Endo, H., Yamaguchi, M., Tamura, T., Nakano, Y., Nishikubo, N., Yoneda, A., et al. (2015). Multiple classes of transcription factors regulate the expression of VASCULAR-RELATED NAC-DOMAIN7, a master switch of xylem vessel differentiation. *Plant & Cell Physiology*, *56*(2), 242–254. https://doi.org/10.1093/pcp/pcu134.

Felipo-Benavent, A., Urbez, C., Blanco-Tourinan, N., Serrano-Mislata, A., Baumberger, N., Achard, P., et al. (2018). Regulation of xylem fiber differentiation by gibberellins through DELLA-KNAT1 interaction. *Development*, *145*(23). https://doi.org/10.1242/dev.164962.

Feng, Z., Sun, X., Wang, G., Liu, H., & Zhu, J. (2012). LBD29 regulates the cell cycle progression in response to auxin during lateral root formation in *Arabidopsis thaliana*. *Annals of Botany*, *110*(1), 1–10. https://doi.org/10.1093/aob/mcs019.

Fornale, S., Lopez, E., Salazar-Henao, J. E., Fernandez-Nohales, P., Rigau, J., & Caparros-Ruiz, D. (2014). AtMYB7, a new player in the regulation of UV-sunscreens in *Arabidopsis thaliana*. *Plant & Cell Physiology*, *55*(3), 507–516. https://doi.org/10.1093/pcp/pct187.

Fraser, C. M., & Chapple, C. (2011). The Phenylpropanoid pathway in Arabidopsis. *The Arabidopsis Book*, *9*. https://doi.org/10.1199/tab.0152.

Frungillo, L., Skelly, M. J., Loake, G. J., Spoel, S. H., & Salgado, I. (2014). S-nitrosothiols regulate nitric oxide production and storage in plants through the nitrogen assimilation pathway. *Nature Communications*, *5*, 5401. https://doi.org/10.1038/ncomms6401.

Geng, H., Wang, M., Gong, J., Xu, Y., & Ma, S. (2021). An Arabidopsis expression predictor enables inference of transcriptional regulators for gene modules. *The Plant Journal*, *107*(2), 597–612. https://doi.org/10.1111/tpj.15315.

Geng, P., Zhang, S., Liu, J., Zhao, C., Wu, J., Cao, Y., et al. (2020). MYB20, MYB42, MYB43, and MYB85 regulate phenylalanine and lignin biosynthesis during secondary cell wall formation. *Plant Physiology*, *182*(3), 1272–1283. https://doi.org/10.1104/pp.19.01070.

Ghelli, R., Brunetti, P., Napoli, N., De Paolis, A., Cecchetti, V., Tsuge, T., et al. (2018). A newly identified flower-specific splice variant of *AUXIN RESPONSE FACTOR8* regulates stamen elongation and Endothecium lignification in Arabidopsis. *Plant Cell*, *30*(3), 620–637. https://doi.org/10.1105/tpc.17.00840.

Glickman, M. H., & Ciechanover, A. (2002). The ubiquitin-proteasome proteolytic pathway: Destruction for the sake of construction. *Physiological Reviews*, *82*(2), 373–428. https://doi.org/10.1152/physrev.00027.2001.

Gong, S. Y., Huang, G. Q., Sun, X., Qin, L. X., Li, Y., Zhou, L., et al. (2014). Cotton *KNL1*, encoding a class II KNOX transcription factor, is involved in regulation of fibre development. *Journal of Experimental Botany*, *65*(15), 4133–4147. https://doi.org/10.1093/jxb/eru182.

Gonzalez, A., Mendenhall, J., Huo, Y., & Lloyd, A. (2009). TTG1 complex MYBs, MYB5 and TT2, control outer seed coat differentiation. *Developmental Biology*, *325*(2), 412–421.

Gonzalez, A., Zhao, M., Leavitt, J. M., & Lloyd, A. M. (2008). Regulation of the anthocyanin biosynthetic pathway by the TTG1/bHLH/Myb transcriptional complex in Arabidopsis seedlings. *The Plant Journal*, *53*(5), 814–827. https://doi.org/10.1111/j.1365-313X.2007.03373.x.

Grant, E. H., Fujino, T., Beers, E. P., & Brunner, A. M. (2010). Characterization of NAC domain transcription factors implicated in control of vascular cell differentiation in *Arabidopsis* and *Populus*. *Planta*, *232*, 337–352.

Gui, J., Luo, L., Zhong, Y., Sun, J., Umezawa, T., & Li, L. (2019). Phosphorylation of LTF1, an MYB transcription factor in *Populus*, acts as a sensory switch regulating lignin biosynthesis in wood cells. *Molecular Plant*, *12*(10), 1325–1337. https://doi.org/10.1016/j.molp.2019.05.008.

Haque, S., Ahmad, J. S., Clark, N. M., Williams, C. M., & Sozzani, R. (2019). Computational prediction of gene regulatory networks in plant growth and development. *Current Opinion in Plant Biology*, *47*, 96–105. https://doi.org/10.1016/j.pbi.2018.10.005.

Hatton, D., Sablowski, R., Vung, M.-H., Smith, C., Schuch, W., & Bevan, M. (1995). Two classes of *cis* sequences contribute to tissue-specific expression of a *PAL2* promoter in transgenic tobacco. *The Plant Journal*, 7(6), 859–876.

Hawe, J. S., Theis, F. J., & Heinig, M. (2019). Inferring interaction networks from multi-omics data. *Frontiers in Genetics*, *10*, 535. https://doi.org/10.3389/fgene.2019.00535.

He, J. B., Zhao, X. H., Du, P. Z., Zeng, W., Beahan, C. T., Wang, Y. Q., et al. (2018). KNAT7 positively regulates xylan biosynthesis by directly activating *IRX9* expression in *Arabidopsis*. *Journal of Integrative Plant Biology*, *60*(6), 514–528. https://doi.org/10.1111/jipb.12638.

Hirai, R., Higaki, T., Takenaka, Y., Sakamoto, Y., Hasegawa, J., Matsunaga, S., et al. (2019). The progression of xylem vessel cell differentiation is dependent on the activity level of VND7 in *Arabidopsis thaliana*. *Plants (Basel)*, *9*(1). https://doi.org/10.3390/plants9010039.

Hu, R., Qi, G., Kong, Y., Kong, D., Gao, Q., & Zhou, G. (2010). Comprehensive analysis of NAC domain transcription factor gene family in *Populus trichocarpa*. *BMC Plant Biology*, *10*, 145.

Huang, D., Wang, S., Zhang, B., Shang-Guan, K., Shi, Y., Zhang, D., et al. (2015). A gibberellin-mediated DELLA-NAC signaling Cascade regulates cellulose synthesis in Rice. *Plant Cell*, *27*(6), 1681–1696. https://doi.org/10.1105/tpc.15.00015.

Hussey, S. G., Mizrachi, E., Creux, N. M., & Myburg, A. A. (2013). Navigating the transcriptional roadmap regulating plant secondary cell wall deposition. *Frontiers in Plant Science*, *4*, 325.

Hussey, S. G., Mizrachi, E., Groover, A., Berger, D. K., & Myburg, A. A. (2015). Genome-wide mapping of histone H3 lysine 4 trimethylation in *Eucalyptus grandis* developing xylem. *BMC Plant Biology*, *15*, 117.

Hussey, S. G., Mizrachi, E., Spokevicius, A. V., Bossinger, G., Berger, D. K., & Myburg, A. A. (2011). *SND2*, a NAC transcription factor gene, regulates genes involved in secondary cell wall development in *Arabidopsis* fibres and increases fibre cell area in *Eucalyptus*. *BMC Plant Biology*, *11*, 173.

Hussey, S. G., Saïdi, M. N., Hefer, C. A., & Grima-Pettenati, A. A. M. J. (2015). Structural, evolutionary and functional analysis of the NAC domain protein family in *Eucalyptus*. *New Phytologist*, *206*(4), 1337–1350.

Jiang, C. K., & Rao, G. Y. (2020). Insights into the diversification and evolution of R2R3-MYB transcription factors in plants. *Plant Physiology*, *183*(2), 637–655. https://doi.org/10.1104/pp.19.01082.

Jiang, P. F., Xu, H., Guan, C. N., Wang, X. X., Wu, A. M., Liu, Y. J., et al. (2020). Functional divergence of *Populus* MYB158 and MYB189 gene pair created by whole genome duplication. *Journal of Systematics and Evolution, (accepted manuscript)*. https://doi.org/10.1111/jse.12595.

Jiao, B., Zhao, X., Lu, W., Guo, L., & Luo, K. (2019). The R2R3 MYB transcription factor MYB189 negatively regulates secondary cell wall biosynthesis in *Populus*. *Tree Physiology*, *39*(7), 1187–1200. https://doi.org/10.1093/treephys/tpz040.

Jin, H., Cominelli, E., Bailey, P., Parr, A., Mehrtens, F., Jones, J., et al. (2000). Transcriptional repression by AtMYB4 controls production of UV-protecting sunscreens in *Arabidopsis*. *The EMBO Journal*, *19*(22), 6150–6161.

Jokipii-Lukkari, S., Sundell, D., Nilsson, O., Hvidsten, T. R., Street, N. R., & Tuominen, H. (2017). NorWood: A gene expression resource for evo-devo studies of conifer wood development. *The New Phytologist*, *216*(2), 482–494. https://doi.org/10.1111/nph.14458.

Kawabe, H., Ohtani, M., Kurata, T., Sakamoto, T., & Demura, T. (2018). Protein S-Nitrosylation regulates xylem vessel cell differentiation in Arabidopsis. *Plant & Cell Physiology*, *59*(1), 17–29. https://doi.org/10.1093/pcp/pcx151.

Kim, H. J., Park, J. H., Kim, J., Kim, J. J., Hong, S., Kim, J., et al. (2018). Time-evolving genetic networks reveal a NAC troika that negatively regulates leaf senescence in *Arabidopsis*. *Proceedings of the National Academy of Sciences of the United States of America*, *115*(21), E4930–E4939. https://doi.org/10.1073/pnas.1721523115.

Kim, W. C., Kim, J. Y., Ko, J. H., Kang, H., & Han, K. H. (2014). Identification of direct targets of transcription factor MYB46 provides insights into the transcriptional regulation of secondary wall biosynthesis. *Plant Molecular Biology*, *85*(6), 589–599. https://doi.org/10.1007/s11103-014-0205-x.

Kim, W. C., Kim, J. Y., Ko, J. H., Kang, H., Kim, J., & Han, K. H. (2014). AtC3H14, a plant-specific tandem CCCH zinc-finger protein, binds to its target mRNAs in a sequence-specific manner and affects cell elongation in *Arabidopsis thaliana*. *The Plant Journal*, *80*(5), 772–784. https://doi.org/10.1111/tpj.12667.

Ko, J.-H., Kim, W.-C., & Han, K.-H. (2009). Ectopic expression of MYB46 identifies transcriptional regulatory genes involved in secondary wall biosynthesis in *Arabidopsis*. *The Plant Journal*, *60*(4), 649–665.

Ko, J.-H., Yang, S. H., Park, A. H., Lerouxel, O., & Han, K.-H. (2007). ANAC012, a member of the plant-specific NAC transcription factor family, negatively regulates xylary fiber development in *Arabidopsis thaliana*. *The Plant Journal*, *50*(6), 1035–1048.

Ko, J. H., Jeon, H. W., Kim, W. C., Kim, J. Y., & Han, K. H. (2014). The MYB46/MYB83-mediated transcriptional regulatory programme is a gatekeeper of secondary wall biosynthesis. *Annals of Botany*, *114*(6), 1099–1107. https://doi.org/10.1093/aob/mcu126.

Kubo, M., Udagawa, M., Nishikubo, N., Horiguchi, G., Yamaguchi, M., Ito, J., et al. (2005). Transcription switches for protoxylem and metaxylem vessel formation. *Genes and Development*, *19*, 1855–1860.

Kulik, A., Wawer, I., Krzywinska, E., Bucholc, M., & Dobrowolska, G. (2011). SnRK2 protein kinases—Key regulators of plant response to abiotic stresses. *OMICS*, *15*(12), 859–872. https://doi.org/10.1089/omi.2011.0091.

Kulkarni, S. R., Vaneechoutte, D., Van de Velde, J., & Vandepoele, K. (2018). TF2Network: Predicting transcription factor regulators and gene regulatory networks in Arabidopsis using publicly available binding site information. *Nucleic Acids Research*, *46*(6), e31. https://doi.org/10.1093/nar/gkx1279.

Kumar, M., Thammannagowda, S., Bulone, V., Chiang, V., Han, K. H., Joshi, C. P., et al. (2009). An update on the nomenclature for the cellulose synthase genes in *Populus*. *Trends in Plant Science*, *14*(5), 248–254. https://doi.org/10.1016/j.tplants.2009.02.004.

Lee, K. H., Du, Q., Zhuo, C., Qi, L., & Wang, H. (2019). LBD29-involved auxin signaling represses NAC master regulators and fiber wall biosynthesis. *Plant Physiology*, *181*(2), 595–608. https://doi.org/10.1104/pp.19.00148.

Lev-Yadun, S., Wyatt, S. E., & Flaishman, M. A. (2005). The inflorescence stem fibers of *Arabidopsis thaliana* revoluta (*ifl1*) mutant. *Journal of Plant Growth Regulation*, *23*(4), 301–306.

Levy, E. D., Landry, C. R., & Michnick, S. W. (2010). Signaling through cooperation. *Science*, *328*, 983–984.

Li, C., Ma, X., Yu, H., Fu, Y., & Luo, K. (2018). Ectopic expression of *PtoMYB74* in poplar and Arabidopsis promotes secondary cell wall formation. *Frontiers in Plant Science*, *9*, 1262. https://doi.org/10.3389/fpls.2018.01262.

Li, E., Bhargava, A., Qiang, W., Friedmann, M. C., Forneris, N., Savidge, R. A., et al. (2012). The class II KNOX gene KNAT7 negatively regulates secondary wall formation in *Arabidopsis* and is functionally conserved in *Populus*. *The New Phytologist*, *194*(1), 102–115. https://doi.org/10.1111/j.1469-8137.2011.04016.x.

Li, E., Wang, S., Liu, Y., Chen, J.-G., & Douglas, C. J. (2011). OVATE FAMILY PROTEIN4 (OFP4) interaction with KNAT7 regulates secondary cell wall formation in *Arabidopsis thaliana*. *The Plant Journal*, *67*(2), 328–341.

Li, H., Dai, X., Huang, X., Xu, M., Wang, Q., Yan, X., et al. (2021). Single-cell RNA sequencing reveals a high-resolution cell atlas of xylem in *Populus*. *Journal of Integrative Plant Biology*, *63*(11), 1906–1921. https://doi.org/10.1111/jipb.13159.

Li, Q., Lin, Y. C., Sun, Y. H., Song, J., Chen, H., Zhang, X. H., et al. (2012). Splice variant of the SND1 transcription factor is a dominant negative of SND1 members and their regulation in *Populus trichocarpa*. *Proceedings of the National Academy of Sciences of the United States of America*, *109*(36), 14699–14704. https://doi.org/10.1073/pnas.1212977109.

Li, S. F., Milliken, O. N., Pham, H., Seyit, R., Napoli, R., Preston, J., et al. (2009). The Arabidopsis MYB5 transcription factor regulates mucilage synthesis, seed coat development, and trichome morphogenesis. *Plant Cell*, *21*(1), 72–89. https://doi.org/10.1105/tpc.108.063503.

Li, W., Lin, Y. C., Li, Q., Shi, R., Lin, C. Y., Chen, H., et al. (2014). A robust chromatin immunoprecipitation protocol for studying transcription factor-DNA interactions and histone modifications in wood-forming tissue. *Nature Protocols*, *9*(9), 2180–2193. https://doi.org/10.1038/nprot.2014.146.

Lin, Y.-C., Li, W., Chen, H., Li, Q., Sun, Y.-H., Shi, R., et al. (2014). A simple improved-throughput xylem protoplast system for studying wood formation. *Nature Protocols*, *9*(9), 2194–2205. https://doi.org/10.1038/nprot.2014.147.

Lin, Y.-C., Li, W., Sun, Y.-H., Kumari, S., Wei, H., Li, Q., et al. (2013). SND1 transcription factor–directed quantitative functional hierarchical genetic regulatory network in wood formation in *Populus trichocarpa*. *The Plant Cell*, *25*, 4324–4341.

Lin, Y. J., Chen, H., Li, Q., Li, W., Wang, J. P., Shi, R., et al. (2017). Reciprocal cross-regulation of VND and SND multigene TF families for wood formation in *Populus trichocarpa*. *Proceedings of the National Academy of Sciences of the United States of America*, *114*(45), E9722–E9729. https://doi.org/10.1073/pnas.1714422114.

Liu, B., Liu, J., Yu, J., Wang, Z., Sun, Y., Li, S., et al. (2021). Transcriptional reprogramming of xylem cell wall biosynthesis in tension wood. *Plant Physiology*, *186*(1), 250–269. https://doi.org/10.1093/plphys/kiab038.

Liu, C., Yu, H., & Li, L. (2019). SUMO modification of LBD30 by SIZ1 regulates secondary cell wall formation in *Arabidopsis thaliana*. *PLoS Genetics*, *15*(1), e1007928. https://doi.org/10.1371/journal.pgen.1007928.

Liu, C., Yu, H., Rao, X., Li, L., & Dixon, R. A. (2021). Abscisic acid regulates secondary cell-wall formation and lignin deposition in *Arabidopsis thaliana* through phosphorylation of NST1. *Proceedings of the National Academy of Sciences of the United States of America, 118*(5). https://doi.org/10.1073/pnas.2010911118.

Liu, H.-I., Wang, G.-C., Feng, Z., & Zhu, J. (2010). Screening of genes associated with dedifferentiation and effect of LBD29 on pericycle cells in *Arabidopsis thaliana. Plant Growth Regulation, 62*(2), 127–136. https://doi.org/10.1007/s10725-010-9495-4.

Liu, L., Missirian, V., Zinkgraf, M., Groover, A., & Filkov, V. (2014). Evaluation of experimental design and computational parameter choices affecting analyses of ChIP-seq and RNA-seq data in undomesticated poplar trees. *BMC Genomics, 15*(Suppl 5), S3. https://doi.org/10.1186/1471-2164-15-S5-S3.

Liu, Y., & Douglas, C. J. (2015). A role for OVATE FAMILY PROTEIN1 (OFP1) and OFP4 in a BLH6-KNAT7 multi-PROTEIN complex regulating secondary cell wall formation in *Arabidopsis thaliana. Plant Signaling & Behavior, 10*(7), e1033126. https://doi.org/10.1080/15592324.2015.1033126.

Liu, Y., Man, J., Wang, Y., Yuan, C., Shi, Y., Liu, B., et al. (2020). Overexpression of *PtrMYB121* positively regulates the formation of secondary cell wall in *Arabidopsis thaliana. International Journal of Molecular Sciences, 21*(20). https://doi.org/10.3390/ijms21207734.

Lois, R., Dietrich, A., Hahlbrock, K., & Schulz, W. (1989). A phenylalanine ammonia-lyase gene from parsley: Structure, regulation and identification of elicitor and light responsive cis-acting elements. *The EMBO Journal, 8*(6), 1641–1648.

Marbach, D., Costello, J. C., Kuffner, R., Vega, N. M., Prill, R. J., Camacho, D. M., et al. (2012). Wisdom of crowds for robust gene network inference. *Nature Methods, 9*(8), 796–804. https://doi.org/10.1038/nmeth.2016.

McCahill, I. W., & Hazen, S. P. (2019). Regulation of cell wall thickening by a medley of mechanisms. *Trends in Plant Science, 24*(9), 853–866. https://doi.org/10.1016/j.tplants.2019.05.012.

McCarthy, R. L., Zhong, R., Fowler, S., Lyskowski, D., Piyasena, H., Carleton, K., et al. (2010). The poplar MYB transcription factors, PtrMYB3 and PtrMYB20, are involved in the regulation of secondary wall biosynthesis. *Plant and Cell Physiology, 51*(6), 1084–1090.

McCarthy, R. L., Zhong, R., & Ye, Z.-H. (2009). MYB83 is a direct target of SND1 and acts redundantly with MYB46 in the regulation of secondary cell wall biosynthesis in *Arabidopsis. Plant and Cell Physiology, 50*(11), 1950–1964.

Meents, M. J., Watanabe, Y., & Samuels, L. A. (2018). The cell biology of secondary cell wall biosynthesis. *Annuls of Botany, 121*, 1107–1125. https://doi.org/10.1093/aob/mcy005.

Mitsuda, N., Iwase, A., Yamamoto, H., Yoshida, M., Seki, M., Shinozaki, K., et al. (2007). NAC transcription factors, NST1 and NST3, are key regulators of the formation of secondary walls in woody tissues of *Arabidopsis. The Plant Cell, 19*, 270–280.

Mitsuda, N., & Ohme-Takagi, M. (2008). NAC transcription factors NST1 and NST3 regulate pod shattering in a partially redundant manner by promoting secondary wall formation after the establishment of tissue identity. *The Plant Journal, 56*(5), 768–778.

Mitsuda, N., Seki, M., Shinozaki, K., & Ohme-Takagi, M. (2005). The NAC transcription factors NST1 and NST2 of *Arabidopsis* regulate secondary wall thickenings and are required for anther dehiscence. *Plant Cell, 17*, 2993–3006.

Miura, K., & Hasegawa, P. M. (2010). Sumoylation and other ubiquitin-like post-translational modifications in plants. *Trends in Cell Biology, 20*(4), 223–232. https://doi.org/10.1016/j.tcb.2010.01.007.

Miyamoto, T., Tobimatsu, Y., & Umezawa, T. (2020). MYB-mediated regulation of lignin biosynthesis in grasses. *Current Plant Biology, 24*. https://doi.org/10.1016/j.cpb.2020.100174.

Nakano, Y., Nishikubo, N., Goué, N., Ohtani, M., Yamaguchi, M., Katayama, Y., et al. (2010). MYB transcription factors orchestrating the developmental program of xylem vessels in *Arabidopsis* roots. *Plant Biotechnology*, *27*, 267–272.

Nakano, Y., Yamaguchi, M., Endo, H., Rejab, N. A., & Ohtani, M. (2015). NAC-MYB-based transcriptional regulation of secondary cell wall biosynthesis in land plants. *Frontiers in Plant Science*, *6*, 288. https://doi.org/10.3389/fpls.2015.00288.

Neale, D. B., Martinez-Garcia, P. J., De La Torre, A. R., Montanari, S., & Wei, X. X. (2017). Novel insights into tree biology and genome evolution as revealed through genomics. *Annual Review of Plant Biology*, *68*, 457–483. https://doi.org/10.1146/annurev-arplant-042916-041049.

Negi, S., Tak, H., & Ganapathi, T. R. (2017). Native vascular related NAC transcription factors are efficient regulator of multiple classes of secondary wall associated genes in banana. *Plant Science*, *265*, 70–86. https://doi.org/10.1016/j.plantsci.2017.09.018.

Newman, L. J., Perazza, D. E., Juda, L., & Campbell, M. M. (2004). Involvement of the R2R3-MYB, *At*MYB61, in the ectopic lignification and dark-photomorphogenic components of the *det3* mutant phenotype. *The Plant Journal*, *37*(2), 239–250.

Ng, L. M., Melcher, K., Teh, B. T., & Xu, H. E. (2014). Abscisic acid perception and signaling: Structural mechanisms and applications. *Acta Pharmacologica Sinica*, *35*(5), 567–584. https://doi.org/10.1038/aps.2014.5.

Noda, S., Koshiba, T., Hattori, T., Yamaguchi, M., Suzuki, S., & Umezawa, T. (2015). The expression of a rice secondary wall-specific cellulose synthase gene, OsCesA7, is directly regulated by a rice transcription factor, OsMYB58/63. *Planta*, *242*(3), 589–600. https://doi.org/10.1007/s00425-015-2343-z.

O'Malley, R. C., Huang, S. S., Song, L., Lewsey, M. G., Bartlett, A., Nery, J. R., et al. (2016). Cistrome and Epicistrome features shape the regulatory DNA landscape. *Cell*, *165*(5), 1280–1292. https://doi.org/10.1016/j.cell.2016.04.038.

Ohashi-Ito, K., Iwamoto, K., & Fukuda, H. (2018). LOB DOMAIN-CONTAINING PROTEIN 15 positively regulates expression of *VND7*, a master regulator of Tracheary elements. *Plant & Cell Physiology*, *59*(5), 989–996. https://doi.org/10.1093/pcp/pcy036.

Ohashi-Ito, K., Oda, Y., & Fukuda, H. (2010). *Arabidopsis* VASCULAR-RELATED NAC-DOMAIN6 directly regulates the genes that govern programmed cell death and secondary wall formation during xylem differentiation. *Plant Cell*, *22*(10), 3461–3473. https://doi.org/10.1105/tpc.110.075036.

Öhman, D., Demedts, B., Kumar, M., Gerber, L., Gorzsás, A., Goeminne, G., et al. (2012). MYB103 is required for *FERULATE-5-HYDROXYLASE* expression and syringyl lignin biosynthesis in *Arabidopsis* stems. *The Plant Journal*, *73*(1), 63–76.

Ohtani, M., Akiyoshi, N., Takenaka, Y., Sano, R., & Demura, T. (2017). Evolution of plant conducting cells: Perspectives from key regulators of vascular cell differentiation. *Journal of Experimental Botany*, *68*(1), 17–26. https://doi.org/10.1093/jxb/erw473.

Ohtani, M., & Demura, T. (2019). The quest for transcriptional hubs of lignin biosynthesis: Beyond the NAC-MYB-gene regulatory network model. *Current Opinion in Biotechnology*, *56*, 82–87. https://doi.org/10.1016/j.copbio.2018.10.002.

Ohtani, M., Nishikubo, N., Xu, B., Yamaguchi, M., Mitsuda, N., Goue, N., et al. (2011). A NAC domain protein family contributing to the regulation of wood formation in poplar. *The Plant Journal*, *67*(3), 499–512. https://doi.org/10.1111/j.1365-313X.2011.04614.x.

Okushima, Y., Fukaki, H., Onoda, M., Theologis, A., & Tasaka, M. (2007). ARF7 and ARF19 regulate lateral root formation via direct activation of *LBD/ASL* genes in *Arabidopsis*. *Plant Cell*, *19*(1), 118–130. https://doi.org/10.1105/tpc.106.047761.

Okushima, Y., Overvoorde, P. J., Arima, K., Alonso, J. M., Chan, A., Chang, C., et al. (2005). Functional genomic analysis of the *AUXIN RESPONSE FACTOR* gene family members in *Arabidopsis thaliana*: Unique and overlapping functions of *ARF7* and *ARF19*. *Plant Cell*, *17*(2), 444–463. https://doi.org/10.1105/tpc.104.028316.

Pascual, M. B., Llebres, M. T., Craven-Bartle, B., Canas, R. A., Canovas, F. M., & Avila, C. (2018). *PpNAC1*, a main regulator of phenylalanine biosynthesis and utilization in maritime pine. *Plant Biotechnology Journal*, *16*(5), 1094–1104. https://doi.org/10.1111/pbi.12854.

Patzlaff, A., McInnis, S., Courtenay, A., Surman, C., Newman, L. J., Smith, C., et al. (2003). Characterisation of a pine MYB that regulates lignification. *The Plant Journal*, *36*(6), 743–754.

Patzlaff, A., Newman, L. J., Dubos, C., Whetten, R. W., Smith, C., McInnis, S., et al. (2003). Characterisation of *Pt*MYB1, an R2R3-MYB from pine xylem. *Plant Molecular Biology*, *53*, 597–608.

Pesch, M., Schultheiss, I., Klopffleisch, K., Uhrig, J. F., Koegl, M., Clemen, C. S., et al. (2015). TRANSPARENT TESTA GLABRA1 and GLABRA1 compete for binding to GLABRA3 in Arabidopsis. *Plant Physiology*, *168*(2), 584–597. https://doi.org/10.1104/pp.15.00328.

Pesquet, E., Zhang, B., Gorzsás, A., Puhakainen, T., Serk, H., Escamez, S., et al. (2013). Non-cell-autonomous postmortem lignification of tracheary elements in *Zinnia elegans*. *The Plant Cell*, *25*(4), 1314–1328.

Ployet, R., Veneziano Labate, M. T., Regiani Cataldi, T., Christina, M., Morel, M., San Clemente, H., et al. (2019). A systems biology view of wood formation in *Eucalyptus grandis* trees submitted to different potassium and water regimes. *The New Phytologist*, *223*(2), 766–782. https://doi.org/10.1111/nph.15802.

Preston, J., Wheeler, J., Heazlewood, J., Li, S. F., & Parish, R. W. (2004). AtMYB32 is required for normal pollen development in *Arabidopsis thaliana*. *The Plant Journal*, *40*(6), 979–995. https://doi.org/10.1111/j.1365-313X.2004.02280.x.

Qian, Y., & Huang, S.-S. C. (2020). Improving plant gene regulatory network inference by integrative analysis of multi-omics and high resolution data sets. *Current Opinion in Systems Biology*, *22*, 8–15. https://doi.org/10.1016/j.coisb.2020.07.010.

Qin, W., Yin, Q., Chen, J., Zhao, X., Yue, F., He, J., et al. (2020). The class II KNOX transcription factors KNAT3 and KNAT7 synergistically regulate monolignol biosynthesis in Arabidopsis. *Journal of Experimental Botany*, *71*(18), 5469–5483. https://doi.org/10.1093/jxb/eraa266.

Qu, G., Peng, D., Yu, Z., Chen, X., Cheng, X., Yang, Y., et al. (2021). Advances in the role of auxin for transcriptional regulation of lignin biosynthesis. *Functional Plant Biology*, *48*(8), 743–754. https://doi.org/10.1071/FP20381.

Quan, M., Liu, X., Du, Q., Xiao, L., Lu, W., Fang, Y., et al. (2021). Genome-wide association studies reveal the coordinated regulatory networks underlying photosynthesis and wood formation in *Populus*. *Journal of Experimental Botany*, *72*(15), 5372–5389. https://doi.org/10.1093/jxb/erab122.

Raes, J., Rohde, A., Christensen, J. H., Peer, Y. V., & Boerjan, W. (2003). Genome-wide characterization of the lignification toolbox in *Arabidopsis*. *Plant Physiology*, *133*, 1051–1071.

Ramsay, N. A., & Glover, B. J. (2005). MYB-bHLH-WD40 protein complex and the evolution of cellular diversity. *Trends in Plant Science*, *10*(2), 63–70. https://doi.org/10.1016/j.tplants.2004.12.011.

Rao, X., Chen, X., Shen, H., Ma, Q., Li, G., Tang, Y., et al. (2019). Gene regulatory networks for lignin biosynthesis in switchgrass (*Panicum virgatum*). *Plant Biotechnology Journal*, *17*(3), 580–593. https://doi.org/10.1111/pbi.13000.

Ren, M., Zhang, Y., Liu, C., Liu, Y., Tian, S., Cheng, H., et al. (2021). Characterization of a high hierarchical regulator, *PtrGATA12*, functioning in differentially regulating secondary wall component biosynthesis in *Populus trichocarpa*. *Frontiers in Plant Science*, *12*, 657787. https://doi.org/10.3389/fpls.2021.657787.

Robe, K., Izquierdo, E., Vignols, F., Rouached, H., & Dubos, C. (2021). The Coumarins: Secondary metabolites playing a primary role in plant nutrition and health. *Trends in Plant Science, 26*(3), 248–259. https://doi.org/10.1016/j.tplants.2020.10.008.

Romano, J. M., Dubos, C., Prouse, M. B., Wilkins, O., Hong, H., Poole, M., et al. (2012). *At*MYB61, an R2R3-MYB transcription factor, functions as a pleiotropic regulator via a small gene network. *New Phytologist, 195*(774–786).

Sablowski, R. W. M., Moyano, E., Culianez-Macia, F. A., Schuch, W., Martin, C., & Bevan, M. (1994). A flower-specific Myb protein activates transcription of phenylpropanoid biosynthetic genes. *The EMBO Journal, 13*(1), 128–137.

Sah, S. K., Reddy, K. R., & Li, J. (2016). Abscisic acid and abiotic stress tolerance in crop plants. *Frontiers in Plant Science,* 7, 571. https://doi.org/10.3389/fpls.2016.00571.

Schuetz, M., Benske, A., Smith, R. A., Watanabe, Y., Tobimatsu, Y., Ralph, J., et al. (2014). Laccases direct lignification in the discrete secondary cell wall domains of protoxylem. *Plant Physiology, 166*(2), 798–807. https://doi.org/10.1104/pp.114.245597.

Serna, L., & Martin, C. (2006). Trichomes: Different regulatory networks lead to convergent structures. *Trends in Plant Science, 11*(6), 274–280. https://doi.org/10.1016/j.tplants.2006.04.008.

Shen, H., Yin, Y., Chen, F., Xu, Y., & Dixon, R. A. (2009). A bioinformatic analysis of *NAC* genes for plant cell wall development in relation to lignocellulosic bioenergy production. *Bioenergy Research, 2*, 217–232.

Shi, D., Jouannet, V., Agusti, J., Kaul, V., Levitsky, V., Sanchez, P., et al. (2021). Tissue-specific transcriptome profiling of the Arabidopsis inflorescence stem reveals local cellular signatures. *Plant Cell, 33*(2), 200–223. https://doi.org/10.1093/plcell/koaa019.

Shi, R., Wang, J. P., Lin, Y.-C., Li, Q., Sun, Y.-H., Chen, H., et al. (2017). Tissue and cell-type co-expression networks of transcription factors and wood component genes in *Populus trichocarpa*. *Planta, 245*(5), 927–938. https://doi.org/10.1007/s00425-016-2640-1.

Smith, R. A., Schuetz, M., Karlen, S. D., Bird, D., Tokunaga, N., Sato, Y., et al. (2017). Defining the diverse cell populations contributing to lignification in Arabidopsis stems. *Plant Physiology, 174*(2), 1028–1036. https://doi.org/10.1104/pp.17.00434.

Smith, R. A., Schuetz, M., Roach, M., Mansfield, S. D., Ellis, B., & Samuels, L. (2013). Neighboring parenchyma cells contribute to *Arabidopsis* xylem lignification, while lignification of interfascicular fibers is cell autonomous. *Plant Cell, 25*(10), 3988–3999. https://doi.org/10.1105/tpc.113.117176.

Soler, M., Camargo, E. L. O., Carocha, V., Cassan-Wang, H., Clemente, H. S., Savelli, B., et al. (2015). The *Eucalyptus grandis* R2R3-MYB transcription factor family: Evidence for woody growth-related evolution and function. *New Phytologist, 206*, 1364–1377.

Soler, M., Plasencia, A., Lepikson-Neto, J., Camargo, E. L., Dupas, A., Ladouce, N., et al. (2016). The woody-preferential gene *EgMYB88* regulates the biosynthesis of phenylpropanoid-derived compounds in wood. *Frontiers in Plant Science,* 7, 1422. https://doi.org/10.3389/fpls.2016.01422.

Song, Q., Lee, J., Akter, S., Rogers, M., Grene, R., & Li, S. (2020). Prediction of condition-specific regulatory genes using machine learning. *Nucleic Acids Research, 48*(11), e62. https://doi.org/10.1093/nar/gkaa264.

Soyano, T., Thitamadee, S., Machida, Y., & Chua, N.-H. (2008). *ASYMMETRIC LEAVES2-LIKE19/LATERAL ORGAN BOUNDARIES DOMAIN30* and *ASL20/LBD18* regulate tracheary element differentiation in *Arabidopsis*. *The Plant Cell, 20*, 3359–3373.

Speicher, T. L., Li, P. Z., & Wallace, I. S. (2018). Phosphoregulation of the plant cellulose synthase complex and cellulose synthase-like proteins. *Plants (Basel),* 7(3). https://doi.org/10.3390/plants7030052.

Spitz, F., & Furlong, E. E. M. (2012). Transcription factors: From enhancer binding to developmental control. *Nature Reviews Genetics, 13*, 613–626.

Sulis, D. B., & Wang, J. P. (2020). Regulation of lignin biosynthesis by post-translational protein modifications. *Frontiers in Plant Science, 11*, 914. https://doi.org/10.3389/fpls.2020.00914.

Sun, Y., Jiang, C., Jiang, R., Wang, F., Zhang, Z., & Zeng, J. (2021). A novel NAC transcription factor from *Eucalyptus*, EgNAC141, positively regulates lignin biosynthesis and increases lignin deposition. *Frontiers in Plant Science, 12*, 642090. https://doi.org/10.3389/fpls.2021.642090.

Sundell, D., Street, N. R., Kumar, M., Mellerowicz, E. J., Kucukoglu, M., Johnsson, C., et al. (2017). AspWood: High-spatial-resolution transcriptome profiles reveal uncharacterized modularity of wood formation in *Populus tremula. Plant Cell, 29*(7), 1585–1604. https://doi.org/10.1105/tpc.17.00153.

Takata, N., Awano, T., Nakata, M. T., Sano, Y., Sakamoto, S., Mitsuda, N., et al. (2019). *Populus* NST/SND orthologs are key regulators of secondary cell wall formation in wood fibers, phloem fibers and xylem ray parenchyma cells. *Tree Physiology, 39*(4), 514–525. https://doi.org/10.1093/treephys/tpz004.

Takata, N., & Taniguchi, T. (2015). Expression divergence of cellulose synthase (*CesA*) genes after a recent whole genome duplication event in *Populus. Planta, 241*, 29–42.

Talbert, P. B., Adler, H. T., Parks, D. W., & Comai, L. (1995). The *REVOLUTA* gene is necessary for apical meristem development and for limiting cell divisions in the leaves and stems of *Arabidopsis thaliana. Development, 121*(9), 2723–2735. https://doi.org/10.1242/dev.121.9.2723.

Tan, T. T., Endo, H., Sano, R., Kurata, T., Yamaguchi, M., Ohtani, M., et al. (2018). Transcription factors VND1-VND3 contribute to Cotyledon xylem vessel formation. *Plant Physiology, 176*(1), 773–789. https://doi.org/10.1104/pp.17.00461.

Tang, X., Zhuang, Y., Qi, G., Wang, D., Liu, H., Wang, K., et al. (2015). Poplar PdMYB221 is involved in the direct and indirect regulation of secondary wall biosynthesis during wood formation. *Scientific Reports, 5*, 12240. https://doi.org/10.1038/srep12240.

Taylor-Teeples, M., Lin, L., Lucas, M.d., Turco, G., Toal, T. W., Gaudinier, et al. (2015). An *Arabidopsis* gene regulatory network for secondary cell wall synthesis. *Nature, 517*, 571–575.

Teng, S., Keurentjes, J., Bentsink, L., Koornneef, M., & Smeekens, S. (2005). Sucrose-specific induction of anthocyanin biosynthesis in Arabidopsis requires the MYB75/PAP1 gene. *Plant Physiology, 139*(4), 1840–1852. https://doi.org/10.1104/pp.105.066688.

Terada, S., Kubo, M., Akiyoshi, N., Sano, R., Nomura, T., Sawa, S., et al. (2021). Expression of peat moss *VASCULAR RELATED NAC-DOMAIN* homologs in *Nicotiana benthamiana* leaf cells induces ectopic secondary wall formation. *Plant Molecular Biology, 106*(3), 309–317. https://doi.org/10.1007/s11103-021-01148-6.

Tobimatsu, Y., & Schuetz, M. (2019). Lignin polymerization: How do plants manage the chemistry so well? *Current Opinion in Biotechnology, 56*, 75–81. https://doi.org/10.1016/j.copbio.2018.10.001.

Tong, H., Chen, H., & Williams, C. M. (2021). Gene regulatory network of secondary cell wall biosynthesis during VND7 induced *de novo* xylem formation. *International Journal of Bioscience, Biochemistry and Bioinformatics, 11*(4), 74–81. https://doi.org/10.17706/ijbbb.2021.11.4.74-81.

Tuskan, G. A., DiFazio, S., Jansson, S., Bohlmann, J., Grigoriev, I., Hellsten, U., et al. (2006). The genome of black cottonwood, *Populus trichocarpa* (Torr. & gray). *Science, 313*(5793), 1596–1604.

Valdivia, E. R., Herrera, M. T., Gianzo, C., Fidalgo, J., Revilla, G., Zarra, I., et al. (2013). Regulation of secondary wall synthesis and cell death by NAC transcription factors in the monocot *Brachypodium distachyon. Journal of Experimental Botany, 64*(5), 1333–1343.

Van de Peer, Y., Mizrachi, E., & Marchal, K. (2017). The evolutionary significance of polyploidy. *Nature Reviews. Genetics*, *18*(7), 411–424. https://doi.org/10.1038/nrg.2017.26.

Vanholme, R., Demedts, B., Morreel, K., Ralph, J., & Boerjan, W. (2010). Lignin biosynthesis and structure. *Plant Physiology*, *153*, 895–905.

Vogt, T. (2010). Phenylpropanoid biosynthesis. *Molecular Plant*, *3*(1), 2–20. https://doi.org/10.1093/mp/ssp106.

Wang, H., Avci, U., Nakashima, J., Hahn, M. G., Chen, F., & Dixona, R. A. (2010). Mutation of WRKY transcription factors initiates pith secondary wall formation and increases stem biomass in dicotyledonous plants. *Proceedings of the National Academy of Science*, *107*(51), 22338–22343.

Wang, H., Zhao, Q., Chen, F., Wang, M., & Dixon, R. A. (2011). NAC domain function and transcriptional control of a secondary cell wall master switch. *The Plant Journal*, *68*(6), 1104–1114.

Wang, J. P., Chuang, L., Loziuk, P. L., Chen, H., Lin, Y. C., Shi, R., et al. (2015). Phosphorylation is an on/off switch for 5-hydroxyconiferaldehyde O-methyltransferase activity in poplar monolignol biosynthesis. *Proceedings of the National Academy of Sciences of the United States of America*, *112*(27), 8481–8486. https://doi.org/10.1073/pnas.1510473112.

Wang, L., Lu, W., Ran, L., Dou, L., Yao, S., Hu, J., et al. (2019). R2R3-MYB transcription factor MYB6 promotes anthocyanin and proanthocyanidin biosynthesis but inhibits secondary cell wall formation in *Populus tomentosa*. *The Plant Journal*, *99*(4), 733–751. https://doi.org/10.1111/tpj.14364.

Wang, Q., Dai, X., Pang, H., Cheng, Y., Huang, X., Li, H., et al. (2021). BEL1-like homeodomain protein BLH6a is a negative regulator of CAl5H2 in Sinapyl alcohol Monolignol biosynthesis in poplar. *Frontiers in Plant Science*, *12*, 695223. https://doi.org/10.3389/fpls.2021.695223.

Wang, S., Yang, H., Mei, J., Liu, X., Wen, Z., Zhang, L., et al. (2019). Rice Homeobox protein KNAT7 integrates the pathways regulating cell expansion and wall stiffness. *Plant Physiology*, *181*(2), 669–682. https://doi.org/10.1104/pp.19.00639.

Wang, X. C., Wu, J., Guan, M. L., Zhao, C. H., Geng, P., & Zhao, Q. (2020). *Arabidopsis* MYB4 plays dual roles in flavonoid biosynthesis. *The Plant Journal*, *101*(3), 637–652. https://doi.org/10.1111/tpj.14570.

Wang, Y., Yu, W., Ran, L., Chen, Z., Wang, C., Dou, Y., et al. (2021). DELLA-NAC interactions mediate GA signaling to promote secondary cell wall formation in cotton stem. *Frontiers in Plant Science*, *12*, 655127. https://doi.org/10.3389/fpls.2021.655127.

Xiao, R., Zhang, C., Guo, X., Li, H., & Lu, H. (2021). MYB transcription factors and its regulation in secondary cell wall formation and lignin biosynthesis during xylem development. *International Journal of Molecular Sciences*, *22*(7). https://doi.org/10.3390/ijms22073560.

Xie, M., Muchero, W., Bryan, A. C., Yee, K., Guo, H. B., Zhang, J., et al. (2018). A 5-Enolpyruvylshikimate 3-phosphate synthase functions as a transcriptional repressor in *Populus*. *Plant Cell*, *30*(7), 1645–1660. https://doi.org/10.1105/tpc.18.00168.

Xie, M., Zhang, J., Singan, V. R., McGranahan, M. J., LaFayette, P. R., Jawdy, S. S., et al. (2020). Identification of functional single nucleotide polymorphism of *Populus trichocarpa* PtrEPSP-TF and determination of its transcriptional effect. *Plant Direct*, *4*(1), e00178. https://doi.org/10.1002/pld3.178.

Xie, M., Zhang, J., Tschaplinski, T. J., Tuskan, G. A., Chen, J. G., & Muchero, W. (2018). Regulation of lignin biosynthesis and its role in growth-defense tradeoffs. *Frontiers in Plant Science*, *9*, 1427. https://doi.org/10.3389/fpls.2018.01427.

Xu, B., Ohtani, M., Yamaguchi, M., Toyooka, K., Wakazaki, M., Sato, M., et al. (2014). Contribution of NAC transcription factors to plant adaptation to land. *Science*, *343*(6178), 1505–1508.

Xu, C., Cao, H., Xu, E., Zhang, S., & Hu, Y. (2018). Genome-wide identification of Arabidopsis LBD29 target genes reveals the molecular events behind auxin-induced cell reprogramming during callus formation. *Plant & Cell Physiology*, *59*(4), 744–755. https://doi.org/10.1093/pcp/pcx168.

Xu, C., Shen, Y., He, F., Fu, X., Yu, H., Lu, W., et al. (2019). Auxin-mediated aux/IAA-ARF-HB signaling cascade regulates secondary xylem development in *Populus*. *The New Phytologist*, *222*(2), 752–767. https://doi.org/10.1111/nph.15658.

Xu, W., Dubos, C., & Lepiniec, L. (2015). Transcriptional control of flavonoid biosynthesis by MYB-bHLH-WDR complexes. *Trends in Plant Science*, *20*(3), 176–185. https://doi.org/10.1016/j.tplants.2014.12.001.

Yamaguchi, M., Kubo, M., Fukuda, H., & Demura, T. (2008). VASCULAR-RELATED NAC-DOMAIN7 is involved in the differentiation of all types of xylem vessels in *Arabidopsis* roots and shoots. *The Plant Journal*, *55*, 652–664.

Yamaguchi, M., Nadia, G., Igarashi, H., Ohtani, M., Nakano, Y., Mortimer, J. C., et al. (2010). VASCULAR-RELATED NAC-DOMAIN6 (VND6) and VND7 effectively induce transdifferentiation into xylem vessel elements under control of an induction system. *Plant Physiology*, *153*, 906–914.

Yang, C., Xu, Z., Song, J., Conner, K., Barrena, G. V., & Wilson, Z. A. (2007). *Arabidopsis MYB26/MALE STERILE35* regulates secondary thickening in the endothecium and is essential for anther dehiscence. *The Plant Cell*, *19*, 534–548.

Yang, L., Zhao, X., Ran, L., Li, C., Fan, D., & Luo, K. (2017). PtoMYB156 is involved in negative regulation of phenylpropanoid metabolism and secondary cell wall biosynthesis during wood formation in poplar. *Scientific Reports*, *7*, 41209. https://doi.org/10.1038/srep41209.

Yang, L., Zhao, X., Yang, F., Fan, D., Jiang, Y., & Luo, K. (2016). PtrWRKY19, a novel WRKY transcription factor, contributes to the regulation of pith secondary wall formation in *Populus trichocarpa*. *Scientific Reports*, *6*, 18643. https://doi.org/10.1038/srep18643.

Yao, S., Chen, P., Yu, Y., Zhang, M., Wang, D., Liu, J., et al. (2021). *PmMYB4*, a transcriptional activator from *Pinus massoniana*, regulates secondary cell wall formation and lignin biosynthesis. *Forests*, *12*, 1618. https://doi.org/10.3390/f12121618.

Yoshida, K., Sakamoto, S., Kawai, T., Kobayashi, Y., Sato, K., Ichinose, Y., et al. (2013). Engineering the *Oryza sativa* cell wall with rice NAC transcription factors regulating secondary wall formation. *Frontiers in Plant Science*, *4*, 383.

Yu, H., & Gerstein, M. (2006). Genomic analysis of the hierarchical structure of regulatory networks. *Proceedings of the National Academy of Sciences of the USA*, *103*(40), 14724–14731.

Yu, J., Zhou, C., Li, D., Li, S., Jimmy Lin, Y. C., Wang, J. P., et al. (2022). A PtrLBD39-mediated transcriptional network regulates tension wood formation in *Populus trichocarpa*. *Plant Commun*, *3*(1), 100250. https://doi.org/10.1016/j.xplc.2021.100250.

Yu, Y., Hu, R., Wang, H., Cao, Y., He, G., Fu, C., et al. (2013). MlWRKY12, a novel *Miscanthus* transcription factor, participates in pith secondary cell wall formation and promotes flowering. *Plant Science*, *212*, 1–9. https://doi.org/10.1016/j.plantsci.2013.07.010.

Zhang, B., Chopra, D., Schrader, A., & Hulskamp, M. (2019). Evolutionary comparison of competitive protein-complex formation of MYB, bHLH, and WDR proteins in plants. *Journal of Experimental Botany*, *70*(12), 3197–3209. https://doi.org/10.1093/jxb/erz155.

Zhang, B., & Schrader, A. (2017). TRANSPARENT TESTA GLABRA 1-dependent regulation of flavonoid biosynthesis. *Plants (Basel), 6*(4). https://doi.org/10.3390/plants6040065.

Zhang, F., Gonzalez, A., Zhao, M., Payne, C. T., & Lloyd, A. (2003). A network of redundant bHLH proteins functions in all TTG1-dependent pathways of *Arabidopsis*. *Development, 130*(20), 4859–4869. https://doi.org/10.1242/dev.00681.

Zhang, J., Xie, M., Tuskan, G. A., Muchero, W., & Chen, J. G. (2018). Recent advances in the transcriptional regulation of secondary cell wall biosynthesis in the woody plants. *Frontiers in Plant Science, 9*, 1535. https://doi.org/10.3389/fpls.2018.01535.

Zhang, Q., Luo, F., Zhong, Y., He, J., & Li, L. (2020). Modulation of NAC transcription factor NST1 activity by XYLEM NAC DOMAIN1 regulates secondary cell wall formation in Arabidopsis. *Journal of Experimental Botany, 71*(4), 1449–1458. https://doi.org/10.1093/jxb/erz513.

Zhang, Q., Xie, Z., Zhang, R., Xu, P., Liu, H., Yang, H., et al. (2018). Blue light regulates secondary cell wall thickening via MYC2/MYC4 activation of the *NST1*-directed transcriptional network in Arabidopsis. *Plant Cell, 30*(10), 2512–2528. https://doi.org/10.1105/tpc.18.00315.

Zhang, S., Yang, J., Li, H., Chiang, V. L., & Fu, Y. (2021). Cooperative regulation of flavonoid and lignin biosynthesis in plants. *Critical Reviews in Plant Sciences, 40*(2), 109–126. https://doi.org/10.1080/07352689.2021.1898083.

Zhao, C., Avci, U., Grant, E. H., Haigler, C. H., & Beers, E. P. (2008). XND1, a member of the NAC domain family in *Arabidopsis thaliana*, negatively regulates lignocellulose synthesis and programmed cell death in xylem. *The Plant Journal, 53*(3), 425–436.

Zhao, C., Lasses, T., Bako, L., Kong, D., Zhao, B., Chanda, B., et al. (2017). XYLEM NAC DOMAIN1, an angiosperm NAC transcription factor, inhibits xylem differentiation through conserved motifs that interact with RETINOBLASTOMA-RELATED. *The New Phytologist, 216*(1), 76–89. https://doi.org/10.1111/nph.14704.

Zhao, Q., Nakashima, J., Chen, F., Yin, Y., Fu, C., Yun, J., et al. (2013). *LACCASE* is necessary and nonredundant with *PEROXIDASE* for lignin polymerization during vascular development in *Arabidopsis*. *Plant Cell, 25*(10), 3976–3987. https://doi.org/10.1105/tpc.113.117770.

Zhao, Y., Sun, J., Xu, P., Zhang, R., & Li, L. (2014). Intron-mediated alternative splicing of WOOD-ASSOCIATED NAC TRANSCRIPTION FACTOR1B regulates cell wall thickening during fiber development in *Populus* species. *Plant Physiology, 164*, 765–776.

Zheng, L., Chen, Y., Ding, D., Zhou, Y., Ding, L., Wei, J., et al. (2019). Endoplasmic reticulum-localized UBC34 interaction with lignin repressors MYB221 and MYB156 regulates the transactivity of the transcription factors in *Populus tomentosa*. *BMC Plant Biology, 19*(1), 97. https://doi.org/10.1186/s12870-019-1697-y.

Zhong, R., Burk, D. H., & Ye, Z.-H. (2001). Fibers. A model for studying cell differentiation, cell elongation, and cell wall biosynthesis. *Plant Physiology, 126*, 477–479.

Zhong, R., Cui, D., & Ye, Z. H. (2018). Secondary cell wall biosynthesis. *New Phytologist, 221*(4), 1703–1723. https://doi.org/10.1111/nph.15537.

Zhong, R., Demura, T., & Ye, Z. H. (2006). SND1, a NAC domain transcription factor, is a key regulator of secondary wall synthesis in fibers of *Arabidopsis*. *Plant Cell, 18*(11), 3158–3170.

Zhong, R., Kandasamy, M. K., & Ye, Z. H. (2021). XND1 regulates secondary wall deposition in xylem vessels through the inhibition of VND functions. *Plant & Cell Physiology, 62*(1), 53–65. https://doi.org/10.1093/pcp/pcaa140.

Zhong, R., Lee, C., Haghighat, M., & Ye, Z. H. (2021). Xylem vessel-specific SND5 and its homologs regulate secondary wall biosynthesis through activating secondary

wall NAC binding elements. *The New Phytologist, 231*(4), 1496–1509. https://doi.org/10.1111/nph.17425.
Zhong, R., Lee, C., McCarthy, R. L., Reeves, C. K., Jones, E. G., & Ye, Z.-H. (2011). Transcriptional activation of secondary wall biosynthesis by rice and maize NAC and MYB transcription factors. *Plant and Cell Physiology, 52*(10), 1856–1871.
Zhong, R., Lee, C., & Ye, Z.-H. (2010a). Global analysis of direct targets of secondary wall NAC master switches in *Arabidopsis*. *Molecular Plant, 3*(6), 1087–1103.
Zhong, R., Lee, C., & Ye, Z. H. (2010b). Functional characterization of poplar wood-associated NAC domain transcription factors. *Plant Physiology, 152*(2), 1044–1055. https://doi.org/10.1104/pp.109.148270.
Zhong, R., Lee, C., Zhou, J., McCarthy, R. L., & Ye, Z.-H. (2008). A battery of transcription factors involved in the regulation of secondary cell wall biosynthesis in *Arabidopsis*. *The Plant Cell, 20*, 2763–2782.
Zhong, R., McCarthy, R. L., Haghighat, M., & Ye, Z. H. (2013). The poplar MYB master switches bind to the SMRE site and activate the secondary wall biosynthetic program during wood formation. *PLoS One, 8*(7), e69219. https://doi.org/10.1371/journal.pone.0069219.
Zhong, R., McCarthy, R. L., Lee, C., & Ye, Z.-H. (2011). Dissection of the transcriptional program regulating secondary wall biosynthesis during wood formation in poplar. *Plant Physiology, 157*, 1452–1468.
Zhong, R., Richardson, E. A., & Ye, Z.-H. (2007a). The MYB46 transcription factor is a direct target of SND1 and regulates secondary wall biosynthesis in *Arabidopsis*. *The Plant Cell, 19*, 2776–2792.
Zhong, R., Richardson, E. A., & Ye, Z.-H. (2007b). Two NAC domain transcription factors, SND1 and NST1, function redundantly in regulation of secondary wall synthesis in fibers of *Arabidopsis*. *Planta, 225*(6), 1603–1611. https://doi.org/10.1007/s00425-007-0498-y.
Zhong, R., Taylor, J. J., & Ye, Z. H. (1997). Disruption of interfascicular fiber differentiation in an *Arabidopsis* mutant. *The Plant Cell, 11*(12), 2159–2170.
Zhong, R., & Ye, Z.-H. (1999). *IFL1*, a gene regulating interfascicular fiber differentiation in *Arabidopsis*, encodes a homeodomain-leucine zipper protein. *The Plant Cell, 11*, 2139–2152.
Zhong, R., & Ye, Z.-H. (2012). MYB46 and MYB83 bind to the SMRE sites and directly activate a suite of transcription factors and secondary wall biosynthetic genes. *Plant and Cell Physiology, 53*(2), 368–380.
Zhong, R., & Ye, Z. H. (2010). The poplar PtrWNDs are transcriptional activators of secondary cell wall biosynthesis. *Plant Signaling & Behavior, 5*(4), 469–472.
Zhong, R., & Ye, Z. H. (2015). The Arabidopsis NAC transcription factor NST2 functions together with SND1 and NST1 to regulate secondary wall biosynthesis in fibers of inflorescence stems. *Plant Signaling & Behavior, 10*(2), e989746. https://doi.org/10.4161/15592324.2014.989746.
Zhong, R., Yuan, Y., Spiekerman, J. J., Guley, J. T., Egbosiuba, J. C., & Ye, Z. H. (2015). Functional characterization of NAC and MYB transcription factors involved in regulation of biomass production in switchgrass (*Panicum virgatum*). *PLoS One, 10*(8), e0134611. https://doi.org/10.1371/journal.pone.0134611.
Zhou, J., Lee, C., Zhong, R., & Ye, Z.-H. (2009). MYB58 and MYB63 are transcriptional activators of the lignin biosynthetic pathway during secondary cell wall formation in *Arabidopsis*. *The Plant Cell, 21*(1), 248–266. https://doi.org/10.1105/tpc.108.063321.
Zhou, J., Zhong, R., & Ye, Z. H. (2014). Arabidopsis NAC domain proteins, VND1 to VND5, are transcriptional regulators of secondary wall biosynthesis in vessels. *PLoS One, 9*(8), e105726. https://doi.org/10.1371/journal.pone.0105726.

Zhou, M., Sun, Z., Wang, C., Zhang, X., Tang, Y., Zhu, X., et al. (2015). Changing a conserved amino acid in R2R3-MYB transcription repressors results in cytoplasmic accumulation and abolishes their repressive activity in *Arabidopsis*. *The Plant Journal*, *84*(2), 395–403. https://doi.org/10.1111/tpj.13008.

Zhou, M., Zhang, K., Sun, Z., Yan, M., Chen, C., Zhang, X., et al. (2017). LNK1 and LNK2 corepressors interact with the MYB3 transcription factor in phenylpropanoid biosynthesis. *Plant Physiology*, *174*(3), 1348–1358. https://doi.org/10.1104/pp.17.00160.

Zhu, T., Nevo, E., Sun, D., & Peng, J. (2012). Phylogenetic analyses unravel the evolutionary history of NAC proteins in plants. *Evolution*, *66*(6), 1833–1848.

Zhu, Y., & Li, L. (2021). Multi-layered regulation of plant cell wall thickening. *Plant & Cell Physiology*. https://doi.org/10.1093/pcp/pcab152.

Zimmermann, I. M., Heim, M. A., Weisshaar, B., & Uhrig, J. F. (2004). Comprehensive identification of *Arabidopsis thaliana* MYB transcription factors interacting with R/B-like BHLH proteins. *The Plant Journal*, *40*, 22–34.

CHAPTER NINE

Regulation of secondary cell wall lignification by abiotic and biotic constraints

Ines Hadj Bachir[a], Raphael Ployet[b], Chantal Teulières[a], Hua Cassan-Wang[a], Fabien Mounet[a,*], and Jacqueline Grima-Pettenati[a]

[a]Laboratoire de Recherche en Sciences Végétales, Université de Toulouse, CNRS, UPS, Auzeville-Tolosane, France

[b]Department of Biochemistry, Genetics and Microbiology, Forestry and Agricultural Biotechnology Institute (FABI), University of Pretoria, Pretoria, South Africa

*Corresponding author: e-mail address: fabien.mounet@univ-tlse3.fr

Contents

Abstract

The appearance of lignin, 400 million-years ago, has been crucial for the successful land colonization by plants. This complex phenolic polymer which confers hydrophobicity and resistance to degradation to secondary cell walls, is also neosynthesized in response to both biotic and abiotic stresses. Lignin biosynthesis during development is regulated by a complex hierarchical transcriptional gene regulatory network. Notably, distinct

Advances in Botanical Research, Volume 104
ISSN 0065-2296
https://doi.org/10.1016/bs.abr.2022.03.008

abiotic stresses can co-opt different genes of this network, leading to different lignification patterns. Co-optation of this developmental network is a mean to facilitate adaptation to stress by promoting functional adaptation. Supporting this hypothesis, a growing number of transcription factors known to be involved in the regulation of secondary cell wall, were shown to integrate both developmental and environmental signals leading to an increased tolerance to abiotic stresses, such as drought or salinity. On the other hand, some transcription factors known to be involved in stress response were shown to impact lignin biosynthesis as part of the functional adaptation performed by plants to cope with natural constraints. In this review, we focus on the most recent findings on the cross-talk between stress-signaling pathways and regulation of lignin biosynthesis in response to abiotic stress and to a lesser extent to biotic stresses. We consider the transcriptional regulation level which has received much attention and the emerging roles of posttranscriptional and posttranslational regulations in the control of lignin biosynthesis in response to stresses. We end discussing future outlines and challenges to fill gaps in this emerging and exciting research field.

1. Introduction

Lignin, the second most abundant biopolymer after cellulose, represents as much as 30% of the total biomass produced on earth (Boerjan, Ralph, & Baucher, 2003; Boudet, Kajita, Grima-Pettenati, & Goffner, 2003). This complex phenolic heteropolymer massively deposited in the secondary cell walls (SCW) of xylem, confers mechanical support, imperviousness, water conductivity and resistance to biodegradation. Lignin is indeed one of the most difficult biopolymers to degrade due to the nature and heterogeneity of its linkages (Vanholme, Demedts, Morreel, Ralph, & Boerjan, 2010). Thanks to these unique properties, lignin has played a crucial role in the emergence of vascular plants and the colonization of terrestrial habitats (Weng & Chapple, 2010).

1.1 Lignification is increased in response to various biotic and abiotic stresses

Besides the crucial roles that lignin plays in plant development, it is also known for several decades that induced lignification is one plant defense response to environmental cues such as wounding, mechanical stress or pathogen attack (for reviews see Cabane, Afif, & Hawkins, 2012; LeGall et al., 2015; Vance, Kirk, & Sherwood, 1980 and reference therein). Increased lignification is indeed a common response to biotic and abiotic stress (Dong & Lin, 2021; Moura, Bonine, de Oliveira Fernandes, Dornelas, & Mazzafera, 2010 and references therein) that contribute to plant adaptation and tolerance to stresses. For instance, reduced lignification increases xylem vulnerability to embolism, whereas higher lignification in xylem can improve drought resistance in plants (Pereira, Domingues-Junior, Jansen, Choat, & Mazzafera, 2018). Salt stress enhances lignin biosynthesis that solidifies the cell wall (Chun et al., 2019). Similarly, enhanced lignification could contribute to cell wall strengthening, preventing cell collapse caused by freezing temperatures (LeGall et al., 2015). Abnormal SCW lignification was associated with the collapse of sieve transport element in *Arabidopsis* C4H mutants (Polo et al., 2020) and in trees submitted to severe drought stress (Donaldson, 2002).

1.2 Lignin is a highly flexible phenolic polymer during development and response to stresses

Lignin is generated by oxidative radical coupling polymerization of mainly three *p*-hydroxycinnamyl alcohol precursors also called monolignols, which differ by the degree of methoxylation of the phenolic hydroxyl group in their aromatic ring (Boerjan et al., 2003). The three main hydroxycinnamyl alcohols (*p*-coumaryl, coniferyl and sinapyl alcohols) give rise to *p*-hydroxyphenyl (H), guaiacyl (G) and syringyl (S) units, respectively (Ralph, 2010). These C6-C3 units are interconnected by a series of ether and carbon-carbon linkages in various bonding patterns (Boudet, Lapierre, & Grima-Pettenati, 1995). The most frequent interunit β-O-4 are the targets of depolymerization processes, while the carbon-carbon and 4-O-5 bonds are more resistant to degradation. The respective proportion of H, G, S lignin has a major impact on the chemical nature and the frequency of the different types of inter-monomeric linkages which in turn affect the resistance to degradation of the polymer. In the last two decades, other monomers deriving from the monolignol synthetic pathway were shown to be incorporated in lignin including hydroxycinnamic acids and aldehydes, coniferyl and sinapyl acetates (Ralph, 2010; Vanholme, De Meester, Ralph, & Boerjan, 2019) highlighting the chemical flexibility of lignin. This high flexibility was further extended to monomers not derived from the monolignol pathway. For instance, tricin which derives from the shikimate pathway was shown to be a true lignin monomer in monocots acting as a nucleation site initiating lignin polymer chains (Lan et al., 2015).

Lignin content and monomeric composition vary greatly within and between species, among tissues and organs of the same plant and even in individual cell layers (Lourenço et al., 2016; Plomion, Leprovost, & Stokes, 2001). For this reason, we think it is most appropriate to use the plural "lignins" as suggested by Boudet et al. (1995). Most gymnosperms lignins do not contain S units, while angiosperms possess both G and S units (Boudet et al., 1995). H lignins are always poorly represented as compared to G and S units. Noteworthy, lignins synthesized in response to stresses have been shown to display distinct structural features from those of developmental lignins (Cesarino, 2019; Lange, Lapierre, & Sandermann, 1995 and references therein). This suggests that the neosynthesized lignins possess different physical properties that allow them to play a role in stress responses. For instance, in a number of cases, it was shown that defense-induced lignins appear to have increased levels of H units compared with lignins found in vascular cell walls (Doster & Bostock, 1988; Hammerschmidt, Bonnen, Bergstrom, & Baker, 1985; Lange et al., 1995; Ride, 1975; Robertsen & Svalheim, 1990). In response to cold (Hausman, Evers, Thiellement, & Jouve, 2000; Ployet et al., 2018), drought or salt (Byrt, Munns, Burton, Gilliham, & Wege, 2018), modifications of the structure and composition of lignin polymer were observed. The syringyl/guaiacyl (S/G) ratio is increased in response to salt stress in roots and stems in maize (Oliveira et al., 2020), to cold stress in *Eucalyptus* xylem cells (Ployet et al., 2018) but is decreased in response to drought in young shoots of *Populus trichocarpa* (Hori et al., 2020). In response to salinity and osmotic stress, the pattern of ferulic acid (FA) incorporation within the SCW of maize and wheat vascular tissues was linked to an increase in lignin content and S/G ratio (Oliveira et al., 2020). FA is esterified to arabinoxylans and participates with lignin monomers in oxidative coupling pathways to generate ferulate-polysaccharide-lignin complexes that cross-link the cell wall. The structure of the lignin produced *de novo* seems to largely depend on the type and intensity of the stress and on the plant species (Cesarino, 2019). Furthermore, the functional relevance for lignin polymer modifications in response to biotic and abiotic constraints is not fully understood. It should be pointed out that the tissue localization of "defense" lignins is also different from those of structural lignins (reviewed recently in Cesarino, 2019).

1.3 The complex monolignol pathway involves multigenic families

Lignin resistance to degradation which is crucial for plant development and defense against environmental cues, is on the other hand, a major obstacle for industrial processes such as pulp and paper and lignocellulosic biofuel production (Wang et al., 2018). It also decreases forage digestibility by cattle (Barrière, Courtial, Soler, & Grima-Pettenati, 2015). Large research efforts have been devoted to engineering lignin amount and composition to reduce biomass recalcitrance (Chanoca, de Vries, & Boerjan, 2019). This has been a driving force to decipher the lignin biosynthetic pathway whose topology has been revisited several times in the last decades. Indeed, the analyses of transgenic plants or mutants downregulated for lignin biosynthetic genes revealed that the monolignol biosynthetic pathway was more complex and reticulated than initially thought (reviewed in Boerjan et al., 2003; Boudet et al., 2003; Grima-Pettenati & Goffner, 1999; Humphreys & Chapple, 2002; Ralph et al., 2004; Vanholme et al., 2010). More recently, co-expression network analyses have allowed the discovery of new alternative routes such as that involving the caffeoyl shikimate esterase (CSE; Vanholme et al., 2013). In brief, 11 enzymatic reactions are implicated in the synthesis of monolignols, which involves the general phenylpropanoid pathway starting with the deamination of phenylalanine and leading to the production of hydroxy-cinnamoyl CoA esters. The enzymes involved in this short sequence of reactions are phenylalanine ammonia-lyase (PAL), cinnamate 4-hydroxylase (C4H) and 4-coumarate CoA ligase (4CL). The hydroxycinnamoyl-CoA esters are not only the precursors of lignin but also of a wide range of end products including flavonoids, anthocyanins and condensed tannins, which vary according to species, cell type and environmental signals (Dixon & Paiva, 1995). In order to produce monolignols, hydroxycinnamoyl-CoA esters undergo successive hydroxylation and O-methylation of their aromatic rings (Boerjan et al., 2003) involving the following enzymatic activities: hydroxycinnamoyl-transferase (HCT); CSE; *p*-coumarate 3-hydroxylase (C3H); caffeoyl-CoA 3-O-methyltransferase (CCoAOMT); ferulate 5-hydroxylase (F5H) and caffeate/5-hydroxyferulate O-methyl-transferase (COMT). CSE catalyzes the conversion of caffeoyl shikimate into caffeate and its activity, combined with that of 4CL to produce caffeoyl-CoA, bypasses the second/reverse reaction of HCT (Vanholme et al., 2013). The conversion of the side-chain carboxyl to an alcohol group is catalyzed successively by cinnamoyl CoA reductase (CCR) and cinnamyl alcohol dehydrogenase (CAD), two enzymes considered to be the most specific of the monolignol biosynthesis pathway (Lacombe et al., 1997). The monolignols are synthesized in the cytoplasm and translocated to the apoplast, where they are dehydrogenated to monolignols radicals by the action of laccases and peroxidases (Berthet et al., 2011; Zhao et al., 2013). These monolignols radicals then couple with each other in a combinatorial fashion generating a variety of chemical bonds. It is worth noting that most of the lignin biosynthesis genes belong to gene families (Carocha et al., 2015) and that the role of individual gene family members is not clearly established. Family members involved in vascular lignification have been identified thanks to their expression patterns, promoter analyses, analyses of transgenics and enzymatic properties (Carocha et al., 2015; Raes, Rohde, Christensen, Van de Peer, & Boerjan, 2003). Members of some gene families have been identified as stress responsive genes such as *AtCCR2* from *Arabidopsis* (Lauvergeat et al., 2001). In contrast to *AtCCR1* involved in constitutive lignification, *AtCCR2* which is poorly expressed during development, is strongly and transiently induced during the incompatible interaction with *Xanthomonas campestris* pv. *campestris* leading to a hypersensitive response. *AtCCR2* was proposed to be involved in the biosynthesis of phenolics whose accumulation may lead to resistance (Lauvergeat et al., 2001) (Table 1).

2. Objective

In this review, we focus on the most recent findings on the cross-talk between stress signaling and regulation of lignin biosynthesis in response to abiotic stress and to a lesser extent to biotic stresses. We first consider the transcriptional regulation level, describing briefly the transcriptional regulatory network regulating lignin biosynthesis during development and further showing that abiotic stresses can co-opt the SCW regulatory network. We provide a survey of (i) the stress-associated transcription factors that can modulate the lignin pathway to enhance stress resistance and (ii) the SCWs-transcription factors recruited by stress signaling pathways whose overexpression leads to better tolerance to environmental constraints. The emerging roles of posttranscriptional and posttranslational regulations in the control of lignin biosynthesis in response to stresses will also be presented. We end discussing future outlines and challenges to fill gaps in this emerging and exiting research field.

3. Transcriptional regulation of the lignin pathway

3.1 A complex transcriptional network regulates lignin biosynthesis

The biosynthesis of lignified secondary cell walls involves hundreds of genes whose expression needs to be tightly regulated in time and space. The coordination between transcriptional regulation of synthesis for each polymer (lignins, cellulose, hemicelluloses) is complex and vital to cell function. In the last two decades, tremendous progress had been made in our understanding of the transcriptional regulation of the deposition of lignified SCWs

Table 1 Transcriptional and posttranscriptional regulators involved at the crossroads between stress signaling pathways and lignin deposition.

Gene alias	Species	Gene accession	Closest ortholog(s) in At	Effect on lignification	Effect on stress resistance	Reference
Abiotic stress-signaling TFs regulating CW lignification						
At SHN2	*Arabidopsis thaliana*	AT5G11190	—	+ Lignin	+ Drought	Ambavaram, Krishnan, Trijatmiko, & Pereira, 2011
Os ERF71	*Oryza sativa*	LOC4340383	RAP2.12 (AT1G53910)	+ Lignin	+ Drought	Lee et al., 2016
Os ERF83	*Oryza sativa*	LOC4334864	ERF1 (AT3G23240)	+ Lignin	+ Drought	Jung et al., 2021
Dc DREB1A	*Daucus carota*	—	CBF3 (AT4G25480)	+ Lignin	+ Drought	Li et al., 2020
Ba DBL1	*Bryum argenteum*	QDB64566	—	+ Lignin	+ Salt/osmotic	Liang et al., 2021
Egu CBF14	*Eucalyptus gunnii*	Eucgr. A02834†	AT4G25490	+ Lignin	+ Cold	Cao et al., 2020
Md MYB88	*Malus domestica*	KY569647	MYB88 (AT2G02820)	+ Lignin	+ Drought	Geng et al., 2018
Md MYB124	*Malus domestica*	KY569648	MYB124 (AT1G14350)	+ Lignin	+ Drought	Geng et al., 2018
Os TF1L	*Oryza sativa*	LOC4345188	AT4G00730	+ Lignin	+ Drought	Bang et al., 2019
Vl bZIP30	*Vitis vinifera*	LOC100266389	AREB3/DPBF3/ bZIP66 (AT3G56850) DBF4/EEL/bZIP12 (AT2G41070)	+ Lignin	+ Drought	Tu et al., 2020
Pu C3H35	*Populus ussuriensis*	MW792496	OXS2/C3H30 (AT2G41900)	+ Lignin	+ Drought	Li et al., 2022

Continued

Table 1 Transcriptional and posttranscriptional regulators involved at the crossroads between stress signaling pathways and lignin deposition.—cont'd

Gene alias	Species	Gene accession	Closest ortholog(s) in At	Effect on lignification	Effect on stress resistance	Reference
TFs involved in CW lignification co-opted by abiotic stresses						
NAC family						
SND1	*Arabidopsis thaliana*	AT1G32770	—	+ Lignin	+ Salt	Jeong et al., 2018
Bp NAC012	*Betula platyphylla*	KT344119	SND1/NST3/ ANAC012 (AT1G32770)	+ Lignin	+ Salt/osmotic	Hu, Zhang, & Yang, 2019
XND1	*Arabidopsis thaliana*	AT5G64530	—	− SCW	− Drought	Tang et al., 2018
Ag NAC1	*Apium graveolens* L.	MT847625	ANAC0002/ATAF1 (AT1G01720) ANAC102 (AT5G63790) ATAF2 (AT5G08790) ANAC032 (AT1G77450)	+ Lignin	+ Salt	Duan et al., 2020
MYB family						
Ej MYB1	*Eriobotrya japonica*	KF767453	MYB63 (AT1G79180) MYB58 (AT1G16490)	+ Lignin	Induced by cold	Xu et al., 2014
Bpl MYB46	*Betula platyphylla*	KP711284	MYB46 (AT5G12870) MYB83 (AT3G08500)	+ Lignin	+ Salt/osmotic	Guo et al., 2017

Md MYB46	*Malus domestica*	MD03G1176000†	MYB83 (AT3G08500) MYB46 (AT5G12870)	+ Lignin	+ Salt/osmotic	Chen et al., 2019
Pto MYB170	*Populus tomentosa*	KY114929	MYB61 (AT1G09540)	+ Lignin	+ Drought	Xu et al., 2017
Eg MYB1	*Eucalyptus gunnii*	Eucgr.G01774†	MYB4 (AT4G38620)	− Lignin	+ Drought	Legay et al., 2010, Soler et al., unpublished
Bp MYB4	*Betula platyphylla*	KA257119	MYB4 (AT4G38620)	− Lignin	+ Salt/osmotic	Yu et al., 2021
Biotic stress-signaling TFs regulating CW lignification						
At MYB46	*Arabidopsis thaliana*	AT5G12870	—	+ SCW	+ *B. cinerea*	Ramirez et al., 2011
At MYB15	*Arabidopsis thaliana*	AT3G23250	—	+ Lignin	+ Biotic stress	Chezem, Memon, Li, Weng, & Clay, 2017; Kim et al., 2020
Os MYB30	*Oryza sativa*	LOC4330027	MYB15 (AT3G23250)	+ Lignin	+ Biotic stress	Kishi-Kaboshi, Seo, Takahashi, & Hirochika, 2018
Os MYB55	*Oryza sativa*	BAF15232	MYB15 (AT3G23250)	+ Lignin	+ Biotic stress	Kishi-Kaboshi et al., 2018
Posttranscriptional regulation (sRNAs) of CW lignification						
Ptr-miR473	*Populus trichocarpa*			+ Lignin	Induced by UV-B	Lu et al., 2005
miRNA828	*Ipomoea batatas*			+ Lignin	Induced by wounding	Lin, Lin, Lin, Chen, & Jeng, 2012

Continued

Table 1 Transcriptional and posttranscriptional regulators involved at the crossroads between stress signaling pathways and lignin deposition.—cont'd

Gene alias	Species	Gene accession	Closest ortholog(s) in At	Effect on lignification	Effect on stress resistance	Reference
sRNA8105	*Ipomoea batatas*			+ Lignin	Induced by wounding	Lin et al., 2013
Zm miR528	*Zea mays*			− LIGNIN	− Lodging	Sun et al., 2018
miR397	*Setaria viridis*			− Lignin	− Salt	Nguyen, Brown, Pegler, Eamens, & Grof, 2020
Vun - miR408-3p	*Vigna unguiculata*			+ Lignin	+ Drought/salt	Mishra, Sahu, & Shaw, 2022
Bm - miR172c-5p	*Bacopa monnieri*			− Lignin	− Drought	Jeena, Joshi, & Shukla, 2021
Posttranslational regulation of CW lignification						
ZML2	*Zea mays*	LOC100277669	GATA28/TIFY2A (AT1G51600)	− Lignin	Degraded by wounding	Velez-Bermudez et al., 2015
Zm MYB11	*Zea mays*	LOC541744	MYB4 (AT4G38620)	− Lignin	Degraded by wounding	Velez-Bermudez et al., 2015
Egr H1.3	*Eucalyptus grandis*	Eucgr.I02364†	H1.3 (AT2G18050)	− Lignin		Soler et al., 2017
Eg MYB1	*Eucalyptus gunnii*	Eucgr.G01774†	MYB4 (AT4G38620)	− Lignin	+ Drought	Legay et al., 2010; Soler et al., unpublished

Pd MPK6 (Map kinase)	*Populus*	Potri.007G139800†	MPK6 (AT2G43790)			Gui et al., 2019
LTF1 (MYB)	*Populus*	KP770010	MYB4 (AT4G38620)	– Lignin	Degraded by wounding	Gui et al., 2019
Pto UBC34	*Populus tomentosa*	MH708242	UBC34 (AT1G17280)	+ Lignin	Induced by salt and heat	Zheng et al., 2019
Pto MYB221	*Populus tomentosa*	—	—	– Lignin		Zheng et al., 2019
Pto MYB156	*Populus tomentosa*	KT990214	MYB4 (AT4G38620)	– Lignin		Zheng et al., 2019
SnRK2	*Arabidopsis thaliana*	AT3G50500 AT5G66880 AT4G33950	—		Induced by ABA	Liu, Yu, Rao, Li, & Dixon, 2021
NST1	*Arabidopsis thaliana*	AT2G46770	—	+ Lignin		Liu et al., 2021

When available, accessions from GenBank or Phytozome V13 (genes marked with †) are listed. For each gene, the fourth column provides the closest *Arabidopsis thaliana* ortholog(s), mentioned in the references cited, or obtained by using BLASTP of NCBI database (best hits >80% of query protein identity).

(for reviews see Camargo, Ployet, Cassan-Wang, Mounet, & Grima-Pettenati, 2019; Grima-Pettenati, Soler, Camargo, & Wang, 2012; Hussey, Mizrachi, Creux, & Myburg, 2013; Schuetz, Smith, & Ellis, 2013; Wang & Dixon, 2012 and reference therein). In *Arabidopsis*, a complex hierarchical regulatory network has been proposed to control SW deposition in which the Transcription Factors (TF) NAM/ATAF/CUC (NAC) such as NST1, SND1, NST2, VND6 and VND7 act as first-level master switches. MYeloBlastosis (MYB) transcription factors (MYB46, MYB83 and MYB55) and NAC TF (SND3 and XND1) (Yamaguchi et al., 2011; Zhong, Lee, & Ye, 2010) act as second-level master switches to regulate a battery of downstream transcription factors and secondary cell wall biosynthesis genes. Noteworthy, among the TF involved in the regulation of the biosynthesis of the polymers of the secondary cell wall, only three MYB were reported to regulate specifically the lignin biosynthesis pathway in *Arabidopsis*: *At*MYB58, *At*MYB63 and *At*MYB85 (Zhong & Ye, 2009). MYB58 and MYB63 have been shown to directly activate nearly all of the genes in the monolignol pathway (Zhou, Lee, Zhong, & Ye, 2009). Overexpression of *MYB58* or its co-ortholog *MYB63* specifically activated the monolignol pathway and lignin accumulation at the expense of biomass production (Zhou et al., 2009). Most of the transcriptional regulatory network underlying the lignified SCW is conserved in other plants (Ohtani & Demura, 2019) but differences have been highlighted between *Arabidopsis* and woody angiosperms for instance (Camargo et al., 2019; Nakano, Yamaguchi, Endo, Rejab, & Ohtani, 2015 and references therein). In addition, the role of lignin-specific SG3-type R2R3-MYBs (MYB58/63) in vascular cell wall lignification seems to be conserved in dicot plants (Zhong & Ye, 2009), but it is less clear in monocots. The orthologs of *MYB58/63* in sorghum (*Sorghum bicolor*) and maize (*Zea mays*), *SbMYB60* and *ZmMYB19/Zm1* only induced marginal increases in lignin content, respectively (Scully et al., 2018). Moreover, recent discoveries of new transcription factors in *Arabidopsis*, and other plant species, suggest that the regulatory network of lignin biosynthesis in plants extends beyond the established NAC-MYB network (Behr, Guerriero, Grima-Pettenati, & Baucher, 2019; Rao & Dixon, 2018).

3.2 Abiotic stresses can co-opt the SCW transcriptional regulatory network

The landmark interactome study of Taylor-Teeples et al. (2015) has identified hundreds of novel regulators and provided considerable insight into the developmental regulation of xylem cell differentiation. The approach consisting in a large-scale yeast one-hybrid analysis of secondary cell wall-related transcription factors expressed in *Arabidopsis* roots, revealed that the complexity of the transcriptional regulation of the SCW was under-evaluated. This study highlighted a complex protein-DNA network between *Arabidopsis* transcription factors and metabolic genes with gene expression regulated by a series of feed-forward loops. It also showed that E2Fc can regulate the secondary wall biosynthesis network acting either as a transcriptional repressor or as an activator. E2Fc has the ability to affect the expression levels of many other transcription factors, including VND7, VND6, MYB46 as well as cellulose, hemicelluloses and lignin-associated genes in a dose-dependent manner. As E2Fc is also involved in regulating cell division through its interaction with the RETINOBLASTOMA-RELATED protein (RBR), it may function to coordinate cell division and cell differentiation during plant development (Ohtani & Demura, 2019). Importantly, the authors filtered the gene network to include only genes that were differentially expressed in response to salt stress or iron deprivation in the root vasculature of *Arabidopsis*. They showed that REVOLUTA (REV), the most upstream regulator in response to iron deprivation, regulates lignin biosynthesis. The expression of VND7 and MYB46, the most upstream regulators in response to salt stress, was greatly increased but the expression of the lignin genes was not altered. Notably, VND7 was ectopically expressed leading to an additional strand of metaxylem in roots exposed to high salinity. This demonstrated that distinct abiotic stresses can co-opt different genes of the xylem regulatory network. Co-optation of a developmental network is a mean to facilitate adaptation to stress by promoting functional adaptation (Taylor-Teeples et al., 2015).

In the last decade, transcription factors known to be involved in stress response were shown to impact lignin biosynthesis as part of the functional adaptation performed by plants to cope with natural constraints. Moreover, a growing number of transcription factors known to be involved in the regulation of SCW formation were shown to integrate both developmental and environmental signals leading to an increased tolerance to abiotic stresses, such as drought and salinity.

3.3 Abiotic stress-signaling transcription factors can modulate lignin deposition

Abiotic stress signaling in plants involves first the perception of the stress through sensors, which triggers signaling cascades within the cell involving protein kinases (Chen et al., 2021). These signaling pathways result finally in the activation of stress responsive genes by transcription factors that will set up a wide range of cell protection mechanisms. Stress responsive genes include many different genes coding for chaperones, detoxification enzymes, ionic transporters, osmolytes biosynthesis enzymes, aquaporins (Zhu, 2016) that will enable the plant to withstand the cellular consequences of environmental constraints. Besides these cellular protection mechanisms, several reviews have pointed out cell wall modifications occurring in response to abiotic stresses that contribute to plant tolerance toward these stresses (Coleman, Brunner, & Tsai, 2021; Houston, Tucker, Chowdhury, Shirley, & Little, 2016; LeGall et al., 2015; Rui & Dinneny, 2020; Tenhaken, 2015). We focus here on recent studies showing that stress response signaling pathways can directly regulate SCW lignification.

APETALA2/ETHYLENE RESPONSIVE FACTOR (AP2/ERF) TFs, which are pivotal regulators of several stress responses, among which drought response (Xie et al., 2019), have been demonstrated to target genes belonging to the lignin biosynthesis pathway leading to further increased plant tolerance to drought. In *Arabidopsis thaliana*, the AP2/ERF *At*SHN2, involved in wax/cutin deposition and drought tolerance, can modify lignin deposition by the direct regulation of NAC and MYB TFs (Ambavaram et al., 2011). In rice (*Oryza sativa*), overexpression of *OsERF71* confers higher drought tolerance associated with an upregulation of genes involved in lignin biosynthesis, as well as an increase of lignin content in the roots. The authors have shown that *Os*ERF71 binds directly to the promoter of *OsCCR1*, a key enzyme of the lignin biosynthesis pathway (Lee et al., 2016). Similarly, overexpression of *OsERF83* leads to increased drought tolerance and is associated with upregulation of lignin biosynthesis genes in rice (Jung et al., 2021). A DEHYDRATION-RESPONSIVE ELEMENT BINDING (DREB) TF from carrot (*Dc*DREB1A) has been shown to increase lignification through the induction of lignin biosynthesis genes while being a positive regulator of drought tolerance (Li et al., 2020). During osmotic stress, a DREB transcription factor from the moss *Bryum argenteum* (*Ba*DBL1) upregulates genes related to lignification pathway, enhancing lignin content and plant tolerance (Liang et al., 2021). Overexpression of *Eucalyptus grandis DREB/CBF14* transcription factor led to an increased lignification of xylem cells accompanied by an increase of the S/G ratio (Cao et al., 2020), mimicking the phenotype obtained with trees submitted to long-term chilling temperatures (Ployet et al., 2018). Besides, in the *CBF*-overexpressing lines, lignin biosynthesis genes were upregulated (Cao et al., 2020), suggesting an interplay between cold-response CBF-signaling pathway and lignin biosynthesis pathway. However, it remains unclear whether CBF transcription factors directly regulate the expression of SCW related genes by direct binding of their promoters.

Other families of transcription factors involved in drought response have also been pointed out for their crosstalk with the lignification pathway. In apple (*Malus domestica*), *Md*MYB88 and *Md*MYB124, previously described for their role in freezing tolerance have been shown to bind directly the promoters of *MdVND6* (key regulator of xylem vessel differentiation and SCW deposition) and *MbMYB46* (SCW master regulator) under drought conditions, leading to an increased lignification of root cell walls (Geng et al., 2018). In rice, overexpression of *OsTF1L*, a TF belonging to the HD-Zip transcription factor family, known for their role in plant growth as well as stress response, leads to increased drought tolerance along with accumulation of lignin in shoots. *Os*TF1L has been shown to bind directly to the promoters of lignin biosynthesis genes, thus enhancing lignin deposition and contributing to water deficit tolerance (Bang et al., 2019). In grapevine, a b-ZIP transcription factor, *Vl*bZIP30, regulates drought tolerance and lignin accumulation by direct binding of the promoters of stress-responsive and SCW-related genes, *VvNAC17* and *VvPRXN1* (Tu et al., 2020). In *Populus ussuriensis*, *Pu*C3H35, a CCCH-type zinc finger protein known for its role in plant growth and stress response, is responsible for lignin deposition in roots under drought stress by direct binding in the promoter of EARLI1 (Li et al., 2022), which has been described in *Arabidopsis* as a positive regulator of CCR1, a key enzyme in lignin biosynthesis (Shi et al., 2011). Altogether, these examples constitute further proof of the capacity of stress-signaling-pathways to co-opt the lignification pathway in order to contribute to better abiotic stress tolerance.

3.4 SCW-TFs recruited by abiotic stress-signaling pathways

Members of the NAC and the MYB TF family involved in the regulation of SCW were shown to be co-opted by stress signaling pathways:

SND1, a member of the NAC TF family well-known to be a level-one master positive regulator of SCW was shown recently to have a dual function (Jeong et al., 2018). Compared to wild type, the *snd1* knockout mutant seedlings accumulated less anthocyanin and exhibited low tolerance to salt stress. The authors demonstrated that SND1 binds not only to the promoter of *Myb46* (a second-level master regulator of SCW), but also to the promoter of the *ABI4* gene, thereby reducing ABA levels under normal growth conditions. Similar findings were found for its ortholog, *BpNAC012* in birch (Hu et al., 2019). Chromatin immunoprecipitation (ChIP) assays revealed that *Bp*NAC012 could bind both to the *cis*-elements CGT[G/A] and SNBE site in the promoters of abiotic stress-responsive genes and SCW-associated genes, respectively. Similarly, the overexpression of *AgNAC1* from celery (*Apium graveolens* L.), a positive regulator of lignin biosynthesis was recently shown to enhance plants' resistance to salt stress (Duan et al., 2020). Through a genome-wide association analysis, XYLEM NAC DOMAIN 1 (XND1), a transcriptional repressor regulating the expression of genes involved in both PCD and secondary wall formation (Zhao, Avci, Grant, Haigler, & Beers, 2008) was identified as a negative regulator of *Arabidopsis* root hydraulic conductivity (Tang et al., 2018). Phenotyping of *xnd1* mutants and natural *XND1* variants show that XND1 modulates root hydraulic conductivity through action on xylem formation and potential indirect effects on aquaporin function and that it diminishes drought stress tolerance.

The expression of *EjMYB1* from Loquat (*Eriobotrya japonica*), a co-ortholog of the lignin-specific activators *AtMYB58/63* in *Arabidopsis*, is induced at low-temperature, concomitantly to an increase of the lignin content in the fruit (Xu et al., 2014). *Ej*MYB1 directly activates lignin genes by binding to AC elements found in their promoters, leading to flesh lignification (Xu et al., 2014). *BplMYB46* from *Betula platyphylla* (ortholog of *AtMYB46*) was one of the first examples highlighting the crosstalk between the abiotic stress and secondary cell wall biosynthesis pathways in trees (Guo et al., 2017). Overexpression of *BplMYB46* not only increases lignin deposition, secondary cell wall thickness and the SW biosynthesis genes (including phenylalanine ammonia lyse, *PAL*), but also improves salt and osmotic tolerance in transgenic birch plants (Guo et al., 2017). *Bpl*MYB46 enhances abiotic stress tolerance in birch, *via* increasing scavenging of reactive oxygen species (ROS) and regulating expression of key ROS-scavenging enzymes: superoxide dismutase (SOD) and peroxidase (POD). Guo et al. (2017) further demonstrated that BplMYB46 binds to the TC-box, GT-box and E-box motifs in addition to the MYBCORE and AC-box, present in lignin biosynthesis genes. The TC-box and GT-box motifs located in the promoters of SOD and POD, respectively are associated with stress tolerance. Similarly, *Md*MYB46 could enhance the salt and osmotic stress tolerance in transgenic *Arabidopsis* and apple (*Malus* × *domestica* Borkh.). *Md*MYB46 not only promotes the biosynthesis of secondary cell wall and deposition of lignin by directly binding to the promoter of lignin biosynthesis genes (*CAD, COMT, CCR*) but also enhances stress tolerance, by activating stress-responsive genes belonging to the AP2/ERF family and dehydration-responsive genes through direct binding to their promoters. Therefore, *MdMYB46* could enhance the salt and osmotic stress tolerance in apple not only by activating lignin biosynthesis pathway, but also by directly activating genes involved in stress-signal transduction (Chen et al., 2019). The respective roles of the two paralogs *PtoMYB170* and *PtoMYB216* from *Populus tomentosa* (co-orthologs of *AtMYB61*) were examined by Tian et al. (2013) and Xu et al. (2017). Using overexpression in poplar, they first showed that *PtoMYB216* and *PtoMYB170* have conserved functions in regulating lignin biosynthesis. Both TFs specifically activate the expression of the upstream genes in the lignin biosynthetic pathway which results in ectopic deposition of lignin in cells that are normally nonlignified (Tian et al., 2013). However, the CRISPR/Cas9 mutant of *PtoMYB170* displays pendent stem phenotype resulting from reduced lignin deposition and more flexible xylem fibers. This strong phenotype suggests that the functional redundancy between *Pto*MYB216 and *Pto*MYB170 is weak because *Pto*MYB216 cannot replace the function of *Pto*MYB170 in lignin deposition (Xu et al., 2017). Interestingly, heterologous expression of *PtoMYB170* in transgenic *Arabidopsis* enhanced stomatal closure in the dark and resulted in drought tolerance of the transgenic plants through reduced water loss. *PtoMYB170* was specifically expressed in guard cells of transgenic *Arabidopsis* while *PtoMYB216* was not. These findings suggest a neofunctionalization of one of the two paralogs which acquired a new function in enhancing drought tolerance. After *Bpl*MYB46 (Guo et al., 2017), *Pto*MYB170 was the second example of tree SCW-TF able to integrate both developmental and stress signals. We have found that poplar plants overexpressing the *Eucalyptus* transcription factor *EgMYB1*, known to repress lignin biosynthesis (Legay et al., 2010), were far more resistant to drought than wild-type plants (unpublished). *Bp*MYB4 from *B. platyphylla*, a close ortholog of *Egr*MYB1 was shown to have the same repressive activity on lignin biosynthesis. It is also induced by several abiotic stresses and promotes ROS scavenging through the positive regulation of resistance-related genes like POD and SOD (Yu et al., 2021).

3.5 Transcription factors involved in the cross talk between biotic stress and lignification

The crosstalk between lignin biosynthesis and defense is extremely complex (Xie et al., 2018). It has been widely observed that lignin deposition and lignin biosynthetic genes are induced during the attack of various pathogens. However, contra-intuitively, "weakening" the cell wall by knocking out essential genes involved in lignin biosynthesis sometimes leads to enhanced resistance toward specific pathogens. The consequences of SCW modifications on pathogenic infection and their role in plant resistance to pathogens have been extensively reviewed in Miedes, Vanholme, Boerjan, and Molina (2014). We consider here only examples of modulation of SCW-transcription factors.

Shortly after *At*MYB46 was identified as a second-level master regulator of secondary cell wall biosynthesis in vascular tissues, it was also shown to be pivotal for mediating disease susceptibility to the necrotrophic fungal pathogen *Botrytis cinerea* (Ramirez et al., 2011). *Arabidopsis myb46* mutants (with impaired lignin biosynthesis) displayed enhanced disease resistance, which was thought to be caused by cell wall integrity damage-triggered high immunity level (Ramirez et al., 2011). Following infection, *At*MYB46 is able to modulate the induction of the pathogen-induced *Ep5C* gene, a type III cell wall-bound peroxidase, through binding to a new *cis*-element located in its promoter region. Different *myb46* mutants exhibit increased disease resistance to *B. cinerea*, a phenotype accompanied by selective transcriptional reprogramming of a set of genes encoding cell wall proteins and enzymes, of which extracellular type III peroxidases (Ramirez et al., 2011). Another MYB transcription factor, *At*MYB15, was shown to be a part of the genetic pathway controlling defense-induced synthesis of G lignin units and basal immunity (Chezem et al., 2017). Treatment with bacterial flagellin FLG22 induced *MYB15* expression and resulted in ectopic lignification in WT plants but not in *myb15* mutants. In addition, *myb15* mutants showed increased susceptibility to *Pseudomonas syringae*, whereas an induced expression of *MYB15* restored basal resistance of *myb15* mutants to wild-type levels. MYB15 directly binds to the secondary wall MYB-responsive element consensus sequence, which encompasses the AC elements, to drive synthesis of G lignin. This TF is also necessary for the basal synthesis of the preformed defense metabolite scopoletin (Chezem et al., 2017). Interestingly, the co-orthologs of *At*MYB15 in rice, *Os*MYB30 and *Os*MYB55, were also shown to contribute to plant immunity by directly activating monolignol genes and inducing higher accumulation of FA, which is an important constituent of grass cell walls (Kishi-Kaboshi et al., 2018).

4. Posttranscriptional regulation of lignin biosynthesis under stress

The posttranscriptional control of lignin deposition has been demonstrated for developmental lignin. It involves alternative splicing, microRNAs, long coding RNAs (reviewed in Zhang, Tuskan, Tschaplinski, Muchero, & Chen, 2020 for poplar).

Increasing evidence suggests that stress lignins are also submitted to posttranscriptional regulation by small RNAs (Cesarino, 2019). A growing number of studies support a role for small RNAs in the regulation of the phenylpropanoid metabolism during stress responses. Small RNAs (sRNAs) are divided into two major classes, microRNAs (miRNAs) and small interfering RNAs (siRNAs), according to their biogenesis (Mallory & Vaucheret, 2006). Small RNAs participate in many molecular mechanisms and signaling transduction pathways that are associated with plant stress tolerance by repressing expression of their target genes.

In *P. trichocarpa*, the mechanical stress responsive Ptr-miR473 was predicted to target a UV-B-resistant gene (*UVR8*) believed to positively regulate phenylpropanoid metabolism associated with C4H for the production of cell wall phenolic compounds, such as lignin and flavonoids (Lu et al., 2005). In sweet potato (*Ipomoea batatas*) overexpression of a wounding-responsive miRNA828 enhanced lignin deposition and H2O2 accumulation through repressed expression of *IbMYB* and *IbTLD* (Lin et al., 2012). These TFs function in the repression of the phenylpropanoid pathway and in the activation of antioxidant enzymes, respectively. Therefore, miR828 induction results in derepression of phenylpropanoid genes and in lower activity of antioxidant enzymes, leading to the accumulation of lignin and H2O2 as a part of defense mechanisms (Lin et al., 2012). The posttranscriptional regulation of wounding-triggered lignin deposition in sweet potato also involves wounding-related small RNA-8105 (Lin et al., 2013). The 22 and 24 nt sRNA8105s bind to the first intron of *IbMYB1* before splicing to mediate RNA cleavage and DNA methylation of *IbMYB1*, respectively. The interaction of sRNA8105 with *IbMYB1* RNA also triggers production of secondary siRNAs to decrease expression of *IbMYB1* family genes. sRNA8105-mediated repression of the IbMYB1 family genes upon wounding induces the expression of monolignol biosynthetic genes and leads to increased lignin deposition (Lin et al., 2013). In maize, the monocot-specific miR528 was shown to regulate lignin biosynthesis under luxurious nitrogen fertilization, affecting lodging resistance (Sun et al., 2018). Nitrogen luxury induces the expression of ZmmiR528 that, in turn, negatively regulates the expression of the lignin-related laccase genes *Zm*LAC3 and *Zm*LAC5, leading to lower lignin levels, decreased S/G ratio and reduced lodging resistance (Sun et al., 2018). Overexpression of microRNA397 (miR397) from the C4 monocot *Setaria viridis* in *Arabidopsis* led to repression of LACCASE (LAC) target gene expression, reduced lignin content and increased sensitivity to salt stress (Nguyen et al., 2020). Transgenic cowpea lines overexpressing artificial vun-miR408-3p displayed enhanced drought and salinity tolerance. Since *LACCASE 12* and *3* were identified as the potential target of *Vun*-miR408-3p, the authors hypothesized that the regulation of lignin content is an essential trait related to the drought tolerance in cowpea (Mishra et al., 2022). Recently, Jeena et al. (2021) showed that *Bm*-miR172c-5p is induced under drought stress and regulates lignin biosynthesis in *Bacopa monnieri*, by targeting the *F5H* gene. Overexpression of *Bm*-miR172c-5p precursor in *B. monnieri* represses *F5H* gene, leading to reduced lignification and secondary xylem thickness. By contrast, the mutation in *Bm*-miR172c-5p showed increased lignification and increased secondary xylem thickness, leading to better physiological response in drought stress conditions (Jeena et al., 2021).

5. Posttranslational regulation of lignin biosynthesis under stress

Several recent examples demonstrate that posttranslational mechanisms involving phosphorylation, translocation, degradation of transcription factors mediated through protein-protein interactions, are involved in the regulation of lignin biosynthesis under stress.

Three maize (*Z. mays*) lignin repressors, MYB11, MYB31, and MYB42 were shown to participate in wound-induced lignification by interacting with ZML2, a protein belonging to the TIFY family (Velez-Bermudez et al., 2015). The three R2R3-MYB factors and ZML2 act as repressors of a set of lignin genes. In particular, MYB11 and ZML2 bind simultaneously to the AC-rich and GAT(A/C) *cis*-elements present in the promoter of the *COMT* gene. Upon wounding and methyl jasmonate treatments, MYB11 and ZML2 proteins are degraded and *COMT* transcription is induced. The authors proposed a molecular regulatory mechanism in which the jasmonate-signaling cascade triggers the degradation of the MYB/ZML complex, resulting in the derepression of a set of lignin genes and, consequently, in the activation of lignin deposition in response to wounding (Velez-Bermudez et al., 2015). Soler et al. (2017) have shown that the *Eucalyptus* transcription factor *Eg*MYB1, known to repress lignin biosynthesis (Legay et al., 2010), interacts specifically with a linker histone variant, *Eg*H1.3. Linker histones can function as "hubs" to recruit many different proteins that affect the posttranslational modifications of core histones to change chromatin conformation and modify gene expression at specific *loci* (Over & Michaels, 2014). The interaction between *Eg*MYB1 and *Eg*H1.3 enhances the repression of *Eg*MYB1's target genes, strongly limiting the amount of lignin deposited in xylem cell walls. The authors suggested that a complex between *Eg*MYB1 and *Eg*H1.3 integrates developmental signals to prevent premature or inappropriate lignification of SCW, providing a mechanism to fine-tune the differentiation of xylem cells in time and space. Several indirect evidences suggest that the protein complex *Eg*MYB1 and *Eg*H1.3 could also integrate stress signals: (i) *H1.3* was originally discovered as a drought responsive gene in *Arabidopsis* (Ascenzi & Gantt, 1999), (ii) *EgMYB1*'s closer ortholog in *Arabidopsis* is *AtMYB4* whose expression is regulated by different environmental conditions such as wounding and UVB light (Jin et al., 2000), (iii) many stress-responsive genes are induced in *Arabidopsis* plants overexpressing both *Eg*MYB1and *Eg*H1.3 (Soler et al., unpublished) and (iv) poplar plants overexpressing *EgMYB1* are drought resistant (unpublished results). Interestingly, LIGNIN BIOSYNTHESIS-ASSOCIATED TF 1 (LTF1) from *Populus,* a putative ortholog of *At*MYB4 and *Eg*MYB1, was shown to be phosphorylated by the MAP kinase MPK6. Activation of *Pd*MPK6 in response to wounding promoted LTF1 phosphorylation in wood cells. In its unphosphorylated state, LTF1 functions as a repressor of lignin biosynthesis. When LTF1 becomes phosphorylated by *Pd*MPK6 in response to external stimuli such as wounding, it undergoes degradation through a proteasome pathway, resulting in activation of lignification. Because the MPK6 signaling module can be activated by various biotic and abiotic environmental stimuli such as cold, light, and drought, the authors proposed that LTF1 phosphorylation could act as a sensory switch to regulate lignin in response to stresses (Gui et al., 2019). Zheng et al. (2019) showed that a *P. tomentosa* E2 UBIQUITIN-CONJUGATING ENZYME 34 (*Pto*UBC34), is involved in the posttranslational regulation of lignin-associated transcriptional repressors *Pto*MYB221 and *Pto*MYB156. *Pto*UBC34 is localized at the endoplasmic reticulum (ER) membrane where it interacts with *Pto*MYB221 and *Pto*MYB156. This specific interaction allows for the translocation of TFs *Pto*MYB221 and *Pto*MYB156 to the ER and reduces their repression activity in a *Pto*UBC34 abundance-dependent manner. Noteworthy, *Pto*UBC34 is induced by treatment with sodium chloride and heat shock suggesting that this mechanism can operate in response to stresses. Very recently, Liu et al. (2021) have shown in *Arabidopsis* that ABA, a stress-signaling hormone, can regulate lignin biosynthesis through direct interaction between SnrK2, a central kinase in the ABA signaling pathway, and NST1, a master regulator of SCW formation. Once NST1 is phosphorylated by SnrK2 at Ser316 residue, it is activated and able to bind the promoters of SCW-related downstream target genes (Liu et al., 2021). Loss of function of NST1 in the *snd1* mutant background results in lack of lignified SCWs in the interfascicular fiber region of the stem, and the Ser316Ala mutant of NST1 fails to complement this phenotype and ABA-induced lignin pathway gene expression.

6. Concluding remarks

Recent studies highlight the crosstalk between the regulation of lignin biosynthesis and responses to stresses which occurs not only at the transcriptional level but also at the posttranscriptional and posttranslational levels. A growing number of SCW-related transcription factors are able to integrate both developmental and environmental signals and their overexpression led to an increased tolerance to abiotic stresses. On the other hand, some stress-signaling transcription factors were shown to impact lignin biosynthesis as part of the functional adaptation performed by plants to cope with environmental cues. The cell-tissue localization and the biochemical and molecular features of stress lignins need however to be further explored to better understand their roles in adaptation to stresses. Emerging-wide "omics" technologies revealed that the complexity of the transcriptional regulation of lignification during development goes far beyond the established NAC-MYB transcriptional regulatory network and that the network structure exhibits flexible changes induced by stress conditions (reviewed in Ohtani & Demura, 2019). Several additional regulatory factors have recently been found to overlay onto the NAC-MYB regulatory module, acting as molecular hubs that integrate developmental and environmental information into lignin biosynthesis. The concept of transcriptional hubs coordinating carbon and energy flux into growth, defense, and lignin biosynthesis is just emerging based on studies in *Arabidopsis* (Xie et al., 2018).

These molecular hubs are attractive targets for manipulation when aiming to control lignification in harmony with plant development and environmental adaptation (Ohtani & Demura, 2019). Moreover, as outlined by Chanoca et al. (2019), most lignin-engineering studies have been performed in controlled conditions and therefore do not provide sufficient insight into the interaction of the engineered plant with environmental factors. Indeed, the performance of lignin-engineered plants appears to be highly influenced by the environmental conditions.

Acknowledgments

The authors acknowledge the University of Toulouse, the Centre National de la Recherche Scientifique (CNRS) and the French Laboratory of Excellence (project TULIP, ANR-10-LABX-41; ANR-11-IDEX-0002-02) for their support. R.P. and I.H.B. were supported by a PhD grant from the Ministère de l'Enseignement Supérieur et de la Recherche.

References

Ambavaram, M. M. R., Krishnan, A., Trijatmiko, K. R., & Pereira, A. (2011). Coordinated activation of cellulose and repression of lignin biosynthesis pathways in rice. *Plant Physiology, 155*(2), 916. https://doi.org/10.1104/PP.110.168641.

Ascenzi, R., & Gantt, J. S. (1999). Subnuclear distribution of the entire complement of linker histone variants in *Arabidopsis thaliana. Chromosoma, 108*(6), 345–355. https://doi.org/10.1007/S004120050386.

Bang, S. W., Lee, D. K., Jung, H., Chung, P. J., Kim, Y. S., Choi, Y. D., et al. (2019). Overexpression of *Os*TF1L, a rice HD-Zip transcription factor, promotes lignin biosynthesis and stomatal closure that improves drought tolerance. *Plant Biotechnology Journal, 17*(1), 118–131. https://doi.org/10.1111/PBI.12951.

Barrière, Y., Courtial, A., Soler, M., & Grima-Pettenati, J. (2015). Toward the identification of genes underlying maize QTLs for lignin content, focusing on colocalizations with lignin biosynthetic genes and their regulatory MYB and NAC transcription factors. *Molecular Breeding, 35*(3), 87. https://doi.org/10.1007/s11032-015-0275-8.

Behr, M., Guerriero, G., Grima-Pettenati, J., & Baucher, M. (2019). A molecular blueprint of lignin repression. *Trends in Plant Science, 24*(11), 1052–1064. https://doi.org/10.1016/J.TPLANTS.2019.07.006.

Berthet, S., Demont-Caulet, N., Pollet, B., Bidzinski, P., Cézard, L., Le Bris, P., et al. (2011). Disruption of LACCASE4 and 17 results in tissue-specific alterations to lignification of *Arabidopsis thaliana* stems. *The Plant Cell, 23*(3), 1124–1137. https://doi.org/10.1105/tpc.110.082792.

Boerjan, W., Ralph, J., & Baucher, M. (2003). Lignin biosynthesis. *Annual Review of Plant Biology, 54*, 519–546. https://doi.org/10.1146/annurev.arplant.54.031902.134938.

Boudet, A. M., Kajita, S., Grima-Pettenati, J., & Goffner, D. (2003). Lignins and lignocellulosics: A better control of synthesis for new and improved uses. *Trends in Plant Science, 8*(12), 576–581. https://doi.org/10.1016/J.TPLANTS.2003.10.001.

Boudet, A. M., Lapierre, C., & Grima-Pettenati, J. (1995). Biochemistry and molecular biology of lignification. *The New Phytologist, 129*(2), 203–236. https://doi.org/10.1111/j.1469-8137.1995.tb04292.x.

Byrt, C. S., Munns, R., Burton, R. A., Gilliham, M., & Wege, S. (2018). Root cell wall solutions for crop plants in saline soils. *Plant Science*, *269*, 47–55. https://doi.org/10.1016/J.PLANTSCI.2017.12.012.

Cabane, M., Afif, D., & Hawkins, S. (2012). Lignins and abiotic stresses. *Advances in Botanical Research*, *61*, 219–262. https://doi.org/10.1016/B978-0-12-416023-1.00007-0.

Camargo, E. L. O., Ployet, R., Cassan-Wang, H., Mounet, F., & Grima-Pettenati, J. (2019). Digging in wood: New insights in the regulation of wood formation in tree species. *Advances in Botanical Research*, *89*, 201–233. https://doi.org/10.1016/BS.ABR.2018.11.007.

Cao, P. B., Ployet, R., Nguyen, C., Dupas, A., Ladouce, N., Martinez, Y., et al. (2020). Wood architecture and composition are deeply remodeled in frost sensitive eucalyptus overexpressing CBF/DREB1 transcription factors. *International Journal of Molecular Sciences*, *21*(8). https://doi.org/10.3390/IJMS21083019.

Carocha, V., Hefer, C., Cassan-wang, H., Fevereiro, P., Alexander, A., Paiva, J. A. P., et al. (2015). Genome-wide analysis of the lignin toolbox of *Eucalyptus grandis*. *The New Phytologist*, *206*, 1297–1313. https://doi.org/10.1111/nph.13313.

Cesarino, I. (2019). Structural features and regulation of lignin deposited upon biotic and abiotic stresses. *Current Opinion in Biotechnology*, *56*, 209–214. https://doi.org/10.1016/J.COPBIO.2018.12.012.

Chanoca, A., de Vries, L., & Boerjan, W. (2019). Lignin engineering in forest trees. *Frontiers in Plant Science*, *10*, 912. https://doi.org/10.3389/fpls.2019.00912.

Chen, X., Ding, Y., Yang, Y., Song, C., Wang, B., Yang, S., et al. (2021). Protein kinases in plant responses to drought, salt, and cold stress. *Journal of Integrative Plant Biology*, *63*(1), 53–78. https://doi.org/10.1111/JIPB.13061.

Chen, K., Song, M., Guo, Y., Liu, L., Xue, H., Dai, H., et al. (2019). *Md*MYB46 could enhance salt and osmotic stress tolerance in apple by directly activating stress-responsive signals. *Plant Biotechnology Journal*, *17*(12), 2341–2355. https://doi.org/10.1111/PBI.13151.

Chezem, W. R., Memon, A., Li, F. S., Weng, J. K., & Clay, N. K. (2017). SG2-type R2R3-MYB transcription factor MYB15 controls defense-induced lignification and basal immunity in *Arabidopsis*. *The Plant Cell*, *29*(8), 1907–1926. https://doi.org/10.1105/TPC.16.00954.

Chun, H. J., Baek, D., Cho, H. M., Lee, S. H., Jin, B. J., Yun, D. J., et al. (2019). Lignin biosynthesis genes play critical roles in the adaptation of Arabidopsis plants to high-salt stress. *Plant Signaling & Behavior*, *14*(8). https://doi.org/10.1080/15592324.2019.1625697.

Coleman, H. D., Brunner, A. M., & Tsai, C. J. (2021). Synergies and entanglement in secondary cell wall development and abiotic stress response in trees. *Frontiers in Plant Science*, *12*, 460. https://doi.org/10.3389/fpls.2021.639769.

Dixon, R. A., & Paiva, N. L. (1995). Stress-induced phenylpropanoid metabolism. *The Plant Cell*, 7(7), 1085–1097. https://doi.org/10.1105/tpc.7.7.1085.

Donaldson, L. A. (2002). Abnormal lignin distribution in wood from severely drought stressed *Pinus radiata* trees. *IAWA Journal*, *23*(2), 161–178. https://doi.org/10.1163/22941932-90000295.

Dong, N. Q., & Lin, H. X. (2021). Contribution of phenylpropanoid metabolism to plant development and plant–environment interactions. *Journal of Integrative Plant Biology*, *63*(1), 180–209. https://doi.org/10.1111/JIPB.13054.

Doster, M. A., & Bostock, R. M. (1988). The effect of temperature and type of medium on oospore production by *Phytophthora syringae*. *Mycologia*, *80*(1), 77–81. https://doi.org/10.1080/00275514.1988.12025501.

Duan, A. Q., Tao, J. P., Jia, L. L., Tan, G. F., Liu, J. X., Li, T., et al. (2020). *Ag*NAC1, a celery transcription factor, related to regulation on lignin biosynthesis and salt tolerance. *Genomics*, *112*(6), 5254–5264. https://doi.org/10.1016/J.YGENO.2020.09.049.

Geng, D., Chen, P., Shen, X., Zhang, Y., Li, X., Jiang, L., et al. (2018). *Md*MYB88 and *Md*MYB124 enhance drought tolerance by modulating root vessels and cell walls in apple. *Plant Physiology*, *178*(3), 1296–1309. https://doi.org/10.1104/PP.18.00502.

Grima-Pettenati, J., & Goffner, D. (1999). Lignin genetic engineering revisited. *Plant Science*, *145*(2). https://doi.org/10.1016/S0168-9452(99)00051-5.

Grima-Pettenati, J., Soler, M., Camargo, E. L. O., & Wang, H. (2012). Transcriptional regulation of the lignin biosynthetic pathway revisited: New players and insights. *Advances in Botanical Research*, *61*, 173–218. https://doi.org/10.1016/B978-0-12-416023-1.00006-9.

Gui, J., Luo, L., Zhong, Y., Sun, J., Umezawa, T., & Li, L. (2019). Phosphorylation of LTF1, an MYB transcription factor in populus, acts as a sensory switch regulating lignin biosynthesis in wood cells. *Molecular Plant*, *12*(10), 1325–1337. https://doi.org/10.1016/J.MOLP.2019.05.008.

Guo, H., Wang, Y., Wang, L., Hu, P., Wang, Y., Jia, Y., et al. (2017). Expression of the MYB transcription factor gene *Bpl*MYB46 affects abiotic stress tolerance and secondary cell wall deposition in *Betula platyphylla*. *Plant Biotechnology Journal*, *15*(1), 107–121. https://doi.org/10.1111/pbi.12595.

Hammerschmidt, R., Bonnen, A. M., Bergstrom, G. C., & Baker, K. K. (1985). Association of epidermal lignification with nonhost resistance of cucurbits to fungi. *Canadian Journal of Botany*, *63*(12), 2393–2398. https://doi.org/10.1139/B85-342.

Hausman, J. F., Evers, D., Thiellement, H., & Jouve, L. (2000). Compared responses of poplar cuttings and *in vitro* raised shoots to short-term chilling treatments. *Plant Cell Reports*, *19*(10), 954–960. http://link.springer.de/link/service/journals/00299/bibs/0019010/00190954.htm.

Hori, C., Yu, X., Mortimer, J. C., Sano, R., Matsumoto, T., Kikuchi, J., et al. (2020). Impact of abiotic stress on the regulation of cell wall biosynthesis in *Populus trichocarpa*. *Plant Biotechnology*, *37*(3), 273. https://doi.org/10.5511/PLANTBIOTECHNOLOGY.20.0326A.

Houston, K., Tucker, M. R., Chowdhury, J., Shirley, N., & Little, A. (2016). The plant cell wall: A complex and dynamic structure as revealed by the responses of genes under stress conditions. *Frontiers in Plant Science*, 7(8), 1–18. https://doi.org/10.3389/fpls.2016.00984.

Hu, P., Zhang, K., & Yang, C. (2019). *Bp*NAC012 positively regulates abiotic stress responses and secondary wall biosynthesis. *Plant Physiology*, *179*(2), 700–717. https://doi.org/10.1104/PP.18.01167.

Humphreys, J. M., & Chapple, C. (2002). Rewriting the lignin roadmap. *Current Opinion in Plant Biology*, *5*(3), 224–229. https://doi.org/10.1016/S1369-5266(02)00257-1.

Hussey, S. G., Mizrachi, E., Creux, N. M., & Myburg, A. A. (2013). Navigating the transcriptional roadmap regulating plant secondary cell wall deposition. *Frontiers in Plant Science*, *4*(8), 1–21. https://doi.org/10.3389/fpls.2013.00325.

Jeena, G. S., Joshi, A., & Shukla, R. K. (2021). *Bm*-miR172c-5p regulates lignin biosynthesis and secondary xylem thickness by altering the ferulate 5 hydroxylase gene in *Bacopa monnieri*. *Plant & Cell Physiology*, *62*(5), 894–912. https://doi.org/10.1093/PCP/PCAB054.

Jeong, C. Y., Lee, W. J., Truong, H. A., Trinh, C. S. N., Jin, J. Y., Kim, S., et al. (2018). Dual role of SND1 facilitates efficient communication between abiotic stress signalling and normal growth in *Arabidopsis*. *Scientific Reports*, *8*(1), 1–10. https://doi.org/10.1038/s41598-018-28413-x.

Jin, H., Cominelli, E., Bailey, P., Parr, A., Mehrtens, F., Jones, J., et al. (2000). Transcriptional repression by *At*MYB4 controls production of UV-protecting sunscreens in *Arabidopsis*. *The EMBO Journal*, *19*(22), 6150–6161. https://doi.org/10.1093/emboj/19.22.6150.

Jung, S. E., Bang, S. W., Kim, S. H., Seo, J. S., Yoon, H. B., Kim, Y. S., et al. (2021). Overexpression of *Os*ERF83, a vascular tissue-specific transcription factor gene, confers drought tolerance in rice. *International Journal of Molecular Sciences*, *22*(14), 7656. https://doi.org/10.3390/IJMS22147656.

Kim, S. H., Lam, P. Y., Lee, M. H., Jeon, H. S., Tobimatsu, Y., & Park, O. K. (2020). The *Arabidopsis* R2R3 MYB transcription factor MYB15 is a key regulator of lignin biosynthesis in effector-triggered immunity. *Frontiers in Plant Science*, *11*, 1456. https://doi.org/10.3389/FPLS.2020.583153/BIBTEX.

Kishi-Kaboshi, M., Seo, S., Takahashi, A., & Hirochika, H. (2018). The MAMP-responsive MYB transcription factors MYB30, MYB55 and MYB110 activate the HCAA synthesis pathway and enhance immunity in rice. *Plant & Cell Physiology*, *59*(5), 903–915. https://doi.org/10.1093/PCP/PCY062.

Lacombe, E., Hawkins, S., Van Doorsselaere, J., Piquemal, J., Goffner, D., Poeydomenge, O., et al. (1997). Cinnamoyl CoA reductase, the first committed enzyme of the lignin branch biosynthetic pathway: Cloning, expression and phylogenetic relationships. *The Plant Journal*, *11*(3), 429–441. http://www.ncbi.nlm.nih.gov/pubmed/9107033.

Lan, W., Lu, F., Regner, M., Zhu, Y., Rencoret, J., Ralph, S. A., et al. (2015). Tricin, a flavonoid monomer in monocot lignification. *Plant Physiology*, *167*(4), 1284–1295. https://doi.org/10.1104/pp.114.253757.

Lange, B. M., Lapierre, C., & Sandermann, H. (1995). Elicitor-induced spruce stress lignin (structural similarity to early developmental lignins). *Plant Physiology*, *108*(3), 1277–1287. https://doi.org/10.1104/PP.108.3.1277.

Lauvergeat, V., Lacomme, C., Lacombe, E., Lasserre, E., Roby, D., & Grima-Pettenati, J. (2001). Two cinnamoyl-CoA reductase (CCR) genes from *Arabidopsis thaliana* are differentially expressed during development and in response to infection with pathogenic bacteria. *Phytochemistry*, *57*(7), 1187–1195. http://www.ncbi.nlm.nih.gov/pubmed/11430991.

Lee, D. K., Jung, H., Jang, G., Jeong, J. S., Kim, Y. S., Ha, S. H., et al. (2016). Overexpression of the *Os*ERF71 transcription factor alters rice root structure and drought resistance. *Plant Physiology*, *172*(1), 575–588. https://doi.org/10.1104/PP.16.00379.

LeGall, H., Philippe, F., Domon, J., Gillet, F., Pelloux, J., & Rayon, C. (2015). Cell wall metabolism in response to abiotic stress. *Plants*, *4*(1), 112–166. https://doi.org/10.3390/plants4010112.

Legay, S., Sivadon, P., Blervacq, A.-S., Pavy, N., Baghdady, A., Tremblay, L., et al. (2010). *Eg*MYB1, an R2R3 MYB transcription factor from eucalyptus negatively regulates secondary cell wall formation in *Arabidopsis* and poplar. *The New Phytologist*, *188*(3), 774–786. https://doi.org/10.1111/j.1469-8137.2010.03432.x.

Li, T., Huang, Y., Khadr, A., Wang, Y. H., Xu, Z. S., & Xiong, A. S. (2020). *Dc*DREB1A, a DREB-binding transcription factor from *Daucus carota*, enhances drought tolerance in transgenic *Arabidopsis thaliana* and modulates lignin levels by regulating lignin-biosynthesis-related genes. *Environmental and Experimental Botany*, *169*, 103896. https://doi.org/10.1016/J.ENVEXPBOT.2019.103896.

Li, D., Yang, J., Pak, S., Zeng, M., Sun, J., Yu, S., et al. (2022). *Pu*C3H35 confers drought tolerance by enhancing lignin and proanthocyanidin biosynthesis in the roots of Populus ussuriensis. *The New Phytologist*, *233*(1), 390–408. https://doi.org/10.1111/NPH.17799.

Liang, Y., Li, X., Yang, R., Gao, B., Yao, J., Oliver, M. J., et al. (2021). *Ba*DBL1, a unique DREB gene from desiccation tolerant moss *Bryum argenteum*, confers osmotic and salt stress tolerances in transgenic *Arabidopsis*. *Plant Science*, *313*, 111047. https://doi.org/10.1016/J.PLANTSCI.2021.111047.

Lin, J. S., Lin, C. C., Li, Y. C., Wu, M. T., Tsai, M. H., Hsing, Y. I. C., et al. (2013). Interaction of small RNA-8105 and the intron of *Ib*MYB1 RNA regulates *Ib*MYB1 family genes through secondary siRNAs and DNA methylation after wounding. *The Plant Journal*, *75*(5), 781–794. https://doi.org/10.1111/TPJ.12238.

Lin, J. S., Lin, C. C., Lin, H. H., Chen, Y. C., & Jeng, S. T. (2012). MicroR828 regulates lignin and H2O2 accumulation in sweet potato on wounding. *The New Phytologist*, *196*(2), 427–440. https://doi.org/10.1111/J.1469-8137.2012.04277.X.

Liu, C., Yu, H., Rao, X., Li, L., & Dixon, R. A. (2021). Abscisic acid regulates secondary cell-wall formation and lignin deposition in *Arabidopsis thaliana* through phosphorylation of NST1. *Proceedings of the National Academy of Sciences of the United States of America*, *118*(5), e2010911118. https://doi.org/10.1073/PNAS.2010911118/-/DCSUPPLEMENTAL.

Lourenço, A., Rencoret, J., Chemetova, C., Gominho, J., Gutiérrez, A., Del Río, J. C., et al. (2016). Lignin composition and structure differs between xylem, phloem and phellem in *Quercus suber* L. *Frontiers in Plant Science*, 7(8), 1612. https://doi.org/10.3389/fpls.2016.01612.

Lu, S., Sun, Y., Shi, R., Clark, C., Li, L., & Chiang, V. L. (2005). Novel and mechanical stress—Responsive microRNAs in *Populus trichocarpa* that are absent from *Arabidopsis*. *The Plant Cell*, *17*(8), 2186–2203. https://doi.org/10.1105/tpc.105.033456.

Mallory, A. C., & Vaucheret, H. (2006). Functions of microRNAs and related small RNAs in plants. *Nature Genetics*, *38*, S31–S36. https://doi.org/10.1038/NG1791.

Miedes, E., Vanholme, R., Boerjan, W., & Molina, A. (2014). The role of the secondary cell wall in plant resistance to pathogens. *Frontiers in Plant Science*, *5*(8), 358. https://doi.org/10.3389/fpls.2014.00358.

Mishra, S., Sahu, G., & Shaw, B. P. (2022). Integrative small RNA and transcriptome analysis provides insight into key role of miR408 towards drought tolerance response in cowpea. *Plant Cell Reports*, *41*(1). https://doi.org/10.1007/S00299-021-02783-5.

Moura, J. C. M. S., Bonine, C. A. V., de Oliveira Fernandes Viana, J., Dornelas, M. C., & Mazzafera, P. (2010). Abiotic and biotic stresses and changes in the lignin content and composition in plants. *Journal of Integrative Plant Biology*, *52*(4), 360–376. https://doi.org/10.1111/j.1744-7909.2010.00892.x.

Nakano, Y., Yamaguchi, M., Endo, H., Rejab, N. A., & Ohtani, M. (2015). NAC-MYB-based transcriptional regulation of secondary cell wall biosynthesis in land plants. *Frontiers in Plant Science*, *6*(5), 288. https://doi.org/10.3389/fpls.2015.00288.

Nguyen, D. Q., Brown, C. W., Pegler, J. L., Eamens, A. L., & Grof, C. P. L. (2020). Molecular manipulation of microRNA397 abundance influences the development and salt stress response of *Arabidopsis thaliana*. *International Journal of Molecular Sciences*, *21*(21), 1–24. https://doi.org/10.3390/IJMS21217879.

Ohtani, M., & Demura, T. (2019). The quest for transcriptional hubs of lignin biosynthesis: Beyond the NAC-MYB-gene regulatory network model. *Current Opinion in Biotechnology*, *56*, 82–87. https://doi.org/10.1016/J.COPBIO.2018.10.002.

Oliveira, D. M., Mota, T. R., Salatta, F. V., Sinzker, R. C., Končitíková, R., Kopečný, D., et al. (2020). Cell wall remodeling under salt stress: Insights into changes in polysaccharides, feruloylation, lignification, and phenolic metabolism in maize. *Plant, Cell & Environment*, *43*(9), 2172–2191. https://doi.org/10.1111/PCE.13805.

Over, R. S., & Michaels, S. D. (2014). Open and closed: The roles of linker histones in plants and animals. *Molecular Plant*, 7(3), 481–491. https://doi.org/10.1093/MP/SST164.

Pereira, L., Domingues-Junior, A. P., Jansen, S., Choat, B., & Mazzafera, P. (2018). Is embolism resistance in plant xylem associated with quantity and characteristics of lignin? *Trees*, *32*(2), 349. https://doi.org/10.1007/s00468-017-1574-y.

Plomion, C., Leprovost, G., & Stokes, A. (2001). Wood formation in trees. *Plant Physiology*, *127*(12), 1513–1523. https://doi.org/10.1104/pp.010816.1.

Ployet, R., Soler, M., Carocha, V., Ladouce, N., Alves, A., Rodrigues, J.-C., et al. (2018). Long cold exposure induces transcriptional and biochemical remodelling of xylem secondary cell wall in eucalyptus. *Tree Physiology*, *38*(3), 409–422. https://doi.org/10.1093/treephys/tpx062.

Polo, C. C., Pereira, L., Mazzafera, P., Flores-Borges, D. N. A., Mayer, J. L. S., Guizar-Sicairos, M., et al. (2020). Correlations between lignin content and structural robustness in plants revealed by X-ray ptychography. *Scientific Reports, 10*(1), 1–11. https://doi.org/10.1038/s41598-020-63093-6.

Raes, J., Rohde, A., Christensen, J. H., Van de Peer, Y., & Boerjan, W. (2003). Genome-wide characterization of the lignification toolbox in *Arabidopsis*. *Plant Physiology, 133*(3), 1051–1071. https://doi.org/10.1104/pp.103.026484.role.

Ralph, J. (2010). Hydroxycinnamates in lignification. *Phytochemistry Reviews, 9*(1), 65–83. https://doi.org/10.1007/s11101-009-9141-9.

Ralph, J., Lundquist, K., Brunow, G., Lu, F., Kim, H., Schatz, P. F., et al. (2004). Lignins: Natural polymers from oxidative coupling of 4-hydroxyphenyl-propanoids. *Phytochemistry Reviews, 3*(1), 29–60. https://doi.org/10.1023/B:PHYT.0000047809.65444.A4.

Ramirez, V., Agorio, A., Coego, A., Garcia-Andrade, J., Hernandez, M. J., Balaguer, B., et al. (2011). MYB46 modulates disease susceptibility to *Botrytis cinerea* in *Arabidopsis*. *Plant Physiology, 155*(4), 1920–1935. https://doi.org/10.1104/pp.110.171843.

Rao, X., & Dixon, R. A. (2018). Current models for transcriptional regulation of secondary cell wall biosynthesis in grasses. *Frontiers in Plant Science, 9*. https://doi.org/10.3389/FPLS.2018.00399.

Ride, J. P. (1975). Lignification in wounded wheat leaves in response to fungi and its possible role in resistance. *Physiological Plant Pathology, 5*(2), 125–134. https://doi.org/10.1016/0048-4059(75)90016-8.

Robertsen, B., & Svalheim, Ø. (1990). The nature of lignin-like compounds in cucumber hypocotyls induced by α-1,4-linked oligogalacturonides. *Physiologia Plantarum, 79*(3), 512–518. https://doi.org/10.1111/J.1399-3054.1990.TB02111.X.

Rui, Y., & Dinneny, J. R. (2020). A wall with integrity: Surveillance and maintenance of the plant cell wall under stress. *The New Phytologist, 225*(4), 1428–1439. https://doi.org/10.1111/NPH.16166.

Schuetz, M., Smith, R., & Ellis, B. (2013). Xylem tissue specification, patterning, and differentiation mechanisms. *Journal of Experimental Botany, 64*(1), 11–31. https://doi.org/10.1093/jxb/ers287.

Scully, E. D., Gries, T., Palmer, N. A., Sarath, G., Funnell-Harris, D. L., Baird, L., et al. (2018). Overexpression of *Sb*Myb60 in *Sorghum bicolor* impacts both primary and secondary metabolism. *The New Phytologist, 217*(1), 82–104. https://doi.org/10.1111/nph.14815.

Shi, Y., Zhang, X., Xu, Z. Y., Li, L., Zhang, C., Schläppi, M., et al. (2011). Influence of EARLI1-like genes on flowering time and lignin synthesis of *Arabidopsis thaliana*. *Plant Biology, 13*(5), 731–739. https://doi.org/10.1111/J.1438-8677.2010.00428.X.

Soler, M., Plasencia, A., Larbat, R., Pouzet, C., Jauneau, A., Rivas, S., et al. (2017). The eucalyptus linker histone variant *Eg*H1.3 cooperates with the transcription factor *Eg*MYB1 to control lignin biosynthesis during wood formation. *The New Phytologist, 213*, 287–299. https://doi.org/10.1111/nph.14129.

Sun, Q., Liu, X., Yang, J., Du, Q., Wang, H., Li, W. X., et al. (2018). MicroRNA528 affects lodging resistance of maize by regulating lignin biosynthesis under nitrogen-luxury conditions. *Molecular Plant, 11*(6), 806–814. https://doi.org/10.1016/j.molp.2018.03.013.

Tang, N., Shahzad, Z., Lonjon, F., Loudet, O., Vailleau, F., & Maurel, C. (2018). Natural variation at XND1 impacts root hydraulics and trade-off for stress responses in *Arabidopsis*. *Nature Communications, 9*(1), 1–12. https://doi.org/10.1038/s41467-018-06430-8.

Taylor-Teeples, M., Lin, L., de Lucas, M., Turco, G., Toal, T. W., Gaudinier, A., et al. (2015). An *Arabidopsis* gene regulatory network for secondary cell wall synthesis. *Nature, 517*, 571–575. https://doi.org/10.1038/nature14099.

Tenhaken, R. (2015). Cell wall remodeling under abiotic stress. *Frontiers in Plant Science*, *5*(1), 1–9. https://doi.org/10.3389/fpls.2014.00771.

Tian, Q., Wang, X., Li, C., Lu, W., Yang, L., Jiang, Y., et al. (2013). Functional characterization of the poplar R2R3-MYB transcription factor *Pto*MYB216 involved in the regulation of lignin biosynthesis during wood formation. *PLoS One*, *8*(10), e76369. https://doi.org/10.1371/journal.pone.0076369.

Tu, M., Wang, X., Yin, W., Wang, Y., Li, Y., Zhang, G., et al. (2020). Grapevine *Vlb*ZIP30 improves drought resistance by directly activating *Vv*NAC17 and promoting lignin biosynthesis through the regulation of three peroxidase genes. *Horticulture Research*, 7(150), 1–15. https://doi.org/10.1038/s41438-020-00372-3.

Vance, C. P., Kirk, T. K., & Sherwood, R. T. (1980). Lignification as a mechanism of disease resistance. *Annual Review of Phytopathology*, *18*, 259–288. https://doi.org/10.1146/ANNUREV.PY.18.090180.001355.

Vanholme, R., Cesarino, I., Rataj, K., Xiao, Y., Sundin, L., Goeminne, G., et al. (2013). Caffeoyl shikimate esterase (CSE) is an enzyme in the lignin biosynthetic pathway in *Arabidopsis*. *Science*, *341*(6150), 1103–1106. https://doi.org/10.1126/SCIENCE.1241602.

Vanholme, R., Demedts, B., Morreel, K., Ralph, J., & Boerjan, W. (2010). Lignin biosynthesis and structure. *Plant Physiology*, *153*(3), 895–905. https://doi.org/10.1104/pp.110.155119.

Vanholme, R., De Meester, B., Ralph, J., & Boerjan, W. (2019). Lignin biosynthesis and its integration into metabolism. *Current Opinion in Biotechnology*, *56*, 230–239. https://doi.org/10.1016/J.COPBIO.2019.02.018.

Velez-Bermudez, I.-C., Salazar-Henao, J. E., Fornale, S., Lopez-Vidriero, I., Franco-Zorrilla, J.-M., Grotewold, E., et al. (2015). A MYB/ZML complex regulates wound-induced lignin genes in maize. *The Plant Cell*, *2*, 1–16. https://doi.org/10.1105/tpc.15.00545.

Wang, H. Z., & Dixon, R. A. (2012). On–off switches for secondary cell wall biosynthesis. *Molecular Plant*, *5*(2), 297–303. https://doi.org/10.1093/MP/SSR098.

Wang, J. P., Matthews, M. L., Williams, C. M., Shi, R., Yang, C., Tunlaya-Anukit, S., et al. (2018). Improving wood properties for wood utilization through multi-omics integration in lignin biosynthesis. *Nature Communications*, *9*(1), 1579. https://doi.org/10.1038/s41467-018-03863-z.

Weng, J.-K., & Chapple, C. (2010). The origin and evolution of lignin biosynthesis. *The New Phytologist*, *187*(2), 273–285. https://doi.org/10.1111/j.1469-8137.2010.03327.x.

Xie, Z., Nolan, T., Jiang, H., Tang, B., Zhang, M., Li, Z., et al. (2019). The AP2/ERF transcription factor TINY modulates brassinosteroid-regulated plant growth and drought responses in *Arabidopsis*. *The Plant Cell*, *31*(8), 1788–1806. https://doi.org/10.1105/TPC.18.00918.

Xie, M., Zhang, J., Tschaplinski, T. J., Tuskan, G. A., Chen, J. G., & Muchero, W. (2018). Regulation of lignin biosynthesis and its role in growth-defense tradeoffs. *Frontiers in Plant Science*, *9*, 1427. https://doi.org/10.3389/FPLS.2018.01427.

Xu, C., Fu, X., Liu, R., Guo, L., Ran, L., Li, C., et al. (2017). *Pto*MYB170 positively regulates lignin deposition during wood formation in poplar and confers drought tolerance in transgenic *Arabidopsis*. *Tree Physiology*, *37*(12), 1713–1726. https://doi.org/10.1093/treephys/tpx093.

Xu, Q., Yin, X. R., Zeng, J. K., Ge, H., Song, M., Xu, C. J., et al. (2014). Activator-and repressor-type MYB transcription factors are involved in chilling injury induced flesh lignification in loquat via their interactions with the phenylpropanoid pathway. *Journal of Experimental Botany*, *65*(15). https://doi.org/10.1093/jxb/eru208.

Yamaguchi, M., Mitsuda, N., Ohtani, M., Ohme-Takagi, M., Kato, K., & Demura, T. (2011). VASCULAR-RELATED NAC-DOMAIN 7 directly regulates the expression of a broad range of genes for xylem vessel formation. *The Plant Journal*, *66*(4), 579–590. https://doi.org/10.1111/j.1365-313X.2011.04514.x.

Yu, Y., Liu, H., Zhang, N., Gao, C., Qi, L., & Wang, C. (2021). The *Bp*MYB4 transcription factor from *Betula platyphylla* contributes toward abiotic stress resistance and secondary cell wall biosynthesis. *Frontiers in Plant Science*, *11*, 2242. https://doi.org/10.3389/fpls.2020.606062.

Zhang, J., Tuskan, G. A., Tschaplinski, T. J., Muchero, W., & Chen, J. G. (2020). Transcriptional and post-transcriptional regulation of lignin biosynthesis pathway genes in *Populus*. *Frontiers in Plant Science*, *11*(5), 652. https://doi.org/10.3389/fpls.2020.00652.

Zhao, C., Avci, U., Grant, E. H., Haigler, C. H., & Beers, E. P. (2008). XND1, a member of the NAC domain family in *Arabidopsis thaliana*, negatively regulates lignocellulose synthesis and programmed cell death in xylem. *The Plant Journal*, *53*(3), 425–436. https://doi.org/10.1111/J.1365-313X.2007.03350.X.

Zhao, Q., Nakashima, J., Chen, F., Yin, Y., Fu, C., Yun, J., et al. (2013). LACCASE is necessary and nonredundant with PEROXIDASE for lignin polymerization during vascular development in *Arabidopsis*. *Plant Physiology*, *25*, 1–13. https://doi.org/10.1105/tpc.113.117770.

Zheng, L., Chen, Y., Ding, D., Zhou, Y., Ding, L., Wei, J., et al. (2019). Endoplasmic reticulum-localized UBC34 interaction with lignin repressors MYB221 and MYB156 regulates the transactivity of the transcription factors in *Populus tomentosa*. *BMC Plant Biology*, *19*(1), 19–97. https://doi.org/10.1186/S12870-019-1697-Y.

Zhong, R., Lee, C., & Ye, Z.-H. (2010). Functional characterization of poplar wood-associated NAC domain transcription factors. *Plant Physiology*, *152*(2), 1044–1055. http://www.plantphysiol.org/cgi/content/abstract/152/2/1044.

Zhong, R., & Ye, Z.-H. (2009). Transcriptional regulation of lignin biosynthesis. *Plant Signaling & Behavior*, *4*(11), 1028–1034. https://doi.org/10.4161/psb.4.11.9875.

Zhou, J., Lee, C., Zhong, R., & Ye, Z.-H. (2009). MYB58 and MYB63 are transcriptional activators of the lignin biosynthetic pathway during secondary cell wall formation in *Arabidopsis*. *The Plant Cell*, *21*(1), 248–266. https://doi.org/10.1105/tpc.108.063321.

Zhu, J. K. (2016). Abiotic stress signaling and responses in plants. *Cell*, *167*(2), 313–324. https://doi.org/10.1016/j.cell.2016.08.029.

FSC
www.fsc.org
MIX
Papier | Fördert
gute Waldnutzung
FSC® C083411

Kolibri 360 GmbH
Ferdinand-Jühlke-Straße 7
99095 Erfurt, Deutschland
produktsicherheit@kolibri360.de